Hans-Jürgen Butt and
Michael Kappl

Surface and Interfacial
Forces

Related Titles

Butt, H.J., Kappl, M.

Physics and Chemistry of Interfaces

2006

ISBN: 978-3-527-40629-6

Friedbacher, G., Bubert, H. (eds.)

Surface and Thin Film Analysis

A Compendium of Principles, Instrumentation, and Applications

2010

ISBN: 978-3-527-32047-9

Förch, R., Schönherr, H., Jenkins, A. T. A. (eds.)

Surface Design: Applications in Bioscience and Nanotechnology

2009

ISBN: 978-3-527-40789-7

Tadros, T. F. (ed.)

Colloid Stability

The Role of Surface Forces - Part II / Colloids and Interface Science Vol. 2

2007

ISBN: 978-3-527-31503-1

Brandrup, J., Immergut, E. H., Grulke, E. A. (eds.)

Polymer Handbook

1999

ISBN: 978-0-471-16628-3

Evans, D. F., Wennerström, H.

The Colloidal Domain

Where Physics, Chemistry, and Biology Meet

1999

ISBN: 978-0-471-24247-5

Garbassi, F., Morra, M., Occhiello, E.

Polymer Surfaces

From Physics to Technology. Revised and Updated Edition

1998

ISBN: 978-0-471-97100-9

Hans-Jürgen Butt and Michael Kappl

Surface and Interfacial Forces

WILEY-VCH

WILEY-VCH Verlag GmbH & Co. KGaA

The Authors

Hans-Jürgen Butt
Max-Planck-Institut für Polymerforschung
Mainz, Germany
butt@mpip-mainz.mpg.de

Michael Kappl
Max-Planck-Institut für Polymerforschung
Mainz, Germany
kappl@mpip-mainz.mpg.de

Description of Cover-illustration:
Top left: Scanning electron micrograph of clusters of silica particles deposited onto a silicon wafer. The particles have a diameter of 1.9 μm and resemble an example of how surface force can dominate the behavior of small scale systems. Even though these structures look fragile, shaking such a sample could not deform them, since inertial forces will be smaller than the van der Waals forces between these particles. Top right: Scanning electron micrograph of red blood cells (courtesy of John Minarcik, Department of Laboratory Medicine & Pathology, University of Alberta). Bottom left: Photograph of a Tokay gecko (Gekko gecko) adhering to a glass plate. Bottom right: Scanning electron micrograph of the fine structure of the gecko foot. Each gecko toe consists of hundred thousands of small keratin hairs called setae. Each seta is further divided into several hundred subunits, the so-called spatulae with a diameter of about 200 nm. This structure allows the gecko to form intimate contact with surfaces and utilize van der Waals and capillary forces to adhere to surfaces. Both gecko images were kindly provided by S. Gorb, Max Planck Institute for Metals Research, Stuttgart.

All books published by Wiley-VCH are carefully produced. Nevertheless, authors, editors, and publisher do not warrant the information contained in these books, including this book, to be free of errors. Readers are advised to keep in mind that statements, data, illustrations, procedural details or other items may inadvertently be inaccurate.

Library of Congress Card No.: applied for

British Library Cataloguing-in-Publication Data
A catalogue record for this book is available from the British Library.

Bibliographic information published by the Deutsche Nationalbibliothek
The Deutsche Nationalbibliothek lists this publication in the Deutsche Nationalbibliografie; detailed bibliographic data are available on the Internet at http://dnb.d-nb.de.

© 2010 WILEY-VCH Verlag GmbH & Co. KGaA, Weinheim

Typesetting   Thomson Digital, Noida, India

Cover Design   Spieszdesign, Neu-Ulm

Printed on acid-free paper

ISBN: 978-3-527-40849-8

Contents

Surface and Interfacial Forces. Hans-Jürgen Butt and Michael Kappl
Copyright © 2010 WILEY-VCH Verlag GmbH & Co. KGaA, Weinheim
ISBN: 978-3-527-40849-8

Symbols and Abbreviations

Many symbols are not unique for a certain physical quantity but are used two or even three times. We use the symbols as they are usually used in the relevant literature. Since the scope of this book includes many disciplines and thus different scientific communities, multiple usage of symbols is unavoidable. In molecular chemistry and physics, for instance, μ is the dipole moment while in engineering μ symbolizes the coefficient of friction.

a	contact radius (m), activity (mol/L)
a_0	Molecular radius (m)
A	Area (m^2)
A_H	Hamaker constant (J)
b	Slip length, distance between grafting sites (m)
c	Number concentration (number of molecules per m^3) or amount concentration ($mol\,m^{-3}$, or $mol\,l^{-1} = M$), mean cosine of contact angles
c_m	Concentration in mass per unit volume ($kg\,m^{-3}$)
d_{cc}	Center-to-center distance between two spheres (m)
D	Distance (m)
D_d	Diffusion coefficient ($m^2\,s^{-1}$)
D_0	Interatomic spacing used to calculate adhesion (m), typically $1.7\,\text{Å}$
$\tilde{D}$	Dimensionless normalized distance
E	Electric field strength ($V\,m^{-1}$), Young's modulus (Pa), surface elasticity ($N\,m^{-1}$)
E^*	Reduced Young's modulus (Pa)
F_{adh}	Adhesion force (N)
F_F	Friction force (N)
F_L	Load (N)
f	Force per unit area ($N\,m^{-2}$)
f^*	Different kinds of dimensionless correction functions
G	Gibbs energy (J)
G_m, G_m^0	Molar Gibbs energy and standard molar Gibbs energy ($J\,mol^{-1}$)
H	Enthalpy (J)
h	Height of a liquid with respect to a reference level (m), Planck constant, film thickness (m)

Surface and Interfacial Forces. Hans-Jürgen Butt and Michael Kappl
Copyright © 2010 Wiley-VCH Verlag GmbH & Co. KGaA
ISBN: 978-3-527-40849-8

K	Spring constant $(\mathrm{N\,m^{-1}})$
K_c	Spring constant of AFM cantilever $(\mathrm{N\,m^{-1}})$
k_c	Bending modulus of a membrane (J)
k_s	Segment elasticity $(\mathrm{N\,m^{-1}})$
L	Center-to-center distance between two spherical particles (m)
L_c	Contour length of a polymer chain (m)
L_0	Thickness of an undisturbed polymer brush (m)
l_r	Length of repeat unit in a polymer (m)
l_s	Segment length of polymer chain also called Kuhn length (m)
M	Torque (N m)
M_w	Molar mass $(\mathrm{kg\,mol^{-1}})$
M_r	Molar mass of repeat unit of a polymer $(\mathrm{kg\,mol^{-1}})$
m	Mass (kg), molecular mass (kg per molecule)
N	Number of molecules (dimensionless or mol), number of segments in a linear polymer chain
n	Refractive index, integer number
P	Pressure (Pa)
P_c	Capillary pressure caused by the curvature of an interface (Pa)
P^V	Equilibrium vapor pressure of a vapor in contact with a liquid with a curved surface (Pa)
P_0	Equilibrium vapor pressure of a vapor in contact with a liquid having a planar surface (Pa)
Q	Electric charge (A s), heat (J), quality factor of a resonator
R	Gas constant
R_1, R_2	Radii of spherical particles (m)
R^*	reduced radius (m)
R_g	Radius of gyration of a polymer (m)
R_p	Radius of a spherical particle (m)
R_0	Size of a polymer chain (m)
r	Radius (m), radial coordinate in cylindrical or spherical coordinates
r_b, r_c, r_d	Radius of a bubble, a capillary, and a drop, respectively (m)
r_1, r_2	Two principal radii of curvature of a liquid (m)
S	Entropy $(\mathrm{J\,K^{-1}})$, number of adsorption binding sites per unit area (mol $\mathrm{m^{-2}}$), spreading coefficient $(\mathrm{N\,m^{-1}})$
T	Temperature (K)
T_Θ	Theta temperature (K)
t	Time (s)
U	Internal energy (J), applied or measured electric potential (V)
V	Volume $(\mathrm{m^3})$ or free energy of interaction between two molecules or particles (J)
V_m	Molar volume $(\mathrm{m^3\,mol^{-1}})$
v	Velocity $(\mathrm{m\,s^{-1}})$, excluded volume parameter $(\mathrm{m^3})$
v_0	Sliding or rolling velocity $(\mathrm{m\,s^{-1}})$
V^A	Free energy for the interaction between two surfaces per unit area $(\mathrm{J\,m^{-2}})$
W_{adh}	adhesion energy per unit are a $(\mathrm{J/m^2})$

x, y, z	Cartesian coordinates (m), reduced electric potential, and the reduced distance $y = x/(2L_0)$
Z	Valency of an ion
α	Polarizability ($C\,m^2\,V^{-1}$), factor defined by Eq. (4.23)
γ	Surface tension ($N\,m^{-1}$). Specifically, γ_L, γ_S and γ_{SL} are the surface tensions of a liquid–vapor, a solid and solid-liquid interface, respectively
γ^s	Surface energy of a solid (J)
$\dot{\gamma}$	Shear rate (s^{-1})
Γ	Grafting density of polymer ($mol\,m^{-2}$ or m^{-2})
δ	Thickness of the hydration layer (m), indentation (m)
ε	Relative permittivity
ς	Zeta potential (V)
η	Viscosity (Pa s)
Θ	Contact angle (deg)
$\varkappa$	Inverse Debye length (m^{-1}) (Eqs. (4.8) and (4.11))
$\varkappa_c$	Capillary constant (m) (Eq. (5.7))
λ	Decay length or wavelength of light (m)
λ_c	Characteristic length scale, critical wavelength of fluctuations (m)
λ_D	Debye length (m)
λ_K	$= \gamma V_m / RT$, Kelvin length (m)
λ_{ev}	penetration depth of evanescent wave (m)
μ	Chemical potential ($J\,mol^{-2}$), dipole moment (C m), friction coefficient
μ_k, μ_s	Dimensionless coefficient of kinetic and static friction, respectively
μ_r	Coefficient of rolling friction (m)
μ_T	Tabor parameter
μ_M	Maugis parameter
ν	Frequency (Hz), Poisson ratio (dimensionless)
Π	Disjoining pressure (Pa)
ϱ	Mass density ($kg\,m^{-3}$)
ϱ_0	Number density of molecules next to a wall (molecules per m^3)
ϱ_n	Molecular density (molecules per m^3)
ϱ_e	Electric charge density ($C\,m^{-3}$)
ϕ	Volume fraction
σ	Surface charge density ($C\,m^{-2}$)
σ_A	Surface area per molecule (m^2)
ξ	Coordinate in the gap between two half-spaces (m)
ξ_m	Matsubara angular frequencies (Hz)

Preface

Two decades have passed since the book by Jacob Israelachvili *"Intermolecular and Surface Forces"* appeared. During this period, our knowledge on interfacial forces has significantly, improved partially due to new experimental tools and improved simulation capabilities. For example, the invention of scanning probe microscopy and the development of optical methods allow us to look at surface forces in much more detail.

Surface forces are relevant in a number of technologies, in particular eco-efficient technologies, for a sustained growth in the face of unbridled exploitation of natural resources, a growing world population, and the expected climate change. For example, a good understanding of particle dispersion is a prerequisite to improve mineral processing and adapt it to the increasing exploitation of raw materials. Food industry, facing a growing demand for healthy food, relies on a good understanding of the stability of emulsions and thus of the interaction of oil drops in water or oily liquids in water. The same is true for oil recovery and waste water treatment. The synthesis of polymers in aqueous emulsions allows an environment-friendly production; again, a good knowledge of the forces that keep the particles dispersed is required.

The number of papers published in journals on colloid and interface science has increased by about seven times during the past 20 years. We suppose that this increase is correlated with an increasing number of active researchers in the field. One reason for this increase is certainly the growth in the world population, the fact that a larger proportion of the world population takes part in the technological progress. Another reason is that several technologies rely more and more on processes at the small scale. One example is the increasing relevance of micro- and nanotechnology, including lab-on-chip technology, microfluids, and biochips. Objects in the micro- and nanoworld are dominated by surface effects rather than gravitation or inertia.

These developments motivated us to write this textbook. It is a general introduction to surface and interfacial forces. Though a basic knowledge of colloid and interface science is helpful, it is not essential because all important concepts have been explained. Certainly, no advanced level of mathematics is required. Looking

Surface and Interfacial Forces. Hans-Jürgen Butt and Michael Kappl
Copyright © 2010 WILEY-VCH Verlag GmbH & Co. KGaA, Weinheim
ISBN: 978-3-527-40849-8

through the pages of this book, you will see a substantial number of equations. Please do not be scared! We preferred to explicitly give all transformations rather than writing "as can easily be seen" and stating the result.

A number of problems with solutions are included to allow private studies. If not mentioned otherwise, the temperature is assumed to be 25 °C. At the end of each chapter, the most important equations, facts, and phenomena are summarized.

This book certainly contains errors. Even after proofreading by different people independently, this is unavoidable. If you find any error, please write us a letter (Max Planck Institute for Polymer Research, Ackermannweg, 55128 Mainz, Germany) or an e-mail (butt@mpip-mainz.mpg.de) so that we can correct it and do not confuse more readers.

We are indebted to several people who helped us collect information, prepare, and critically read this manuscript. In particular, we would like to thank Maria D'Acunzi, Günter Auernhammer, Clemens Bechinger, Elmar Bonaccurso, Derek Y.C. Chan, Vince Craig, Raymond Dagastine, Markus Deserno, Georg Floudas, Stanislav Gorb, Karina Grundke, Vagelis Harmandaris, Manfred J. Hampe, Manfred Heuberger, Katharina Hocke, Roger Horn, Naoyuki Ishida, Gunnar Kircher, Reinhard Miller, Maren Müller, Martin Oettel, Sandra Ritz, Tim Salditt, Tanja Schilling, Doris Vollmer, and Xuehua Zhang.

Mainz, August 2009   *Hans-Jürgen Butt and Michael Kappl*

1
Introduction

The topic of this book is forces acting between interfaces. There is no clear, unique definition of an interfacial force. One possible definition is as follows: Interfacial forces are those forces that originate at the interface. For example, electrostatic double-layer forces are caused by surface charges at the interface. Such a definition would, however, not include van der Waals forces. For van der Waals interaction, the surface atoms do not have a distinct role compared to the bulk atoms. Still, van der Waals forces substantially contribute to the interaction between small particles. One could define surface forces as all interactions that increase proportional to the interfacial area. Then, for certain geometries gravitation should also be included. Gravitation is, however, not described here. On the other hand, hydrodynamic interactions would be excluded because they depend on the specific shape of interacting interfaces and not only on the interfacial area.

We take a pragmatic approach and discuss all forces that are relevant in systems, that have a small characteristic length scale, and whose structure and dynamics are dominated by interfaces rather than gravitation and inertia. In this sense, this book is about the structure and dynamics of system with a small characteristic length scale. At this point we need to specify two terms: "Interface" and "characteristic length scale."

An interface is the area that separates two phases. If we consider the solid, liquid, and gas phases, we immediately get three combinations of interfaces: the solid–liquid, the solid–gas, and the liquid–gas interfaces. The term surface is often used synonymously, although interface is preferred for the boundary between two condensed phases and in cases where the two phases are named explicitly. For example, we talk about a solid–gas interface but a solid surface. Interfaces can also separate two immiscible liquids such as water and oil. These are called liquid–liquid interfaces. Interfaces may even separate two different phases within one component. In a liquid crystal, for example, an ordered phase may coexist with an isotropic phase. Solid–solid interfaces separate two solid phases. They are important for the mechanical behavior of solid materials. Gas–gas interfaces do not exist because gases mix.

Often interfaces and colloids are discussed together. Colloid is a synonym for colloidal system. Colloidal systems are disperse systems in which one phase has dimensions in the order of 1 nm to 1 μm (Figure 1.1). The word "colloid" comes from

Surface and Interfacial Forces. Hans-Jürgen Butt and Michael Kappl
Copyright © 2010 Wiley-VCH Verlag GmbH & Co. KGaA
ISBN: 978-3-527-40849-8

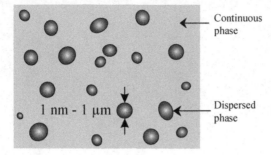

Figure 1.1 Schematic of a dispersion.

Table 1.1 Types of dispersions.

Continuous phase	Dispersed phase	Term	Examples
Gas	Liquid	Aerosol	Clouds, fog, smog, hairspray
	Solid	Aerosol	Smoke, dust, pollen
Liquid	Gas	Foam	Lather, whipped cream, foam on beer
	Liquid	Emulsion	Milk, skin creams
	Solid	Sol	Ink, muddy water, dispersion paint
Solid	Gas	Porous solids[a]	Partially sintered or pressed powders
		Foam	Styrofoam, soufflés
	Liquid	Solid emulsion	Butter
	Solid	Solid suspension	Concrete

a) Porous solids have a bicontinuous structure while in a solid foam the gas phase is clearly dispersed.

the Greek word for glue and was used the first time in 1861 by Graham.[1] He applied it to materials that seemed to dissolve but were not able to penetrate a membrane, such as albumin, starch, and dextrin. A colloidal dispersion is a two-phase system that is uniform on the macroscopic but not on the microscopic scale. It consists of grains or droplets of one phase in a matrix of the other phase.

Different kinds of dispersions can be formed. Most of them have important applications and have special names (Table 1.1). While there are only 3 types of interfaces, we can distinguish 10 types of disperse systems because we have to discriminate between the continuous, dispersing (external) phase and the dispersed (inner) phase. In some cases, this distinction is obvious. Nobody will, for instance, mix up fog with a foam although in both cases a liquid and a gas are involved. In other cases, the distinction between continuous and inner phases cannot be made because both phases might form connected networks. Some emulsions for instance tend to form a bicontinuous phase, in which both phases form an interwoven network.

The characteristic length scale of a system can often be given intuitively. For example, for a spherical particle one would use the radius, for a thin film the thickness. For more complex systems, intuition, however, leads to ambiguous

1) Thomas Graham, 1805–1869. British chemist, professor in Glasgow and London.

results. We suggest to use the ratio of the total volume V divided by the total interfacial area A of a system as the characteristic length scale: $\lambda_c = V/A$. For a sphere of radius r_s, the characteristic length scale is $\lambda_c = r_s/3$. For a thin film, it is equal to the thickness. For a dispersion of spherical particles with a volume fraction ϕ, the total volume of the system is $V = N4\pi r_s^3/(3\phi)$, where N is the number of particles. With a total surface area of $A = N4\pi r_s^2$, we get a characteristic length scale of $\lambda_c = r_s/(3\phi)$.

◼ **Example 1.1**

If aluminum is electrochemically oxidized under appropriate circumstances, porous layers of many microns thickness with regular pores of diameters between 15 and 400 nm can be made [1, 2]. After etching away the aluminum support, free membranes are obtained (Figure 1.2). What is the characteristic length scale of a porous alumina membrane? The volume of a piece of membrane of facial area A_0 and thickness h_0 is $V = h_0 A_0$. With a radius of the cylindrical pores of r_c and a distance between neighboring pores of d, we get a total surface area $A = N2\pi r_c h_0$. Here, N is the number of pores. We neglect the top and bottom surfaces and the sides. For a two-dimensional hexagonal packing of the pores we get $N = A_0/(d^2 \sin 60°)$. This leads to

$$\lambda_c = \frac{\sin 60°}{2\pi} \frac{d^2}{r_c}. \tag{1.1}$$

With typical values $d = 200$ nm and $r_c = 50$ nm, we get $\lambda_c = 110$ nm. Increasing the radius of the pores at constant interporous spacing decreases the characteristic length scale.

2 μm

Figure 1.2 Scanning electron microscope image of a piece of a porous aluminum oxide membrane lying on top of a continuous membrane. (Courtesy of H. Duran.)

Surface forces determine the structure and dynamics of soft systems with small characteristic length scales. This certainly includes nanosystems with $\lambda_c \leq 100$ nm. Depending on the specific situation, it can also include systems with larger length scales. A typical system is a liquid with dispersed particles. If the particles attract each other, they will aggregate. Often, aggregation is not desired and the surfaces are naturally or artificially designed to give rise to repulsive forces. If the particles repel each other, they do not aggregate and the dispersion is stable. The same arguments hold for emulsions, for example, an oil-in-water emulsion. If the oil drops repel each other, they remain dispersed and the emulsion is stable. Otherwise, they coagulate, form larger and larger drops, and eventually form a continuous oil phase.

Our knowledge of surface forces is mainly about relatively stable systems. Typically, these systems are at or close to equilibrium. We would like to point out that surface forces also determine the dynamics of systems far away from equilibrium. This topic is, however, still in its infancy. Due to the increased number of parameters and the fact that systems far away from equilibrium depend on their specific history, they are more complex and difficult to understand.

2
Van der Waals Forces

In this chapter, we will deal with the forces between neutral molecules due to dipolar interactions. This type of forces are called van der Waals forces to honor the contribution of Johannes Diderik van der Waals[1] in the field of the equation of state for gases and liquids. In his famous doctoral thesis *Over de Continuiteit van den Gas – en Vloeistoftoestand* (On the continuity of the gas and liquid state), he showed the necessity of taking into account the finite volumes of the gas molecules as well as the intermolecular forces to establish the relationship between the pressure, volume, and temperature of gases and liquids. Such intermolecular forces can easily be understood on the basis of electrostatics if at least one of the molecules carries a dipole moment. To explain why even nonpolar molecules are able to attract each other – which is obvious from the fact that gases of such molecules do condense to liquids when cooled to sufficiently low temperatures – is more complex and requires the application of quantum theory.

Van der Waals forces are of universal importance since they exist between any combination of molecules and surfaces. Therefore, essentially any textbook dealing with surface phenomena will contain a section on van der Waals forces. The most comprehensive textbook is that of Parsegian [3], which gives introduction to the topic at different levels (from beginner to full theory) and contains a collection of formulas for many different interaction geometries. Two older much cited textbooks by Langbein [4] and Mahanty and Ninham [5] are no longer in print.

2.1
Van der Waals Forces Between Molecules

2.1.1
Coulomb Interaction

Forces between macroscopic objects result from a complex interplay of the interaction between molecules in the two objects and the medium separating

---

1) Johannes Diderik van der Waals, 1837–1923. Dutch physicist, professor in Amsterdam. Nobel Prize in Physics, 1910.

Surface and Interfacial Forces. Hans-Jürgen Butt and Michael Kappl
Copyright © 2010 Wiley-VCH Verlag GmbH & Co. KGaA
ISBN: 978-3-527-40849-8

them. The starting point for an understanding of intermolecular forces is the Coulomb[2] force. The Coulomb force is the electrostatic force between two charges Q_1 and Q_2:

$$F = \frac{Q_1 Q_2}{4\pi\varepsilon\varepsilon_0 D^2} \cdot$$

$$(2.1)$$

The potential energy between two electrical charges that are a distance D apart is

$$V = \frac{Q_1 Q_2}{4\pi\varepsilon\varepsilon_0 D} \cdot$$

$$(2.2)$$

For charges with opposite sign, the potential energy is negative. They reduce their energy when they get closer. If the two charges are in a medium, the permittivity ε is higher than 1 and the electrostatic force is reduced accordingly.

■ **Example 2.1**

The potential energy between Na^+ and Cl^-, 1 nm apart, in a vacuum is

$$V = -\frac{(1.60 \times 10^{-19} \, C)^2}{4\pi \cdot 8.85 \times 10^{-12} \, A\,s\,V^{-1}\,m^{-1} \cdot 10^{-9} \, m} = -2.30 \times 10^{-19} \, J.$$

This is 56 times higher than the thermal energy $k_B T = 4.12 \times 10^{-21}$ J at room temperature.

2.1.2
Monopole–Dipole Interaction

For most molecules, the total electric charge is zero. Still, the electric charge is often not evenly distributed. A molecule can have a more negative side and a more positive side. In carbon monoxide, for example, the oxygen carries more negative charge than the carbon atom. To first order, the electric properties of such molecules are described by the so-called dipole moment. For the most simple case of two opposite charges Q_d and $-Q_d$ a distance d apart, the dipole moment μ is given by $\mu = Q_d\,d$. It is given in units of Cm (Coulomb × meter). In literature, still often the old unit "Debye" is used. 1 Debye is equal to a positive and a negative unit charge being 0.21 Å apart; it is $1\,D = 3.336 \times 10^{-30}$ Cm. The dipole moment is a vector that points from minus to plus.[3] If we do not have two point charges within the molecule, we have to integrate the charge density ϱ_e over the whole volume of the molecule. This leads to the general

---

[2] Charles Augustin Coulomb, 1736–1806. French physicist and engineer.

[3] Chemists often use a symbol $\longmapsto$ to indicate dipole moments, where "+" marks the positive end of the molecule and the arrow thus points in the direction opposite to the dipole moment.

definition of the dipole moment:

$$\vec{\mu} = \int \varrho_e(\vec{r})\, \vec{r}\, dV.$$ (2.3)

Let us return to intermolecular interactions. For the following calculations, we assume the interaction between charges in vacuum, setting $\varepsilon = 0$. If more than two charges are present, the net potential energy of a charge can be calculated by summing up the contributions of all the other charges. This is called the superposition principle. Using this superposition principle, we can calculate the potential energy between a dipole and a single charge Q:

$$V = \frac{QQ_d}{4\pi\varepsilon_0 r_1} - \frac{QQ_d}{4\pi\varepsilon_0 r_2}.$$ (2.4)

Here, r_1 and r_2 are the distances between the single charge and the two charges of the dipole. With

$$r_1 = D + \frac{d}{2}\cos\vartheta \quad \text{and} \quad r_2 = D - \frac{d}{2}\cos\vartheta,$$ (2.5)

we obtain

$$V(D, \vartheta) = \frac{QQ_d}{4\pi\varepsilon_0 D}\left[\frac{1}{1 + \frac{d}{2D}\cos\vartheta} - \frac{1}{1 - \frac{d}{2D}\cos\vartheta}\right].$$ (2.6)

Under the assumption that the distance between charge and dipole is much larger than the extension of the dipole ($D \gg d$), we can approximate

$$\frac{1}{1 + \frac{d}{2D}\cos\vartheta} \approx 1 - \frac{d}{2D}\cos\vartheta \quad \text{and} \quad \frac{1}{1 - \frac{d}{2D}\cos\vartheta} \approx 1 + \frac{d}{2D}\cos\vartheta.$$ (2.7)

Inserting Eqs. (2.7) into (2.6) gives the final result

$$V(D, \vartheta) = -\frac{QQ_d d\cos\vartheta}{4\pi\varepsilon_0 D^2} = -\frac{Q\mu\cos\vartheta}{4\pi\varepsilon_0 D^2}.$$ (2.8)

■ **Example 2.2**

The maximum potential energy ($\vartheta = 0°$) between a water molecule (dipole moment of 1.85 D) with fixed orientation and a Na$^+$ ion 1 nm apart in a vacuum is

$$V_{max} = -\frac{1.60 \times 10^{-19}\,\mathrm{C} \cdot 1.85 \cdot 3.336 \times 10^{-30}\,\mathrm{C\,m}}{4\pi \cdot 8.85 \times 10^{-12}\,\mathrm{A\,s\,V^{-1}\,m^{-1}} \cdot (10^{-9}\,\mathrm{m})^2} = -8.94 \times 10^{-21}\,\mathrm{J}.$$

This is 2.2 times higher than the thermal energy $k_B T = 4.12 \times 10^{-21}$ J at room temperature.

In practice, a molecule with a dipole moment is often mobile. If the dipole is free to rotate and is, for example, close to a positive charge, it tends to rotate until its negative pole points toward the positive charge, which would correspond to the maximum interaction. For a finite temperature $T > 0$, thermal fluctuations tend to enforce a randomly fluctuating orientation and will drive the dipole away from an optimal orientation. For completely random fluctuations, the average interaction would be zero. However, the orientations with lower potential energy will be favored compared to those with higher potential energy. On average, the balance between these two opposing effects will result in a net preferential orientation and dipole and monopole will attract each other. To calculate the net interaction, we have to integrate over all possible orientations. Each orientation is weighted with a Boltzmann factor. The mean interaction potential is

$$V(D) = \frac{\int_0^\pi V(D, \vartheta)\, e^{-\frac{V(D,\vartheta)}{k_B T}} \sin\vartheta\, d\vartheta}{\int_0^\pi e^{-\frac{V(D,\vartheta)}{k_B T}} \sin\vartheta\, d\vartheta} \tag{2.9}$$

To evaluate the integral, we assume small interaction energies $V(D, \vartheta) < k_B T$ and expand

$$e^{-\frac{V}{k_B T}} \approx 1 - \frac{V}{k_B T}.$$

This leads to

$$
\begin{aligned}
V(D) &= \frac{\dfrac{Q\mu}{4\pi\varepsilon D^2} \displaystyle\int_0^\pi \cos\vartheta \left(1 - \dfrac{Q\mu \cos\vartheta}{4\pi\varepsilon D^2 k_B T}\right) \sin\vartheta\, d\vartheta}{\displaystyle\int_0^\pi \left(1 - \dfrac{Q\mu \cos\vartheta}{4\pi\varepsilon_0 D^2 k_B T}\right) \sin\vartheta\, d\vartheta} \\[2em]
&= \frac{\dfrac{Q\mu}{4\pi\varepsilon D^2} \displaystyle\int_0^\pi \cos\vartheta \sin\vartheta\, d\vartheta + \left(\dfrac{Q\mu}{4\pi\varepsilon D^2}\right)^2 \dfrac{1}{k_B T} \displaystyle\int_0^\pi \cos^2\vartheta \sin\vartheta\, d\vartheta}{\displaystyle\int_0^\pi \sin\vartheta\, d\vartheta - \dfrac{Q\mu}{4\pi\varepsilon D^2} \displaystyle\int_0^\pi \cos\vartheta \sin\vartheta\, d\vartheta}.
\end{aligned}
$$

$$\tag{2.10}$$

The four integrals can be calculated:

$$\int_0^\pi \cos\vartheta \sin\vartheta \, d\vartheta = -\int_1^{-1} x \, dx = 0, \tag{2.11}$$

$$\int_0^\pi \cos^2\vartheta \sin\vartheta \, d\vartheta = -\int_0^\pi \cos^2\vartheta \, d\cos\vartheta = -\int_1^{-1} x^2 dx = \frac{2}{3}, \tag{2.12}$$

$$\int_0^\pi \sin\vartheta \, d\vartheta = 2. \tag{2.13}$$

In Eqs. (2.11) and (2.12), we changed the variables:

$$x = \cos\vartheta \quad \text{and} \quad dx = -\sin\vartheta \, d\vartheta. \tag{2.14}$$

Inserting these results into Eq. (2.10) and taking into account that the free energy is half of the interval energy, we obtain the final result:

$$V(D) = -\frac{Q^2\mu^2}{6(4\pi\varepsilon_0)^2 k_B T D^4}. \tag{2.15}$$

We see that for the case of the freely rotating dipole with thermal motion, the interaction decays much faster with distance, namely, proportional to D^{-4} instead of D^{-2}. In addition, the interaction becomes weaker with increasing temperature.

■ **Example 2.3**

Calculate the free energy between a Na^+ ion and a freely rotating water molecule (dipole moment 6.17×10^{-30} C m) 1 nm apart in vacuum at 25 °C.

$$V = -\frac{(1.60 \times 10^{-19} \text{ C})^2 (6.17 \times 10^{-30} \text{ C m})^2}{6(4\pi \cdot 8.85 \times 10^{-12} \text{ A s V}^{-1} \text{ m}^{-1})^2 \cdot 4.12 \times 10^{-21} \text{ J} \cdot (10^{-9} \text{ m})^4}$$

$$= -3.20 \times 10^{-21} \text{ J}.$$

This is slightly less than the thermal energy $k_B T = 4.12 \times 10^{-21}$ J at the same temperature.

2.1.3
Dipole–Dipole Interaction

To calculate the interaction between two dipoles, we again use the superposition principle to add up the interactions between individual charges. The position of the two dipoles with respect to each other is described by the distance and three angles.

Two angles, denoted by ϑ_1 and ϑ_2, describe the angle between the dipoles and the connecting axis. A third one, ϕ, is the angle between the projections of the dipoles to an area perpendicular to the connecting line. The interaction between two fixed dipoles is given by

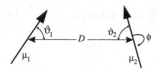

$$V(D, \theta_1, \theta_2, \phi) = \frac{\mu_1\mu_2}{4\pi\varepsilon_0 D^3} (2\cos\vartheta_1\cos\vartheta_2 - \sin\vartheta_1\sin\vartheta_2\cos\phi). \quad (2.16)$$

The maximum attractive interaction is obtained if the dipoles point in the same direction along the axis connecting their centers ($\vartheta_1 = \vartheta_2 = 0°$). It is

$$V_{max}(D) = -\frac{\mu_1\mu_2}{2\pi\varepsilon_0 D^3}. \quad (2.17)$$

■ **Example 2.4**

Calculate the maximum potential energy between two water molecules 1 nm apart in vacuum.

$$V_{max} = -\frac{(6.17 \times 10^{-30}\ \text{C m})^2}{2\pi \cdot 8.85 \times 10^{-12}\ \text{A s V}^{-1}\ \text{m}^{-1} \cdot (10^{-9}\ \text{m})^3} = -6.85 \times 10^{-22}\ \text{J}.$$

This value corresponds to about 17% of the thermal energy $k_B T = 4.12 \times 10^{-21}$ J.

2.1.3.1 Keesom Interaction

When the two dipoles are allowed to rotate freely, we have the same type of balance between preferential orientation of the dipoles and thermal motion as for the case of monopole–freely rotating dipole configuration discussed above. The angular average again has to be calculated by using Eq. (2.9) with the corresponding potential and Boltzmann weighting factors. Two freely rotating dipoles will attract each other because they preferentially orient with their opposite charges facing each other. This thermally averaged dipole–dipole free energy is often referred to as the Keesom energy[4] [6]:

---

4) Wilhelmus Hendrik Keesom, 1876–1956. Dutch physicist, professor in Utrecht and Leiden.

$$V = -\frac{\mu_1^2\mu_2^2}{3(4\pi\varepsilon_0)^2 k_B T D^6} = -\frac{C_{\text{orient}}}{D^6} \cdot \left(\begin{array}{c} \quad\\ \mu_1 \end{array} \overset{\text{---}D\text{---}}{\longleftrightarrow} \begin{array}{c} \quad\\ \mu_2 \end{array} \right) \qquad (2.18)$$

As we can see, the distance dependence has changed from proportional to D^{-3} for the fixed dipoles to D^{-6} for the thermally averaged dipole–dipole interaction. Again, the interaction energy decreases as T^{-1} for increasing temperature.

■ **Example 2.5**

Calculate the free energy between two freely rotating water molecules 1 nm apart in vacuum at room temperature.

$$V = -\frac{(6.17 \times 10^{-30} \text{ C m})^4}{3(4\pi \cdot 8.85 \times 10^{-12} \text{ A s V}^{-1} \text{ m}^{-1})^2 \cdot 4.12 \times 10^{-21} \text{ J} \cdot (10^{-9} \text{ m})^6}$$

$$= -9.5 \times 10^{-24} \text{ J}.$$

This is only 0.2% of the thermal energy $k_B T = 4.12 \times 10^{-21}$ J. Reducing the distance to 0.5 nm increases the potential to 15% of $k_B T$.

All expressions reported so far give the free energies of interaction because they were derived under constant volume conditions. Until now, free energy and internal energy were identical. For the randomly oriented dipole–dipole interaction, entropic effects, namely, the ordering of one dipole by the field of the other dipole, contribute to the free energy. If one dipole approaches another, one-half of the internal energy is taken up in decreasing the rotational freedom of the dipoles as they become progressively more aligned. For this reason, the free energy given in Eq. (2.18) is only half the internal energy. The Gibbs energy is in this case equal to the (Helmholtz) free energy.

2.1.3.2 Debye Interaction
When a charge approaches a molecule without a static dipole moment, all energies considered so far would be zero. Nevertheless, there is an attractive force, which arises from a charge shift in the nonpolar molecule induced by the charge. This induced dipole moment interacts with the charge. The Helmholtz free energy is

$$V = -\frac{Q^2\alpha}{2(4\pi\varepsilon_0)^2 D^4}. \qquad (2.19)$$

Here, α is the polarizability in $\text{C}^2 \text{ m}^2 \text{ J}^{-1}$. The polarizability is defined by $\mu_{\text{ind}} = \alpha E$, where E is the electric field strength. Often it is given as $\alpha/4\pi\varepsilon_0$ in units of Å^{-3}.

■ **Example 2.6**

The polarizability of a water molecule in the gas phase has the value of 1.65×10^{-40} C^2 m^2 J^{-1}. Which dipole moment is induced by a unit charge that is 1 nm away and what is the potential energy between the two?
The electric field of a point charge at a distance D is

$$E = \frac{Q}{4\pi\varepsilon_0 D^2} = 1.44 \times 10^9 \text{ V m}^{-1}.$$

The induced dipole moment is

$$\mu_{ind} = 1.65 \times 10^{-40} \text{ } C^2 \text{ } J^{-1} \text{ } m^2 \cdot 1.44 \times 10^9 \text{ V m}^{-1} = 2.38 \times 10^{-31} \text{ C m}$$

and the potential energy is -1.71×10^{-22} J.

In analogy, a molecule with a static dipole moment will interact with a polarizable molecule by inducing a dipole moment in the polarizable molecule. If the dipoles can freely rotate, the Helmholtz free energy for interaction between a permanent dipole and an induced dipole is

$$V = -\frac{\mu^2\alpha}{(4\pi\varepsilon_0)^2 D^6} = -\frac{C_{ind}}{D^6}. \tag{2.20}$$

This interaction is called the Debye interaction [7]. It will also arise between two identical polarizable molecules that have a permanent dipole moment. In this case, a factor of 2 has to be inserted on the right-hand side of Eq. (2.20).

2.1.3.3 London Dispersion Interaction

All energies considered so far can be calculated using classical physics. However, they fail to explain the attraction between two nonpolar molecules. That such an attraction exists is evident because all gases condense at some temperature. Responsible for this attraction is the so-called London[5] or dispersion force. To calculate the dispersion force, quantum mechanical perturbation theory is required. An impression about the origin of dispersion forces can be obtained by considering an atom with its positively charged nucleus around which electrons circulate with a typically high frequency of 10^{15}–10^{16} Hz. At every instant, the atom is therefore polar. Only the direction of the polarity changes with this high frequency. When two such oscillators approach, they start to influence each other. Attractive orientations have higher probabilities than repulsive ones. As an average, this leads to an attractive force.

The free energy between two molecules with ionization energies $h\nu_1$ and $h\nu_2$ can be approximated by [8]

5) Fritz London, 1900–1954. American physicist of German origin, professor in Durham.

$$V = -\frac{3}{2}\frac{\alpha_1\alpha_2}{(4\pi\varepsilon_0)^2 D^6}\frac{h\nu_1\nu_2}{(\nu_1 + \nu_2)} = -\frac{C_{disp}}{D^6} \tag{2.21}$$

Dispersion interactions increase with the polarizability of the two molecules α_1 and α_2. The optical properties enter in the form of the excitation frequencies. Expression (2.21) only considers one term of a series over dipole transition moments. Usually, this term is by far the most dominant one.

Keesom, Debye, and London contributed much to our understanding of forces between molecules. For this reason, the three different types of dipole interactions are named after them. The van der Waals force is the Keesom plus the Debye plus the London dispersion interaction, thus all the terms that consider dipole–dipole interactions:

$$V_{vdW}(D) = -\frac{C_{vdW}}{D^6} \quad \text{with} \quad C_{vdW} = C_{orient} + C_{ind} + C_{disp}. \tag{2.22}$$

All three terms contain the same distance dependency: the potential energy decreases with $1/D^6$. Usually, the London dispersion term is dominating. This is in most cases true even for polar molecules that interact not only via the Debye and Keesom force but also via the London dispersion forces. In Table 2.1, the contributions of the individual terms for some gases are listed.

Table 2.1 Contributions of the Keesom, Debye, and London potential energy to the total van der Waals interaction between similar molecules as calculated with Eqs. (218),(220), and (2.21) using $C_{total} = C_{orient} + C_{ind} + C_{disp}$.

	μ (D)	$\alpha/4\pi\varepsilon_0$ $(10^{-30}m^3)$	$h\nu$ (eV)	C_{orient}	C_{ind}	C_{disp}	C_{total}	C_{exp}
He	0	0.2	24.6	0	0	1.2	1.2	0.86
Ne	0	0.40	21.6	0	0	4.1	4.1	3.6
Ar	0	1.64	15.8	0	0	50.9	50.9	45.3
CH_4	0	2.59	12.5	0	0	101.1	101.1	103.3
HCl	1.04	2.7	12.8	9.5	5.8	111.7	127.0	156.8
HBr	0.79	3.61	11.7	3.2	4.5	182.6	190.2	207.4
HI	0.45	5.4	10.4	0.3	2.2	364.0	366.5	349.2
$CHCl_3$	1.04	8.8	11.4	9.5	19.0	1058	1086	1632
CH_3OH	1.69	3.2	10.9	66.2	18.3	133.5	217.9	651.0
NH_3	1.46	2.3	10.2	36.9	9.8	64.6	111.2	163.7
H_2O	1.85	1.46	12.6	95.8	10.0	32.3	138.2	176.2
CO	0.11	1.95	14.0	0.0012	0.047	64.0	64.1	60.7
CO_2	0	2.91	13.8	0	0	140.1	140.1	163.6
N_2	0	1.74	15.6	0	0	56.7	56.7	55.3
O_2	0	1.58	12.1	0	0	36.2	36.2	46.0

They are given in units of 10^{-79} J m^6. For comparison, the van der Waals coefficient C_{exp} as derived from the van der Waals equation of state for a gas $(P + a/V_m^2)(V_m - b) = RT$ is tabulated. From the experimentally determined constants a and b, the van der Waals coefficient can be calculated with $C_{exp} = 9ab/(4\pi^2 N_A^3)$ assuming that at very short range the molecules behave like hardcore particles. Dipole moments μ, polarizabilities α, and the ionization energies $h\nu$ of isolated molecules are also listed.

The data in Table 2.1 demonstrate that even for molecules with permanent dipole moment, the London dispersion interaction usually dominates. An exception is the water molecule, where a strong permanent dipole moment is combined with the small size of the molecule. Small molecules usually have a low polarizability.

So far, we have implicitly assumed that the molecules stay so close to each other that the propagation of the electric field is instantaneous. For the London dispersion interaction, however, this is not necessarily true. To illustrate this, let us have a closer look at what happens when two molecules interact. In one molecule, a spontaneous random dipole moment arises, which generates an electric field. The electric field expands with the speed of light. It polarizes the second molecule, whose dipole moment in turn causes an electric field that reaches the first molecule with the speed of light. The process takes place as calculated only if the electric field has enough time to cover the distance D between the molecules before the fluctuating dipole moment has completely changed again. The exchange of interaction takes a time $\Delta t = D/c$, where c is the speed of light. If the first dipole changes faster than Δt, the interaction becomes weaker. The time during which the dipole moment changes is in the order of $1/\nu$. Hence, only if

$$\frac{D}{c} < \frac{1}{\nu}, \tag{2.23}$$

the interaction takes place as considered. The relevant frequencies are those corresponding to the ionization of the molecule, which are typically 3×10^{15} Hz. Thus, for $D > c/\nu \approx 3 \times 10^8 / 3 \times 10^{15}$ m = 10 nm, the van der Waals interaction decreases more steeply (i.e., for molecules with $1/D^7$) than for smaller distances. This effect is known as retardation, and one speaks about retarded van der Waals interaction and forces. We will discuss this in more detail in Section 2.4.

2.2
The Van der Waals Force Between Macroscopic Solids

We move from the interaction between two molecules to the interaction between two macroscopic solids. It was recognized soon after London had published his explanation of the dispersion forces that dispersion interaction could be responsible for the attractive forces acting between macroscopic objects. This idea led to the development of a theoretical description of van der Waals forces between macroscopic bodies based on the pairwise summation of the forces between all molecules in the objects. This concept was developed by Hamaker[6] [9] based on earlier work by Bradley [10] and de Boer [11]. This microscopic approach of Hamaker of pairwise summation of the dipole interactions makes the simplifying assumption that the

6) Hugo Christiaan Hamaker, 1905–1993, Dutch scientist at the Philips Laboratories and professor in Eindhoven.

dipolar interaction between a pair of molecules is not influenced by the presence of a third molecule with a permanent or induced dipole moment. This is strictly fulfilled only in the case of dilute gases but not solids, so we cannot expect precise quantitative predictions from this theory. We will nevertheless start out with the Hamaker approach, since it is instructive to get a qualitative understanding of van der Waals interactions between macroscopic objects.

2.2.1
Microscopic or Hamaker Approach

As derived above, the potential energy of the van der Waals interaction between two molecules A and B is given by

$$V_{AB}(D) = -\frac{C_{AB}}{D^6}. \tag{2.24}$$

The minus sign arises because it is an attractive force. C_{AB} is a material specific constant and equal to C_{total} in Eq. (2.22). It sums up contributions of all three dipole–dipole interactions.

In order to determine the interaction between macroscopic solids, in the first step we calculate the van der Waals energy between molecule A and an infinitely extended body with a planar surface made of molecules B. This is also relevant in understanding the adsorption of gas molecules to surfaces. We sum up the van der Waals energy between molecule A and all molecules in solid B. Practically, this is done by integration of the molecular density ϱ_B over the entire volume of the solid:

$$V_{Mol/plane} = -C_{AB} \iiint \frac{\varrho_B}{D'^6} dV = -C_{AB}\varrho_B \int_0^\infty \int_0^\infty \frac{2\pi r\, dr\, dx}{[(D+x)^2 + r^2]^3}. \tag{2.25}$$

We have used cylindrical coordinates (Figure 2.1) and we have assumed that the density of molecules B in the solid is constant. With $2r\, dr = d(r^2)$, we get

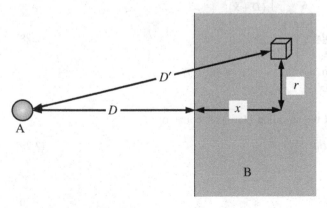

Figure 2.1 Calculating the van der Waals force between a macroscopic body and a molecule.

$$V_{\text{Mol/plane}} = -\pi\varrho_B C_{AB} \int_0^\infty \int_0^\infty \frac{d(r^2)}{[(D+x)^2 + r^2]^3} dx$$

$$= -\pi\varrho_B C_{AB} \int_0^\infty \left[-\frac{1}{2[(D+x)^2 + r^2]^2} \right]_0^\infty dx$$

$$= -\frac{\pi\varrho_B C_{AB}}{2} \int_0^\infty \frac{1}{(D+x)^4} dx \tag{2.26}$$

$$= -\frac{\pi\varrho_B C_{AB}}{2} \left[-\frac{1}{3(D+x)^3} \right]_0^\infty$$

$$= -\frac{\pi\varrho_B C_{AB}}{6D^3}.$$

Here, we already see a remarkable property of macroscopic van der Waals forces: the energy of a molecule and a macroscopic body decreases less steeply than the energy between two molecules. Instead of the D^{-6} dependence, the energy falls off with distance proportional to D^{-3}.

In the second step, we calculate the van der Waals energy between two infinitely extended solids that are separated by a parallel gap of thickness D. Therefore, we use Eq. (2.26) and integrate over all molecules in solid A:

$$V = -\frac{\pi C_{AB}\varrho_B}{6} \iiint \frac{\varrho_A}{(D+x)^3} \, dV = -\frac{\pi C_{AB}\varrho_A\varrho_B}{6} \int_0^\infty \int_{-\infty}^\infty \int_{-\infty}^\infty \frac{dz\,dy\,dx}{(D+x)^3}. \tag{2.27}$$

Here, y and z are the coordinates parallel to the gap and ϱ_A is the density of molecules in material A. Again, for simplicity we assume that A is a homogeneous material and ϱ_A is constant. The integral is infinite because the solids are infinitely large. We have to divide by the area. The van der Waals energy per unit area is

$$V^A = \frac{V}{A} = -\frac{\pi C_{AB}\varrho_A\varrho_B}{6} \int_0^\infty \frac{dx}{(D+x)^3}$$

$$= -\frac{\pi C_{AB}\varrho_A\varrho_B}{6} \left[-\frac{1}{2(D+x)^2} \right]_0^\infty \tag{2.28}$$

$$= -\frac{\pi C_{AB}\varrho_A\varrho_B}{12D^2}.$$

With the definition of the so-called Hamaker constant

$$A_H = \pi^2 C_{AB}\varrho_A\varrho_B, \tag{2.29}$$

we get

$$V^A = -\frac{A_H}{12\pi D^2}. \tag{2.30}$$

The force per unit area is equal to the negative derivative of V^A versus distance:

$$f = -\frac{dV^A}{dD} = -\frac{A_H}{6\pi D^3}. \tag{2.31}$$

In the same way, it is possible to calculate the van der Waals energy between solids having different geometries. One important case is the interaction between two spheres with radii R_1 and R_2, which was also studied in Hamaker's original publication [9]. He found a van der Waals energy of

$$V = -\frac{A_H}{6}\left[\frac{2R_1R_2}{d_{cc}^2-(R_1+R_2)^2} + \frac{2R_1R_2}{d_{cc}^2-(R_1-R_2)^2} + \ln\left(\frac{d_{cc}^2-(R_1+R_2)^2}{d_{cc}^2-(R_1-R_2)^2}\right)\right]. \tag{2.32}$$

Here, d_{cc} is the distance between the centers of the spheres. The distance between the surfaces is $D = d_{cc} - R_1 - R_2$.

If the radii of the spheres are substantially larger than the distance ($D \ll R_1, R_2$), Eq. (2.32) can be simplified (Ref. [12], p. 543):

$$W = -\frac{A_H}{6D} \cdot \frac{R_1R_2}{R_1+R_2}. \tag{2.33}$$

The van der Waals force is the negative derivative:

$$F = -\frac{A_H}{6D^2} \cdot \frac{R_1R_2}{R_1+R_2}. \tag{2.34}$$

For a sphere and a flat, planar surface, the energy and force can be obtained by letting R_2 go to infinity:

$$V = -\frac{A_H R_p}{6D} \quad \text{and} \quad F = -\frac{A_H R_p}{6D^2}. \tag{2.35}$$

For consistency, we renamed the radius of the first sphere to R_p; index "p" for particle.

■ **Example 2.7**

A spherical silica (SiO_2) particle hangs on a planar silica surface caused by the van der Waals attraction of $F = A_H R_p/6D^2$. The van der Waals attraction increases linearly with the radius of the sphere. The gravitational force, which pulls the sphere down, is $4\pi R_p^3 \varrho g/3$. It increases cubically with the radius. As a consequence, the behavior of small spheres is dominated by van der Waals forces, while for large spheres gravity is more important. At which radius is the gravitational force so strong that the sphere detaches?

Use a Hamaker constant of $A_H = 6 \times 10^{-20}$ J = 60 zJ, the density is $\varrho = 2600$ kg m^{-3}. For two solids in contact, we take the distance to be 1.7 Å, which corresponds to a typical interatomic spacing. We use the condition that just before the sphere falls down, the van der Waals force is equal to the gravitational force:

$$\frac{A_H R_p}{6 D^2} = \frac{4}{3} \pi R_p^3 \varrho g \Rightarrow R_p = \sqrt{\frac{A_H}{8\pi D^2 \varrho g}} \Rightarrow$$

$$R_p = \left(\frac{6 \times 10^{-20} \, \text{J}}{8\pi \cdot (1.7 \times 10^{-10} \, \text{m})^2 \cdot 2600 \, \text{kg m}^{-3} \cdot 9.81 \, \text{m s}^{-1}} \right)^{1/2} = 1.8 \, \text{mm}.$$

In reality, the value is certainly lower due to surface roughness and contamination. Roughness and contamination layers increase the effective contact distance.

As already mentioned, the microscopic approach by Hamaker to sum up the molecular contributions is questionable since the pairwise additivity is usually not fulfilled. Higher order interactions could in principle be taken into account as demonstrated for the case of three interacting molecules [13]. With a simple semiclassical discussion of the problem, one finds that the higher order terms are smaller by a factor $(\alpha/4\pi\varepsilon_0 D^3)^N$ for the Nth order [14]. Thus, we can neglect higher order terms if $\alpha/4\pi\varepsilon_0 D^3 \ll 1$. When plugging in typical values, we see that for $D \approx 1$ nm, this conditions is still fulfilled and thus corrections due to higher orders should be reasonably small. However, the Lifshitz theory, which we will discuss below, gives precise predictions for the Hamaker constant, allows to take the influence of an intervening medium into account, and uses easier to access bulk material properties for calculations. It is therefore the method of choice for quantitative calculations.

An important contribution of Hamaker (and de Boer and Bradley) was to recognize that for macroscopic bodies, the distance dependence of the van der Waals forces can be significantly different from the D^{-6} and thus much longer ranged than originally expected from the D^{-6} dependence for single molecules. This finding implies that van der Waals forces between macroscopic objects can reach significant values if an intimate contact (i.e., small D) is achieved.

■ **Example 2.8**

For a typical value of the Hamaker constant of $A_H = 10 \, \text{zJ} = 10^{-20} \, \text{J}$ and a minimum separation of 1.7 Å, we get a force per unit area of strength

$$f = \frac{A_H}{6\pi D^3} = \frac{10^{-20} \, \text{J}}{6\pi(1.7 \times 10^{-10} \, \text{m})^3} = 1.1 \times 10^8 \, \text{Pa}. \tag{2.36}$$

Thus, perfectly smooth and clean surfaces in contact and with a contact area of $1 \, \text{cm}^2$ could carry a load of ≈ 1.1 tons. However, if the distance is increased to 10 nm, the value drops to a mere

$$f = \frac{A_H}{6\pi D^3} = \frac{10^{-20} \, \text{J}}{6\pi(10^{-8} \, \text{m})^3} = 530 \, \text{Pa}, \tag{2.37}$$

which corresponds to a load of only 5 g per $1 \, \text{cm}^2$. This is a reduction by more than five orders of magnitude but still a significant and clearly detectable force.

Figure 2.2 Picture of a Gecko taken through a glass plate (left) and scanning electron microscope images of the setae and spatulae (right). (Images were kindly provided by S. Gorb.)

This example already demonstrates the importance of surface roughness for adhesion forces. For rough surfaces, the distance of closest approach will usually be limited by the height of surface asperities. One possible approach to achieve high adhesion on rough surfaces is the use of soft or adaptive surfaces that can conform with the roughness.

◼ **Example 2.9**

Geckos are able to run up and down walls, and even walk along the ceiling head downward. Therefore, they have evolved one of the most versatile and effective mechanisms to adhere to a surface. Geckos accomplish this by using van der Waals forces [15, 16]. Their feet are covered with millions of tiny foot hair, called setae; typically, they have 14 000 setae per mm^2. These setae run over into many even smaller spatulae (Figure 2.2). The soft nanostructures can easily comply with surfaces and even out the surface roughness. In this way, the Gecko achieves a high total contact area, which leads to van der Waals forces strong enough to carry its weight.

2.2.2
Macroscopic Calculation: Lifshitz Theory

An important step toward a quantitative description of van der Waals forces was the publication of Casimir and Polder in 1948 [17]. Using recent developments in quantum electrodynamics, they calculated the retarded dispersion forces between an atom and a metallic plate in vacuum. In the same year, Casimir also published his well-known paper on the attraction between two parallel perfectly conducting plates in vacuum [18], where he derived the force by considering quantum fluctuations in vacuum (for details see Section 2.6). Casimir was the first to recognize that the optical properties of the metal plates are directly linked to the production of virtual photons and the intrinsic polarizability of the metal plates. This shift in paradigm from the microscopic view of single interacting dipoles to the electromagnetic fluctuations that are connected to the bulk material properties significantly advanced our understanding of van der Waals forces.

From 1954 to 1956, Lifshitz derived the theoretical description for the forces between two parallel plates of dielectric materials across a vacuum [19]. This theory was extended together with Dzyaloshinskii and Pitaevskii between 1959 and 1961 to include the effect of a third dielectric filling the gap between the plates [20]. However, the complicated structure of their solution hindered its widespread acceptance and initially caused doubt of its practical use [21]. A simplified derivation of the van der Waals forces between parallel plates was introduced by van Kampen *et al.* [22] based on a model in which the fluctuations were represented by a sum of harmonic oscillators. Since the bulk modes are independent of distance between the surfaces, only surface modes contribute to the van der Waals force. Based on the van Kampen calculations, Parsegian and Ninham showed in a series of papers in 1970 that the van der Waals forces could be calculated based on available dielectric data [23]. This paned the way for a general quantitative description of van der Waals forces.

Before we go into detail with the equations, let us briefly discuss the main features of the van der Waals forces in the light of the Lifshitz theory. We will use the configuration of two half-spaces made of material 1 and material 2 separated by a distance. From now on we denote that distance by x, rather than D. The symbol D will be reserved for the distance between objects and x is used for the distance between two half-spaces separated by a gap with two parallel infinitely extended planar interfaces. The intervening medium is made of material 3. In such a configuration, the interaction between materials 1 and 2 occurs due to correlations of the electromagnetic fluctuations, which corresponds to the exchange of virtual photons between materials 1 and 2. An intervening material 3 with a refractive index >1 will influence the exchange of the virtual photons. In a simplified picture, the virtual photons have now to travel a longer optical path, which corresponds to a larger physical distance and thus a reduced interaction.

The main finding of the Lifshitz theory was that the equations derived with the Hamaker approach are essentially still valid:

$$V^A(x) = -\frac{A_H}{12\pi x^2}.$$ (2.38)

The calculation of van der Waals forces for simple geometries comes essentially down to determination of the Hamaker constant A_H. However, the Hamaker constant has to be calculated from the bulk dielectric and magnetic properties of the materials. Since for most materials the magnetic contribution is negligible compared to the dielectric, we assume that the magnetic susceptibilities are unity. For equations including magnetic contributions, we refer to Ref. [3]. In Lifshitz theory, the nonretarded Hamaker constant A_{132} for two half-spaces made of materials 1 and 2 separated by an intervening medium 3 is given by

$$A_{132} = \frac{3}{2}k_B T \sum_{m=0}^{\infty}{}' \int_0^{\infty} x \ln(1 - \Delta_{13}\Delta_{23}e^{-x})\,dx,$$ (2.39)

with

$$\Delta_{ij} = \frac{\varepsilon_i(i\omega_m) - \varepsilon_j(i\omega_m)}{\varepsilon_i(i\omega_m) + \varepsilon_j(i\omega_m)} \qquad (2.40)$$

and

$$\omega_m = \frac{2\pi m k_B T}{\hbar} \quad \text{with} \quad m = 1, 2, 3, 4, \ldots \qquad (2.41)$$

being the so-called Matsubara angular frequencies. The prime in the sum over m in Eq. (2.39) denotes that the term with $m = 0$ has to be multiplied with $1/2$. The so-called London dispersion dielectric response function $\varepsilon(i\omega)$ with *imaginary* frequencies $i\omega$ is itself not a physical property of the materials but is related to the real physical dielectric response function $\varepsilon(\omega)$ of the materials

$$\varepsilon(\omega) = \varepsilon'(\omega) + i\varepsilon''(\omega) \qquad (2.42)$$

by the Kramers–Kronig relation

$$\varepsilon(i\omega) = 1 + \frac{2}{\pi} \int_0^\infty \frac{\omega \varepsilon''(\omega)}{\omega^2 + \omega^2} \, d\omega. \qquad (2.43)$$

Note that $\varepsilon(i\omega)$ itself is not a complex but a real quantity.

The lowest value of the photon energy $\hbar\omega_m$ for $m = 1$ given by Eq. (2.41) at room temperature is 0.159 eV, which corresponds to the IR range (frequency 3.84×10^{13} Hz, wavelength $\lambda = 7.8\,\mu m$). Summation of Eq. (2.39) typically has to be done up to values of $m \approx 1000$ or more in order to achieve good convergence of the sum. This corresponds to wavelength up to the far UV. In fact, most of the terms contributing to the sum will come from the UV portion of the spectrum (Figure 2.3). For very high frequencies, all dielectric permittivities approach unity and the difference in Eq. (2.40) becomes zero.

The term with $m = 0$ – commonly called zero frequency contribution – in the sum of Eq. (2.39) contains the dielectric response functions of the materials evaluated at zero frequency. These are identical to the static dielectric constants of the materials. Actually, the zero frequency term corresponds to the sum of the Debye and Keesom interactions. In most cases, the zero frequency term will make only a minor

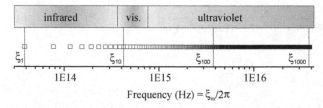

Figure 2.3 Matsubara frequencies (drawn in units of Hz, while ω_m has the unit rad s^{-1} = 2π Hz), at which the terms of the sum in Eq. (2.39) have to be evaluated. All values are for room temperature. The IR and optical range contribute only with a few values. Most contributions come from the UV.

contribution to the total interactions. However, there are exceptions. For substances with large permanent dipole moment, the Debye interaction can become significant. It should also be noted that the Δ_{ij} terms defined by Eq. (2.40), which are used to sum up the total interaction, depend on *differences* in dielectric properties of the materials. This implies that for materials that have similar dielectric properties in the UV range, the static or IR part of the interaction may become important.

For a numerical calculation of A_{132}, an alternative form of Eq. (2.39) can be used [5]:

$$A_{132} = \frac{3}{2}k_B T \sum_{m=0}^{\infty}{}' \sum_{s=1}^{\infty} \frac{[\Delta_{13}(i\omega_m)\Delta_{23}(i\omega_m)]^s}{s^3}. \tag{2.44}$$

The sum over $s = 1, 2, 3, \ldots$ is a quickly converging sum, and for approximate calculation, the $s = 1$ term is already sufficient, which gives the simplified equation

$$A_{132} = \frac{3}{2}k_B T \sum_{m=0}^{\infty}{}' [\Delta_{13}(i\omega_m)\Delta_{23}(i\omega_m)]. \tag{2.45}$$

The difficulty in calculating A_H from the above equations arises from the necessity that the dielectric response function $\varepsilon''(\omega)$ of all materials involved must be known for the full range of relevant frequencies $\nu = \omega/2\pi$, from the infrared to the far ultraviolet. Lack of this information was a critical obstacle for the practical use of the Lifshitz theory. This issue was solved first by using approaches that constructed a function of $\varepsilon(i\omega)$ by interpolating the limited amount of spectral data available. Today, new experimental methods enable a much more detailed determination of $\varepsilon(i\omega)$ resulting in more precise values of A_H. These methods are primarily vacuum ultraviolet (VUV) spectroscopy [24] and valence electron energy loss spectroscopy (VEELS) that can be carried out in electron microscopes [25].

Let us discuss the most important approaches to obtaining values of A_H for given spectroscopic data.

Simple Spectral Method [23] In the simple spectral method, a model dielectric response function is used. It combines a Debye relaxation term to describe the response at microwave frequencies with a sum of terms of classical form of Lorentz electron dispersion (corresponding to a damped harmonic oscillator model) for the frequencies from IR to UV:

$$\varepsilon(i\omega) = 1 + \frac{C_{mw}}{1 + \omega/\omega_{mw}} + \sum_j \frac{C_j}{1 + (\omega/\omega_j)^2 - \chi_j\omega} \tag{2.46}$$

$$\approx 1 + \frac{C_{mw}}{1 + \omega/\omega_{mw}} + \sum_j \frac{C_j}{1 + (\omega/\omega_j)^2}, \tag{2.47}$$

where ω_{mw} is derived from the characteristic relaxation wavelength, which is typically in the microwave range. C_{mw} is simply the difference $n_0 - n_{mw}$, where n_0 is the static refractive index and n_{mw} is the refractive index at microwave frequencies and the damping term χ_j is assumed to be small. The values of C_j and ω_j are obtained using the relation for the refractive index n:

$$n(\omega)^2 - 1 = \left[n(\omega)^2 - 1\right] \frac{\omega^2}{\omega_j^2} + C_j. \tag{2.48}$$

By plotting $n^2 - 1$ versus $(n^2 - 1)\omega^2$ – the so-called Cauchy plot – both quantities are directly obtained from the slope and intersection of a line fit.

For many materials, the simple spectral model gives reasonable estimates of the Hamaker constant when the microwave term is omitted and a single oscillator term is used. An example for a material with a more complex spectral response is water. The original parameterization of Ninham with a microwave term and just one oscillator term for IR and UV each [23] was improved in several steps to a model including a relaxation term, five IR, and six UV oscillator terms [26, 27]. Two alternative models closer to a full spectral analysis for water have recently been proposed by Dagastine *et al.* [28] and Fernandez-Varea and Garcia-Molina [29].

Tabor–Winterton Approximation [30, 31] This approximation was originally introduced by Tabor and Winterton for the interpretation of their SFA measurements. It uses the following approximations to describe the spectral response:

- Contributions from the infrared part of the spectrum are ignored, since the ultraviolet part is expected to contribute most.
- Absorption in the UV occurs within a narrow frequency band and can therefore be described by a single oscillator model.
- The resonance frequency in the UV is assumed to be the same for all three materials and equal to the plasma frequency ν_e of about 3×10^{15} Hz.
- Absorption in the UV is estimated from the refractive index in the range of visible light $n^2 = \varepsilon_{vis}$.

This corresponds to a dielectric response function

$$\varepsilon(i\omega) = 1 + \frac{n^2 - 1}{1 + \omega^2/(2\pi\nu_e)^2}. \tag{2.49}$$

Using Eq. (2.44) and taking only the value for $s = 1$ into account (higher values are increasingly smaller since the Δ_{ij} are smaller than 1), the Hamaker constant for the interaction of materials 1 and 2 over an intervening material 3 can be calculated by replacing the sum over m with an integral for $m > 1$. One obtains the useful relation [31]

$$A_{132} \approx \frac{3}{4} k_B T \left(\frac{\varepsilon_1 - \varepsilon_3}{\varepsilon_1 + \varepsilon_3}\right)\left(\frac{\varepsilon_2 - \varepsilon_3}{\varepsilon_2 + \varepsilon_3}\right)$$
$$+ \frac{3h\nu_e}{8\sqrt{2}} \frac{(n_1^2 - n_3^2)(n_2^2 - n_3^2)}{\sqrt{n_1^2 + n_3^2} \cdot \sqrt{n_2^2 + n_3^2} \cdot \left(\sqrt{n_1^2 + n_3^2} + \sqrt{n_2^2 + n_3^2}\right)}. \tag{2.50}$$

For the case of two identical materials 1 interacting over an intervening material 3, this equation further simplifies to

$$A_{131} = \frac{3}{4} k_B T \frac{\varepsilon_1 - \varepsilon_3}{\varepsilon_1 + \varepsilon_3} + \frac{3\pi h \nu_e}{8\sqrt{2}} \frac{\left(n_1^2 - n_3^2\right)^2}{\left(n_1^2 + n_3^2\right)^{3/2}}. \tag{2.51}$$

In Eqs. (2.50) and (2.51), the first term corresponds to the static ($\omega = 0$) dielectric response, which is due to the Keesom and Debye interaction and the second term corresponds to the London dispersion interaction. Except for materials with high values of the dielectric constant (e.g., water), the first term will be much smaller than the second one. In their original paper, Tabor and Winterton ignored this term since in their case medium 3 was vacuum ($n_3 = 1$).

Full Spectral Method [32]   The full spectral method is based on a comprehensive spectral characterization of all involved materials. This involves as first step the measurement of optical properties over a broad spectral range from IR to far UV; alternatively, an *ab initio* band structure calculation is also possible. One possible approach is to measure reflectance over the whole frequency range using IR and VUV spectroscopy. The measured reflected amplitude $\mathcal{R}(h\nu)$ is used to obtain the so-called reflection phase $\phi(h\nu)$ using the corresponding Kramers–Kronig relation

$$\phi(h\nu) = -\frac{2h\nu}{\pi} \mathcal{P} \int_0^\infty \frac{\ln \mathcal{R}(h\nu')}{(h\nu')^2 - (h\nu)^2} \, dh\nu'. \tag{2.52}$$

The symbol $\mathcal{P}$ denotes that the principal value of the integral has to be taken (omitting the divergency at $\nu' = \nu$).

The difficulty in this calculation is that measured reflection results are often not easily available over the whole frequency range. Therefore, an interpolation of the data to values of low (0 eV) and high (≈ 1000 eV) energy values may be required [33]. With known values of $\mathcal{R}(h\nu)$ and $\phi(h\nu)$, $\varepsilon''(\omega)$ and $\varepsilon''(\omega)$ are obtained from the equations for the complex refractive index $n = n' + in''$ and complex dielectric function $\varepsilon = \varepsilon' + i\varepsilon''$:

$$\frac{n' - 1 + in''}{n' + 1 + in''} = \mathcal{R}(h\nu) \exp\left(i\phi(h\nu)\right), \tag{2.53}$$

$$\varepsilon' + i\varepsilon'' = (n' + in'')^2. \tag{2.54}$$

The values of $\varepsilon''(\omega)$ can then serve as input for Eq. (2.43) to obtain the dielectric response function $\varepsilon(i\omega)$ for each material. Finally, the Hamaker constant is calculated using Eq. (2.44).

The full spectral method is much more involved than the other approaches. However, once the dielectric response function for a certain material has been determined, it can be stored in a spectroscopic database that will then allow fast and automated calculations of Hamaker constants for different material combinations. For a detailed discussion on this topic, see Ref. [34]. By the use of VEELS in TEM, it is even possible to resolve variations in the dielectric properties on the nanometer scale and calculate from these spectra the dispersion interaction across grain boundaries, which were found to amount to up to 10% of the total grain boundary energies [35].

A comparison of the different methods for calculations of A_{132}, where medium 3 was vacuum [36], showed that the Tabor–Winterton approach tended to overestimate the values of A_{132} for materials with high refractive index $n > 1.8$ but gives reasonable values for materials with $n < 1.8$. If sufficient spectral information is available, using the full spectral method is certainly the first choice. For the case of molecular clusters, even *ab initio* calculations of Hamaker constants are feasible [37].

However, in case of limited spectral data or as a quick estimate, Eqs. (2.50) and (2.51) are very helpful since they allow the calculation of the Hamaker constant from the refractive indices, dielectric permittivities, and UV absorption frequency. A list of these values for some selected materials is given in Table 2.2.

■ **Example 2.10**

Calculate the Hamaker constant for the interaction of amorphous silicon oxide (SiO$_2$) with silicon oxide across water at 20 °C. According to Table 2.2, we insert $\varepsilon = 78.5$ and $n = 1.33$ for water, $\varepsilon = 3.82$ and $n = 1.46$ for silicon oxide, and $\nu_e = 3.4 \times 10^{15}$ Hz for the mean absorption frequency, into Eq. (2.50):

$$A_H = 0.304 \times 10^{-21} \, \text{J} \cdot \left(\frac{3.82 - 78.5}{3.82 + 78.5} \right)^2$$

$$+ 59.7 \times 10^{-20} \, \text{J} \cdot \frac{(1.46^2 - 1.33^2)^2}{(1.46^2 + 1.33^2) \cdot \left(2\sqrt{1.46^2 + 1.33^2} \right)}$$

$$= 0.250 \times 10^{-21} \, \text{J} + 5.10 \times 10^{-21} \, \text{J} = 5.35 \times 10^{-21} \, \text{J}.$$

Equation (2.50) not only allows us to calculate the Hamaker constant, but also allows us to easily predict whether we can expect attraction or repulsion. An attractive van der Waals force corresponds to a positive sign of the Hamaker constant, repulsion corresponds to a negative Hamaker constant. Van der Waals forces between similar materials are always attractive. This can easily be deduced from Eq. (2.50) using $\varepsilon_1 = \varepsilon_2$ and $n_1 = n_2$: the Hamaker constant is positive, which corresponds to an attractive force. If two different media interact across vacuum ($\varepsilon_3 = n_3 = 1$), or practically a gas, the van der Waals force is also attractive. Van der Waals forces between different materials across a condensed phase can be repulsive. Repulsive van der Waals forces occur when medium 3 is more strongly attracted to medium 1 than medium 2. Repulsive forces were, for instance, measured for the interaction of silicon nitride with silicon oxide in diiodomethane [40]. Repulsive van der Waals forces can also occur across thin films on solid surfaces. In the case of thin liquid films on solid surfaces, there is often a repulsive van der Waals force between the solid–liquid and the liquid–gas interface [41], which stabilizes the film (see Section 7.6). This prevents flotation.

Table 2.2 Permittivity ε, refractive index n, and main absorption frequency ν_e in the UV for various solids, liquids, and polymers at 20 °C (Refs. [31, 38, 39], handbooks, and own measurements).

Material	ε	n	ν_e (10^{15} Hz)
Al_2O_3 (alumina)	9.3–11.5	1.75	3.2
C (diamond)	5.7	2.40	2.7
$CaCO_3$ (calcium carbonate, average)	8.2	1.59	3.0
CaF_2 (fluorite)	6.7	1.43	3.8
$KAl_2Si_3AlO_{10}(OH)_2$ (muscovite mica)	5.4	1.58	3.1
KCl (potassium chloride)	4.4	1.48	2.5
NaCl (sodium chloride)	5.9	1.53	2.5
Si_3N_4 (silicon nitride, amorphous)	7.4	1.99	2.5
SiO_2 (quartz)	4.3–4.8	1.54	3.2
SiO_2 (silica, amorphous)	3.82	1.46	3.2
TiO_2 (titania, average)	114	2.46	1.2
ZnO (zinc oxide)	11.8	1.91	1.4
Acetone	20.7	1.359	2.9
Chloroform	4.81	1.446	3.0
n-Hexane	1.89	1.38	4.1
n-Octane	1.97	1.41	3.0
n-Hexadecane	2.05	1.43	2.9
Ethanol	25.3	1.361	3.0
1-Propanol	20.8	1.385	3.1
1-Butanol	17.8	1.399	3.1
1-Octanol	10.3	1.430	3.1
Toluene	2.38	1.497	2.7
Water	78.5	1.333	3.6
Polyethylene	2.26–2.32	1.48–1.51	2.6
Polystyrene	2.49–2.61	1.59	2.3
Poly(vinyl chloride)	4.55	1.52–1.55	2.9
Poly(tetrafluoroethylene)	2.1	1.35	4.1
Poly(methyl methacrylate)	3.12	1.50	2.7
Poly(ethylene oxide)		1.45	2.8
Poly(dimethyl siloxane)	2.6–2.8	1.4	2.8
Nylon 6	3.8	1.53	2.7
Bovine serum albumin	4.0		2.4–2.8

The above analysis applies to insulating materials. For electrically conductive materials such as metals, the dielectric constant is infinite and equations such as Eq. (2.47) do no longer apply. In this case, we can approximate the dielectric response function of the metal by

$$\varepsilon(i\omega) = 1 + \frac{2\pi\nu_e^2}{\omega^2}, \tag{2.55}$$

where ν_e is the so-called plasma frequency of the electron gas; it is typically 5×10^{15} Hz. Using Eq. (2.55), instead of Eq. (2.45), leads to the approximate

Hamaker constant

$$A_H \approx \frac{3}{16\sqrt{2}} h v_e \approx 4 \times 10^{-19} \, J \tag{2.56}$$

for two metals interacting across a vacuum. Thus, the Hamaker constants of metals and metal oxides can be more than an order of magnitude higher than those of nonconducting media. Much effort to achieve detailed modeling of the (retarded) van der Waals forces between metals has been made in recent years in the research on the Casimir force (see Section 2.6).

While a calculation of Hamaker constants using the full spectral information should give the most reliable results, this information is often not available, and of course experimental proof of theoretical calculations is always desirable. The experimental determination of Hamaker has followed several approaches. Direct measurement of van der Waals forces (see Section 2.5) has been carried out using the SFA and more often AFM in combination with the colloid probe technique. The latter allows more flexibility in the choice of materials. By fitting the experimentally observed forces, Hamaker constants can be derived. Van der Waals interactions between colloids can be estimated from sedimentation experiments using ultracentrifugation [42] or field flow fragmentation [43], from flocculation kinetics [44] or the critical yield stress in rheology [45]. Other methods that are based on the interaction with defined test substances are inverse gas chromatography [46, 47], immersion calorimetry [48], and contact angle measurements for the determination of surface energies (see Section 2.23).

In Table 2.3, nonretarded Hamaker constants are listed for different material combinations. Hamaker constants, calculated from spectroscopic data, are found in many publications [29, 36, 38, 49–54]. Reviews are given in Refs [34, 39].

2.2.2.1 Combining Relations for Hamaker Constants

There are several equations that can be used to calculate Hamaker constants between different combinations of materials from known Hamaker constants. A useful one is

$$A_{132} \approx A_{12} - A_{32} - A_{13} + A_{33}, \tag{2.57}$$

where A_{ij} denotes the Hamaker constant for materials i and j interacting across vacuum. By applying this relation to A_{131}, we obtain

$$A_{131} \approx A_{11} + A_{33} - 2A_{13} \approx A_{313}. \tag{2.58}$$

A second relation (based on the geometric mean) gives

$$A_{12} \approx \sqrt{A_{11} A_{22}}. \tag{2.59}$$

Substituting Eq. (2.59) into Eq. (2.57) results in

$$A_{132} \approx \left(\sqrt{A_{22}} - \sqrt{A_{33}} \right) \left(\sqrt{A_{11}} - \sqrt{A_{33}} \right). \tag{2.60}$$

Table 2.3 Hamaker constants for medium 1 interacting with medium 2 across medium 3.

Medium 1	Medium 2	Medium 3	A_H calc.	A_H exp.	References
Mica	Vacuum	Mica	89 ± 12	158 ± 25	[36, 39, 56–58]
Mica	Water	Mica	9 ± 4	22 ± 3	[29, 36, 39, 55]
Mica	Vacuum	Si_3N_4		64 ± 7	[59]
Mica	Water	Si_3N_4	24.5		[39]
Mica	Ethanol	Si_3N_4		116	[60]
Mica	Vacuum	Ag	330		[57]
Mica	Vacuum	SiO_2			
Mica	Water	SiO_2	12		[61]
SiO_2	Vacuum	SiO_2	68 ± 3	61	[36, 39, 48, 53, 54, 63]
SiO_2 (cr)	Vacuum	SiO_2 (cr)	95	87	[48, 54]
SiO_2	Water	SiO_2	4.6 ± 1.6		[29, 36, 39, 54]
SiO_2 (cr)	Water	SiO_2 (cr)	14 ± 3		[39, 54]
SiO_2	Dodecane	SiO_2	1.4		[39]
SiO_2	Al_2O_3	SiO_2	25		[54]
SiO_2	Water	Vacuum	-16		[54]
SiO_2 (cr)	Water	Vacuum	-23		[54]
SiO_2	Vacuum	Si_3N_4		66 ± 27	[59]
SiO_2	Vacuum	PTFE	62	76	[53]
SiO_2	Vacuum	Ag	184	132	[53]
SiO_2	Vacuum	Cu	167	142	[53]
SiO_2	Vacuum	W		1.3 ± 0.2	[64]
SiO_2	Vacuum	TiN	120	88	[53]
SiO_2	Vacuum	PS		120 ± 20	[65]
SiO_2	Water	Protein (BR, BSA)	6.7	5–10	[62, 66]
Si_3N_4	Vacuum	Si_3N_4	174 ± 6		[36, 39]
Si_3N_4	Water	Si_3N_4	48 ± 3		[29, 36, 39]
Si_3N_4	SiO_2	Si_3N_4	36 ± 3		[36, 54]
Si_3N_4	Diiodomethane	Si_3N_4	10		[40]
Si_3N_4	Bromonaphthalene	Si_3N_4	28		[40]
Si_3N_4	Diiodomethane	SiO_2	-8		[40]
Si_3N_4	Bromonaphthalene	SiO_2	-2		[40]
Si_3N_4	Vacuum	Ag		373 ± 89	[59]
Si_3N_4	Vacuum	PS		190 ± 20	[65]
Si_3N_4	Water	PS		80 ± 15	[65]
SiC	Vacuum	SiC	247	71	[39, 48]
SiC	Water	SiC	108		[39]
Si	Vacuum	Si	234		[63]
Si (am)	Water	Si (am)	88		[29]
Al_2O_3	Vacuum	Al_2O_3	145	165	[36, 39, 48]
Al_2O_3	Water	Al_2O_3	36 ± 2		[29, 36, 39, 52]
Al_2O_3	SiO_2	Al_2O_3	19		[36]
Al_2O_3	SiO_2	Vacuum	-42		[54]
Al_2O_3	Methanol	Al_2O_3	84		[52]
Al_2O_3	1-Propanol	Al_2O_3	76		[52]
Al_2O_3	2-Propanol	Al_2O_3	149		[52]
Al_2O_3	1-Butanol	Al_2O_3	75		[52]
Al_2O_3	2-Butanol	Al_2O_3	85		[52]

Table 2.3 (*Continued*)

Medium 1	Medium 2	Medium 3	A_H calc.	A_H exp.	References
Al_2O_3	Benzene	Al_2O_3	77		[52]
Al_2O_3	Toluene	Al_2O_3	79		[52]
TiO_2	Vacuum	TiO_2	167 ± 14	138	[36, 39, 48]
TiO_2	Water	TiO_2	57 ± 3	37	[29, 36, 39, 67, 68]
TiO_2	SiO_2	Vacuum	-54		[54]
ZnO	Vacuum	ZnO	92		[39]
ZnO	Water	ZnO	19		[39]
$BaTiO_3$	Vacuum	$BaTiO_3$	180		[39]
$BaTiO_3$	Water	$BaTiO_3$	93 ± 12		[39, 52]
$BaTiO_3$	Methanol	$BaTiO_3$	143		[52]
$SrTiO_3$	Vacuum	$SrTiO_3$	210 ± 50		[39, 54]
$SrTiO_3$	Water	$SrTiO_3$	92 ± 30		[39, 54]
$SrTiO_3$	Water	Vacuum	-61 ± 20		[54]
$CaCO_3$	Vacuum	$CaCO_3$		138	[48]
$CaCO_3$	Water	$CaCO_3$	13		[29]
Cement	Water	Cement	≈ 16		[69]
CaF_2	Vacuum	CaF_2	69		[39]
CaF_2	Water	CaF_2	4.4 ± 0.5		[29, 39]
CdS	Vacuum	CdS	114		[39]
CdS	Water	CdS	35 ± 2		[29, 39]
PbS	Vacuum	PbS	82		[39]
KBr	Vacuum	KBr	56		[39]
KCl	Vacuum	KCl	55		[39]
NaCl	Vacuum	NaCl	65		[39]
NaF	Vacuum	NaF	41		[39]
MgO	Vacuum	MgO	121		[39]
MgO	Water	MgO	21 ± 1		[29, 39]
Ag	Vacuum	Ag	488	387	[53]
Ag	Water	Ag	149	20–100	[52, 70]
Ag	Methanol	Ag	175		[52]
Ag	1-Propanol	Ag	168		[52]
Ag	2-Propanol	Ag	225		[52]
Ag	1-Butanol	Ag	167		[52]
Ag	2-Butanol	Ag	175		[52]
Ag	Benzene	Ag	166		[52]
Ag	Toluene	Ag	168		[52]
Ag	Vacuum	PTFE	164	132	[53]
Au	Vacuum	Au	332		[58]
Au	Water	Au		44 ± 15	[71, 72, 73, 74, 75]
Au	Ethanol	Au		40	[76]
Cu	Vacuum	Cu	402	275	[53]

(*Continued*)

Table 2.3 (*Continued*)

Medium 1	Medium 2	Medium 3	A_H calc.	A_H exp.	References
Cu	Vacuum	PTFE	149	122	[53]
Pt	Vacuum	Pt		180	[78]
Pt	Vacuum	Graphite		320	[78]
Pt	Vacuum	Ni		530	[78]
Pt	Vacuum	Ti		390	[78]
Pt	Vacuum	Al		380	[78]
Pt	Vacuum	Cu		65	[78]
Diamond	Vacuum	Diamond	296		[39]
Diamond	Water	Diamond	134 ± 4		[29, 39]
PS	Vacuum	PS	71 ± 5		[63, 79, 80]
PS	Water	PS	7.7	10	[79, 81]
PI	Vacuum	PI	60		[80]
PVC	Vacuum	PVC	75		[80]
PE	Vacuum	PE	51		[80]
PTFE	Vacuum	PTFE	47 ± 6	43	[48, 51, 53]
PTFE	Water	PTFE	4.9 ± 1.3		[51]
PS	Water	Protein (BSA)	6.2		[62]
PVC	Water	Protein (BSA)	9.3		[62]
PMMA	Water	Protein (BSA)	7.9		[62]
PTFE	Water	Protein (BSA)	0.73		[62]
Protein (BSA)	Water	Protein (BSA)	12.5		[62]
Cellulose	Vacuum	Cellulose	58		[77]
Cellulose ester	Vacuum	Cellulose ester		68–84	[82]
Cellulose	Water	Cellulose	8		[77]
PL bilayer	Water	PL bilayer	7.5	See Section 10.3	
Water	Vacuum	Water	36.6		[58]
Cyclopentane	Vacuum	Cyclopentane	29.9		[51]
n-Pentane	Vacuum	n-Pentane	25.9		[51]
n-Hexane	Vacuum	n-Hexane	27.6		[51]
n-Heptane	Vacuum	n-Heptane	30.0		[51]
n-Octane	Vacuum	n-Octane	43.5		[80]
Cyclopentane	Water	Cyclopentane	3.9		[51]
n-Pentane	Water	n-Pentane	4.4		[51]
n-Hexane	Water	n-Hexane	4.1		[51]
n-Heptane	Water	n-Heptane	3.9		[51]
Diiodomethane	Vacuum	Diiodomethane	75		[80]
Benzene	Vacuum	Benzene	49.5		[80]

We did not discriminate between gases and vacuum and consistently wrote "vacuum" in both cases. All values are given in zJ $= 10^{-21}$ J. BSA: bovine serum albumin; BR: bacteriorhodopsin; PS: polystyrene; PI: poly(isoprene); PVC: poly(vinyl chloride); PTFE: poly(tetrafluoroethylene); PMMA: poly(methyl methacrylate); am: amorphous; cr: crystalline. Errors reflect the variation from different references or the errors given in a reference. If no errors are given, the value is based on only one reference without reported error.

For the symmetric situation, we get

$$A_{131} \approx \left(\sqrt{A_{11}} - \sqrt{A_{33}} \right)^2$$

and

$$A_{232} \approx \left(\sqrt{A_{22}} - \sqrt{A_{33}} \right)^2,$$

which together with Eq. (2.58) results in

$$A_{132} \approx \sqrt{A_{131} A_{232}}. \tag{2.61}$$

So, if we know the Hamaker constant of material 1 interacting across medium 3 with itself, A_{131}, and we know the Hamaker constant of material 2 interacting across medium 3 with itself, A_{232}, then we can estimate the Hamaker constant for the interaction between material 1 and material 2 across medium 3, A_{132}. However, all these equations should be used with caution and only for cases where the London dispersion interaction makes the dominating contribution to the total van der Waals force. Otherwise, one might obtain values that can be off by an order of magnitude and may even have the wrong sign [87]. Therefore, calculation of the Hamaker constant using one of the above approximations is recommended only if no sufficient spectral information is available.

2.2.3
Surface Energy and Hamaker Constant

Van der Waals forces play an important role in adhesion phenomena and cohesive energies of materials. For molecular solids we can derive a simple relation between the surface energy and the Hamaker constant; molecular solids are solids in which the molecules attract each other solely by van der Waals interaction, for example, in a wax. We can in first approximation estimate the surface energy of molecular crystals using the following gedankenexperiment: A crystal is cleaved in two parts. These parts are separated to an infinite distance (Figure 2.4). The work required per unit area to overcome the van der Waals attraction is $V^A = A_H / 12\pi D_0^2$, where D_0 is the distance between two atoms. Upon cleavage, two fresh surfaces are formed. With the surface energy γ_S, the total work required for creation of the two surfaces is $2\gamma_S \cdot A$. Equating the results leads to

$$\gamma_S = \frac{A_H}{24\pi D_0^2}. \tag{2.62}$$

However, such calculation should be seen with caution. The application of a continuum view of matter (manifested by the use of a Hamaker constant) at distances at which a molecular "graininess" is expected to become important is certainly optimistic. Furthermore, rearrangement of surface atoms upon contact or separation might occur. Nevertheless, it was found that in many cases reasonable estimates of the surface energy can be obtained when an interatomic distance of 1.7 Å is used.

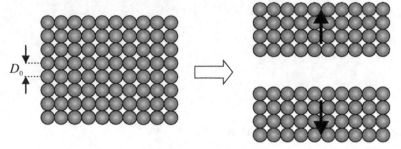

Figure 2.4 Cleaving a molecular crystal to calculate the surface energy of a solid.

■ **Example 2.11**

The Hamaker constant of helium–vacuum–helium is 5.7×10^{-22} J. Calculating the surface energy with Eq. (2.62) and an atomic distance of 1.6 Å leads to 0.26 mJ m^{-2}. This value is in good agreement with measured values for liquid helium of 0.12–0.35 mJ m^{-2}. For Teflon, we calculate a surface energy of 16–28 mJ m^{-2} [51] using an atomic spacing of 1.7 Å and a Hamaker constant of $3.4-6.0 \times 10^{-20}$ J.

A frequently used method to measure the Hamaker constant of solids is based on the idea that the interaction of a solid surface with nonpolar liquids will mainly occur through the London dispersion interaction. One can therefore define a dispersive surface tension γ_S^D that contains only the dispersive contribution. The value of γ_S^D can be determined from contact angle measurements with nonpolar liquids using the Fowkes equation [83]

$$\cos \theta = 2\sqrt{\frac{\gamma_S^D}{\gamma_L^D}} - 1. \tag{2.63}$$

Here, θ is the contact angle (see Section 5.3) and γ_L^D is the surface tension of the nonpolar liquid. By plotting $\cos \theta$ versus $1/\sqrt{\gamma_L^D}$ for several nonpolar liquids, one obtains γ_S^D from the slope of a linear fit. The Hamaker constant A_{1V1} for material 1 interacting with itself over a gap in vacuum is

$$A_{1V1} = 24\pi D_0^2 \gamma_S^D. \tag{2.64}$$

2.3
The Derjaguin Approximation

In Section 2.1, we had calculated the van der Waals force between two spheres and between two planar surfaces. What if the two interacting bodies do not have such a simple geometry? We could try to do an integration similar to the one that was carried out for the two spheres. This, however, might be very difficult and lead to long expressions. The Derjaguin[7] approximation is a simple way to overcome this

7) Boris Vladimirovich Derjaguin, 1902–1994. Russian physicochemist, professor in Moscow.

problem. This approximation is valid only if the characteristic decay length of the surface force is small in comparison to the curvature of the surfaces. The Derjaguin approximation is not only applied to van der Waals forces but can also in general be used to relate the energy per unit area for a parallel gap between two half-spaces to the force between arbitrary bodies.

2.3.1
The General Equation

The Derjaguin approximation relates the energy per unit area between two planar surfaces V^A that are separated by a gap of width x to the energy between two bodies of arbitrary shape V that are at a distance D (Figure 2.5):

$$V(D) = \int_D^\infty V^A(x)\, dA. \tag{2.65}$$

The integration runs over the entire surface of the solid. Please note that here A is the cross-sectional area. Often, we have to deal with rotational symmetric configurations. Then, it is reasonable to integrate in cylindrical coordinates:

$$V(D) = 2\pi \int_0^\infty V^A(x(r))\, r\, dr. \tag{2.66}$$

In many cases, the following expression is more useful:

$$V(D) = \int_D^\infty V^A(x) \frac{dA}{dx}\, dx. \tag{2.67}$$

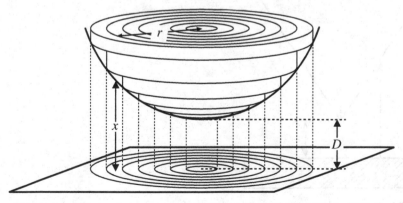

Figure 2.5 Schematic of Derjaguin's approximation for a rotational symmetric body interacting with a planar surface.

Derjaguin used this approach to calculate the interaction between two ellipsoids [84]. The same approximation was also introduced in 1977 by Blocki *et al.* [85] for calculating interaction forces between nuclei of atoms. They coined the term "proximity forces" in their publication. While the term "Derjaguin approximation" is still the standard term in surface science, the term "proximity force approximation" has become popular among physicists in the field of nuclear physics and the Casimir force (see Section 2.6).

■ **Example 2.12**

Calculate the van der Waals force between a cone with an opening angle α and a planar surface (Figure 2.6). The cross-sectional area is given by

$$A = \pi[(x-D)\tan \alpha]^2 \quad \text{for} \quad x \geq D,$$

which leads to

$$\frac{dA}{dx} = 2\pi \tan^2\alpha(x-D).$$

For the force, we can use a similar equation as for the energy:

$$F(D) = \int_D^\infty f(x)\frac{dA}{dx}dx = -\int_D^\infty \frac{A_H}{6\pi x^3} \cdot 2\pi \cdot \tan^2\alpha \cdot (x-D) \cdot dx$$

$$= -\frac{A_H\tan^2\alpha}{3} \cdot \int_D^\infty \frac{x-D}{x^3}dx,$$

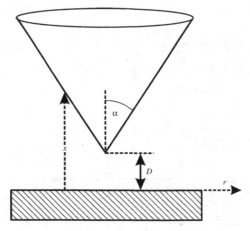

Figure 2.6 Interaction between a cone and a planar surface.

$$F(D) = -\frac{A_H \tan^2\alpha}{3} \left[-\frac{1}{x} + \frac{D}{2x^2} \right]_D^\infty = \frac{A_H \tan^2\alpha}{3} \cdot \left(-\frac{1}{D} + \frac{1}{2D} \right)$$

$$= -\frac{A_H \tan^2\alpha}{6D}.$$

A special and important case is the interaction between two identical spheres. It is relevant for understanding the stability of dispersions. For the case of two spheres of equal radius R_p, the parameters x and r are related by (Figure 2.7)

$$x(r) = D + 2R_p - 2\sqrt{R_p^2 - r^2} \Rightarrow dx = \frac{2r}{\sqrt{R_p^2 - r^2}} dr \Rightarrow 2r\,dr = \sqrt{R_p^2 - r^2} \cdot dx.$$

$$(2.68)$$

If the range of the interaction is substantially smaller than R_p, then we only need to consider the outer caps of the two spheres and only the contributions with a small r are effective. We can simplify

$$2r\,dr = \sqrt{R_p^2 - r^2} \cdot dx \approx R_p\,dx \qquad (2.69)$$

and integral (2.67) becomes

$$V(D) = \pi R_p \int_D^\infty V^A(x)\,dx. \qquad (2.70)$$

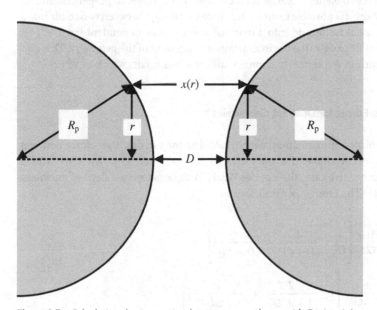

Figure 2.7 Calculating the interaction between two spheres with Derjaguin's approximation.

From the potential energy, we can calculate the force between two spheres:

$$F = -\frac{dV}{dD} = -\pi R_p \frac{d}{dD} \left(\int_D^\infty V^A(x)\,dx \right) = \pi R_p V^A(D)$$ (2.71)

since $V^A(\infty) = 0$. Thus, assuming that the range of the interaction is substantially smaller than R_p, there is a simple relationship between the potential energy per unit area and the force between two spheres.

As one example, we calculate the van der Waals force between two identical spheres of radius R_p from the van der Waals energy per unit area $V^A = -A_H/12\pi x^2$ (Eq. (2.38)). Using Derjaguin's approximation, we can directly write

$$F = -\frac{A_H R_p}{12 D^2},$$ (2.72)

which agrees with Eq. (2.34) for $R_1 = R_2 = R_p$.

Equation (2.71) refers to the interaction between two spheres. For a sphere that is at a distance D from a planar surface, we get a similar relation:

$$F = 2\pi R_p V^A(D).$$ (2.73)

White [86] extended the approximation to solids with arbitrary shape.

Derjaguin's approximation has a fundamental consequence. In general, the force or energy between two bodies depends on the shape, on the material properties, and on the distance. Now it is possible to divide the force (or energy) between two solids into a purely geometrical factor and into a material and distance-dependent term $V^A(x)$. Thus, it is possible to describe the interaction independent of the geometry. This also gives us a common reference, to compare different materials, which is $V^A(x)$.

2.3.2
Van der Waals Forces for Different Geometries

Using Derjaguin's approximation, we can calculate the van der Waals force between objects of different shape (for an extended list, see Ref. [3]). We start with a particularly instructive case, the van der Waals force between two slabs of thickness d (Figure 2.8a). The energy per unit area is

$$\frac{V}{A} = \frac{A_{132}}{12\pi} \left[\frac{1}{D^2} - \frac{2}{(D+d)^2} + \frac{1}{(D+2d)^2} \right]$$

$$= \frac{A_{132}}{12\pi D^2} \left[1 - \frac{2}{(1+d/D)^2} + \frac{2}{(1+2d/D)^2} \right].$$ (2.74)

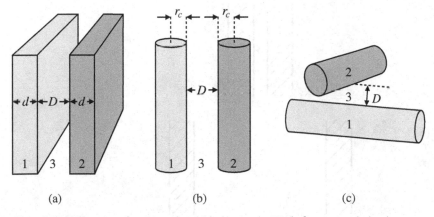

Figure 2.8 Different configurations for which the van der Waals force was calculated.

Equation (2.74) tells us how far the dispersion interaction effectively reaches into the material: for the case $D \ll d$, the second and third terms in the square brackets will vanish and the interaction will be the same as if the slabs had infinite thickness. This means that the van der Waals interaction between two bodies with parallel planar surface will occur essentially between surface layers with a thickness of the order of the separation D. So for very small values of D, only a very thin surface layer will contribute. As a consequence, the van der Waals interaction between layered materials will have complex dependence on distance. As an example, we will take the interaction between two half-spaces of material 1 coated with a thin layer of material 2, separated by a gap filled with material 3. The corresponding Hamaker constant A_{12321}, which should better be called Hamaker function, will depend on the gap thickness D. For a very small gap width $D \ll d$, the van der Waals interaction will essentially occur between the two layers of material 2 and $A_{12321}(D \approx 0)$ will be equal to A_{232}, just as if we had the interaction of two half-spaces of material 2 across material 3. For distances much larger than the film thickness of material 2, the interaction will reach far into the material 1 and $A_{12321}(D \gg 0)$ will approach the value of A_{131}, just as if the layers of material 2 would not exist (Figure 2.9).

The equation for the interaction between two slabs can also be helpful for calculations for other geometries.

■ **Example 2.13**

Calculate the free energy between two cubes of side length L. Two opposite faces are assumed to be oriented parallel.

Knowing the interaction between two slabs of thickness d (Eq. (2.74)) and the cross-sectional area of interaction $A = L^2$, we obtain

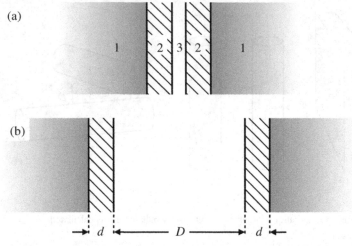

Figure 2.9 Dependence of effective Hamaker constant on the separation distance between two half-spaces coated with a thin layer. Case (a): for $D \ll d$, $A_{12321} \rightarrow A_{232}$. Case (b): for $D \gg d$, $A_{12321} \rightarrow A_{131}$.

$$V(D) = \frac{A_{132} L^2}{12\pi D^2} \left[1 - \frac{2}{(1 + L/D)^2} + \frac{2}{(1 + 2L/D)^2} \right].$$

For $L \gg D$, this reduces to $V(D) = A_{132} L^2 / (12\pi D^2)$.

For two parallel, infinitely long cylinders both of equal radius r_c (Figure 2.8b) and with a radius much larger than the distance, the free energy per unit length l is

$$\frac{V(D)}{l} = \frac{A_{132}}{24 D^{3/2}} \sqrt{r_c}. \tag{2.75}$$

If the cylinders are oriented perpendicular to each other (Figure 2.8c), the free energy for $r_c \gg D$ is

$$V(D) = \frac{A_{132}}{6D} r_c. \tag{2.76}$$

2.4
Retarded Van der Waals Forces

As already mentioned in Section 2.1.3, the D^{-2} dependence for the van der Waals force between parallel plates is valid only for short distances D. When D is larger than

5–10 nm, the finite speed of the virtual photons leads to retardation. The full equations of the Lifshitz theory already contain this effect, but a discussion on the complete Lifshitz theory is beyond the scope of this chapter and the interested reader may refer to Ref. [3]. We will follow that book for a short qualitative discussion of the topic for the example of half-spaces of materials 1 and 2 interacting over a gap filled with material 3.

A rough approximation to understand qualitatively the transition from the non-retarded van der Waals force to the retarded regime can be derived using Eq. (2.45). Note that only the London dispersion contribution is affected by retardation. When retardation comes into play, the Hamaker constant A_{123} becomes, in fact, a "Hamaker function" $A_{123}(x)$ that will depend on separation x. Under the simplifying assumption that the speed of light is the same in all three materials, this Hamaker function then can be written as [3]

$$A_{132} = \frac{3}{2} k_B T \sum_{m=0}^{\infty}{}' \Delta_{13}(i\omega_m)\Delta_{23}(i\omega_m) R_m(r_m) \tag{2.77}$$

with the same definitions as for Eq. (2.39) and the additional damping function

$$R_m(r_m) = (1 + r_m)e^{-r_m} \quad \text{with} \quad r_m = \frac{2x\sqrt{\varepsilon_3}}{c}\omega_m \tag{2.78}$$

that describes the change of the Hamaker constant with separation D due to retardation. For the interaction between two identical materials, we obtain ($\Delta_{23} = \Delta_{13}$)

$$A_{131} = \frac{3}{2} k_B T \sum_{0}^{\infty}{}' (\Delta_{13}(i\omega_m))^2 R_m(r_m) \tag{2.79}$$

and ω_m again representing the Matsubara frequencies. The dimensionless values r_m are given as the product of a time constant that is equal to twice the travel time for a photon over the gap filled with material 3 times the mth Matsubara frequency. In other words, r_m is the ratio of the travel time of the photons, $2x\sqrt{\varepsilon_3}/c$, and the fluctuation lifetime $1/\omega_m$. In the limit $\to 0$, r_m will go to zero and $R_m \to 1$. Equation (2.77) will become identical to Eq. (2.45). The convergence of the sum is in this case ensured due to the decay of $(\Delta_{13}(i\omega_m))^2$ at high frequencies. For the case $D \to \infty$, also r_m will go to infinity and we get $R_m = r_m\exp(-r_m) \to 0$. Thus, for large distances the London dispersion contribution will decay to zero and only the Keesom and Debye interactions will survive. This implies that at very large distances the van der Waals force between two plates will become proportional to D^{-2} again.

In the intermediate range, the behavior is more complex. The damping function R_m is low (i.e., leads to a stronger damping) for high values of m, which means for higher values of the frequencies ω_m. This behavior is due to the shorter fluctuation lifetime (i.e., the time the oscillation takes to change the dipole orientation) $1/\omega_m$ at higher frequencies. The exponential damping factor $\exp(-r_m)$ will make the high-frequency components (corresponding to higher values of m) in the sum of Eq. (2.79) die out first with increasing distance x. This means that in the case of retardation, the sum in Eq. (2.79) is essentially completed when m has reached the value m_{max} for which $r_m = 1$:

$$m_{\max} = \frac{\hbar c}{4\pi k_B T x \sqrt{\varepsilon_3}}. \tag{2.80}$$

Assuming $\varepsilon_3 \approx 1$, we get $m_{\max} = 6.23 \times 10^{-7} \mathrm{m}/x$. For $x = 10\,\mathrm{nm}$, we get $m_{\max} = 62$, which corresponds to contributions up to a wavelength of $\lambda_{\max} = 126$ nm. For $x = 50\,\mathrm{nm}$, we get $m_{\max} = 12$ corresponding to $\lambda_{\max} = 652$ nm and for $x = 100\,\mathrm{nm}$ we get $m_{\max} = 6$ and $\lambda_{\max} = 1.3\,\mu\mathrm{m}$.

Often, the distance dependence of the van der Waals force between parallel plates is written as

$$V^A(D) = -\frac{A_{132}}{12\pi x^p} \tag{2.81}$$

to express the effect of retardation as a change in the exponent p of the power law instead of using a Hamaker function. In this case, the value of p must increase with the onset of retardation to a value larger than 2 but return to a value of 2 at very large separations. At short separations, we have the unretarded van der Waals force with an exponent of $p = 2$. At separations of 2–5 nm, the exponent starts to increase due to retardation. In this range, both the decrease of $(\Delta_{13}(i\omega_m))^2$ and R_m with increasing ω_m contribute. With further increase of distance, one reaches the point where $(\Delta_{13}(i\omega_m))^2$ becomes independent of m since it has reached its low frequency limit and only the damping factor R_m remains relevant. In this range, we have

$$R_m(r_m) = (1 + r_m)e^{-r_m} \approx r_m e^{-r_m}. \tag{2.82}$$

Since $r_m \propto mD$, this equation can be rewritten as

$$R_m(r_m) = mCDe^{-mCD}, \tag{2.83}$$

where C is a constant. The sum in Eq. (2.79) can then be rewritten using Eq. (2.83) and converting the sum to an integral:

$$A_{131} = (\Delta_{13}(i\omega_1))^2 \int mCxe^{-mCx}\,\mathrm{d}m = (\Delta_{13}(i\omega_1))^2 \frac{1}{Cx} \int x' De^{-x'}\,\mathrm{d}x'. \tag{2.84}$$

The prefactor of the integral introduces an additional factor of $1/x$ that leads to an exponent of $p = 3$ in this range. For even higher separations (more than the thermal wavelength of $7.8\,\mu\mathrm{m}$ for $m = 1$), even the fluctuations at the lowest Matsubara frequency have been screened out and the exponent goes back to $p = 2$.

Exact calculations of the retarded van der Waals force are usually carried out numerically using the full Lifshitz theory in combination with spectral data of the materials. For some special geometries as sphere–sphere interaction, better analytical approximations for R_m than Eq. (2.78) have been developed. For discussion of these approximations, see Ref. [89].

2.4.1
Screening of Van der Waals Forces in Electrolytes

In many applications, we are interested in the van der Waals force across an aqueous medium. Then, an important question concerning the effect of dissolved ions on the van der Waals interaction arises. Since ions are free to move, they should also respond to the electromagnetic fluctuations that are responsible for the London dispersion interactions. However, since the mobility of ions in water is rather low compared to the high frequencies involved in the London dispersion forces, only the lowest frequencies will contribute. A comparison of the typical diffusion time of ions in aqueous solutions soon shows that within the time frame of even the lowest Matsubara frequency $\omega_1 = 3.84 \times 10^{13}$ Hz, the ions will not move significantly. Therefore, the ions will couple only to the zero-frequency part of the van der Waals forces, that is, the Keesom and Debye interactions. Since water molecules have a large static dipole moment, Keesom and Debye interactions in aqueous systems may be relatively high compared to other solvents, where usually the London dispersion interaction will be dominant.

An electric field due to dipoles emanating from a material in a salt solution will be screened according to the Gouy–Chapman theory in an exponential decay fashion typical for the electric double layer (see Chapter 4). The decay length will be given by the Debye length λ_D (Eq. (4.8)). For the interaction between two parallel surfaces over a gap of width x filled with electrolyte solution, the field from surface 1 will have decayed to $\exp(-x/\lambda_D)$ when it has reached the other surface. The dielectric response of surface 2 will be shielded the same way. This will lead to damping of the interaction that we can describe by a simple correction factor:

$$V^A(x) = V_0^A(x)\left(1 + \frac{2x}{\lambda_D}\right)e^{-2x/\lambda_D}. \tag{2.85}$$

Here, $V_0^A(x)$ is the van der Waals interaction without screening. For a 0.1 M NaCl solution, the Debye length is $\lambda_D = 0.95$ nm. For a distance between the surfaces of 2 nm, the Keesom and Debye interaction will be reduced to less than 8% of the unscreened interaction. The screening of the static contributions by ions is especially relevant for biological systems, where the interactions may be dominated by the Keesom and Debye interactions and physiological solutions typically contain more than 0.1 M salt. For systems where the London dispersion forces are dominating, screening will have a minor effect.

■ **Example 2.14**

For the interaction of lipid bilayers across a layer of water, a Hamaker constant of 7.5×10^{-21} J is calculated. A value of only 3×10^{-21} J was measured. One reason is probably a reduction of the Keesom and Debye by the presence of ions [88].

2.5
Measurement of Van der Waals Forces

Even when taking into account that van der Waals forces for macroscopic objects fall slower than the fast D^{-6} decay law for single molecules, the direct measurement of van der Waals forces is demanding with respect to precise distance control and surface roughness. Surface roughness hinders close approach of the surfaces and is difficult to take into account unless the exact structure of the surface on the nanometer scale is known. This is why all of the early measurements of van der Waals forces were carried out at relatively large distances in the retarded regime. A second critical issue is separating van der Waals forces from possible other force contributions. For measurements in vacuum or dry gases, electrostatic forces are of special concern and have to be avoided or taken into account.

The first precise measurements of van der Waals forces started in the 1950s. Derjaguin and Abricossova [90, 91] "weighed" the force between a quartz sphere and a quartz plate for distances between 100 and 1000 nm with a special microbalance that included an active feedback mechanism. Their results were in approximate agreement with the Lifshitz theory. At about the same time, Overbeek and Sparnaay [92, 93] in the Netherlands tried to measure van der Waals forces between parallel glass plates. The force was detected by deflection of a mechanical spring and distance by interferometry. The problem to align the plates perfectly parallel and to get rid of all residual surface charges that lead to electrostatic force was obviously not completely solved – they observed forces that were about 500 times larger than those of Derjaguin. Furthermore, the force decayed with distance less steeply than expected from Lifshitz theory. These discrepancies led to a hot debate (insightful reading how hot it got in Ref. [94]) that was finally resolved through later experiments by Kitchener and Prosser [95] and Black *et al.* [96]. They confirmed the original findings of Derjaguin.

With the advent of the surface forces apparatus (SFA, see Section 3.1) that employs molecularly smooth mica surfaces, Tabor and Winterton [30] were the first to achieve measurements of surface forces down to separations of 5 nm and observe the change from retarded to non-retarded interaction. Shortly after, Israelachvili and Tabor [56] recorded van der Waals forces for separations from 2 to 130 nm by using different springs in the SFA and recording the position of the jump to contact instability, where the gradient of the attractive van der Waals interaction equals the spring constant of the system. Reasonable agreement with the predictions from the Lifshitz theory was found and further improved in a reanalysis of the results [97].

Wittmann and Splittgerber just "had a look" at van der Waals forces in 1971 [98]. They used a quartz block with highly polished surface and a thin quadratic quartz plate that was polished on both sides and had a ridge of ≈ 160 nm at one end that acted as a spacer. They placed the plate onto the block such that the ridge acted as a spacer at one end, while the other end was in direct contact. Optical interferometry was used to measure the bow of the plate due to the surface forces acting between plate and block and the transition between nonretarded forces at 5 nm and retarded van der Waals forces at 100 nm distance was observed from the curvature shape. Hunklinger *et al.* [99] used the dynamic interaction between a plate and a vibrating spherical lens both made of borosilicate glass to probe the van der Waals forces for distances between 80 nm and 1.2 μm and determined the Hamaker constant with an accuracy of $\approx 20\%$. Derjaguin *et al.* [100] measured the forces between crossed platinum or quartz fibers at distances of 10–100 nm and found exponents of 1.93–2.01 and 3.01–3.23 for the non-retarded and retarded van der Waals interaction, respectively. The attractive force between two chromium surfaces was measured in the distance range of 132–670 nm by van Blokland and Overbeek [101]. Results agreed with Lifshitz calculations based on a free electron gas model to describe the dispersion interaction of metals. Coakley and Tabor [57] employed the SFA in the same type of "jump-in" experiments as Israelachvili to study the van der Waals force between mica surfaces covered with calcium stearate or silver films. More recent experiments with a SFA type of setup (using capacitive sensors for distance detection) allowed direct recording of full force versus distance curves without instabilities down to distances of 20 nm. An agreement within 5% with Lifshitz theory was found for mica [102] and Pyrex glass [103] surfaces.

With the introduction of the atomic force microscope (AFM) and the colloid probe technique (see Section 3.2), it became easier to record force versus distance curves for a wide variety of materials. An early measurement of the van der Waals forces with the AFM was done by Hutter and Bechhoefer in 1994 [105] between a Si_3N_4 tip and a mica surface for distances between 9 and 50 nm in air. The crossover from the nonretarded to the retarded regime could clearly be distinguished from the fits in the ranges between 9–16 and 20–50 nm, where exponents of -2.19 and -2.92 were obtained for force versus distance. The lower limit of force–distance curves with the AFM – as for all spring deflection-based devices – is given by the jump-in stability that occurs when the attractive force gradient exceeds the spring constant.

The use of colloid probes instead of AFM tips offers the advantage of a well-defined contact geometry and larger interaction forces. Palasantzas et al. [104] measured the force between a gold-coated colloid probe and a gold-coated surface for distances of 12–100 nm. For separations less than 18 nm, they observed the expected distance dependence of $F_{vdW} \propto D^{-2}$. For larger separations, the exponent showed a transition to a value of -2.5, which corresponds to distances below 100 nm, the transition to the fully retarded regime is not yet reached.

A general problem of the direct measurement of van der Waals forces (as for attractive surface forces in general) is the jump-in occurring when the force gradient exceeds the spring constant. This instability can be suppressed by using a magnetic feedback as demonstrated by Ashby et al. [106], who measured vdW forces between gold surfaces coated with CH_3- and OH-terminated thiols. Very high Hamaker constants of ≈ 100 zJ were found, indicating that the interaction was dominated by the underlying gold substrate. Another way to circumvent the snap-in problem is the use of dynamic AFM modes. Operating the AFM using either amplitude modulation (AM) or frequency modulation (FM) allows the reconstruction of force versus distance curves from damping or frequency shift curves. Examples are the interaction between a Si_3N_4 tip and a gold surface measured by AM-AFM [107] down to a distance of 4 nm. A good agreement of the power law dependence and magnitude of the Hamaker constant with theory was found. Guggisberg et al. [108] employed FM-AFM to measure surface forces between Si tips and Cu(111) surfaces in UHV. After compensation of electrostatic forces, fits of the van der Waals forces in the range of 1–5 nm gave excellent agreement with theory and allowed separation between van der Waals forces and chemical forces that dominate the interaction for distances below 1 nm. The most precise direct measurements of retarded van der Waals forces have been carried out in experimental studies of the Casimir force since 1998 and are discussed in Section 2.6.

Repulsive van der Waals forces can occur for the right combinations of materials. First indirect proofs for repulsive van der Waals forces came from the absence of heteroflocculation in dispersions of graphite and PTFE in glycerol [109] and from the engulfment behavior of particles at solidification fronts [110]. A detailed study on the equilibrium thickness of liquid ^{4}He films on copper substrates due to repulsive van der Waals forces showed excellent agreement with the Lifshitz theory [111]. First direct force measurements of repulsive van der Waals forces were carried out with the AFM. Repulsive van der Waals forces were found between a PTFE surface and gold spheres [112] using different nonpolar liquids and for silica and α-alumina spheres in cyclohexane [113]. Switching between attraction and repulsion was demonstrated for an Si_3N_4 AFM tip in diiodomethane and 1-bromonapthalene when changing the substrate from Si_3N_4 to SiO_2 [40].

2.6
The Casimir Force

2.6.1
Casimir Forces Between Metal Surfaces

When Overbeek and Verwey studied the stability of colloids at the Philips Research Laboratories in Eindhoven, they observed that at large distances the interaction between colloids decays more steeply than expected by the London theory of van der Waals forces. Overbeek recognized that for distances larger than the wavelength corresponding to the typical absorption frequencies of atoms, retardation might occur, which would make the interaction to decay faster than R^{-6} for larger distances between the two molecules and might be explained by their experimental results. However, the exact calculation of the distance dependence of the van der Waals force taking retardation effects into account cannot be derived on the basis of such a simple picture but has to follow a full treatment by means of quantum electrodynamics.

Such calculations were first carried out by their colleagues Casimir and Polder [17], who derived equations for the interaction of a single molecule with a perfectly conducting wall at distance D. They obtained

$$V_{\text{Mol/plane}}(D) = -\frac{3\alpha\hbar c}{8\pi D^4} \tag{2.86}$$

and for two molecules with polarizabilities α_1 and α_2 at distance D

$$V(D) = -\frac{23\alpha_1\alpha_2\hbar c}{4\pi D^7}. \tag{2.87}$$

When Casimir discussed his results with Niels Bohr in Copenhagen, Bohr suggested that these interactions might have to do something with quantum mechanical zero-point energy. This comment inspired Casimir to explore a different theoretical approach that starts from the interactions of the molecules with the photons present due to zero-point energy inside a cavity. He obtained the result that two parallel conducting plates will feel an attractive force simply due to the modifications of the zero-point fluctuations of the electromagnetic field within the cavity between them. In his publication from 1948, Casimir [18] considered a cubic cavity of volume L^3 bounded by perfectly conducting walls with a perfectly conducting square plate with side L placed in this cavity parallel to the xy face either (1) at a small distance D from the xy face (2) or at a much larger distance (e.g., $L/2$). The quantity $1/2\sum\hbar\omega$ where the summation extends over all possible resonance frequencies ω of the cavities is divergent in both cases and devoid of physical meaning, but the difference between these sums in the two situations,

$$\frac{1}{2}\left(\sum\hbar\omega\right)_1 - \frac{1}{2}\left(\sum\hbar\omega\right)_2,$$

will obtain a well-defined value and this value corresponds to the Casimir interaction between the plate and the xy face. With this approach, Casimir derived a formula that describes the force per unit area between two parallel metallic plates of area A at a distance D in vacuum:

$$f(D) = -\frac{\pi^2\hbar c}{240\,D^4}. \tag{2.88}$$

Casimir himself noted in his paper: "Although the effect is small, an experimental confirmation seems not unfeasible and might be of a certain interest."

This "certain interest" has grown significantly during the last decades especially since 1990. Experiments of Lamoreaux [114] and Mohideen [115] reached a precision that allowed a critical comparison with existing theoretical models of the Casimir force. From the standpoint of applications, Casimir forces can become a critical issue in nanotechnology and MEMS, where they can lead to stiction or collapse of components [116]. From the side of fundamental science, high-precision measurements of Casimir forces have become a benchmark for the existence of new hypothetical fundamental forces [117] that are often summarized as "non-Newtonian gravitation." This fancy name was coined due to the alternative experimental approach to detect such forces as a deviation from the classical Newtonian laws of gravitation. There are a number of recent review articles on the topic of Casimir forces [118, 119].

At this point we should clarify that the Casimir force is not a really new type of force. It is simply another term for a special case of the van der Waals forces, namely, the retarded van der Waals force between metallic surfaces. While the terms "retarded van der Waals force" or "retarded London dispersion force" are prevalent in the physical chemistry and colloid community, the term "Casimir force" or "Casimir–Polder force" has become popular in the physics community. This means that in principle the Lifshitz theory is applicable to describe the Casimir forces. The problem with using Lifshitz theory for ideal metals is the fact that for these the dielectric constant diverges ($\varepsilon \rightarrow \infty$) and therefore the Lifshitz theory breaks down. However, for real metals, the use of the Lifshitz theory is possible with corresponding dielectric models of the metals.

The simple form of Eq. (2.88) is appealing, especially when compared to the more complicated equations of the Lifshitz theory. However, one should not forget that the derivation of Eq. (2.88) was done under assumptions that are usually not fulfilled in practical situations:

- Temperature was assumed to be 0 K.
- The metal surfaces were assumed to be perfect conductors.
- The metal surfaces were assumed to be perfectly flat.
- The equation is valid only for the arrangement of perfectly parallel plates.

We briefly discuss the relevance of the above assumptions for real experiments on Casimir forces.

The Effect of Temperature   This effect is in principle already included in the Lifshitz theory. For distances in the order of $\approx 1\,\mu m$ that are much smaller than the thermal wavelength $\lambda_T = \hbar c / k_B T$, which corresponds to 7.6 μm at room temperature, there is only a weak influence of temperature. For $x > \lambda_T$, temperature will have a significant influence. This is related to the fact that at such large distance all higher frequency fluctuations have been damped out due to retardation and the thermal fluctuations will become the dominating term in the summation over all frequencies in the Lifshitz theory.

The influence of finite temperature on the Casimir force has been calculated [120–122] and some experiments have been carried out [123]. Consensus has, however, not been reached. This is partially due to the fact that an experimental verification is difficult since van der Waals forces at large distances are weak. For that reason, they are usually also insignificant, and our poor understanding is more an academic problem.

Finite Conductivity of Real Metals   Real metals will always have a finite conductivity. Therefore, at very high frequencies they will no longer act as perfect mirrors but even become transparent. This leads to corrections in the order of 10–20% for separations larger than $\approx 1\,\mu m$. The calculations can be carried out within the Lifshitz theory either by using the plasma frequency model or by using the full spectral method explained above [124]. A related phenomenon is the finite penetration depth of electric fields into a real metal surface (skin depth). This has to be kept in mind when using metal-coated surfaces in force measurements. When the thickness of the metal layer becomes comparable to the skin depth, the Casimir force will decrease [125]. A general problem in calculations is the still limited availability of spectroscopic data for metals over the required spectral range. This is even exacerbated by the fact that in many experiments, thin films of metals are used that were deposited by sputtering or evaporation. The conductive properties of such films are known to deviate from bulk properties and in principle one should use the same samples for force measurements and determination of the spectral properties for calculations based on the Lifshitz theory [126].

Surface Roughness Except for very small, single-crystalline surfaces, roughness cannot be completely avoided. The effect of roughness is expected to be most prominent at short distances. With increasing separation of the surfaces, nanoscale roughness will become more and more averaged out. One possible approach for calculating the Casimir force between surfaces with a given surface roughness is to use a pairwise summation of the retarded interatomic potentials, as in the microscopic Hamaker approach, and dividing the result by a renormalization parameter, which is used to approximately correct for the nonadditivity of the forces [127]. For the case of stochastic roughness with an amplitude $\delta_{1,2}$ much smaller than the distance between the parallel surfaces 1 and 2, perturbation calculations up to the fourth order were used [127] to obtain the equation

$$f_{CR}(x) = f_C(x) \left\{ 1 + 10 \left[\left(\frac{\delta_1}{x} \right)^2 + \left(\frac{\delta_1}{x} \right)^2 \right] + 105 \left[\left(\frac{\delta_1}{x} \right)^2 + \left(\frac{\delta_2}{x} \right)^2 \right]^2 \right\},$$

(2.89)

where f_{CR} and f_C are the Casimir force per unit area for two perfectly flat parallel surfaces and for the rough surfaces, respectively. For the case of a sphere interacting with a plane, the prefactors change:

$$F_{CR}(D) = F_C(D) \left\{ 1 + 6 \left[\left(\frac{\delta_1}{D} \right)^2 + \left(\frac{\delta_1}{D} \right)^2 \right] + 45 \left[\left(\frac{\delta_1}{D} \right)^2 + \left(\frac{\delta_2}{D} \right)^2 \right]^2 \right\}.$$

(2.90)

A special case is the interaction of periodically structured surfaces that can give rise to a lateral Casimir force [128].

Casimir Force for Nontrivial Geometries As in the case of the normal van der Waals forces, the Derjaguin approximation can be used to calculate the Casimir force for geometries other than that of parallel plates. Note that in many papers on the Casimir force, this approach is called "proximity force approximation" due to historic reasons (see Section 2.3). It was shown that the error introduced by this approximation should be smaller than $0.4\ D/R$ for $D < 300$ nm [129]. Full calculations without approximation have been done for some configurations, for example, sphere/plate [130], but these are usually cumbersome. An alternative approach for approximate calculations was introduced by Jaffe and Scardicchio [131]. It is based on classical ray optics. The results obtained using this method were found to be closer to the full calculations than those from the proximity force approximation (for a review, see Ref. [13]).

Measurements of the Casimir forces are especially demanding since the forces are weak. An early attempt to verify Eq. (2.88) experimentally was carried out by Overbeek and Spaarnay [133] in 1958. However, the error in their measurement was too large. While there had been several later attempts to quantify Casimir forces, their precision was not yet convincing until the modern measurements by Lamoreaux [114]. Lamoreaux used a torsion pendulum to measure the interaction between a gold-coated spherical lens ($\approx$ 12 cm radius) and a flat plate for distances between 0.6 and 6 μm. The torsion pendulum consisted of the gold-coated flat plate on one arm and a plate that formed the center electrode of dual parallel capacitors on the other arm. The spherical lens was mounted on a piezoactuator to vary the distance between sphere and plate. The torsion of the pendulum due to the Casimir force was detected by the change in capacitance and by using a feedback loop. As a feedback, a voltage was applied to the capacitors to exactly counterbalance the Casimir force. An important step in the data analysis was the subtraction of electrostatic forces, which amounted to about 80% of the total interaction at closest separation. The experimental error at closest separation was estimated to be of the order of 5%. The following corrections were later calculated by Bordag *et al.* [134]. The finite conductivity of the gold surfaces should reduce the Casimir force by 23%. The thermal corrections that led to a higher value compared to $T = 0$ were calculated to be 2.7% at $D = 1\,\mu$m but 174% at $D = 6\,\mu$m. The contribution of roughness was estimated to be 30%, assuming a ratio of roughness amplitude to distance of 0.1.

One year later Mohideen and coworkers [115] used the colloid probe AFM technique to probe the Casimir force between a metalcoated planar surface and a metal coated sphere. After several improvements [135], they could achieve a measurement over the range of distance starting from 62 to 400 nm with an experimental error of less than 1% at the lowest distance (Figure 2.10). This level of precision allows a quantitative comparison with theory. In this case, the influences of surface roughness, finite temperature, and finite conductivity of the metal have been taken into account.

A problem is the correct determination of zero distance, especially in the presence of surface roughness and deformation of the surfaces in contact. The problem of surface roughness was addressed by Ederth [136] by using especially flat, template-stripped gold surfaces. The gold surface was further coated with a thiol monolayer to prevent surface contamination. In his SFA-like setup, the interaction of crossed cylinders down to distances of 20 nm could be measured with 1% precision. In addition, gold surfaces in the analysis of later experiments were usually taken as bare gold even when experiments were carried out in air, where gold is known to readily adsorb a carbonaceous contamination layer that is much less defined than the thiol monolayer. van Zwol *et al.* [126] combined AFM measurements between gold-coated surfaces with ellipsometric measurements of the optical properties of the gold layers to obtain correct input data for the theoretical modeling of their results. Using stiffer

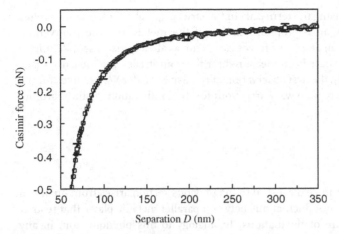

Figure 2.10 Measurement of the Casimir force between a gold-coated sphere and gold-coated plate. (From Ref. [135] with permission from U. Mohideen.)

AFM cantilever to shift the jump-in, they could record the forces down to separations of 12 nm. However, experimental accuracy was lower than in the previous studies. A modern approach to measure the Casimir force using the original parallel plate geometry was first realized by Bressi *et al.* [137], who measured the change in resonance frequency of a cantilever approaching a flat surface. This experiment was carried out inside an SEM to allow precise parallel alignment. The parallel plate configuration was implemented by using micromechanical torsion devices that were operated either statically [138] or dynamically [139] and have led to the highest precision measurements of the Casimir force with an experimental error of only 0.19% at the closest separation of 160 nm and 9.0% at the largest separation of 750 nm.

As mentioned above, van der Waals forces can become repulsive for the right combination of materials but should always be attractive across vacuum. Thus, Casimir forces should always be attractive. There has been some discussion if they might become repulsive by using hypothetic materials with extraordinary magnetic properties [140], but will in fact be hard to achieve with realistic materials [141]. We also mentioned that van der Waals and Casimir forces between two identical materials should be always attractive, independent of the intervening material. With the advent of the so-called metamaterials, that is, materials that have a *negative* refractive index, this is not necessarily true anymore [142]. By using metamaterials as intervening medium, Casimir forces between metallic surfaces could in principle become repulsive [143]. However, the experimental verification might be hard to achieve in practice [144] because metamaterials are structured on the 100 nm length scale. Practically, this effect should therefore be irrelevant.

Recently, there have been measurements of retarded van der Waals forces between metal surfaces in liquid [145]. The forces were termed Casimir forces, although we prefer using the term only in the strict original definition that applies only in vacuum. By an appropriate choice of the metals and the intervening medium, repulsive van der Waals forces can occur, as already discussed. Repulsive van der Waals forces have been measured for the nonretarded regime earlier (see Section 2.5). Recently, the first observation of repulsive van der Waals forces for the retarded, long-range regime was carried out for the combination of silica–bromo-benzene–gold [146].

2.6.2
Critical Casimir Force

The original Casimir force discussed in the previous section arises from quantum electromagnetic vacuum fluctuations between parallel metallic plates that restrict the possible spectrum of fluctuations. In analogy to this phenomenon, in any system with long-range forces, the energy of the ground state will depend on the boundary conditions. Any change in confinement will therefore lead to change in ground-state energy. Thus, a force will be connected to such a change. Just like the quantum fluctuations in the original Casimir force, classical fluctuations of matter due to thermal motion can lead to such a force. One example are the thermal density fluctuations in a liquid due to the molecular motion. These fluctuations will diverge near a second-order phase transition and will thus become long ranged. These diverging fluctuations are usually denoted as "critical fluctuations" and the temperature of the phase transition is called "critical temperature" T_c. One well-known example of critical temperature T_c is that of the fluid–gas phase transition. Above T_c, there exists only a single phase, while below T_c, fluid and gas phase can coexist. When approaching T_c, strong long-range fluctuations in the system are observed, as if the system were switching between the two states. Two other examples are ^{4}He at the superfluid transition and binary liquids close to their critical point.

It was already predicted by Fisher and de Gennes [147] that suppression of the critical fluctuations by confinement between two closely separated walls should lead to a force between the plates. Unlike the classical Casimir force, this so-called critical Casimir force strongly depends on temperature and should be largest when approaching the critical temperature T_c (for an introduction, see Ref. [148]). The length scale of the fluctuations is described by means of the so-called bulk order parameter ξ. It changes with temperature according to

$$\xi = \xi_0 \left(1 - \frac{T}{T_c}\right)^{-1} \tag{2.91}$$

Here, ω_0 is an amplitude that corresponds to a typical length scale of the intermolecular pair potential. The critical exponent $b \approx 0.65$ is an universal quantity; its precise value depends on the internal symmetry of the system. The free energy per unit area of a system confined by two parallel walls at distance x at temperature T can be expressed [147] as

$$V^A(x) = \frac{k_B T_c}{x^2} f^*(x/\xi),$$
(2.92)

where $f^*(x)$ is a universal scaling function that depends only on the symmetry and boundary conditions of the system [149]. Unfortunately, an analytical expressions for $f^*(x)$ cannot be derived for the three-dimensional case. However, it was shown recently that it can be derived numerically using Monte Carlo simulations [150, 151].

First indirect experimental observations of the critical Casimir force were made by Chan and Garcia [152]. They measured the thickness of ^{4}He films on a copper substrate and detected a thinning of the films close to the critical point of transition to superfluidity, indicating an attractive critical Casimir force. For a ^{3}He/^{4}He mixture close to the tricritical point, the same authors found a repulsive critical Casimir force, which caused film thickening on the copper substrate [153] (for a later, refined theoretical analysis, see Ref. [154]). The tricritical point is the point in the phase diagram where the superfluidity transition line terminates at the top coexistence line of ^{3}He/^{4}He.

Whether the critical Casimir force is attractive or repulsive depends on the boundary conditions. If the order parameter vanishes at both surfaces (symmetric boundary conditions), the resulting critical Casimir forces is attractive. For nonsymmetric boundary conditions (i.e., if the order parameter remains finite at one of the interfaces while it vanishes at the other), the critical Casimir force is repulsive. The critical Casimir force for a binary liquid mixture of methylcyclohexane and perfluoromethylcyclohexane was studied by Fukuto et al. [155]. The increase in thickness of such a liquid film on a silicon wafer close to critical point was measured using X-ray reflectivity and agreed with theoretical models.

Direct measurement of the critical Casimir force was achieved recently by Hertlein et al. [156] using TIRM (see Section 3.4) to probe the forces between small spheres and a planar surface in binary mixtures of water and 2,6-lutidine. Close to the critical temperature of the binary mixture, they observed long-range forces of $\approx$ 600 fN. By using a polystyrene sphere with strongly charged surface (preferentially wetted by water) or without charges (preferentially wetted by lutidine) and by chemical treatment of the silica substrate by NaOH (rendering it hydrophilic) or hexamethyldisilaxane (making it hydrophobic), the sign of the forces could be controlled. For symmetric boundary conditions (both sides preferring the same liquid) attraction was observed, whereas for asymmetric surfaces, repulsion occurred. For symmetric conditions, the fluctuation spectrum is even more strongly suppressed than for surfaces where no preferential adsorption is occurring, which leads to an amplification of the attraction. For asymmetric boundary conditions, the concentration fluctuations originating from the two different surfaces favor different species. When the two surfaces approach each other, the overlap of the surface-induced fluctuations with preference for different species hinders the occurrence of fluctuations of one species with the correlations length ξ, which leads to a net repulsion. For both situations, experimental potential energy versus distance profiles were in excellent agreement with the theoretical predictions of Ref. [150].

Critical Casimir forces are rather weak and therefore hard to detect. What might still make them interesting for future applications is the fact that they are based on a very universal concept, can be controlled by temperature, and can be either attractive or repulsive depending on surface treatment.

2.7
Summary

- The van der Waals interaction between molecules is the sum of the Keesom, Debye, and London dispersion interactions. The Keesom interaction describes the average interaction between freely rotating dipoles. The Debye interaction describes the interaction between a molecule with permanent dipole moment and the induced dipole moment of a polarizable molecule. The London dispersion interaction arises from quantum mechanical charge fluctuations. In most cases, the London dispersion interaction gives the largest contribution to the total van der Waals interaction. The van der Waals interaction energy between molecules is proportional to D^{-7}.
- Between macroscopic bodies, the distance dependence of the van der Waals interaction depends on the geometry of the objects. Generally, it decays less steeply than between single molecules, for example, proportional to D^{-2} for two planar surfaces.
- The van der Waals interaction between different materials depends on their differences in optical properties and can be characterized by the so-called Hamaker constant. Hamaker constants either can be calculated from dielectric and spectroscopic information of the materials using the Lifshitz theory or can be determined experimentally.
- Between two bodies of the same material, the van der Waals interaction is always attractive independent of the intervening medium. An intervening medium reduces the van der Waals interaction. For certain combination of three different materials, repulsive van der Waals forces can occur. This is important for the stability of thin films.
- The Derjaguin approximation (also called proximity force approximation) allows the calculation of the van der Waals interaction between macroscopic bodies with complex geometries from the knowledge of the interaction potential between planar surfaces, as long as the radii of curvature of the objects are large compared to the separation between them.
- For distances larger than 5–10 nm, the finite speed of light leads to a reduction of the London dispersion interaction. This effect is called retardation. Retarded van der Waals forces exhibit a distance dependence with a power law exponent that is larger by 1 than for the unretarded case.
- A special case of the retarded van der Waals forces is the so-called Casimir force. It is the attractive force between two metallic surfaces across vacuum that arises from quantum fluctuations of the vacuum.

- Long-range fluctuations in matter close to a critical point (phase transition) can lead to long-range forces, which are called critical Casimir forces.

2.8
Exercises

2.1. In atomic force microscopy, the tip shape is often approximated by a parabolic shape with a certain radius of curvature R at the end. Calculate the van der Waals force for a parabolic tip versus distance. We only consider nonretarded contributions. Assume that the Hamaker constant A_H is known.

2.2. The van der Waals interaction between two parallel slabs approaches that of two half-spaces if the slab thickness is much larger than the separation between the slabs. How thick must a slab be to have a van der Waals interaction that deviates less than 10% or less than 1% from that of a half-space, when the separation of the slabs is $D = 1$ nm or $D = 10$ nm?

2.3. Explain by a simple argument why Hamaker constants of hydrocarbons with longer chains are higher than those for shorter chains (see Table 2.3).

3
Experimental Methods

The direct measurement of surface forces is challenging due to their short-ranged nature. One has therefore to combine a sensitive detection of forces with a precise control of distance on the subnanometer scale. A critical quantity in experiments on force versus distance relations is the distance of closest approach that is actually achieved in an experiment. Surface roughness and contaminations are serious issues when trying to establish intimate contact between two objects. In contrast to friction forces, which were already studied by Leonardo da Vinci, systematic studies of surface forces were not done before the beginning of the twentieth century. For an overview of the history of the development of devices to measure surface forces, see Refs [157, 158].

Early attempts to measure surface forces were carried out by Tomlinson in 1928. He studied adhesion forces between crossed glass fibers or glass fibers and glass spheres [159]. He introduced the advantageous interaction geometry of crossed cylinders, which avoids the complication in controling relative orientation of the surfaces as in the case of parallel plates. In subsequent experiments, Bradley [10] measured the adhesion between quartz spheres. However, these early measurements did not include a precise determination of separation distance. A first step in that direction were the experiments by Lord Rayleigh [160], who studied the work necessary to peel a thin glass slide from a glass plate while monitoring the separation by interference fringes.

In the 1950s, Derjaguin and Abricossova of the Russian Academy of Science, Moscow, and Overbeek and Sparnaay, from the University of Utrecht, the Netherlands, tried to verify the theoretical predictions of Lifshitz on the distance dependence of van der Waals forces. Derjaguin used a specially constructed microbalance with electromagnetic feedback to measure the force between a quartz sphere and a quartz plate for distances between 100 and 1000 nm [90, 91]. Their results were in approximate agreement with the theoretical predictions for van der Waals forces by Lifshitz. Overbeek and Sparnaay [92] measured van der Waals forces between parallel glass plates by deflection of a mechanical spring while observing the distance by interferometry. Their experiments were complicated by having to keep the plates precisely parallel. They measured a ≈ 500 times larger value for the interaction energy compared to Derjaguin's results and a distance dependence of the van der

Surface and Interfacial Forces. Hans-Jürgen Butt and Michael Kappl
Copyright © 2010 Wiley-VCH Verlag GmbH & Co. KGaA
ISBN: 978-3-527-40849-8

Waals force that was less steep than expected in theory. Later experiments by Kitchener and Prosser, Imperial College of London [95], and by the group of Overbeek and Sparnaay [96] using a sphere-plate geometry confirmed the original findings by Derjaguin and explained the discrepancy in the early experiments by Overbeek and Sparnaay as originating from residual electrostatic charges.

A significant breakthrough in the experimental study of surface forces was the introduction of the surface forces apparatus, which will be described in the next section.

3.1
Surface Forces Apparatus

The surface forces apparatus (SFA) developed by Tabor, Winterton [30], and Israelachvili [56] contains two crossed silica cylinders with a radius of curvature of roughly 1–2 cm to which thin sheets of mica are glued to obtain atomically flat surfaces (inset of Figure 3.1). One mica-coated cylinder is mounted to a piezoactuator, which is used to change the distance between the two cylinders when recording force versus distance curves.

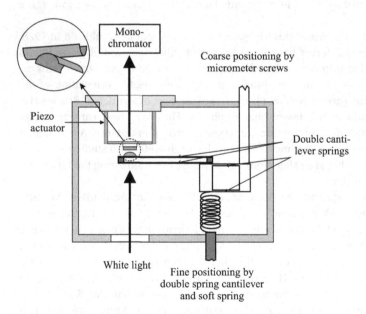

Figure 3.1 Schematic of a surface forces apparatus. Two mica sheets are glued to silica half cylinders to form a crossed cylinder geometry (inset on upper left). Their surfaces are brought in close proximity by micrometer screws and a double-cantilever spring mechanism and can then moved in and out of contact by a piezoactuator. The distance between the mica sheets is measured using optical interferometry and the force is deduced from the observed deflection of the second, much softer double-cantilever spring.

Distance control between the surfaces typically consists of a combination of several stages. For coarse approach, micrometer screws are used. In another stage, fine movement can be achieved by a mechanical design that combines a stiff double-cantilever spring with a much softer spring and a micrometer screw. When the soft spring is compressed with the micrometer screw at certain distance, the stiffer spring will deflect by a fraction of that distance that is given by the ratio of the spring constants. With this trick, deflection of the stiffer spring can be controlled on the 1 nm scale. Finally, a piezoactuator allows movements with subnanometer precision. The second mica surface is mounted on a lever arm of known and adjustable spring constant. To minimize sample tilting and twisting, this lever arm also follows a double-cantilever spring design.

The separation between the two surfaces is optically measured by using multiple beam interferometry (MBI) [161]. First the piezoactuator movement is calibrated at large separation, where the lever arm is not deflected. At such large distances, the MBI signal changes linearly with the actuator movement (measured in some type of encoder units). At close distance, the lever arm starts to deflect due to surface forces and the optical MBI reading is no longer a linear function of the actuator movement. With the known spring constant k of the lever arm, the surface force can simply be calculated using Hook's law:

$$F = k(D - D'). \tag{3.1}$$

Here, D is the actual distance between the surfaces as measured by MBI and D' is the position calculated from encoder readings assuming zero deflection of the lever arm. Spring constants range between 10 and 10^6 N m^{-1} and can often be adjusted within that range by using a clamp that allows to vary the effective cantilever length. The lowest value would correspond to a force resolution of 10^{-8} N. The most useful range of spring constants lies between 100 and 1000 N m^{-1}, where the lower value is chosen to reach a reasonably high maximal force that can be achieved within the range of the piezoactuator and the upper value is related to the maximum acceptable force error due to instrumental drift [162]. An important issue that applies not only to the SFA but also to all systems using spring deflection-based force measurements is the fact that if the force gradient gets higher than the spring constant, an instability occurs. For example, when the gradient of the attractive van der Waals forces upon approaching a surface exceeds the spring constant, the surfaces will jump into contact and during this jump no information about the interaction potential is obtained.

It should be emphasized that using the MBI method the absolute distance is measured between the surfaces. This distance measurement can be extremely precise (better than 1 Å), is not affected by instrumental drift, and gives an absolute value of separation distance D. It includes the determination of absolute zero distance, which means it can be detected if the mica sheets get into molecular contact or are still separated by adsorbed layers. In contrast, the forces acting between the surfaces are deduced from the difference between measured absolute distance D and movement D' of the actuator. Thus, the precision of the force measurement depends on precise knowledge of the spring constant and the true actuator movement. The calibration of the lever arm spring constant in the SFA is straightforward due to its macroscopic dimensions. It can, for example, be done by placing defined masses on the lever arm while measuring the deflection. The precision of the piezoactuator is typically in the order of 1 Å. However, the precision of the actuator movement will be influenced by instrumental drift, which can become a serious issue if not properly accounted for. For an in-depth discussion of sources and consequences of instrumental drift in the SFA, see Refs [162, 163].

3.1.1
Mica

The choice of the clay mineral muscovite mica for interacting surfaces in the SFA is attributed to its unique combination of properties. Clay minerals are formed by two building blocks [164]: tetrahedrons of oxygen with Si^{4+} ions in their centers and octahedrons of oxygen with Al^{3+} or Mg^{2+} in their centers. The tetrahedrons share oxygens and form hexagonal rings. Some oxygen atoms form hydroxyls, in particular when the clay is filled with Ca^{2+}. This pattern can be repeated *ad infinitum* to form flat tetrahedral sheets. Similarly, the octahedrons are linked to form octahedral layers. The tetrahedral and octahedral sheets can be stacked on top of each other in various forms to build different kinds of clays. In muscovite mica, each sheet is composed of three layers (Figure 3.2). The top and bottom layers are formed by hexagons filled with Si^{4+}. The intermediate layer is octahedral and each octahedron is filled with Al^{3+} or Mg^{2+}. The sheets are held together by potassium cations in between. Since this electrostatic binding is relatively weak, mica can easily be cleaved along the (001) plane, obtaining an atomically flat surface over many mm^2. Furthermore, mica is flexible, has a high shear and tensile strength, is chemically stable, and is inert to most liquids. The only disadvantage of mica is that it is birefringent, which slightly complicates evaluation of the FECO patterns (see Figure 3.2).

As interacting surfaces in the SFA, $\approx 0.3-4\,\mu m$ thin sheets of freshly cleaved mica are used. Mica is transparent and flexible enough to easily align onto the silica cylinders. The preparation of these mica sheets is usually done from a large block of muscovite mica by cleaving off thin sheets. Cleavage of defect-free sheets can be achieved by inserting the sharp tip of pointed tweezers. Once thinned down to the desired thickness by repeated cleavage, smaller rectangular sheets ($\approx 10 \times 10\,mm^2$ in size) are cut out by melt cutting using a hot platinum wire. Melt cutting is used to avoid generation of mica flakes that readily deposit on the freshly cleaved face. However, great care has to be taken to avoid contamination of the mica surface with platinum nanoparticles (for a discussion, see Ref. [165] and references therein). Alternatively, careful cutting with precision surgical scissors was found to produce contamination-free surfaces if properly carried out [166]. The freshly prepared sheets are placed face down on a backing sheet of freshly cleaved mica for storage and silver evaporation to avoid contamination. The backside of the mica sheets is sputter coated with a silver film of $\approx 55\,nm$ thickness as a reflective coating. This thickness gives a good compromise between sharpness of the interference fringes, which is higher for thicker, better reflecting layers, and their intensity, which becomes smaller with increasing layer thickness. In addition, at a thickness of 55 nm, slight variations in

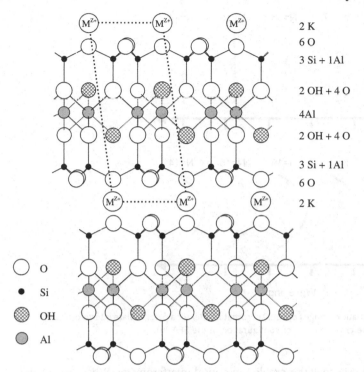

2 K
6 O
3 Si + 1Al

2 OH + 4 O
4Al
2 OH + 4 O

3 Si + 1Al
6 O
2 K

○ O
● Si
◉ OH
● Al

Figure 3.2 Atomic structure of muscovite mica ($KAl_2Si_3AlO_{10}(OH)_2$) in side view. A unit cell is indicated by the dashed line. The number of atoms of a certain species in a layer for two unit cells is indicated at the top right. As an average in the third molecular layer from the top, one silicon atom is replaced by an aluminum atom.

layer thickness do not significantly influence reflectivity, transmittance, and phase change and thus the position of the fringes [167].

The mica sheets are finally glued onto the silica cylinders with the silvered face against the glue. For measurements in air or aqueous solution, usually Epikote 1004 epoxy resin is used, which was tested to pose no contamination risk [168]. The silica cylinders are simply precoated with a thin layer of molten glue. When the mica sheet is placed on top of the liquid glue, the highly flexible sheets spontaneously adapt to the curvature of the cylinders. An alternative preparation route is to do the final cleavage of the mica sheets by peeling with an adhesive tape after mounting them onto the glass cylinders [169]. The disadvantage of this method is that the thickness of the two mica sheets will be different, which will make interpretation of the MBI pattern more complex.

3.1.2
Multiple Beam Interferometry

The surface separation between the mica surfaces is measured using multibeam interferometry. The multilayer structure formed by the silver layers, mica sheets, and

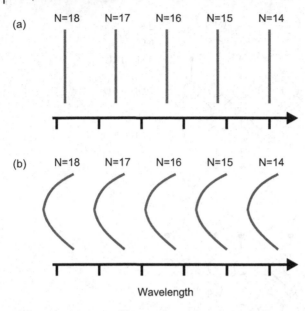

(a) N=18   N=17   N=16   N=15   N=14

(b) N=18   N=17   N=16   N=15   N=14

Wavelength

Figure 3.3 Schematic of fringes of equal chromatic order for a parallel plate interferometer (a) and as obtained for the crossed cylinder configuration in the SFA (b).

gap in between essentially resembles an optical interferometer. When illuminated with white light from one side, constructive interference will occur only if the total increase in the phase of a plane wave that is going through one reflection cycle on both mirrors matches N times 360°, where N is an integer number. Only for wavelengths that fulfill this condition, light will be transmitted. The transmitted light is sent to the entrance slit of a spectrograph. For the case of parallel mirrors, one would obtain a pattern of fringes in the focal plane of the spectrograph (Figure 3.3a). The number N is called the "chromatic" order and the fringes are therefore denoted as "fringes of equal chromatic order" or "FECO." The silver mirrors in the SFA are not parallel but crossed cylinders, a configuration that is equivalent to a sphere-plate geometry. This implies that for each point along a cross section the optical distance between the mirrors will be different and the wavelength at which constructive interference occurs will shift continuously with distance from the center. Therefore, the FECO pattern observed in an SFA is curved (Figure 3.3b).

In reality, the FECO pattern of an SFA will look more complex for several reasons. First, odd- and even-order fringes are affected differently by the refractive index of the gap medium. Second, birefringence of the mica leads to doubling of the fringes. This effect can be suppressed by using mica sheets of the same thickness and orienting them so that their optical axes are perpendicular to each other so as to cancel out the effect of birefringence (alternatively one may maximize the effect to obtain two clearly separated fringes). Third, reflections at the inner surfaces will lead to additional secondary (reflections within each mica layer) and gap fringes (reflections with the gap). Since these fringes originate from the inner surfaces, they contain information

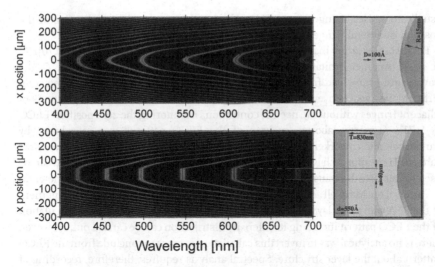

Figure 3.4 Examples of the FECO patterns as observed in a real SFA experiment. Top: with a gap of 10 nm between the mica surfaces. The curved shape resembles a cross section of the sphere-plate geometry equivalent to the crossed cylinders. Bottom: with surfaces in contact. The flat portion of the fringes resembles the width of the contact zone. Schematics of the layer structure of the silvered mica sheets for each situation are shown on the left. (Images kindly provided by M. Heuberger.)

about the surface topography. However, these fringes are hardly visible for the highly reflective silver coatings (reflectivity of $\approx 98\%$). By using thinner, less reflective silver layers, they can in principle be used to obtain topographic information [167]. Figure 3.4 shows examples of real FECO patterns observed in an SFA.

For the most simple and common case of a symmetric layer structure with mica sheets of equal thickness separated by a gap of width D with refractive index n, an analytical solution for extracting n and D from the observed FECO pattern exists [170]. This linearized approximation is valid for gap distances of up to ≈ 200 nm and allows to determine D with a resolution of 0.1 nm. At the start of an SFA measurement, the mica surfaces are brought into contact to determine wavelength λ_0 and chromatic order N_0 of a lead fringe for zero distance. During the experiment, the distance D is then calculated from the wavelength shift $\lambda - \lambda_0$ of this lead fringe using the following equations:

$$D(\lambda) \approx \frac{(\lambda - \lambda_0)N_0}{2n_{\text{mica}}} \quad \text{for odd} \quad N_0, \tag{3.2}$$

$$D(\lambda) \approx \frac{(\lambda - \lambda_0)N_0 n_{\text{mica}}}{2n^2} \quad \text{for even} \quad N_0, \tag{3.3}$$

Here, n_{mica} and n are the refractive indices of mica and the medium in the gap, respectively. Note that the positions of odd fringes depend only on the gap distance D whereas positions of even fringes will also depend on the refractive index. This is why usually an odd lead fringe is used. By combining the two equations and analyzing the position of several fringes, D and n can both be determined.

For distances larger than 200 nm, a different, less accurate approximation [170] can be used. Another limitation of this approach is that for large changes of D, the wavelength of the lead fringe will move out of the visible range making detection difficult. One way to overcome this limitation is to evaluate the relative positions of adjacent fringes without the need of continuous monitoring the zero position FECO λ_0 [171]. A more elaborate analysis of the FECO patterns was introduced by Vanderlick and coworkers [172, 173] on the basis of spectral analysis of the whole pattern. Their approach is based on the multilayer matrix method [174], which can be used to calculate the transmission characteristics of any multilayer interferometer on the basis of the Maxwell equations. This means for a given experimental arrangement of the SFA with silver layers, mica sheets, adsorbed layers, and gap, a full calculation of the FECO pattern including its intensity distribution can be carried out. However, there is no analytical way to invert this calculation process to conclude from the FECO pattern about the layer structure. Spectral analysis requires, therefore, recording of the full intensity versus wavelength information of the FECO pattern and the numerical matching of this pattern with spectra calculated by the multilayer matrix method. This method was optimized by the group of Heuberger [175] for real-time operation in the SFA (see following discussion).

Historically, Bailey and Courtney-Pratt [176] were the first to combine crossed cylinder mica sheets as molecularly smooth surfaces and FECO for distance measurement in 1955. They used this approach to map out the contact area between mica surfaces at different loads and shear forces. The first SFA was built by Tabor and Winterton in 1968 [30] and subsequently by Tabor and Israelachvili in 1972 [56]. In 1976, the newly designed SFA Mk 1 by Israelachvili and Adams [55, 177] was introduced that allowed measurements in liquids or vapors. With the SFA Mk 2, an adjustable double-cantilever spring extended the range of measurable forces and an optional friction device [644] allowed to measure shear forces of molecular thin films at defined loads. Between 1985 and 1989 Israelachvili and McGuiggan developed the SFA Mk 3I to overcome several limitations of the SFA Mk 2 [179]. The new system had a more compact design for higher attainable stiffness and lower thermal drift, an improved control system for mechanical movement of surfaces, and was more easy to clean.

Developments by other groups included the construction of an SFA with a glass chamber by Klein [180] for measurements in liquids and the SFA Mk 4 by Parker *et al.* [181]. The latter was based on the SFA Mk 2 but was more easy to assemble and use. The so-called extended SFA or eSFA was developed by Heuberger *et al.* [175, 182] on the basis of the SFA Mk 3. To significantly reduce instrumental drift, a thorough analysis of sources of errors [162] and thermal drift [163] was carried out and used to optimize construction and temperature control [182]. The second improvement is the use of fast spectral correlation [175] to make use of the full FECO pattern to numerically obtain separation distance D with a precision of ≈ 25 pm as well as the refractive index of the gap medium at the rate of several hertz. Alternatively, distance D can be monitored with higher time resolution (1 kHz) by detecting the shift of a single fringe via a split photodiode [183]. Recently, it was also shown that video rate visualization of gap width and refractive index is possible using a combination of optical correlator and CCD camera [184].

3.1.3
Friction Force Measurements

The SFA was not only a major breakthrough in the measurement of forces normal to the surface but it also became a powerful tool to study friction on the molecular level, especially for confined liquid films or adsorbed molecular layers. When confining liquids in the SFA down to a few molecular layers, the mica sheets will deform and flatten out in the contact region due to the relatively soft layer of glue. This leads to parallel surfaces over a contact area of 10–100 μm. To allow friction force measurements, two additions were necessary: the possibility to shear the two mica surfaces relative to each other and to detect the corresponding shear force. These requirements were fulfilled using different designs (for a detailed review see Ref. [185]). The first system introduced by the Israelachvili group used a double-cantilever spring consisting of piezoelectric bimorphs to shear the lower surface. The upper surface was held by vertical leaf springs with integrated strain gauges to detect their bending due to the friction force. It is also possible to shear the whole upper holder with a mechanical drive that is less precise than the bimorphs but allows 5 mm of travel range [186]. The Granick group used a design where the upper surface is held by a pair of vertical leaf springs that were connected to piezoelectric bimorphs. One bimorph was used to move the upper surface via one leaf spring, the second bimorph acted as a sensor of the friction force via the second leaf spring [187]. The group of Klein used a piezoelectric tube scanner for vertical and lateral movement of the upper mica surface. The tube scanner itself is mounted onto a platform that is held by vertical leaf springs. Deflection of these springs is detected by capacitive sensor [188]. Another design that allows shearing of the lower surface in two dimensions was introduced by Qian *et al.* [189], and recently a new system that allows large-scale (500 μm) sliding motion with closed loop control of normal load was introduced [190]. A very different approach to study friction at very high oscillation frequencies was taken by Berg *et al.* [191] who combined an SFA setup with a quartz crystal microbalance.

3.1.4
Surface Modification

The use of molecularly smooth mica surfaces is a conceptual strength of the SFA. At the same time, it is a severe limitation since it is the only surface that can be directly studied. There have been several approaches to overcome this limitation and allow the use of different surfaces. Evaporation of thin metal layers onto mica leads to increased roughness of the surfaces even when template stripping [192–194] is used. Most simple, but less controlled, is surface modification by adsorption, preferentially, of molecules that self-assemble into monomolecular layers. Alternatively, chemical modification of the mica surfaces by silanes can be achieved. Finally, mica can be replaced by other transparent materials, which will, however, not be atomically flat. By redesigning the SFA, also nontransparent surfaces can become a possible target of investigation. If only one of the two interacting surfaces is exchanged, one can still use reflective MBI for distance measurement [195]. With both surfaces nontransparent, a different detection scheme has to be introduced, which also means giving up the independent measurement of absolute surface separation. Common to all these approaches is the high demand of clean handling and operation to avoid contamination within the relatively large contact area.

3.2
Atomic Force Microscope

The atomic force microscope (AFM) was invented in 1986 by Binnig *et al.* [196]. Originally, it was developed for imaging the topography of surfaces. In the AFM, a sharp tip that sits at the end of a microfabricated cantilever is brought in contact with the sample surface, the tip is then raster-scanned over the surface by moving either the tip or the sample (Figure 3.5). The deflection of the cantilever due to the surface topography is measured using a laser beam that is reflected from the backside of the cantilever onto a split photodiode. Height traces of the single scan lines are then put together to a three-dimensional surface topography by software on the control computer. Imaging resolution will depend on several factors. Using piezoactuators makes the movement of the sample possible with angstrom precision, and the optical lever technique allows a subangstrom resolution for detection of the cantilever deflection. Typical tip radii for standard AFM probes are in the order of 5–50 nm.

In many cases, imaging resolution critically depends on the tip–sample interaction. For soft surfaces, contact pressure of the tip will lead to sample deformation and reduced resolution. Understanding the forces acting between the tip and the sample

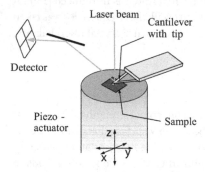

Figure 3.5 Schematic of an AFM setup. The sample is in contact with a sharp tip that sits at the end of microfabricated cantilever and is raster-scanned by a piezoactuator. The deflection of the cantilever due to the surface topography is measured by a laser beam that is reflected from the backside of the detector onto a photodetector.

was therefore an important step in understanding AFM operation and image contrast mechanism. An example was the finding by Weisenhorn *et al.* [197] that under ambient conditions, capillary forces will commonly dominate the forces between the tip and the sample and thus imaging in water allows to dramatically reduce surface forces. First force-versus-distance experiments to study surface forces in liquid were carried out in 1991 [198, 199].

Since then, AFM force measurements have become a prominent tool in surface science, covering a broad range of topics (review in Ref. 200). The unique capability of the AFM to acquire forces locally and with high sensitivity makes it possible to get information about the interactions of even a single-molecular pair. This kind of experiments is known as "single molecule force spectroscopy" and has been applied mainly to the field of polymers and biomolecules (for reviews see Refs [201–203]). Especially in the case of biological interactions, single molecule force spectroscopy has become an important tool for investigation of receptor–ligand interaction at the single–molecule level.

3.2.1
Force Measurements with the AFM

In an AFM measurement, the sample is moved up and down by the piezotranslator, while measuring the deflection of the cantilever. In principle, almost any kind of sample can be investigated by AFM, but for fundamental studies, preferentially planar surfaces with low surface roughness such as mica, HOPG, or silicon wafers are used. Problems due to sample roughness and contamination are greatly reduced in the AFM compared to other surface force techniques since the sample only needs to be smooth on a scale comparable to the radius of curvature at the end of the tip. Surface quality can be directly checked by imaging the surface and allows to select clean and smooth areas of the sample. Atomic force microscopes can be operated in air, different gases, vacuum, or liquid. Different environmental cells, in which the kind of gas and the temperature can be adjusted, are commercially available. To acquire force curves in liquid, different types of liquid cells are employed. Typically, liquid cells consist of a special transparent cantilever holder that forms the upper lid of the cell, an O-ring sealing the cell from the side, and the sample surface forming the bottom of the cell.

In a typical AFM experiment, the piezoelectric translator moves with constant velocity up and down so that its position versus time can be described by a triangular function while the detector signal of the photodiode is recorded. The outcome of such a measurement is a measure of the cantilever deflection, Z_c, versus position of the piezotranslator, Z_p, normal to the sample surface. A schematic example is given in Figure 3.6. For large distances between probe and surface, no surface forces are acting and the deflection will be zero (1). When the probe gets close to the surface, the cantilever will start to deflect due to surface forces (e.g., bend down due to attractive van der Waals forces) (2). In cases where the gradient of the attractive force exceeds the spring constant of the cantilever, the probe will jump in contact with the surface (dotted line). From this point onward, probe and surface will move in parallel (constant compliance region, 3) when assuming a hard contact. Upon reversal of

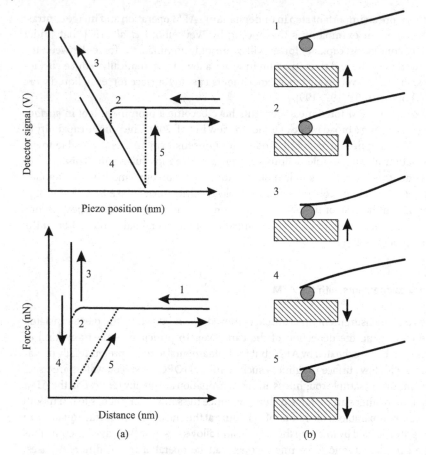

Figure 3.6 Schematic of an AFM measurement. (1) Zero force region, where probe and surface are far from each other; (2) "snap-in" to surface due to attractive force; (3) approach part of the constant compliance region, where probe and surface move up in parallel; (4) retract part of the constant compliance region, where probe and surface move down in parallel and adhesion leads to negative bending; and (5) "jump-out" occurs when the restoring force of the cantilever exceeds the adhesion force. (a) Detector signal versus piezoactuator position. (b) Force versus distance curve derived from (a).

the direction, the cantilever will again move in parallel with the sample (4). If adhesive forces between probe and sample surface exist, the cantilever will bend downward (4) until the restoring force of the cantilever exceeds the adhesion force and the cantilever snaps back to its equilibrium position (5, dotted line).

The "jump to contact" (2) and "jump-out" (5) instabilities have the consequence that not all parts of the interaction potential can be reconstructed from the measurement. Such instabilities can be avoided or suppressed by using stiffer cantilevers, but this would be at the cost of reduced sensitivity. To obtain a real force-versus-distance curve (called "force curve"), Z_c and Z_p have to be converted to force and distance. First, a line fit is done on the zero force region to subtract any offset in the detector signal. Then, a linear fit of the constant compliance region is done to obtain the slope that is equal to the conversion factor between detector signal in Volts and cantilever deflection in nanometers. The tip–sample separation D is calculated by adding the deflection to the piezotranslator position $D = Z_p + Z_c$ and the force F is obtained by multiplying the deflection Z_c with the spring constant K_c of the cantilever.

A critical point in the evaluation of AFM force curves is the determination of zero distance, meaning the point where the probe starts to touch the surface. While in the SFA, the analysis of the FECO pattern gives clear information on any gap remaining between the mica surfaces, there is no independent check for absolute distance in the AFM. This can lead to ambiguities either due to adsorbed layers or due to deformation of soft surfaces. One attempt to overcome this limitation was the combination of AFM with TIRM [204] (see Section 3.4), where the scattering signal from a colloid probe was used for absolute determination of separation distance.

3.2.2
AFM Cantilevers

The cantilever is in fact a key element of the AFM and its mechanical properties are largely responsible for its performance. The development of production processes for AFM cantilevers made of silicon or silicon nitride using standard photolithography [205, 206] allowed cost-effective and reliable manufacturing and has certainly driven the commercial success of AFM. Today, AFM cantilevers with integrated tips are mostly fabricated from silicon or silicon nitride in the shape of a single rectangular beam or a triangle (Figure 3.7). Both are covered with a native oxide layer of 1–2 nm thickness.

The mechanical properties of cantilevers are characterized by the spring constant K_c and the resonance frequency. Both can in principle be calculated from the material properties and dimensions of the cantilever. The spring constant of a rectangular cantilever with length L, width w, and thickness d (Figure 3.7) made

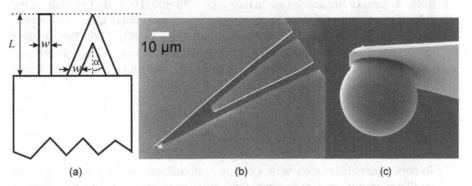

| (a) | (b) | (c) |

Figure 3.7 (a) Schematic top view of a rectangular and triangular AFM cantilever with length L, width w, and opening angle α; (b) SEM image of an AFM cantilever with sharp tip; and (c) colloid probe prepared by sintering a polystyrene particle to a tippless AFM cantilever.

from a material with Young's modulus E can be calculated by using thin plate theory [208]:

$$K_c = \frac{Ewd^3}{4L^3}. \qquad (3.4)$$

A good cantilever should have a high sensitivity. High sensitivity in Z_c is achieved with low spring constants or low d/L ratio. Hence, in order to have a large deflection at small force cantilevers should be long and thin. In addition, the design of a suitable cantilever is influenced by other factors:

- External vibrations, such as vibrations of the building, the table, or acoustical noise, which are usually in the low-frequency regime, are less transmitted to the cantilever, when the resonance frequency v_0 of the cantilever

$$v_0 = \frac{1.875^2}{2\pi\sqrt{12}} \frac{d}{L^2} \sqrt{\frac{E}{\varrho_c}}. \qquad (3.5)$$

  is as high as possible. Here, ϱ_c is the density of the cantilever material. The equation is valid for a rectangular cantilever. A high resonance frequency is also important to be able to scan fast because the resonance frequency limits the time resolution [205, 207].
- Cantilevers often have different top and bottom faces due to reflective coatings or processing during manufacturing. Temperature changes, adsorption of substances, or electrochemical reactions in liquid environment can change the surface stresses of both faces differently. This leads to a bending of the cantilever [206, 210]. Practically, these changes in surface stress lead to an unpredictable drift of the cantilever deflection that disturbs force measurements. To reduce drift the ratio d/L should be high.

These requirements lead to the conclusion that AFM cantilevers should be small. Only short and thin cantilevers are soft, have a high sensitivity, and a high resonance frequency. Typical cantilevers are few 100 μm long, some 10 μm wide, and 0.5–3 μm thick. Resonance frequencies are in the range of 10–400 kHz in air. Especially in the field of single-molecule force spectroscopy, several researchers aim to make even smaller cantilevers with higher resonance frequency [211, 212]. The smallest cantilevers were 10 μm long, 0.1–0.3 μm thick, and 3–5 μm wide. Further size reduction would, however, make it more and more difficult to fabricate cantilevers and to focus the laser beam onto such small structures.

3.2.3
Calibration of the Spring Constant

To obtain quantitative force versus distance information, precise knowledge of the cantilever spring constant is a prerequisite. In principle, values of the cantilever spring constants are given by the manufacturers, but these values are often not very reliable. Therefore, calibration of spring constants is an important part of quantitative AFM measurements. Several methods are employed. They can be grouped into dimensional methods, that use the dimensions and material properties, static methods that apply a known force and measure the resulting deflection, and dynamic ones that are based on the resonance frequency of the cintilever.

Spring constants of AFM cantilevers can be calculated from the dimensions and material properties of the cantilever according to Eq. (3.4). For triangular cantilevers, as a first approximation, it can be viewed as the combination of two parallel rectangular beams. If w is the width of one arm, the spring constant is 205.

$$K_c = \frac{Ewd^3}{2L^3}. \tag{3.6}$$

More complex analytical equations have been deduced by several authors [207, 213–215] that take into account deviations both from the ideal V-shape and from the thin plate theory. Numerical calculations using finite element allow an even more realistic simulation of shapes and bending behavior [207, 214, 216, 217]. For a recent discussion of these different dimensional approaches and their refinement, see Refs [218, 219].

Calculation of spring constant from cantilever dimensions and material properties suffers from two main problems – exact determination of cantilever thickness is difficult and Young's modulus of the thin layer of cantilever material may differ from the bulk value – and thus, calculated spring constants were found to significantly deviate from actually measured ones [207, 215]. The possible existence of reflective coatings on the backside of the cantilever further complicates the calculations. Therefore, experimental determination for each single cantilever is usually employed.

One method is the thermal noise method introduced by Hutter and Bechhoefer [220]. It is implemented in many commercial AFMs. The thermal noise method is based on the equipartition theorem of statistical mechanics that states that the mean thermal energy of any harmonic system at temperature T is equal to $k_B T/2$ per degree of freedom. For the small thermal oscillation amplitudes (≈ 0.1 nm), the AFM cantilever can be seen as a harmonic oscillator with spring constant k_c. The mean square deflection of the cantilever due to thermal fluctuations must fulfill the following condition:

$$\langle x^2 \rangle = \frac{k_B T}{K_c}. \tag{3.7}$$

Practically, the thermal noise of the cantilever is recorded by measuring the deflection of the cantilever for some time while it is free in air far from any surface. From recorded signals, the power spectrum $P(\nu)$, which is a plot of x^2 versus frequency, is calculated by fast Fourier transform. The total value of $\langle x^2 \rangle$ would then be given by the integral of the power spectrum:

$$\langle x^2 \rangle = \int\limits_0^\infty P(\nu) d\nu = \frac{k_B T}{K_c}. \tag{3.8}$$

The first resonance peak is fitted after subtracting background noise. Since thermal excitation acts as a white noise driving force, the power spectrum should follow the response function of a simple harmonic oscillator:

$$P(\nu) = \frac{A\nu_0^4}{(\nu^2 - \nu_0^2) + (\nu\nu_0/Q)^2}. \tag{3.9}$$

Here, ν_0 and Q are the resonance frequency and quality factor of the first resonance peak. By combining Eqs. (3.8) and (3.9) and solving the integral, we obtain

$$K_c = \frac{2k_B T}{\pi A \nu_0 Q}. \tag{3.10}$$

However, the value of K_c obtained from fitting just the lowest peak will overestimate the spring constant, since part of the measured deflection will be due to higher oscillation modes with each of the modes having the mean thermal energy $k_B T/2$. From an analysis of all vibration modes and their relative contributions, correction factors have been deduced for rectangular [221] and V-shaped cantilevers [222, 223]. A second important effect is that deflection is usually detected with the optical lever technique. The optical lever technique measures the inclination at the end of the cantilever rather than its deflection. Deflection and inclination are proportional to each other, but the proportionality factor depends on whether the end of the cantilever is free or in contact with a surface. At the same deflection, the inclination of the first vibration mode is lower than the inclination of a cantilever with a vertical end load. Conversely, if the signal caused by the first vibration mode is similar to the signal of the same cantilever with a vertical end load, the deflection for the first vibration mode is higher. Together these effects lead to correction factors of 0.81 for rectangular and 0.78 for V-shaped cantilevers. Detailed instructions on how to practically implement the thermal noise calibration method including all necessary correction factors are given in Ref. [224].

Cleveland $et\ al.$ [225, 226] introduced another calibration method, which includes dimensional data but without the need to know the cantilever thickness. The Cleveland method requires knowledge of cantilever length L and width w, elastic modulus E, and density ϱ of the cantilever material. From the measured resonance frequency ν_0, one can calculate the spring constant:

$$K_c = 2w(\pi L \nu_0)^3 \sqrt{\frac{\varrho_c^3}{E}}. \tag{3.11}$$

One possible source of error with this method are the material characteristics E and ϱ_c. For thin films, they might deviate from the bulk literature values, depending on how the cantilevers have been produced. Therefore, Cleveland $et\ al.$ have shown in the same paper that alternatively one could measure the change in resonance frequency upon addition of a mass M at the end of the cantilever and then calculate the spring constant from

$$K_c = (2\pi)^2 \frac{M}{1/v_M^2 - 1/v_0^2}. \tag{3.12}$$

Here, v_M and v_0 are the resonance frequencies with and without added mass. However, attaching the spheres is cumbersome and potentially destructive, and as an alternative, addition of a thin gold layer [227] or small droplets by an inkjet nozzle [228] has been used. The Sader method circumvents these problems by determining spring constants from the resonance frequency and the quality factor Q alone, when both the width w and length L of the rectangular cantilever and the damping behavior of the surrounding fluid are known. The spring constant can be calculated according to

$$K_c = 0.1906\varrho w^2 L Q \Gamma_i(Re)(2\pi v_0)^2, \tag{3.13}$$

where ϱ is the density of the fluid and $\Gamma_i(Re)$ is the imaginary part of the so-called hydrodynamic function (plotted in Ref. [226]) that depends on the Reynolds number Re and takes into account the viscosity η of the fluid.

$$Re = \frac{\varrho w^2 2\pi v_0}{4\eta}. \tag{3.14}$$

Although in principle this method could be applied in any fluid, uncertainties become higher for stronger hydrodynamic interaction [229]. For that reason, the Sader method should preferentially be applied in air. Comparisons of the thermal noise and Sader method by several authors gave an uncertainty of 5–10% for obtained spring constant for both methods [230, 231, 233].

Static methods of spring constant calibration apply a defined force while measuring the resulting deflection of the cantilever. This can be done by adding a defined mass to the end of the cantilever [234], applying a hydrodynamic drag force on the cantilever by linear movement in a liquid [235], detecting the hydrodynamic force for a colloid probe approaching a wall [236, 237], a fit of DLVO forces between a silica sphere and a silica surface [238], or use of a nanoindenter [239, 240].

A simple possibility is the use of a reference cantilever with known spring constant K_{ref} [241, 242, 244]. Therefore, the reference cantilever is used as sample and the cantilever that should be calibrated is mounted in the AFM. Then, force curves are taken on a hard substrate and on top of the end of the reference cantilever. From the slopes of the force curves S_{ref} and S_{subs} taken on the solid substrate and the reference cantilever, respectively, one can simply obtain the spring constant from

$$K_c = K_{ref} \frac{S_{subs} - S_{ref}}{S_{ref}}. \tag{3.15}$$

If spring constant of reference cantilever was determined precisely, most of the error stems from variation in alignment of cantilevers. Cumpson *et al.* [243] have addressed this issue by the use of a long (1.6 mm) reference cantilever with alignment marks. The use of piezoresistive reference cantilever eliminates the need of photo-detector calibration [245] and would even allow their use as traceable SI transfer standards [246].

3.2.4
Microfabricated Tips and Colloidal Probes

For quantitative AFM force experiments not only the cantilever spring constant but also the tip geometry needs to be known. A standard procedure is to use a scanning electron microscope (SEM) to image the tip. However, charging of the tip during imaging in case of the insulating silicon nitride may be a problem unless low-voltage or low-vacuum SEMs are used. Image resolution is typically in the order of 5 nm, which may not be good enough to resolve the shape of very sharp tips. In addition, SEMs are not calibrated in the normal direction. Therefore, SEM imaging is used for semiquantitative tip shape and tip wear characterization.

Tip wear is one of the major problems in force measurements. The assumption that the surfaces of tip and sample do not change during a force experiment is probably in most cases wrong. Computer simulations showed that a transfer of atoms between the two surfaces is likely as they get into contact [247]. Chung *et al.* [248] could show that during the first approach significant structural changes of the tip apex can occur, even at low forces.

Higher resolution can be achieved with transmission electron microscopy (TEM), where even the crystalline shape of the tip could be resolved [248], but this is achieved at the price of much higher experimental effort and is not done routinely. An alternative method for tip shape characterization is to image a sharp structure and reconstruct the tip shape from the image [249–251]. In this case, one relies on a defined structure of the sharp object and the characterization itself might cause damage to the tip. Therefore, characterizing the tip size and shape for routine applications is an unsolved problem, neither is the resolution satisfactory nor are the methods noninvasive.

To overcome the problem of tip shape for nanoscale AFM tips, the so-called colloid probe technique was introduced [198, 199] (for reviews, see Refs [252–254]). By attaching smooth and spherical particles to the end of (tipless) AFM cantilevers, one obtains a probe with defined geometry and surface chemistry (Figure 3.7c). Attachment of the particles is usually done under the control of an optical microscope with a micromanipulator. A tiny amount of glue is placed onto the very end of the cantilever and then the colloidal probe is brought in contact with this spot. This can be achieved either by placing glue and particle by the use of thin wires [255, 256] or a micropipette [257] or by moving the cantilever and first touching a small glue droplet and then the particle to be mounted [258, 259]. If a micromanipulator is not available, an AFM itself may be employed by using its sample stage and integrated video microscope [260, 261]. For some materials such as glass [262] or polymers [263], sintering instead of gluing was demonstrated to exclude the risk of probe contamination.

However, the colloid probe technique is not limited to particles. For biological studies, strategies to attach single spores [264], bacteria-coated beads [265], or single cells [266, 267] have been developed. To study forces in emulsions [268] or flotation cells such as oil drops [269, 270, 696] and bubbles have been attached to cantilevers [271].

A limitation of the colloid probe technique, however, is the minimum particle size that can be reproducibly attached by using optical microscopy. For spheres smaller than 1 μm, it becomes difficult to correctly position the particle at the very end of the cantilever to avoid touching the substrate with the edge of the cantilever. In this respect, the name colloid probe is somewhat misleading since colloidal particles are usually smaller than 1 μm. Recently, there have been attempts to attach nanoparticles to the end of AFM tips either by wet chemistry [272] or by epoxy-coating of tips and dipping them into a powder [273].

In cases where AFM tips have been functionalized with special chemical end groups, they can be used to map out chemical interactions between tip and surfaces using either the force volume mode or the friction force microscopy. For both types of experiments, the term "chemical force microscopy" has been introduced (reviewed in Ref. [274]).

3.2.5
Friction Forces

Soon after the introduction of the AFM, it was recognized that small modifications allowed the detection of friction forces [275]. For this purpose, a scan direction perpendicular to the long axis of the cantilever is used. A friction force between the tip and the substrate will then lead to a torsion of the cantilever and this in turn induces a lateral movement of the laser beam on the detector. By using a quadrant photodiode as shown in Figure 3.8, this lateral shift can be recorded and allows simultaneous recording of surface topography and friction [276]. This mode of operation is called friction force microscopy (FFM) or lateral force microscopy (LFM).

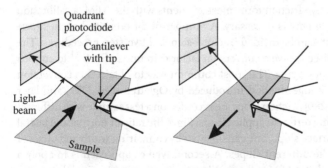

Figure 3.8 Schematic of the friction force microscopy operation of an AFM. When using a scan direction perpendicular to the long axis of the cantilever, friction between tip and surfaces leads to a lateral deflection of the laser beam that is recorded by a quadrant photodiode.

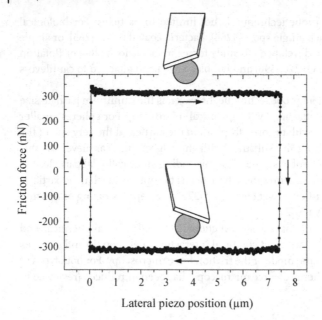

Figure 3.9 Example of the friction force obtained in an AFM friction force measurement [295]. The friction force between probe and surface induces a tilt of the cantilever that is recorded as a lateral movement of the laser spot on the photodetector. The upper and lower parts of the so-called friction loop correspond to the trace and retrace scan, respectively. The width of the friction loop (i.e., the difference between trace and retrace) is proportional to the friction force.

A typical lateral deflection signal recorded for a single left/right scan cycle along a scan line is plotted in Figure 3.9. The upper and lower parts of this so-called friction loop result from trace (scanning left) and retrace (scanning right back on the same line). The width of the friction loop, which means the difference of the average values from trace and retrace, is proportional to the friction force.

To carry out quantitative friction force measurements with the AFM, a calibration of the lateral force sensitivity is necessary. A number of different techniques have been introduced, which can be divided into one-step and two-step procedures. The one-step procedures directly give a conversion factor α in units of $N\,V^{-1}$ to convert detector voltage to friction force. The most common way to obtain this conversion factor is the so-called wedge method introduced by Ogletree *et al.* [277] and later refined by others [278–280]: friction loops are recorded on a test sample with different well-defined slopes using different applied loads. Since the ratio of normal and lateral forces changes with surface slope, one can obtain α by comparing width and offset of the friction loops on the different slopes. A second type of approach is to apply a known normal force with the AFM to the cantilever at an off-center position by using a tip mounted on a surface [281], by attachment of a lever arm perpendicular to the end of the cantilever [282, 283].

A third possibility is to apply a defined friction or torque while recording the detector response. This was achieved using the Lorentz force on a conductive cantilever in a magnetic field [284], using glass fibers of known stiffness [285, 286], or using especially designed micromechanical devices [287, 288].

The two-step processes consist of (1) determination of the torsional spring constant K_L of the cantilever and (2) the lateral sensitivity of the detection system. With the known height of the AFM tip or diameter of the colloid probe, the friction force can then be calculated from the measured detector signal.

If the dimensions of the cantilever are known, the torsional spring constant can be calculated [214, 219, 289]. For rectangular cantilevers, calibration methods equivalent to the Cleveland and Sader methods for the normal spring constant have been developed [290]. Another approach uses the application of a defined vertical force at the end of the cantilever with a defined lateral offset [281]. The lateral sensitivity can be obtained from a lateral movement of the detector by a defined distance in case of known geometry of cantilever position and laser alignment [291, 292] or by placing a mirror in the beam path that can be tilted by defined angles [281]. In principle, the lateral sensitivity could also be deduced from the initial slope of the friction loop [293]; however, this is valid only if the contact stiffness is high enough [294]. In the case of colloid probes, these can be pushed by a defined distance against a vertical step [295, 296], and in the case of AFM tips, a calibration grating with vertical steps of defined height can be used to cause defined lateral bending while recording the detector signal [297].

3.2.6
Force Maps

Instead of taking force–distance curves only on selected points of the sample, one can also acquire force-versus-distance curves at every point corresponding to a pixel of the AFM image. Since the tip is scanned along the surface and also moves in the z-direction normal to the surface, the term "force volume mode" has been coined for this mode of operation. From the array of force–distance curves, the spatial variation of interactions throughout the sample surface can be obtained. This is usually done by postprocessing of the force data, resulting in two-dimensional maps of physicochemical sample properties. Since lateral movement of the AFM tip can be done in a retracted position in force volume mode, lateral forces on the sample during scanning can be avoided.

A fundamental problem in force volume mode is the relatively long acquisition time that can easily amount to tens of minutes and can give rise to excessive drift. This and the memory usage for storing all data points limit the resolution of the force maps to typically 64×64 data points compared to the typical 512×512 pixels for standard AFM images. A critical point that has hindered the widespread use of force volume is automated data evaluation, which is currently not part of standard AFM software of the commercial suppliers. Since some thousand force curves have to be evaluated for each data set, sophisticated software routines have to be developed by the users for automated analysis to efficiently use force volume mode.

3.2.7
Dynamic Modes

Recently, it has even become possible to probe the interaction between single atoms by using frequency modulation AFM (FM-AFM) (for a review see Ref. [298]. This special AFM mode was introduced by Albrecht *et al.* [299] to allow noncontact imaging of surfaces in ultrahigh vacuum up to atomic resolution images that could be obtained with this method. In FM-AFM, one uses active excitation of the resonance of the AFM cantilever with a fixed oscillation amplitude at its resonance frequency v_0. When the tip is brought in close proximity to the surface, the presence of a force gradient will correspond to an additional spring constant

$$\Delta K = \frac{dF}{dz},\tag{3.16}$$

which leads to a change in resonance frequency by

$$\Delta v = v_0 \frac{\Delta K}{K_c},\tag{3.17}$$

An active feedback circuit is used to sense this change in resonance frequency and allow scanning of the surface at constant frequency shift Δv, which corresponds to a scan at constant potential gradient. In 2001, Lantz *et al.* [300] were able to measure forces between single atoms on a silicon tip and a silicon surface, and in 2002 Giessibl *et al.* could probe the friction forces between single silicon atoms [301]. While originally developed for UHV conditions, it was recently shown that operation of FM-AFM is even possible in liquid [302] and can be used for quantitative force measurements in solution [303].

3.3
Optical Tweezers

In optical tweezers, the interaction between light and matter is used to control the movement of micrometer-sized objects by a single laser beam. Optical trapping of particles in a liquid was first demonstrated by Ashkin *et al.* [304], who showed that dielectric particles in the size range of 10 μm down to 25 nm could be trapped in aqueous solution using a single laser beam. Soon after they demonstrated that this technique also allows the trapping and manipulation of biological objects such as cells [305], which has much inspired their use in biological applications [306–308] as well as in the fields of colloid science [309, 310] and microrheology [311]. Although optical tweezers are almost exclusively used in liquids, where hydrodynamic drag facilitates stable trapping, there has been recent effort to establish this technique also for aerosols [312, 313]. For reviews on optical tweezers, we refer to Ref. [314] that gives a very detailed description of the technique and Refs [315, 316] for recent trends and developments.

Optical tweezers are based on the premise that a particle placed into the focus of a collimated laser beam will feel two optical forces. The first one is the scattering force. It points in the direction of the light propagation and is proportional to the light intensity. The second one is the gradient force. For a dielectric particle with higher refractive index than the surrounding medium, this force is directed toward higher light intensity (i.e., toward the focal spot) and is proportional to the intensity gradient. Therefore, the particle will be trapped close to the focal spot if the gradient force is large enough.

For particles with a large diameter compared to the laser wavelength λ, this gradient force can be explained by simple geometric optics (Figure 3.10); this is within the framework of Mie theory. The rays A and B are diffracted by the optically more dense particle. Conservation of momentum for the diffracted rays leads to a net force on the particle that is directed toward the focus of the laser beam. Reflected rays contribute only little to the force. When the particle diameter is much smaller than wavelength λ, wave optics has to be used and the theory for Rayleigh scattering can be applied to calculate the scattering and gradient force. Rayleigh theory treats the scattering particle as a single point dipole. In this picture, the scattering force arises from absorption and re-emission of light from the dipole and is given by

$$F_S = \frac{128 I_0 \pi^5 R_p^6 n_1}{3c\lambda^4} \cdot \frac{(n_2/n_1)^2 - 1}{(n_2/n_1)^2 + 2}. \tag{3.18}$$

I_0 is the laser intensity, R_p is the particle radius, c is the speed of light, and n_1 and n_2 are the refractive indices of medium and particle, respectively. The gradient force

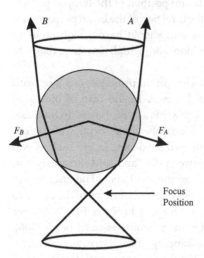

Figure 3.10 Optical gradient force on a particle with refractive index higher than the surrounding. Conservation of momentum for the diffracted rays leads to a net force that pulls the particle into the focus.

arises from interaction of the induced dipole with the intensity gradient ∇I_0:

$$F_G = \frac{2\pi R_p^3}{c} \cdot \frac{(n_2/n_1)^2 - 1}{(n_2/n_1)^2 + 2} \cdot \nabla I_0. \qquad (3.19)$$

Both, Mie and Rayleigh scattering can give a qualitative understanding of the working principle of optical tweezers. In most applications, particles with diameters of 0.2–5 μm are used, which fall just in between these two limiting ranges. For a quantitative evaluation of trapping potentials, complex numerical calculations are unavoidable. Results of such calculations agree with experimental results [317, 318]. Close to the trap center, the particle will feel a quadratic potential corresponding to a linear spring. For this situation, the trap stiffness, (i.e., the restoring force on the particle toward the trap center due to a deviation from the trap center) can simply be characterized by a spring constant K_t. The range over which the trap still behaves as a linear spring is limited to about 150 nm and the maximum distance a trapped particle can move within losing it from the trap is in the order of 400 nm. Trapping is only possible if the scattering force does not exceed the gradient force. In addition, the trapping force has to be stronger than thermal motions of the particle. While the first condition will be easier to match with smaller particles, the second one will limit the minimum size of particles that can be trapped. This makes trapping of nanoparticles demanding. A few examples of optical tweezer studies of nanoparticles are reviewed in Ref. [319].

To maximize the gradient force, high numerical aperture objectives should be used to focus the laser beam to a diffraction limited spot. Note that for a given optical setup, scattering force and gradient force will both increase linearly with laser intensity. As a consequence of the scattering force, the equilibrium position of the trapped particle will be located slightly beyond the focal spot in the direction of the beam propagation. To achieve a low noise trapping, the optical system should have a high pointing stability to avoid fluctuations of the trap position and a high power stability to minimize force fluctuations.

An important issue is the choice of laser wavelength. In principle, any laser with sufficient beam stability and laser power can be used. In the case of biological samples, however, radiation damage by the laser power is a critical issue. In this case, a wavelength in the near infrared is most suitable due to the low absorption of water in the range of 800–1200 nm. Typically, diode-pumped solid-state lasers such as Nd: YAG, Nd:YLF, or Nd:YVO$_4$ with wavelengths between 1047 and 1064 nm are used in a Gaussian TEM$_{00}$ mode that allows to minimize the focal spot. The laser power required will depend on the precise application, especially on the size of the object to be trapped and the desired trap stiffness. A typical force of 1 pN per 10 mW of laser power in the specimen plane has been reported for micrometer-sized beads [306]. The lowest laser power that still allowed stable trapping of a 0.53 μm particle in an optimized setup was reported as 0.6 mW [320]. The force range of optical tweezers of 0.025 pN [321] to 100 pN [322] is a perfect match for single-molecule studies.

To apply a force with optical tweezers to single molecules, typically a spherical particle of ≈ 1 μm diameter is trapped with the laser. The bead will act as a handle to exert forces onto the molecule that is additionally fixed either to a surface or to a second bead. In the latter case, the second bead may be held by a second optical trap or by aspiration with a micropipette.

In many cases, optical tweezers are built with an inverted optical microscope that gives a convenient platform providing the focusing optics, a measurement chamber, and imaging capabilities. The main additional components are the laser, a beam shaping and steering optics, and a detection system for the bead position (Figure 3.11). The laser beam is sent through a beam expander to widen its diameter. This leads to a cutoff of the shoulders of the Gaussian beam profile by the entrance aperture of the microscope objective, which results in a more stable trapping. The reason for this can be seen from Figure 3.10. The outer part of the focused laser beam will contribute most to the gradient force, whereas the central part will hardly be diffracted and will mainly contribute to the scattering force. Therefore, widening the beam above the size of the objective entrance aperture can maximize the relative

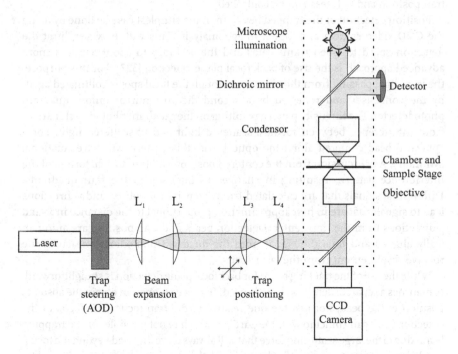

Figure 3.11 Schematic of an optical tweezer setup. It consists of an optical video microscope (components within dotted line) and the components needed for the laser trap. The laser beam is widened by a beam expander and coupled into the microscope by a dichroic mirror. The beam is focused on a diffraction limited spot by the high numerical aperture objective to form the laser trap for the bead. The light scattered by the bead is collected by the condenser and reflected onto the detector by a second dichroic mirror to monitor the bead position. Positioning of the laser trap can be done by either a displacement of lens (L_3, slow) or an acoustooptical deflector (AOD, fast).

intensity of the extremal rays and thus optimize trapping efficiency. The objective itself must be a high numerical aperture objective (NA = 1.2–1.4) to obtain a steep intensity gradient. This limits the effective z working distance and thus the maximum distance the trapped object can be above the glass bottom of the measurement cell.

Beam steering can be achieved by several means. One possible approach is to use a beam steering mirror and a simple telescope to reimage the mirror plane onto the entrance aperture of the objective. A tilting of the mirror will lead to a lateral movement of the focal spot without changing focus and intensity. This element might simultaneously act as a dichroic mirror to separate the incoming laser light from the imaging light path going in the opposite direction from the collimator to the CCD camera. Alternatively, one of the lenses of a telescope in the beam path can be repositioned. A much faster steering control is possible by acoustooptical deflectors. These consist of a transparent crystal in which a diffraction grating is induced by an acoustic wave. The deflection angle is controlled via the acoustic wavelength and the intensity can be controlled by the acoustic wave amplitude, which means that both trap position and stiffness can be controlled.

Position detection of the trapped bead can in the simplest case be done by using the CCD video camera and digital image analysis. This will, however, limit the detection speed to video frame rates and the accuracy to about 5 nm. A more advanced approach is the use of back focal plane detection [323]. For this purpose, the laser light passing from the objective through the focal spot is collimated again by the condenser and reflected by a second dichroic mirror onto a quadrant photodetector. The intensity pattern resulting on the quadrant photodetector arises from interference between forward-scattered light and unscattered light. For a spherical bead centered along the optical axis, this pattern will have rotational symmetry. Any excursion from the centered position will lead to a distortion of the interference pattern, resulting in changes of intensity in the four quadrants. Differential signals that are calculated from the pairs for both x- and y-directions lead to signals that are in first approximation proportional to the distance in x- and y-directions from the trap center. Detection beam and trap position are automatically aligned and thus the signal from the quadrant photodiode measures the relative displacement from the laser focus.

While the use of the tracking beam for backfocal plane detection is straightforward, sometimes a separate detection laser is used. In this way, one can obtain the absolute position of the bead, not just the one relative to the trap focus. The focus of the detection laser can coincide with the particle (which is not possible for the trapping beam due to the large scattering force that will always drive the bead beyond the focus) resulting in highest sensitivity. The beam of the detection laser can also be less sharply focused, allowing a larger detection range. While position detection in the xy plane is straightforward, it is more difficult in axial direction along the laser beam. The back-focal plane detection approach also allows, in principle, to deduce the vertical position of the trapped bead to be recorded since the total laser light intensity hitting the four quadrants of the photodetector will depend on the vertical position due to interference between the forward-scattered light and the light directly passing onto the detector [324].

3.3.1
Calibration

Whenever optical tweezers are used not only for manipulation but also for quantitative measurements, a calibration of distance and force detection must be carried out. The most straightforward distance calibration is to apply defined forced movements to the particle in steps of known distance across the detector region while recording the detector output. This approach can be implemented either by moving a fixed bead using defined movements of the sample stage or by defined displacements of a bead caught in a steerable trap that has been calibrated beforehand, for example, by video tracking. Alternatively, the thermal motion of a bead with known size in the trap can be analyzed [323]. The power spectrum of the thermal motion of a trapped bead in $(nm^2\,Hz^{-1})$ is given by [306]

$$S_x(\nu) = \frac{k_B T}{\pi^2 \beta (\nu_0^2 - \nu^2)}. \tag{3.20}$$

Here, ν_0 is the roll-off frequency (the frequency where the power spectrum values had dropped to half its asymptotic low frequency value) and β is the hydrodynamic drag coefficient. For a sphere with radius R_p in a liquid with viscosity η, it is given by $\beta = 6\pi \eta R_p$. The uncalibrated power spectrum S_V measured by the detector in $(V^2\,Hz^{-1})$ is connected with the calibrated one by

$$S_V(\nu) = a_{det}^2 S_x(\nu). \tag{3.21}$$

Here, a_{det} is the linear sensitivity of the detector in $(V\,nm^{-1})$. For frequencies $\nu \gg \nu_0$, the quantity $S_x(\nu)\nu^2$ will approach a constant value of $k_B T/\pi^2\beta$. By plotting $S_V(\nu)\nu^2$ versus ν and taking the plateau value in the limit of $\nu \gg \nu_0$, one obtains the calibration factor ϱ simply from

$$\varrho = \frac{S_V(\nu)\nu^2}{S_x(\nu)\nu^2} = \frac{S_V(\nu)\nu^2\pi^2\beta}{k_B T}. \tag{3.22}$$

The advantage of this method is that no forced movement of the bead is required. It is, however, valid only for small displacements where linear dependence between displacement and detector signal is still valid, and one must make sure that detection bandwidth is high enough to obtain an undistorted power spectrum.

Force calibration is commonly not directly done, but trap stiffness is determined. Then, the force is calculated by multiplying trap stiffness by particle displacement from the trap center. The first method involves again measuring the power spectrum of thermal motion of a bead in the harmonic potential, which is described by a Lorentzian as given by Eq. (3.20). By fitting the power spectrum and using the relation

$$v_0 = \frac{K_t}{2\pi\beta}, \tag{3.23}$$

the trap stiffness K_t can be calculated.

Several issues have to be considered when this method is used. The first is the fact that for particles close to a surface, the hydrodynamic drag coefficient β will be increased due to wall effects and depend on surface separation D. For movements parallel to the surface, this can be taken into account using [325, 326] (see Eqs. (6.46) and (6.50))

$$\beta(D) = \frac{6\pi\eta R_p}{1 - \frac{9}{16}\left(\frac{R_p}{D}\right) + \frac{1}{8}\left(\frac{R_p}{D}\right)^3 - \frac{45}{256}\left(\frac{R_p}{D}\right)^4 - \frac{1}{16}\left(\frac{R_p}{D}\right)^5}. \tag{3.24}$$

For calibration of vertical stiffness, the same procedure is carried out for the vertical detector signal but using a different correction formula for β derived by Brenner [327]. The second issue is connected to the bandwidth of the quadrant photodiode, which should be at least one order of magnitude higher than v_0. This condition may not always be fulfilled especially if silicon photodiodes are used in combination with infrared lasers. A detailed discussion on this topic can be found in Ref. [328].

A second possibility to calibrate trap stiffness is to use the equipartition theorem. The mean square displacement of a particle in a harmonic potential with stiffness K_t is given by the equipartition theorem as

$$\frac{1}{2}k_B T = \frac{1}{2}K_t\langle x^2\rangle. \tag{3.25}$$

So, by measuring the variance $\langle x^2\rangle$ of the bead displacement, the stiffness is obtained without the need to know particle shape, height above the surface, or viscosity of the medium. However, it requires distance calibration of the detector and sufficiently high detection bandwidth as before.

The third and most direct method is to apply a defined viscous drag force on the particle and detect its response, for example, by moving the stage with a triangular waveform with amplitude A_0 and frequency v. This will result in a trajectory of the bead within one period described by

$$x(t) = \frac{A_0\beta v}{2K_t}\left[1 - \exp\left(-\frac{K_t}{\beta}t\right)\right]. \tag{3.26}$$

Here, the asymptotic value $A_0\beta f/2k_t$ can be used to estimate trap stiffness. This method does not require high detection bandwidth and allows mapping of the linear range of the trap where k_t is constant. Recently, Tolic-Norrelykke *et al.* [329] used a combination of sinusoidal drag force and power spectrum to obtain both distance and force calibration without the need of independent information on the hydrodynamic drag coefficient. A similar approach has been introduced to allow calibration in viscoelastic media such as the cytoplasm, where the viscoelastic properties of the medium are not known a priori [330].

3.3.2
Multiple Traps

With optical tweezers more than one object can be simultaneously trapped. First implementations used beam splitters [331] or rapid switching between different positions with a single beam [332]. Interference patterns between two laser beams may be used to trap particles at the intensity maxima [333, 334]. A more flexible way to create fixed arrays of optical tweezers is to use holographic beam splitters that can even be constructed as phase-only diffractive elements, which allows to direct all laser power to the trapping points [335]. A breakthrough based on this principle was the introduction of spatial light modulators (SLMs) as holographic elements [336]. Spatial light modulators are two-dimensional arrays of pixels that allow to impose a defined phase shift for each pixel by varying the optical path length, for example, by variation of the orientation of a liquid crystal. By using computer-controlled spatial light modulators with numerical calculation of the required holograms, not only lateral movement of the trap position can be achieved but also the focal power can be controlled and the axial position can be shifted relative to the focal plane of the microscope objective. This allows full three-dimensional motion of many independent optical traps [337, 338] and creation of three-dimensional crystals [339] and microstructures [340] over a range that is limited by the out-of-focus performance of the microscope and the spatial resolution of the spatial light modulator [341].

Another extension of the classical optical tweezers is the introduction of torque generation using the angular momentum carried by light. An overview of the different ways to apply torque by using optical means is given In ref. [342]. However, up to now such experiments were mainly proof of principle, and torque application with optical tweezers has not yet reached widespread use compared to magnetic tweezers.

3.4
Total Internal Reflection Microscopy

Total internal reflection microscopy (TIRM) was introduced in 1987 by Prieve *et al.* [343]. TIRM allows to probe the interaction of a single microsphere with a transparent flat plate. In a TIRM experiment, a microsphere is allowed to sediment toward the plate. The technique relies on repulsive forces between sphere and plate. This repulsion will typically result from electric double layer or steric forces. They keep the sphere from getting into contact with the plate. Thermal fluctuations will constantly change the precise distance. The distance between sphere and plate is monitored by the light intensity scattered from the particle when illuminated by an evanescent wave and can be determined with a resolution of ≈ 1 nm. By recording the fluctuations in vertical position of the sphere due to Brownian motion, the potential energy of interaction and the diffusion coefficient of the sphere can be deduced. For overviews of the technique, see Refs [344, 345].

A basic TIRM setup consists of a laser, a prism that either forms the bottom of the measurement cell or is attached via an index matching material to the glass bottom of the cell, and an inverted optical microscope equipped with a sensitive photon detector to collect the scattered laser light (Figure 3.12).

TIRM is based on the following effect: If we shine light onto an interface coming from an optically dense medium to an optically less dense medium, it is totally reflected provided the angle is below a certain critical angle θ_c. This angle is given by $\sin \theta_c = n_2/n_1$, where n_1 and n_2 are the refractive indices of the dense and less dense medium, respectively. Far away from the interface, we will not observe any light. The light intensity does, however, not abruptly decrease to zero at the interface but it can penetrate into the less dense medium for a short distance. This is called the evanescent field. The intensity of the evanescent light field decays exponentially. The decay length λ_{ev} is given by

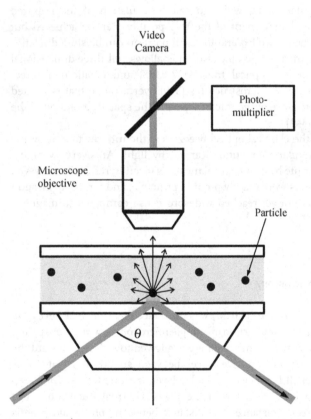

Figure 3.12 Schematic of a TIRM setup. A laser is reflected from the bottom of the measurement cell at an angle above the critical angle to achieve total internal reflection of the light. Light of the evanescent wave that is scattered by a colloidal particle is collected by a microscope objective and detected by a photomultiplier. From observed intensity fluctuations, the Brownian motion of the particle within the combined surface and gravitational potential can be determined and thus a calculation of the surface potential is possible.

$$\lambda_{ev} = \frac{\lambda}{4\pi n_1 \sqrt{\sin^2\theta - \sin^2\theta_c}}. \tag{3.27}$$

Here, λ is the laser wavelength. The actual angle of incidence θ of the laser beam has to be above the critical angle. Equation (3.27) shows that we can adjust the penetration depth λ_{ev} by changing the angle of incidence.

■ **Example 3.1**

At a wavelength of 633 nm zinc crown glass has a refractive index of $n_1 = 1.52$. The critical angle with water ($n_2 = 1.33$) is $\theta_c = 61°$. For an angle of incidence of $\theta = 65°$, the decay length of the evanescent field is $\lambda_{ev} = 140$ nm. For $\theta_c = 85°$, it decreases to $\lambda_{ev} = 70$ nm.

For the dependence of the scattered intensity $I(z)$ on distance z, one assumes that $I(z)$ decays with distance z of the particle from the planar surface in the same way as the intensity of the evanescent wave itself [346]:

$$I(z) = I_0 e^{-z/\lambda_{ev}} \tag{3.28}$$

Here, I_0 is the intensity at the surface. Together with the exponential dependence, this leads to a distance resolution of about 1 nm [344].

The optical microscope can be used to find a particle within the cell and move it above the laser spot by the microscope stage. Then, the measurement is started by recording the scattered light intensity with millisecond time resolution for an extended time ($\approx$15 min). Due to the Brownian motion, the particle will not stay at the potential minimum $U_{min} = U(z_m)$, where z_m is the height above the planar surface where the potential energy is minimal. The probability density $P(z)$ to find the particle at a given distance z above the surface is given by a Boltzmann distribution:

$$P(z) = P(z_m)\exp\left(-\frac{U(z) - U_{min}}{k_B T}\right), \tag{3.29}$$

with $P(z_m)$ being the probability density at the potential minimum. The recorded intensity versus time series is converted into a histogram $N(I)$. Here, N is the number of time intervals, in which an intensity I was observed. When recording a sufficiently high number of data points, this histogram will converge to the probability density distribution $P(I(z))$. The measured probability density distribution $P(I(z))$ is related to the theoretical distribution $P(z)$ via

$$P(z)dz = P(I)dI = N(I)\frac{dI(z)}{dz}. \tag{3.30}$$

By inserting Eq. (3.28) into Eq. (3.30), we obtain

$$P(z)dz = -N(I)\frac{I(z)}{\lambda_{ev}}. \tag{3.31}$$

Combining Eqs. (3.29) and (3.31) leads to

$$N(I(z_m))\frac{I(z_m)}{\lambda_{ev}} \cdot e^{-\frac{U(z)-U(z_m)}{k_B T}} = N(I(z))\frac{I(z)}{\lambda_{ev}}. \tag{3.32}$$

By defining $U(z_m) = 0$ and rearranging Eq. (3.32), we finally obtain a relation between the measured data and the potential energy:

$$\frac{U(z-z_m)}{k_B T} = \ln\left[\frac{N(I(z_m))I(z_m)}{N(I(z))I(z)}\right] = \ln\left[\frac{N(I(z_m))}{N(I(z))}\right] + \frac{z-z_m}{\lambda_{ev}}. \tag{3.33}$$

Equation (3.33) gives information only in terms of the relative positions $z-z_m$. To obtain the absolute distance, one has to determine I_0. In the case of electric double-layer repulsion, this can be achieved by exchanging the solution against one with high salt concentration to suppress repulsion and thus bring the particle in contact with the surface, to which it will attach by the van der Waals forces. Particle attachment can be verified by the strong decrease in fluctuation of intensity. The direct measurement of I_0 then allows to calculate the distance z_m from

$$z_m = \lambda_{ev} \ln \frac{I_0}{I(z_m)}. \tag{3.34}$$

This type of calibration experiment has to be repeated for each particle, since scattering intensities can vary up to one order of magnitude for nominally identical particles. Slight size variations may or may not lead to resonant scattering [347]. The potential energy profile can be determined with a resolution of about $0.1k_B T$, corresponding to forces as small as 10 fN [344]. For low forces, the radiation pressure force from the evanescent wave may no longer be negligible [348]. To avoid lateral drift of the particle and to allow exchange of solutions without flushing the particle from the area of the evanescent wave, optical tweezers can be combined with the basic TIRM setup [349]. While TIRM has mainly been used to study colloidal interactions, its application to cells or liposomes is possible as well [350].

It should be noted that hydrodynamic forces on the sphere, which will be significant close to a wall, do not contribute to the observed equilibrium potential energy profile. Drag forces do not change the equilibrium distribution. They only change the speed of fluctuations. As a result, the TIRM signal can also be used to measure the hydrodynamic drag on the particle by determining its averaged apparent diffusion coefficient from the autocorrelation function of the scattering signal [351]. The disadvantage of this method is that it only probes the diffusion at the potential minimum. Alternatively, one may analyze the distribution of vertical displacements of a particle at short timescales from an arbitrary starting point. For short timescales (≈ 1 ms), the variance of the displacements was found to be proportional to the diffusion coefficient times the displacement time [352].

One limitation of the TIRM method is the poor sampling of energetically unfavorable positions by the particle: according to Eq. (3.29), an increase in potential energy of $7k_B T$ leads to decrease in probability by a factor of 1000. To extend the range of probed particle–substrate separations, an additional force can be applied.

Such attempts have been made with the radiation pressure of a laser beam [353, 349], optical trapping [354], and electrophoresis [355].

Background noise can lead to a shift and broadening of the potential energy profile. For standard experimental conditions with clear solutions and particles with high scattering intensities, signal-to-noise ratio should not be an issue. However, for poorly scattering objects such as cells or for turbid solutions, background scattering can become significant. In such situations, subtraction of background scattering can still lead to valid potential energy profiles [356]. While in traditional TIRM experiments, only a single particle is studied at a time, an extension of the method to ensembles of particles was demonstrated recently [357]. This allows to study not only isolated particle–wall interactions but also forces within the particle ensemble by additional video microscopy.

The exponential dependence between scattering intensity and separation between surface and particle is at the heart of the TIRM method. At the time of the development of the method, no full theoretical calculations for this scattering problem existed, but this type of dependence was shown experimentally [346]. In recent years, the validity of this assumption was analyzed in more detail and significant deviations were found under certain conditions. By combining TIRM and AFM [204], direct measurement of the change in scattering amplitude with distance could be obtained both for AFM tips and for colloid probes. Deviations from the expected exponential decay were found for distances smaller than $3\lambda_{ev}$ [358] in the case of larger particles. Deviations could be minimized by using smaller particles and p-polarized light. Hertlein *et al.* [359] used a dual-wavelength TIRM setup that allowed the detection of the separation distance at two different wavelengths. Thus, they could use one laser wavelength at optimized conditions for independent distance measurements while probing influence of different experimental parameters on the quality of distance measurement at the other wavelength. The observed deviations of the $I(z)$ profiles from the idealized exponential dependence were compared with numerical simulations and excellent agreement was found, opening the possibility to improve distance detection by detailed numerical analysis.

3.5
Magnetic Tweezers

The use of magnetic forces to manipulate particles for studying biological objects dates back to the 1920s. First uses were the movement of magnetic particles in protoplasts [360] or echinoderm eggs [361]. In later experiments, magnetic forces were used to study the viscoelastic properties of the cytoplasm of cells [362–364]. Ziemann *et al.* [365] designed a setup for linear oscillation of a magnetic particle to measure the viscoelastic modulus of actin networks and coined the term "magnetic tweezers" for their setup. In the most simple implementation, magnetic tweezers can be set up by combining two permanent magnets with a superparamagnetic bead (Figure 3.13). The particle will feel an upward pulling force in the direction of the gradient of the magnetic field. In contrast to the situation with optical tweezers, the particle is not trapped in a potential well. The magnetic force has to be counter-balanced, for example, by the force needed to stretch a linker molecule that binds the particle to the substrate, typically a DNA or other biomolecule of interest. The force exerted on the magnetic bead leads to a movement of the bead that is recorded by using an inverted optical microscope with a digital camera.

The beads used in magnetic tweezers are usually superparamagnetic beads since they have no remnant magnetization without external magnetic field and thus do not pose the problem of agglomeration. Superparamagnetic materials consist of small clusters (≈ 10 nm) of a ferromagnetic material (in most cases, iron oxide) embedded in a polymeric matrix. Due to the small size of the ferromagnetic clusters, they can randomly flip their magnetization direction just by thermal excitation. Therefore, they will not exhibit a permanent magnetic moment and behave like a paramagnetic material but exhibit a high magnetic susceptibility due to the ferromagnetism of the ferromagnetic material.

In an external magnetic field of field strength B, a superparamagnetic bead with radius R_p will have a magnetic moment m_m of

$$m_m = \frac{4\pi R_p^3}{\mu_0} \frac{\mu_r - 1}{\mu_r + 2} B \tag{3.35}$$

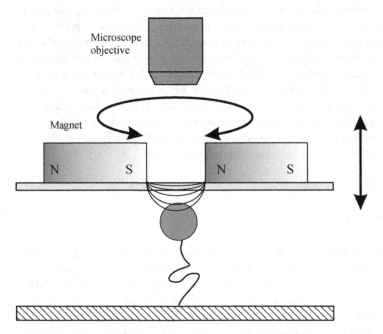

Figure 3.13 Schematic of a basic magnetic tweezer setup. A paramagnetic bead is attached to the bottom of the measurement cell via a linker molecule. The field gradient of the two permanent magnets induces a pulling force on the molecule. The bead can be moved vertically or rotated by moving the magnets. Displacement of the bead is measured by a combination of digital video microscopy and image processing.

as long as the magnetic field is lower than the material-dependent saturation field. The related potential energy of the bead is $U = -\frac{1}{2} m_m B$. The force on the bead is given by

$$F = -\nabla U = \frac{2\pi R_p^3}{\mu_0} \frac{\mu_r - 1}{\mu_r + 2} \nabla(B^2).$$ (3.36)

The selection of the optimum bead size is a compromise. On the one hand, the particles should be as small as possible to minimize force due to Brownian motion. On the other hand, optical video microscopy is used for position tracking and the magnetic force scales with the particle volume (Eq. (3.36)). Typical bead diameters are $0.5\text{–}5\,\mu\text{m}$.

For a perfectly paramagnetic bead, a magnetic field cannot impose any torque. Superparamagnetic beads exhibit a slight polarization anisotropy. This leads to a torque proportional to B in a magnetic field. It tends to orient the particle along the anisotropic axis and can be as high as 1 nN nm. This value is higher than any torque that will be imposed on the bead by the linking molecule and allows to rotate the particle by the magnetic field, by rotating the permanent magnets.

The permanent magnets – typically made of the rare earth magnetic alloy neodymium iron boron – are mounted with a small (< 1 mm) gap between them. The maximum magnetic field is proportional to the inverse of the gap width and the typical length scale is of the order of the gap size. The force acting on the particle can be controlled by moving the magnets up and down and will typically change by about 1 pN per 1 mm of movement. This corresponds to an extremely soft spring constant of 10^{-6} pN nm $^{-1}$. As a result, in magnetic tweezers the change in force over distances of typical bead sizes is negligibly small. This implies that the bead is held by a constant force independent of its precise position, as long as the position of the permanent magnets is not changed. Such a situation is commonly denoted as "force clamp." Therefore, magnetic tweezers are intrinsically force clamp experiments, and since they do not need any complex active force feedback controls, they have "infinite" bandwidth of the force control.

The position of the probe bead is commonly determined by video microscopy and digital image processing. The x- and y- positions can be determined from fits of the bead image with a resolution of ≈ 10 nm [366]. Higher resolution in bead position in all directions can be achieved if the magnetic tweezers are equipped with a laser illumination allowing back-focal plane interferometry [367, 368].

The magnetic force acting on the bead can be calculated from its thermal fluctuations. For a given extension l of the linker molecule under a magnetic force F, one can assume that for small excursions of the bead the system behaves as a linear spring with spring constant F/l. Therefore, the equipartition theorem tells us that

$$F = \frac{k_B T l}{\langle \Delta x^2 \rangle}.$$ (3.37)

Another possibility is measuring the drift speed v_p of a particle in a liquid of known viscosity η [363]. In equilibrium, the magnetic force is counterbalanced by the drag force that can be calculated using Stokes Eq. $(6.11) F = 6\pi \eta R_p \nu_p$. This type of calibration becomes difficult for multipole setups with small distances between the poles as the force will be position-dependent and the distances may be too small for the particles to reach a stationary speed. An alternative approach is then to measure the deflection of a glass cantilever with known spring constant onto which a magnetic bead with know magnetization is attached [369].

Magnetic tweezers based on permanent magnets have a disadvantage that force can be applied only in one direction without any three-dimensional position control. The use of electromagnets for magnetic tweezers (which are therefore sometimes called electromagnetic tweezers) allows easy control of the amplitude of the magnetic field by the current through the coils. Using soft magnetic materials as pole pieces of a coil magnet enables high field gradients but limits the time response due to hysteresis effects. Early designs used only two pole pieces to allow one-dimensional oscillatory movement for rheological applications [365]. Amblard *et al.* [370] used four independent magnetic elements to enable two-dimensional control of the bead position and bead rotation.

The group of Croquette [366] was the first to implement a full three-dimensional control of the bead movement using an array of six electromagnets above the measurement chamber. Even such magnetic tweezers do not create a stable three-dimensional potential minimum to hold the bead but rely on dynamic stabilization by using feedback from the optical detection system to dynamically adjust the field distribution. A more sophisticated system with tetragonal arrangement of pole tips and three-dimensional optical position tracking by back-focal plane scattering similar to that used in optical tweezers was introduced by Fisher *et al.* [367].

A severe limitation of magnetic tweezers is the maximum force that can be achieved. Early setups could only reach maximal forces in the order of 1–200 pN. Higher field gradients would allow either higher forces or use of smaller particles for intracellular applications. One way to accomplish higher field gradients is the use of pole pieces with a sharp tip. Several setups have used this effect to allow higher field gradients by using pole tip arrays with optimized shape [369, 371, 371–373]. The high field gradient at the end of a single pole tip can also be used for three-dimensional movement by scanning the tip in close proximity of the bead [374]. Kollmannsberger and Fabry [375] could achieve forces of up to 100 nN following that concept. Active feedback from the optical tracking was used to reposition the needle to allow the application of a constant force. In some cases, ferromagnetic beads were used to achieve higher forces [372, 376, 377].

Magnetic tweezers are especially well suited for measuring forces in biological systems. They do not have the radiation and heat damage issues like optical tweezers, can act as a perfect force clamp, and easily allow the application of torque. The latter option has made them the preferred tool to study DNA coiling or rotation of molecular motors. Three-dimensional manipulation with magnetic tweezers is less straightforward than for optical tweezers, but dynamic bead control using active feedback is now possible. The lower resolution of position determination in earlier setups has been improved in recent years by adding more sophisticated interference tracking systems.

3.6
Summary

- Much of the fundaments we know about surface forces are based on experiments with the SFA. With the SFA, surface forces are measured between two atomically smooth mica surfaces. Distance is measured interferometrically, which allows absolute determination of separation distance with a resolution of typically 0.1 nm (with down to 25 pm achievable). Its absolute force sensitivity is not as high as in several other methods, but in terms of the usually more relevant force per unit area, its sensitivity is excellent. Lateral (friction) forces can be measured in addition to normal standard force versus distance measurements and have contributed much to our understanding of lubrication by thin films. Additional information such as refractive index and contact area can be obtained. The main reason for the limited number of groups using this instrument is the difficult operation of such a system that needs a very experienced and skillful expert. The large interaction areas demand a contamination-free surface preparation and can lead to significant hydrodynamic forces in highly viscous media, which could make equilibrium measurements hard to achieve.
- The atomic force microscope has become one of the key tools in nanoscience. While the force sensitivity is smaller than for some of the other methods, measurements can be done under a wide range of conditions (from highly viscous media to ultrahigh vacuum) and on almost any material. Its high spatial resolution can result in single molecule or even atom force profiles and force maps that show lateral variation in interaction forces. Forces ranging from 5 pN to some 100 nN resolution can typically be detected or applied. Friction forces can be recorded and related to the atomic structure of the sample surface. Computer-controlled movement of the AFM tip with defined applied load opens up the field of nanolithography and nanomanipulation. The small interaction areas and the possibility to image samples with molecular resolution makes the issue of contamination much less critical.
- Optical tweezers have become a versatile tool for force measurements in liquids, with typical force range of 0.1–100 pN, which makes it an excellent tool to study biological interactions down to the molecular level. Apart from force measurement, they have also found application in the micromanipulation of colloids and cells, especially with the advent of multiple holographic tweezers.
- Total internal reflection microscopy enables the measurement of colloidal forces down to weak forces of 10 fN under conditions of free Brownian motion that may better resemble true colloidal systems compared to other methods where force distance curves are recorded via enforced movement of surfaces. However, its application is limited to transparent surfaces and repulsive interaction potentials.
- Magnetic tweezers allow the measurement of forces down to 10^{-3} pN. Therefore, magnetic tweezers have mainly been applied to the measurement of molecular interactions. One advantage is the possibility to apply a defined torque, which has made them the most prominent tool to study twisting of molecules such as DNA.

3.7
Exercises

3.1. What will be a typical thermal noise of an AFM cantilever at room temperature with a spring constant of (a) $0.1 \, N \, m^{-1}$ and (b) $40 \, N \, m^{-1}$?

3.2. TIRM. How large is the penetration depth of the evanescent wave into water for a TIRM setup using a HeNe laser (wavelength 633 nm) and a coupling prism with a refractive index of 1.6 for an angle of incidence that is $0.1°$, $1°$, $5°$, and $10°$ above the critical angle?

3.3. Assuming that the colloid is stabilized by an electric double-layer force (see Chapter 4), derive an expression that gives the location of the potential minimum and the shape of the potential around this minimum.

4
Electrostatic Double-Layer Forces

4.1
The Electric Double Layer

This chapter is about charged solid surfaces in liquids. The most important liquid is water. Because of its high dielectric constant, water is a good solvent for ions. For this reason, most surfaces in water are charged. Different processes can lead to charging. Ions adsorb to a surface or dissociate from a surface. A protein might, for instance, expose an amino group on its surface. This can become protonated and thus positively charged ($\sim NH_2 + H^+ \rightarrow \sim NH_3^+$). Oxides are often negatively charged in water due to the dissociation of a proton from a surface hydroxyl group ($\sim OH \rightarrow \sim O^- + H^+$). Another way of charging a conducting surface is to apply an external electric potential between this surface and a counterelectrode. This is typically done in an electrochemical cell.

Surface charges cause an electric field. This electric field attracts counterions. The layer of surface charges and counter ions is called "electric double layer." In the simplest model of an electric double layer, the counterions directly bind to the surface and neutralize the surface charges much like in a plate capacitor. In the honor of the work of Ludwig Helmholtz[1] on electric capacitors, it is called the Helmholtz layer. The electric field generated by the surface charges is accordingly limited to the thickness of a molecular layer. On the basis of the Helmholtz picture, one could interpret some basic features of charged surfaces, but the model failed to explain one of the main properties that could be easily measured: the capacitance of an electric double layer.

In the years 1910–1917, Gouy[2] and Chapman[3] went a step further. They took into account a thermal motion of the ions. Thermal fluctuations tend to drive the

1) Hermann Ludwig Ferdinand Helmholtz, 1821–1894. German physicist and physiologist, professor in Königsberg, Bonn, Heidelberg, and Berlin.
2) Louis George Gouy, 1854–1926. French physicist, professor in Lyon.
3) David Leonard Chapman, 1869–1958. English chemist, professor in Manchester and Oxford.

Surface and Interfacial Forces. Hans-Jürgen Butt and Michael Kappl
Copyright © 2010 Wiley-VCH Verlag GmbH & Co. KGaA
ISBN: 978-3-527-40849-8

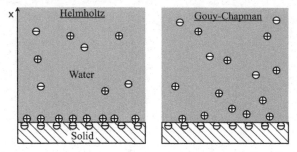

Figure 4.1 Helmholtz and Gouy–Chapman model of the electric double layer.

counterions away from the surface. They lead to the formation of a diffuse layer, which is more extended than a molecular layer. For the simple case of a planar, negatively charged plane, this is illustrated in Figure 4.1. Gouy and Chapman applied their theory on the electric double layer to planar surfaces [378, 379]. Later, Debye[4] and Hückel[5] calculated the potential and ion distribution around spherical surfaces [380].

Both the Gouy–Chapman and Debye–Hückel are continuum theories. They treat the solvent as a continuous medium with a certain dielectric constant, ignoring the molecular nature of the liquid. Also, the ions are not treated as individual point charges but as a continuous charge distribution. For many applications, this is sufficient and the predictions of continuum theory agree with experimental results. Before we finally calculate the free energy of an electric double layer and force between two double layers, we discuss the limitations and problems of the continuum model. At the end of this chapter, electrostatic forces in nonpolar media are described.

4.2
Poisson–Boltzmann Theory of the Diffuse Double Layer

4.2.1
The Poisson–Boltzmann Equation

The aim of this chapter is to calculate the electric potential ψ near a charged planar interface. In general, this potential depends on the distance normal to the surface x. Therefore, we consider a planar solid surface with a homogeneously distributed electric surface charge density σ, which is in contact with a liquid. The surface charge generates a surface potential $\psi_0 = \psi(x = 0)$.

4) Peter Debye, 1884–1966. American physicist of Dutch origin, professor in Zürich, Utrecht, Göttingen, Leipzig, Berlin, and Ithaca. Nobel Prize for chemistry, 1936.

5) Erich Armand Arthur Joseph Hückel, 1896–1980. German chemist and physicist, professor in Marburg.

What is the potential distribution $\psi(x, y, z)$ in the solution? In general, charge density and electric potential are related by the Poisson[6] equation:

$$\nabla^2 \psi = \frac{\partial^2 \psi}{\partial x^2} + \frac{\partial^2 \psi}{\partial y^2} + \frac{\partial^2 \psi}{\partial z^2} = -\frac{\varrho_e}{\varepsilon \varepsilon_0}. \tag{4.1}$$

Here, ϱ_e is the local electric charge density in Cm^{-3}. With the Poisson equation, the potential distribution can be calculated once the exact charge distribution is known. A complication in our case is that the ions in solution are free to move. Before we can apply the Poisson equation, we need to know more about their spatial distribution. This information is provided by Boltzmann[7] statistics. According to the Boltzmann equation, the local ion density is given by

$$c_i = c_i^0 \cdot e^{-W_i / k_B T}, \tag{4.2}$$

where W_i is the work required to bring an ion in solution from infinite distance to a certain position closer to the surface. Equation (4.2) tells us how the local ion concentration c_i of the ith ion species depends on the electric potential at a certain position. For example, if the potential at a certain position in solution is positive, the chance of finding an anion at this position is increased while the cation concentration is reduced.

Now, we assume that only electric work has to be done. We neglect, for instance, that the ion must displace other molecules. In addition, we assume that only a 1:1 salt is dissolved in the liquid. We further assume that the concentration of this background salt is much higher than the concentration of ions, which have dissociated from the surface to build up the surface charge (otherwise the number of anions and cations would not be the same); for low ion concentration, see Ref. [381]. The electric work required to bring a charged cation to a place with potential ψ is $W^+ = e\psi$. For an anion, it is $W^- = -e\psi$. The local anion and cation concentrations c^- and c^+ are related to the local potential ψ by the Boltzmann factor: $c^- = c_0 \cdot e^{e\psi/k_B T}$ and $c^+ = c_0 \cdot e^{-e\psi/k_B T}$. Here, c_0 is the bulk concentration of the salt. The local charge density is

$$\varrho_e = e\left(c^+ - c^-\right) = c_0 e \cdot \left(e^{-\frac{e\psi(x,y,z)}{k_B T}} - e^{\frac{e\psi(x,y,z)}{k_B T}} \right). \tag{4.3}$$

To remind you that the potential depends on the position, we explicitly wrote $\psi(x, y, z)$. Substituting the charge density into the Poisson equation (4.1) leads to

$$\nabla^2 \psi = \frac{c_0 e}{\varepsilon \varepsilon_0} \cdot \left(e^{\frac{e\psi(x,y,z)}{k_B T}} - e^{-\frac{e\psi(x,y,z)}{k_B T}} \right). \tag{4.4}$$

Often, this equation is referred to as the Poisson–Boltzmann equation. It is a partial differential equation of second order, which in most cases has to be solved numerically. Only for some simple geometries can it be solved analytically. One such geometry is a planar surface.

---

6) Denis Poisson, 1781–1840. French mathematician and physicist, professor in Paris.
7) Ludwig Boltzmann, 1844–1906. Austrian physicist, professor in Vienna.

4.2.2
Planar Surfaces

For the simple case of an infinitely extended planar surface, the potential cannot change in the y- and z-directions because of the symmetry and so the differential coefficients with respect to y and z must be zero. We are left with the Poisson–Boltzmann equation that contains only the coordinate normal to the plane x:

$$\frac{d^2\psi}{dx^2} = \frac{c_0 e}{\varepsilon\varepsilon_0} \cdot \left(e^{\frac{e\psi(x)}{k_B T}} - e^{-\frac{e\psi(x)}{k_B T}} \right). \tag{4.5}$$

Before we solve this equation for the general case, it is illustrative and, for many applications, sufficient to treat the special case of low potentials. How does the potential change with distance for low potentials? "Low" means, in a strict sense, $e|\psi| \ll k_B T$. At room temperature, this is $\psi \leq 25\,\text{mV}$. Fortunately, in most applications, the result is valid even for higher potentials, up to 50–80 mV.

For low potentials, we can expand the exponential functions into a series and neglect all but the first (i.e., the linear) term:

$$\frac{d^2\psi}{dx^2} = \frac{c_0 e}{\varepsilon\varepsilon_0} \cdot \left(1 + \frac{e\psi}{k_B T} - 1 + \frac{e\psi}{k_B T} \pm \cdots \right) \approx \frac{2c_0 e^2}{\varepsilon\varepsilon_0 k_B T} \cdot \psi. \tag{4.6}$$

This is sometimes called the "linearized Poisson–Boltzmann equation."[8] The general solution of the linearized Poisson–Boltzmann equation is

$$\psi(x) = C_1 \cdot e^{-\varkappa x} + C_2 \cdot e^{\varkappa x}, \tag{4.7}$$

with

$$\varkappa = \sqrt{\frac{2c_0 e^2}{\varepsilon\varepsilon_0 k_B T}}. \tag{4.8}$$

C_1 and C_2 are constants, which are defined by the boundary conditions. The boundary conditions require that, at the surface, the potential is equal to the surface potential, $\psi(x = 0) = \psi_0$, and that, for large distances from the surface, the potential should disappear $\psi(x \to \infty) = 0$. The second boundary condition guarantees that, for very large distances, the potential becomes zero and does not grow infinitely. It directly leads to $C_2 = 0$. From the first boundary condition, we get $C_1 = \psi_0$. Hence, the potential is given by

$$\psi = \psi_0 \cdot e^{-\varkappa x} \tag{4.9}$$

The potential decreases exponentially. The decay length is given by $\lambda_D = \varkappa^{-1}$. It is called the *Debye length*.

The Debye length decreases with increasing salt concentration. This is plausible because the more ions are in solution the more effective is the screening of surface

---

[8] In the physics community, the linearization is sometimes called "Debye–Hückel approximation." Debye and Hückel used the approximation to describe the electric potential around a sphere.

charge. If we quantify all factors for water at 25 °C, then for a monovalent salt, the Debye length is

$$\lambda_D = \frac{3.04\text{Å}}{\sqrt{c_0 \, 1\,\text{mol}^{-1}}}, \tag{4.10}$$

with the concentration c_0 in M ($=\text{mol}\,\text{l}^{-1}$). For example, the Debye length of a 0.1 M aqueous NaCl solution at 25 °C is 0.96 nm.

In water λ_D cannot be longer than 960 nm. Due to the dissociation of water (according to $2H_2O \rightarrow H_3O^+ + OH^-$), the ion concentration cannot decrease below 1×10^{-7} M. Practically, the Debye length even in distilled water is only a few 100 nm. One reason is dissolved CO_2 that is converted to carbonic acid and is partially dissociated:

$$CO_2 + H_2O \rightleftharpoons H_2CO_3 \rightleftharpoons HCO_3^- + H^+$$

Until now, we have assumed that we are dealing only with the monovalent so-called 1 : 1 salts. If ions of higher valency are also present, the inverse Debye length is given by

$$\varkappa = \sqrt{\frac{e^2}{\varepsilon \varepsilon_0 k_B T} \sum_i c_i^0 Z_i^2}. \tag{4.11}$$

Here, Z_i is the valency of the ith ion sort. Please keep in mind that the concentrations have to be given in particles per m^3.

■ **Example 4.1**

Human blood plasma, that is, blood without red and white blood cells and without thrombocytes, contains 143 mM Na^+, 5 mM K^+, 2.5 mM Ca^{2+}, 1 mM Mg^{2+}, 103 mM Cl^-, 27 mM HCO_3^-, 1 mM HPO_4^{2-}, and 0.5 mM SO_4^{2-}. What is the Debye length? We insert

$c_{Na}^0 = 861 \times 10^{23}\,\text{m}^{-3}$	$Z_{Na} = 1$	$c_K^0 = 30 \times 10^{23}\,\text{m}^{-3}$	$Z_K = 1$
$c_{Ca}^0 = 15 \times 10^{23}\,\text{m}^{-3}$	$Z_{Ca} = 2$	$c_{Mg}^0 = 6 \times 10^{23}\,\text{m}^{-3}$	$Z_{Mg} = 2$
$c_{Cl}^0 = 620 \times 10^{23}\,\text{m}^{-3}$	$Z_{Cl} = -1$	$c_{HCO_3}^0 = 163 \times 10^{23}\,\text{m}^{-3}$	$Z_{HCO_3} = -1$
$c_{HPO_4}^0 = 6 \times 10^{23}\,\text{m}^{-3}$	$Z_{HPO_4} = -2$	$c_{SO_4}^0 = 3 \times 10^{23}\,\text{m}^{-3}$	$Z_{SO_4} = -2$

into Eq. (4.11). Considering that at 36 °C the dielectric constant of water is $\varepsilon = 74.5$, we get a Debye length of 0.78 nm.

All ions come from the dissociation of salts according to $AB \rightarrow A^- + B^+$. For this reason, the sum $\sum_i c_i^0 Z_i$ should always be zero (electroneutrality). Inserting the values from above, we find a surplus of cations of 22 mM. These cations come from the dissociation of organic acids (6 mM) and proteins (16 mM).

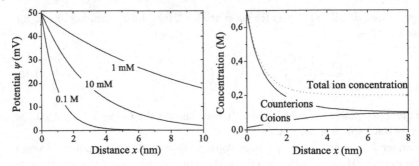

Figure 4.2 (a) Potential versus distance for a surface potential of $\psi_0 = 50\,\text{mV}$ and different concentrations of a monovalent salt in water. (b) Local co- and counterion concentrations are shown for a monovalent salt at a bulk concentration of 0.1 M and a surface potential of 50 mV. In addition, the total concentration of ions, that is, the sum of the co- and counterion concentrations, is plotted.

Figure 4.2 illustrates several features of the diffuse electric double layer. The potential decreases exponentially with increasing distance. This decrease becomes steeper with increasing salt concentration. The concentration of counterions is drastically increased close to the surface. As a result, the total concentration of ions at the surface and thus the osmotic pressure is increased.

4.2.3
The Full One-Dimensional Case

In many practical cases, we can use the low-potential assumption and it leads to realistic results. In addition, it is a simple equation and dependencies like the one on the salt concentration can easily be seen. In some cases, however, we have high potentials and we cannot linearize the Poisson–Boltzmann equation. Now, we treat the general solution of the one-dimensional Poisson–Boltzmann equation and drop the assumption of low potentials. It is convenient to solve the equation with the dimensionless potential $y \equiv e\psi/k_B T$. Please do not mix this up with the spatial coordinate y! In this section, we use the symbol "y" for the dimensionless potential because many textbooks do so. The Poisson–Boltzmann equation for a 1:1 salt becomes

$$\frac{d^2 y}{dx^2} = \frac{c_0 e^2}{\varepsilon\varepsilon_0 k_B T} \cdot (e^y - e^{-y}) = \frac{2c_0 e^2}{\varepsilon\varepsilon_0 k_B T} \cdot \frac{1}{2}(e^y - e^{-y}) = \varkappa^2 \cdot \sinh y \tag{4.12}$$

using $\sinh y = 1/2 \cdot (e^y - e^{-y})$. To solve the differential equation, we multiply both sides by $2 \cdot dy/dx$:

$$2 \cdot \frac{dy}{dx} \cdot \frac{d^2 y}{dx^2} = 2 \cdot \frac{dy}{dx} \cdot \varkappa^2 \cdot \sinh y. \tag{4.13}$$

The left-hand side is equal to $\frac{d}{dx}\left(\frac{dy}{dx}\right)^2$. We insert this and integrate:

$$\int \frac{d}{dx'}\left(\frac{dy}{dx'}\right)^2 dx' = 2\varkappa^2 \cdot \int \frac{dy}{dx'} \cdot \sinh y \cdot dx' \Leftrightarrow$$

(4.14)

$$\left(\frac{dy}{dx}\right)^2 = 2\varkappa^2 \cdot \int \sinh y' \cdot dy' = 2\varkappa^2 \cdot \cosh y + C_1.$$

C_1 is an integration constant. It is determined by the boundary conditions. At large distances, the dimensionless potential y and its derivative dy/dx are zero. Since $\cosh y = 1$ for $y = 0$, this constant is $C_1 = -2\varkappa^2$. It follows that

$$\left(\frac{dy}{dx}\right)^2 = 2\varkappa^2 \cdot (\cosh y - 1) \Rightarrow \frac{dy}{dx} = -\varkappa \cdot \sqrt{2\cosh y - 2}.$$

(4.15)

In front of the square root there is a minus sign because y has to decrease for a positive potential with increasing distance, that is, $y > 0 \Rightarrow dy/dx < 0$. Now, we remember the mathematical identity $\sinh \frac{y}{2} = \sqrt{\frac{1}{2}(\cosh y - 1)}$. Thus,

$$\frac{dy}{dx} = -2\varkappa \cdot \sinh \frac{y}{2}.$$

(4.16)

Separation of variables and integration leads to

$$\frac{dy}{\sinh \frac{y}{2}} = -2\varkappa \cdot dx \Rightarrow \int \frac{dy'}{\sinh \frac{y'}{2}} = -2\varkappa \cdot \int dx' \Rightarrow$$

(4.17)

$$2 \cdot \ln\left(\tanh \frac{y}{4}\right) = -2\varkappa x + 2C_2.$$

C_2 is another integration constant. Written explicitly, we get

$$\ln\left(\frac{e^{y/4} - e^{-y/4}}{e^{y/4} + e^{-y/4}}\right) = -\varkappa x + C_2.$$

(4.18)

Multiplying the denominator and numerator (in brackets) by $e^{y/4}$ leads to

$$\ln\left(\frac{e^{y/2} - 1}{e^{y/2} + 1}\right) = -\varkappa x + C_2.$$

(4.19)

By using the dimensionless surface potential $y_0 = y(x = 0) = e\psi_0/k_B T$ we can determine the integration constant

$$\ln\left(\frac{e^{y_0/2}-1}{e^{y_0/2}+1}\right) = C_2. \tag{4.20}$$

Substituting the results in Eq. (4.19)

$$\ln\left(\frac{e^{y/2}-1}{e^{y/2}+1}\right) - \ln\left(\frac{e^{y_0/2}-1}{e^{y_0/2}+1}\right) = \ln\left[\frac{(e^{y/2}-1)(e^{y_0/2}+1)}{(e^{y/2}+1)(e^{y_0/2}-1)}\right] = -\varkappa x \tag{4.21}$$

$$\Rightarrow e^{-\varkappa x} = \left[\frac{(e^{y/2}-1)(e^{y_0/2}+1)}{(e^{y/2}+1)(e^{y_0/2}-1)}\right].$$

Solving the equation for $e^{y/2}$ leads to the alternative expression

$$e^{y/2} = \frac{e^{y_0/2}+1+(e^{y_0/2}-1)\cdot e^{-\varkappa x}}{e^{y_0/2}+1-(e^{y_0/2}-1)\cdot e^{-\varkappa x}}. \tag{4.22}$$

It is useful to define

$$\alpha = \frac{e^{e\psi_0/2k_BT}-1}{e^{e\psi_0/2k_BT}+1} = \tanh\left(\frac{e\psi_0}{4k_BT}\right). \tag{4.23}$$

Both expressions are identical, which is easily seen when remembering the definition of the tanh function:

$$\tanh z = \frac{e^z-e^{-z}}{e^z+e^{-z}}.$$

Multiplication of both the numerator and denominator by e^z leads to $(e^{2z}-1)/(e^{2z}+1)$. For low surface potentials, $e\psi_0 \ll 4k_BT$, we have $\alpha = e\psi_0/4k_BT$. With α defined in Eq. (4.23), we can express Eq. (4.22) by

$$e^{y/2} = \frac{1+\alpha e^{-\varkappa x}}{1-\alpha e^{-\varkappa x}}. \tag{4.24}$$

Let us compare results obtained with the linearized Poisson–Boltzmann equation (4.9) with the full solution equation (4.24). Figure 4.3 (left) shows the potential calculated for a monovalent salt at a concentration of 20 mM in water. The Debye length is 6.8 nm. For a low surface potential of 50 mV, both results agree well. When the surface potential is increased to 100, 150, or even 200 mV, the full solution leads to lower potentials. At distances below $\approx \lambda_D/2$, the decay is, therefore, steeper than just the exponential decay. This steep decay at small distances becomes progressively more effective at higher and higher surface potentials, which leads to saturation behavior. For example, the potential at a distance of one Debye length can never exceed 40 mV irrespective of the surface potential.

Such saturation behavior should also lead to a saturation in the electrostatic double-layer repulsion. This has been experimentally observed in one case [382] but not in others [383, 384]. The force resulting at high surface potentials is therefore still an open question.

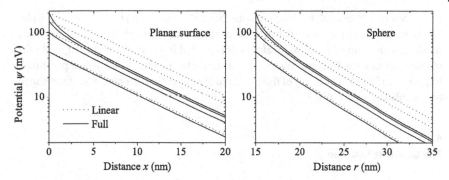

Figure 4.3 Potential versus distance for surface potentials of 50, 100, 150, and 200 mV (from bottom to top) with 2 mM monovalent salt. Left: Planar surface. Results calculated with the full solution equation (4.24) and the solution of the linearized Poisson–Boltzmann equation (4.9) are shown. Right: Potential around a sphere of 15 nm radius. The radial coordinate originating in the center of the particle is plotted rather than the distance from the surface of the particle. Solutions for the linearized (Eq. (4.27)) and the full (Eq. (4.28)) Poisson–Boltzmann are plotted.

4.2.4
The Electric Double-layer Around a Sphere

In many applications, the electric double layer around spherical particles is studied. If the radius of the particle, R_p, is much larger than the Debye length, we can treat the double layer as planar. Otherwise, we have to consider the Poisson–Boltzmann equation for spherical symmetry:

$$\frac{d^2\psi}{dr^2} + \frac{2}{r}\frac{d\psi}{dr} = \frac{c_0 e}{\varepsilon\varepsilon_0}\cdot\left(e^{\frac{e\psi(r)}{k_BT}} - e^{-\frac{e\psi(r)}{k_BT}}\right). \tag{4.25}$$

Here, r is the radial coordinate. The boundary conditions are $\psi(r = R_p) = \psi_0$ for a given surface potential ψ_0 and $\psi = 0$ for $r \to \infty$.

For low potentials, we can linearize the differential equation:

$$\frac{d^2\psi}{dr^2} + \frac{2}{r}\frac{d\psi}{dr} = \varkappa^2\psi. \tag{4.26}$$

This linearized Poisson–Boltzmann equation is solved by [380, 385]

$$\psi(r) = \psi_0 \cdot \frac{R_p}{r} \cdot e^{-\varkappa(r-R_p)}. \tag{4.27}$$

For high potentials, one has to solve the full Poisson–Boltzmann equation in radial coordinates. To our knowledge, no analytical solution of Eq. (4.25) has been reported. In analogy to Eq. (4.22), a good approximation for large $\varkappa R_p$ is [386]

$$e^{y/2} = \frac{1 + \alpha \cdot e^{-\varkappa(r-R_p)} \cdot R_p/r}{1 - \alpha \cdot e^{-\varkappa(r-R_p)} \cdot R_p/r}. \tag{4.28}$$

Other approximate analytical expressions have been derived [387]. As an example, the calculated electric potential around a sphere of $R_p = 15$ nm in an aqueous medium containing 2 mM monovalent salt is plotted in Figure 4.3 on the right. As for planar surfaces, the linear solution overestimates the potential for surface potentials higher than 50 mV. The decay of the potential is steeper than for planar surface due to the additional factor $1/r$.

4.2.5
The Grahame Equation

In many cases, we have an idea about the number of charged groups on a surface. Then, we might want to know the potential. The question is how are surface charge σ and surface potential ψ_0 related? This question is also important because if we know σ versus ψ_0 we can calculate $d\sigma/d\psi_0$. This is basically the capacitance of the double layer and can be measured. The measured capacitance can be compared with the theoretical result to verify the whole theory.

Grahame derived an equation between σ and ψ_0 based on the Gouy–Chapman theory. We can deduce the equation easily from the so-called electroneutrality condition. This condition demands that the total charge, that is, the surface charge plus the charge of the ions in the whole double layer, must be zero. The total charge in the double layer is $\int_0^\infty \varrho_e dx$ and we get [388]

$$\sigma = -\int_0^\infty \varrho_e dx. \tag{4.29}$$

By using the one-dimensional Poisson equation and the fact that at large distances the potential, and thus its gradient, is zero $(d\psi/dx|_{z=\infty} = 0)$, we get

$$\sigma = \varepsilon\varepsilon_0 \int_0^\infty \frac{d^2\psi}{dx^2} \cdot dx = -\varepsilon\varepsilon_0 \frac{d\psi}{dx}\bigg|_{x=0}. \tag{4.30}$$

With $dy/dx = -2\varkappa \cdot \sinh(y/2)$ and

$$\frac{dy}{dx} = \frac{d(e\psi/k_B T)}{dx} = \frac{e}{k_B T} \cdot \frac{d\psi}{dx}, \tag{4.31}$$

we get the Grahame equation:

$$\sigma = \sqrt{8c_0\varepsilon\varepsilon_0 k_B T} \cdot \sinh\left(\frac{e\psi_0}{2k_B T}\right). \tag{4.32}$$

For low potentials, we can expand sinh into a series $(\sinh x = x + x^3/3! + \cdots)$ and ignore all but the first term. That leads to the simple relationship

$$\sigma = \frac{\varepsilon\varepsilon_0 \psi_0}{\lambda_D}. \tag{4.33}$$

■ **Example 4.2**

On the surface of a certain material, there is one ionized group per $(4\,\text{nm})^2$ in aqueous solution containing 10 mM NaCl. What is the surface potential? With a Debye length of 3.04 nm at 25 °C the surface charge density in SI units is $\sigma = 1.60 \times 10^{-19}\text{A s}/16 \times 10^{-18}\,\text{m}^2 = 0.01\text{A s m}^{-2}$. With this we get

$$\psi_0 = \frac{\sigma\lambda_D}{\varepsilon\varepsilon_0} = \frac{0.01\text{A s m}^{-2} \cdot 3.04 \times 10^{-9}\text{m}}{78.4 \cdot 8.85 \times 10^{-12}\text{A s V}^{-1}\text{m}^{-1}} = 0.0438\text{V}. \qquad (4.34)$$

By using the Grahame equation (4.32), we get a surface potential of 39.7 mV.

Figure 4.4 shows the calculated relationship between the surface potential and surface charge for different concentrations of a monovalent salt. We see that for small potentials the surface charge density is proportional to the surface potential. Depending on the salt concentration, the linear approximation (dashed) is valid till $\psi_0 \approx 40-80$ mV. At high salt concentration, more surface charge is required to reach the same surface potential than for a low salt concentration.

4.2.6
Capacity of the Diffuse Electric Double Layer

The differential capacitance between two regions of separated charges is, in general, defined as dQ/dU. Here, Q is the charge on each "electrode" and U is the voltage. The capacity of an electric double layer per unit area is thus

$$C_{GC}^A = \frac{d\sigma}{d\psi_0} = \sqrt{\frac{2e^2 c_0 \varepsilon\varepsilon_0}{k_B T}} \cdot \cosh\left(\frac{e\psi_0}{2k_B T}\right) = \frac{\varepsilon\varepsilon_0}{\lambda_D} \cdot \cosh\left(\frac{e\psi_0}{2k_B T}\right). \qquad (4.35)$$

The index "GC" is a reminder that we calculated the capacitance in the Gouy–Chapman model. For sufficiently small surface potentials, we can expand cosh into a

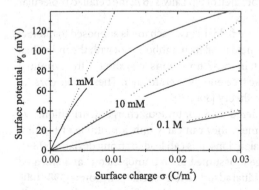

Figure 4.4 Surface potential versus surface charge calculated with the full Grahame equation (4.32), (continuous line) and with the linearized version (4.33), (dotted).

series $(\cosh x = 1 + x^2/2! + x^4/4! + \cdots)$ and consider only the first term. Then, we get

$$C_{GC}^A = \frac{\varepsilon \varepsilon_0}{\lambda_D}. \tag{4.36}$$

It is instructive to compare this to the capacitance per unit area of a plate capacitor $\varepsilon \varepsilon_0/d$. Here, d is the separation between the two plates. We see that the electric double layer behaves like a plate capacitor, in which the distance between the plates is given by the Debye length. The capacity of a double layer – that is the ability to store charge – rises with increasing salt concentration because the Debye length decreases.

To avoid confusion, we should point out that C_{GC}^A as defined above is the differential capacitance. The integral capacitance per unit area is σ/ψ_0. Experimentally, the differential capacitance is easier to measure.

4.3
Beyond Poisson–Boltzmann Theory

4.3.1
Limitations of the Poisson–Boltzmann Theory

In the treatment of the diffuse electric double layer, several assumptions were made [389–392]:

- The finite size of the ions was neglected [393, 394]. In particular, close to the surface, this is a daring assumption because the ion concentration can get very high. For example, if we have a surface potential of 100 mV, the counterion concentration is increased by a factor ≈ 50. At a bulk concentration of 0.1 M, the Poisson–Boltzmann theory predicts a local concentration at the surface of roughly 5 M. Then, the ions have an average distance less than 1 nm. Considering the diameter of an ion with its hydration shell of typically 3–6 Å, the detailed molecular structure should become significant.

- Poisson–Boltzmann theory is a mean field theory. Each ion is supposed to interact only with the average electrostatic field of all its neighbors, not with the individual neighbors. However, under certain conditions ions coordinate their mutual arrangements such as to lower the free energy even further. These ionic correlations are neglected in mean field theory [390, 395].

- The ions in solution were considered as a continuous charge distribution. We ignored their discrete nature, namely, they can carry only a multiple of the unit charge. In particular for di- or trivalent ions, this can lead to strong deviations [394, 396, 397]. Also, the surface charge is assumed to be homogeneous and smeared out. In reality, it is formed by individual adsorbed ions or charged groups [398–400].

- All non-Coulombic interactions and thus ion-specific effects were disregarded. In water, for example, each ion has a hydration shell. If the ions approach each other, these hydration shells overlap and drastically change the interaction (see Section 10.2.1).

- The solvent is supposed to be continuous and we took the permittivity of the medium to be constant. This is certainly a rough approximation because polar molecules are hindered from rotating freely in the strong electric field at the surface. In addition, the high concentration of counterions in the proximity of the surface can drastically change the permittivity.
- Surfaces are assumed to be flat on the molecular scale. In many cases, this is not a reasonable assumption. If we, for instance, consider a biological membrane in a physiological buffer, the ion concentration is roughly 150 mM leading to a Debye length of 0.8 nm. Charges in phospholipids are distributed over a depth of up to 8 Å. In addition, the molecules thermally jump up and down so that the charges are distributed over a depth of almost 1 nm.

Despite these strong assumptions, the Poisson–Boltzmann theory describes electric double layers surprisingly well. The reason is that errors partly compensate each other. Including non-Coulombic interactions leads to an increase of the ion concentration at the surface and a reduced surface potential. On the other hand, taking the finite size of the ions into account leads to a lower ion concentration at the surface and thus an increased surface potential. A reduction of the dielectric permittivity due to the electric field increases its range but at the same time reduces the surface potential because less ions dissociate or adsorb.

Ion-specific effects, however, cannot be explained by the Poisson–Boltzmann theory. Specific effects of divalent cations, for example, are an active field of research. A specific effect of anions was already discovered in 1888. Lewith and Hofmeister observed great differences between the minimum concentrations of various salts required to precipitate a given protein from solution [401, 402]. The precipitation capacity of a salt was largely given by the specific anion. They could arrange the anions in a series according to their precipitation capability. This series is called the Hofmeister series. Later, it was discovered that the same capacity of a specific anion to precipitate proteins was also manifest in very different effects (reviewed in Ref. [403]). The effect is still not fully understood. Two contributions are the different abilities of anions to arrange the water molecules around them and their different van der Waals interactions [404, 405]; anions have stronger van der Waals interactions than cations because of the additional electron.

In summary, for aqueous solutions the Gouy–Chapman theory provides relatively good predictions for monovalent salts at concentrations below 0.2 M and for potentials below 50–80 mV [390]. The fact that the surface charge in reality is not continuously but discretely distributed leads, according to experience, to deviations only with bivalent and trivalent charges. Often, however, the surface charges do not lie precisely in one plane. This is true, for example, for biological membranes. In this case, larger deviations might result.

How can we improve the Poisson–Boltzmann theory? Basically, it is not that difficult to account for one or two of the mentioned defects, but such improvements have limited practical relevance. Since the defects introduced by the assumptions lead to compensating errors, removing a defect might even lead to less realistic results. The most rigorous approach to improve the Poisson–Boltzmann theory is by starting from first principles, applying statistical thermodynamic equations for bulk electrolytes and nonuniform fluids. Excellent reviews about the statistical mechanics of double layers have appeared [406, 407]. Also, computer simulations significantly increased our understanding of electric double layers [389, 390, 408, 409]. The practical significance of statistical mechanics and computer simulations remains, however, somewhat academic because it does not lead to simple analytical formulas that can easily be applied.

There is one relatively simple semiempirical extension of the Gouy–Chapman theory, which can improve the description of charge and potential at the surface. This extension was proposed by Stern.[9]

4.3.2
The Stern Layer

Stern combined the ideas of Helmholtz and that of a diffuse layer [410]. In Stern theory, we take a pragmatic, though somewhat artificial, approach and divide the double layer into two parts: an inner part, the Stern layer, and an outer part, the Gouy or diffuse layer. Essentially, the Stern layer is a layer of ions that is directly adsorbed to the surface and that is immobile [411]. In contrast, the Gouy–Chapman layer consists of mobile ions that obey Poisson–Boltzmann statistics. The potential at the point where the bound Stern layer ends and the mobile diffuse layer begins is the zeta potential (ς potential).

Stern layers can be introduced at different levels of sophistication. In the simplest case, we only consider the finite size effect of the counterions (Figure 4.5). Due to their size, which in water might include their hydration shell, they cannot get infinitely close to the surface but always remain at a certain distance. This distance δ between the surface and the centers of these counterions marks the so-called outer

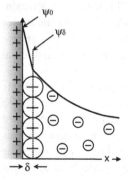

Figure 4.5 Simple version of the Stern layer.

9) Otto Stern, 1888–1969. German physicist, professor in Hamburg. Nobel Prize in physics in 1943.

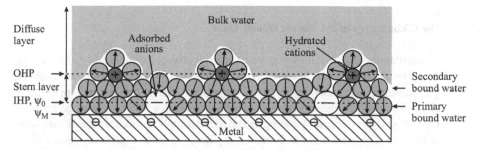

Figure 4.6 Stern layer at a metal surface. Due to the high electrical conductivity, the potential in the metal ψ_M is constant up to the surface. The inner (IHP) and outer (OHP) Helmholtz planes are indicated. In the first layer of primary bound water, the permittivity is typically $\varepsilon = 6$. In the secondary layer of water, it is of the order of $\varepsilon \approx 30$.

Helmholtz plane. It separates the Stern from the Gouy–Chapman layer. For a positively charged surface, this is indicated in Figure 4.5.

At the next level, we also take specific adsorption of ions into account (Figure 4.6). Specifically adsorbed ions bind tightly at a short distance. This distance characterizes the inner Helmholtz plane. In reality, all models can describe only certain aspects of the electric double layer. A good model for the structure of many metallic surfaces in an aqueous medium is shown in Figure 4.6. The metal itself is negatively charged. This can be due to an applied potential or due to the dissolution of metal cations. Often, anions bind relatively strongly, and with a certain specificity, to metal surfaces because anions are more polarizable than cations. Water molecules show a distinct preferential orientation and thus a strongly reduced permittivity. They determine the inner Helmholtz plane.

Next comes a layer of nonspecifically adsorbed counterions with their hydration shell. Still, the permittivity is significantly reduced because the water molecules are not free to rotate. This layer specifies the outer Helmholtz plane. Finally, there is the diffuse layer. A detailed discussion of the structure of the electric double layer at a metal surface is included in Ref. [412].

An important quantity with respect to experimental verification is the differential capacitance of the total electric double layer. In the Stern picture it is composed of two capacitors in series: the capacity of the Stern layer, C_{St}^A, and the capacitance of the diffuse Gouy–Chapman layer. The total capacitance per unit area is given by

$$\frac{1}{C^A} = \frac{1}{C_{St}^A} + \frac{1}{C_{GC}^A}. \tag{4.37}$$

Let us estimate C_{St} using the simple equation for a plate capacitor. The two plates are formed by the surface and by the adsorbed ions. Denoting the radius of the hydrated ions by R_{ion}, the distance is in the order of $R_{ion}/2 \approx 2$ Å. The capacitance per unit area of the Stern layer is $C_{St}^A = 2\varepsilon_{St}\varepsilon_0/r_{ion}$. The permittivity at the surface is reduced and typically of the order of $\varepsilon_{St} \approx 6\text{-}32$ for water. Using a value of $\varepsilon_{St} = 10$ we estimate a capacitance for the Stern layer of $C_{St}^A = 0.44\ \text{F m}^{-2} = 44\ \mu\text{F cm}^{-2}$. Experimental values are typically $10\text{-}100\ \mu\text{F cm}^{-2}$.

4.4
The Gibbs Energy of the Electric Double Layer

To calculate the electrostatic force, we first derive an expression for the Gibbs energy of an electric double layer. The energy of an electric double layer plays a central role in colloid science, for instance, to describe the properties of charged polymers (polyelectrolytes) or the interaction between colloidal particles. Here, we only give results for diffuse layers because it is simpler and in most applications only the diffuse layer is relevant. The formalism is, however, applicable to other double layers as well.

In order to calculate the Gibbs energy of a Gouy–Chapman layer, we split its formation into three steps [413, 415]. In reality, it is not possible to do these steps separately, but we can do the gedankenexperiment without violating any physical principle.

First, the uncharged colloidal particle is brought into an infinitely large solution. This solution contains specifically adsorbing ions and indifferent ions. Surface ions will bind or dissociate from surface groups, driven by chemical forces. It is important to realize that these chemical forces drive the formation of the double layer. To calculate the chemical energy, we have to realize that the dissociation (or binding) of ions does not proceed forever because the more the ions dissociate the higher the electric potential. This potential prevents ions from dissociating further from the surface. The process stops when the chemical energy is equal to the electrostatic energy. The electrostatic energy of one ion of charge Q at this point is simply $Q\psi_0$. The chemical energy of this ion is $-Q\psi_0$. Hence, the Gibbs energy per unit area for the formation of an electric double layer is $-\sigma\psi_0$.

Second, we bring the counterions to the surface. They are supposed to go directly to the surface at $x = 0$. The number of counterions is equal to the number of ions in the diffuse double layer. The first counterions are still attracted by the full surface potential. Their presence, however, reduces the surface potential and the following counterions only notice a reduced surface potential. To bring counterions to the surface, the work $dG = \psi_0' d\sigma$ has to be performed, where ψ_0' is the surface potential at a certain time of charging – or better discharging – process. The total energy we gain (and the double layer loses) is

$$\int_0^\sigma \psi_0' \, d\sigma'. \tag{4.38}$$

In the third step, the counterions are released from the surface. Stimulated by thermal fluctuations, they partially diffuse away from the surface and form the diffuse double layer. The entropy and, at the same time, the energy increases. One can show that the gain in free energy due to an increase in entropy and the loss in free energy due to an increase is inner energy just compensate, so that in the third step no contribution to the Gibbs energy results.

Summing up all contributions, we obtain the total Gibbs energy of the diffuse double layer per unit area:

$$g = -\sigma\psi_0 + \int_0^\sigma \psi_0' d\sigma'. \tag{4.39}$$

Calculus tells us that

$$d(\psi_0' \, \sigma') - \sigma' \, d\psi_0' + \psi_0' \, d\sigma' \Rightarrow \int d(\psi_0'\sigma') = \int \sigma' \, d\psi_0' + \int \psi_0' \, d\sigma'. \tag{4.40}$$

By using this equation, we can write

$$g = -\sigma\psi_0 + \int_0^{\sigma\psi_0} d(\sigma'\psi_0') - \int_0^{\psi_0} \sigma' \, d\psi_0' = -\int_0^{\psi_0} \sigma' \, d\psi_0'. \tag{4.41}$$

The integral can be solved with the help of Grahame's equation (4.32):

$$g = -\int_0^{\psi_0} \sigma d\psi_0' = -\int_0^{\psi_0} \sqrt{8c_0\varepsilon\varepsilon_0 k_B T} \cdot \sinh\left(\frac{e\psi_0'}{2k_B T}\right) \cdot d\psi_0'$$

$$= -\sqrt{8c_0\varepsilon\varepsilon_0 k_B T} \cdot \frac{2k_B T}{e} \cdot \left[\cosh\left(\frac{e\psi_0'}{2k_B T}\right)\right]_0^{\psi_0} \tag{4.42}$$

$$= -8c_0 k_B T \lambda_D \cdot \left[\cosh\left(\frac{e\psi_0}{2k_B T}\right) - 1\right].$$

For low potentials we can use the even simpler relation (4.33) and get

$$g = -\varkappa\varepsilon\varepsilon_0 \cdot \int_0^{\psi_0} \psi_0' d\psi_0' = -\frac{\varkappa\varepsilon\varepsilon_0}{2} \psi_0^2 = -\frac{1}{2}\sigma\psi_0. \tag{4.43}$$

The Gibbs energy of an electric double layer is negative because it forms spontaneously. Roughly, it increases in proportion to the square of the surface potential.

■ **Example 4.3**

Estimate the energy per unit area of an electric double layer for a surface potential of 40 mV in an aqueous solution containing 0.01 M monovalent ions. With

$$g = -\frac{\varkappa\varepsilon\varepsilon_0}{2} \psi_0^2 = -\frac{\varepsilon\varepsilon_0}{2\lambda_D} \psi_0^2$$

and a Debye length of 3.04 nm we get

$$g = -\frac{78.4 \cdot 8.85 \times 10^{-12} \, A\,s\,V^{-1}\,m^{-1}}{2 \cdot 3.04 \times 10^{-9} \, m} \cdot (0.04 \, V)^2 = -0.183 \times 10^{-3} \, Jm^{-2}.$$

Compared to typical surface tensions of liquids, this is small.

4.5
The Electrostatic Double-Layer Force

If two charged surfaces approach each other and the electric double layers overlap, an electrostatic double-layer force arises. This electrostatic double-layer force is essential for the stabilization of dispersion in aqueous media.

Please note that the electrostatic double-layer force is fundamentally different from the Coulomb force. The difference is the presence of free charges (ions) in solution. They screen the electrostatic field emanating from the surfaces.

The interaction between two charged surfaces in liquid depends on the surface charge. Here, we only consider the linear case and assume that the surface potentials are low. If we had to use the nonlinear Poisson–Boltzmann theory, the calculations would become substantially more complex. In addition, only monovalent salts are considered. An extension to other salts can easily be made.

4.5.1
General Equations

When calculating the Gibbs energy per unit area $V^A(x)$ two approaches can be used. Either the change in Gibbs energy of the two double layers during the approach is calculated or the disjoining pressure in the gap is determined. Both approaches lead to the same result. Depending upon conditions and on personal preference, one or the other method is more suitable. Hogg, Healy, and Fuerstenau used the first condition to calculate the energy between two spheres with constant, but different, surface potentials [416]. Parsegian and Gingell chose to determine the osmotic pressure for two different surfaces with different boundary conditions [417].

We start with the first approach and calculate the change in Gibbs free energy for two approaching double layers. The Gibbs energy of one isolated electric double layer per unit area is (Eq. (4.41))

$$-\int_0^{\psi_0} \sigma' \, d\psi_0'. \tag{4.44}$$

For two homogeneous double layers, which are infinitely separated, the Gibbs energy per unit area is twice this value:

$$g^\infty = -2\int_0^{\psi_0} \sigma' \, d\psi_0'. \tag{4.45}$$

If the two surfaces approach each other up to a distance x, the Gibbs energy changes. Now, surface charge and potential depend on the distance:

$$g(x) = -2\int_0^{\psi_0(x)} \sigma'(x) \, d\psi_0'. \tag{4.46}$$

The Gibbs free interaction energy per unit area is [413, 416]

$$V^A(x) = \Delta g = g(x) - g^\infty. \tag{4.47}$$

For the force per unit area, we get

$$f = -\frac{dV^A}{dx}.$$ (4.48)

Why do charged surfaces interact? To get an intuitive understanding, we consider what happens to surface charges when two surfaces are brought closer, holding the potential constant. Remember, surface charge and surface potential are related by (Eq. (4.32))

$$\sigma = -\varepsilon\varepsilon_0 \frac{d\psi}{d\xi}\bigg|_{\xi=0}.$$

The surface charge is (except for two constants) equal to the potential gradient $|d\psi/d\xi|$ at the surface. If two surfaces approach each other, the potential gradient decreases (Figure 4.7). The surface charge density decreases accordingly. If, for instance, AgI particles are brought together, then I^- ions are removed from the negatively charged surface and the surface charge density decreases. During the approach of SiO_2 particles, previously dissociated H^+ ions bind again. Neutral hydroxide groups are formed and the negative surface charge becomes weaker. This reduction of the surface charge increases their Gibbs energy (otherwise the double layer would not have formed). As a consequence the surfaces repel each other.

At this point, it is probably instructive to discuss the use of the symbols D, x, and ξ. D is the shortest distance between two solids of arbitrary geometry. Usually, we use x for the thickness of the gap between two infinitely extended solids. For example, it appears in the Derjaguin approximation because there we integrate over many such hypothetical gaps. ξ is a coordinate describing a position within the gap. At a given gap thickness x, the potential changes with ξ (Figure 4.7). D is the distance between finite, macroscopic bodies.

Alternatively, the interaction can be calculated with the force per unit area f

$$V^A(x) = -\int_\infty^x f(x')dx'.$$ (4.49)

By using Poisson–Boltzmann theory, we can derive a simple expression for the disjoining pressure. For the linear case (low potentials) and for a monovalent salt, the one-dimensional Poisson–Boltzmann equation (Eq. (4.9)) is

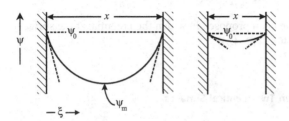

Figure 4.7 Change in the potential distribution when two parallel planar surfaces approach each other. The gap is filled with electrolyte solution.

$$ec_0\left(e^{\frac{e\psi}{k_BT}}-e^{-\frac{e\psi}{k_BT}}\right)-\varepsilon\varepsilon_0\cdot\frac{d^2\psi}{dx^2}=0. \qquad (4.50)$$

The first term is sometimes written as $2ec_0\cdot\sinh\left(e\psi/k_BT\right)$. Integration leads to [417]

$$c_0k_BT\left(e^{\frac{e\psi}{k_BT}}+e^{-\frac{e\psi}{k_BT}}\right)-\frac{\varepsilon\varepsilon_0}{2}\cdot\left(\frac{d\psi}{dx}\right)^2=P. \qquad (4.51)$$

So far, P is only an integration constant. As we see later, it has a physical meaning: P corresponds to the pressure in the gap. The first term describes the osmotic pressure caused by the increased number of particles (ions) in the gap. The second term, sometimes called the Maxwell stress term, corresponds to the electrostatic force caused by the electric field of one surface that affects charges on the other surface and vice versa.

To obtain the force per unit area, we have to realize that the solution in the infinitely extended gap is in contact with an infinitely large reservoir. As the force per unit area f, only the difference of the pressure inside the gap and the pressure in the reservoir is effective. Therefore, the osmotic pressure in the reservoir $2k_BTc_0$ must be subtracted from P to get the force per unit area: $\Pi = P-2k_BTc_0$. Finally, for the force per unit area, we obtain [418]

$$f=c_0k_BT\cdot\left(e^{\frac{e\psi}{k_BT}}+e^{-\frac{e\psi}{k_BT}}-2\right)-\frac{\varepsilon\varepsilon_0}{2}\cdot\left(\frac{d\psi}{d\xi}\right)^2. \qquad (4.52)$$

To determine the force in a specific situation, the potential must first be calculated. This is done by solving the Poisson–Boltzmann equation. In a second step, the force per unit area is calculated. It does not matter for which point ξ we calculate f, the value must be the same for every ξ.

In order to calculate the potential distribution in the gap, we not only need the Poisson–Boltzmann equation but also boundary conditions must be specified. Two common types of boundary conditions are as follows:

- **Constant potential**: Upon approach of the two surfaces, the surface potentials remain constant: $\psi(\xi=0)=\psi_1$ and $\psi(\xi=x)=\psi_2$.
- **Constant charge**: During approach, the surface charge densities σ_1 and σ_2 are constant. The boundary conditions for the potential are (Eq. (4.30))

$$\left.\frac{d\psi}{d\xi}\right|_{\xi=0}=-\frac{\sigma_1}{\varepsilon\varepsilon_0}\quad\text{and}\quad\left.\frac{d\psi}{d\xi}\right|_{\xi=x}=\frac{\sigma_2}{\varepsilon\varepsilon_0} \qquad (4.53)$$

Also, intermediate models have been proposed in which the potential at the surface regulates the surface charge [408, 419–423].

4.5.2
Electrostatic Interaction Between Two Identical Surfaces

An important case is the interaction between two identical parallel surfaces of two infinitely extended solids. It is, for instance, important to understand the coagulation of sols. We can use the resulting symmetry of the electric potential to simplify the calculation. For identical solids, the surface potential ψ_0 on both surfaces is equal. In between, the potential decreases (Figure 4.7). In the middle, the gradient must be zero because of the symmetry, that is, $d\psi(\xi = x/2)/d\xi = 0$. Therefore, the disjoining pressure in the center is given only by the osmotic pressure. Toward the two surfaces, the osmotic pressure increases. This increase is, however, compensated by a decrease in the Maxwell stress term. Since in equilibrium the pressure must be the same everywhere, we have

$$f(x) = c_0 k_B T \left(e^{\frac{e\psi_m}{k_B T}} + e^{-\frac{e\psi_m}{k_B T}} - 2 \right), \tag{4.54}$$

where ψ_m is the electric potential in the middle. f depends on the gap thickness x since ψ_m changes with x. We can thus determine the force from the potential in the middle [425].

For low potentials, we can further simplify this expression. Therefore, we write the exponential functions in a series and neglect all terms higher than the quadratic one:

$$\begin{aligned} f &= k_B T c_0 \left(1 + \frac{e\psi_m}{k_B T} + \frac{1}{2}\left(\frac{e\psi_m}{k_B T}\right)^2 + \cdots + 1 - \frac{e\psi_m}{k_B T} + \frac{1}{2}\left(\frac{e\psi_m}{k_B T}\right)^2 \pm \cdots - 2 \right) \\ &\approx \frac{c_0 e^2}{k_B T} \psi_m^2 = \frac{\varepsilon\varepsilon_0}{2\lambda_D^2} \cdot \psi_m^2. \end{aligned}$$

$$\tag{4.55}$$

It remains to find ψ_m. If the electric double layers of the two opposing surfaces overlap only slightly ($x \gg \lambda_D$), then we can simply superimpose them and approximate

$$\psi_m = 2\psi'(x/2), \tag{4.56}$$

where ψ' is the potential of an isolated double layer. Adding the potentials of the undisturbed surfaces to obtain the potential in the gap is often referred to as the linear superposition approximation. For ψ', we can use various exact functions. For low potentials, we can insert Eq. (4.9) that leads to a repulsive force per unit area of

$$f(x) = \frac{2\varepsilon\varepsilon_0}{\lambda_D^2} \cdot \psi_0^2 \cdot e^{-x/\lambda_D}. \tag{4.57}$$

To calculate the Gibbs interaction energy per unit area, we still have to integrate

$$V^A(x) = -\int_\infty^x \Pi(x')\, dx' = -\frac{2\varepsilon\varepsilon_0\psi_0^2}{\lambda_D^2} \int_\infty^x e^{-x'/\lambda_D}\, dx' = \frac{2\varepsilon\varepsilon_0\psi_0^2}{\lambda_D} \cdot e^{-x/\lambda_D}. \tag{4.58}$$

If we use expression (4.22) for ψ', which is also valid at higher potentials, we get

$$V^A(x) = 64 c_0 k_B T \lambda_D \alpha^2 \cdot e^{-\frac{x}{\lambda_D}} \tag{4.59}$$

with α defined by Eq. (4.23).

An expression valid to even higher potentials than Eq. (4.59) is [413, p. 97], [414, p. 407]

$$V^A(x) = 64c_0 k_B T \lambda_D \alpha^2 \cdot \frac{e^{-x/\lambda_D}}{1 + e^{-x/\lambda_D}}, \tag{4.60}$$

$$f(x) = 64c_0 k_B T \alpha^2 \cdot \frac{e^{-x/\lambda_D}}{(1 + e^{-x/\lambda_D})^2}.$$

To obtain the force per unit area, we applied $f = -dV^A/dx$. Equation (4.60) is valid for constant surface potential. For constant charge boundary conditions, we get [417]

$$V^A(x) = \frac{2\sigma^2 \lambda_D}{\varepsilon\varepsilon_0} \cdot \frac{1 + e^{-x/\lambda_D}}{e^{x/\lambda_D} - e^{-x/\lambda_D}}, \tag{4.61}$$

$$f(x) = \frac{2\sigma^2}{\varepsilon\varepsilon_0} \cdot \frac{2 + e^{x/\lambda_D} + e^{-x/\lambda_D}}{(e^{x/\lambda_D} - e^{-x/\lambda_D})^2}.$$

For low surface potentials and large distances, both equations reduce to

$$V^A(x) = \frac{2\varepsilon\varepsilon_0}{\lambda_D} \cdot \psi_0^2 \cdot e^{-x/\lambda_D} = \frac{2\lambda_D}{\varepsilon\varepsilon_0} \cdot \sigma^2 \cdot e^{-x/\lambda_D}, \tag{4.62}$$

which is equal to Eq. (4.58).

4.5.3
Electrostatic Interaction Between Different Surfaces

Let us finally turn to the more general case of two dissimilar surfaces. We start with two parallel, planar surfaces having constant and dissimilar surface potentials $\psi_1 \neq \psi_2$. It is not possible to find a simple analytical expression using the full Poisson–Boltzmann equation. Depending on the approximations made, different expressions have been proposed [416–418.].

By using the linearized Poisson–Boltzmann equation (4.6), the free energy of interaction per unit area can be calculated to be [416–418.]

$$V^A(x) = \varepsilon\varepsilon_0 \varkappa \frac{2\psi_1\psi_2 - (\psi_1^2 + \psi_2^2) e^{-\varkappa x}}{e^{\varkappa x} - e^{-\varkappa x}}. \tag{4.63}$$

Comparison with exact solutions of the Poisson-Boltzmann equation showed that expression (4.63) gives adequate agreement for low surface potentials [418]. For $e^{\varkappa x} \gg e^{-\varkappa x}$, Eq. (4.63) can be simplified:

$$V^A(x) = \varepsilon\varepsilon_0 \varkappa [2\psi_1\psi_2 e^{-\varkappa x} - (\psi_1^2 + \psi_2^2)e^{-2\varkappa x}]. \tag{4.64}$$

For constant charge, surfaces with carges of equal sign repel each other at all distances. For constant potential surfaces of equal sign in potential also repel each other at large distances. If the surface potentials are not equal for very short distance the force becomes attractive. (Figure 4.8) [426].

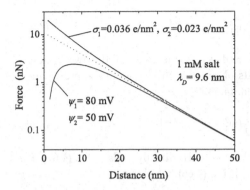

Figure 4.8 Electrostatic double-layer force between a sphere of $R_1 = 3\,\mu\text{m}$ radius and a flat surface in water containing 1 mM monovalent salt. Results were calculated for constant potentials ($\psi_1 = 80\,\text{mV}$, $\psi_2 = 50\,\text{mV}$) with Eq. (4.68). For constant surface charge density ($\sigma_1 = 0.0058\,\text{Cm}^{-2} = 0.036\,\text{e}\,\text{nm}^{-2}$, $\sigma_2 = 0.0036\,\text{Cm}^{-2} = 0.023\,\text{e}\,\text{nm}^{-2}$), we applied Eq. (4.69).

For constant surface charge densities σ_1 and σ_2, various expressions have been proposed, using different approximations. With the linearized Poisson–Boltzmann equation (4.6) one obtains [417, 418, 427]

$$V^A(x) = \frac{1}{\varepsilon\varepsilon_0\varkappa}\frac{2\sigma_1\sigma_2 + (\sigma_1^2 + \sigma_2^2)e^{-\varkappa x}}{e^{\varkappa x} - e^{-\varkappa x}}. \tag{4.65}$$

For $e^{\varkappa x} \gg e^{-\varkappa x}$, we can again simplify:

$$V^A(x) = \frac{1}{\varepsilon\varepsilon_0\varkappa}\left[2\sigma_1\sigma_2 e^{-\varkappa x} + (\sigma_1^2 + \sigma_2^2)e^{-2\varkappa x}\right]. \tag{4.66}$$

Unequal plates of constant charge give repulsion at close distance, even though the plates may have charges of opposite sign. An essential feature of the interaction of double layers at constant charge is that the total charge in the diffuse layer remains constant. As the plates approach each other, this charge is compressed into a decreasing volume. As a result, the ion density and the osmotic pressure increase, giving a repulsive force.

Gregory [418] points out that Eq. (4.66) tends to overestimate the interaction energy, in particular at short distances. He derives a more accurate though slightly more complex equation. The mixed case of a surface with constant charge and one with constant potential has been treated in Ref. [428].

Let us turn from planar, parallel surface to spheres. The interaction between spheres can be calculated in different ways. (a) We can start with expressions (4.63–4.66) and apply the Derjaguin approximation [416, 427–430]. This leads to a good approximation for $R_1, R_2 \gg \lambda_D$ and short distance. (b) Or, we superimpose the potentials around spheres [430–433]. This is a good approximation for small spheres and large distances. (c) Ohshima solved the linearized Poisson–Boltzmann equation in two dimensions analytically for constant potential conditions [434]. As a leading term, he obtained

$$V(D) = 4\pi\varepsilon\varepsilon_0 R^* \left[\psi_1\psi_2 e^{-\kappa D} - \frac{1}{4}(\psi_1^2 + \psi_2^2) e^{-2\kappa d}\right]. \tag{4.67}$$

Here, $R^* = R_1 R_2/(R_1 + R_2)$ with R_1 and R_2 as the radii of the two spheres. Equation (4.67) agrees with the results obtained with Derjaguin's approximation $F(D) = 2\pi R^* V^A(D)$ when inserting Eq. (4.64):

$$F(D) = 4\pi\varepsilon\varepsilon_0\kappa R^* \left[\psi_1\psi_2 e^{-\kappa D} - \frac{1}{2}(\psi_1^2 + \psi_2^2) e^{-2\kappa d}\right]. \tag{4.68}$$

Integration of Eq. (4.68) according to $V = -\int F dD$ directly leads to Eq. (4.67). For constant surface charge density and with Eq. (4.66), we obtain

$$F(D) = \frac{4\pi R^*}{\varepsilon\varepsilon_0\kappa} \left[\sigma_1\sigma_2 e^{-\kappa x} + \frac{1}{2}(\sigma_1^2 + \sigma_2^2)e^{-2\kappa x}\right], \tag{4.69}$$

$$V(D) = \frac{4\pi R^*}{\varepsilon\varepsilon_0\kappa^2} \left[\sigma_1\sigma_2 e^{-\kappa x} + \frac{1}{4}(\sigma_1^2 + \sigma_2^2)e^{-2\kappa x}\right]. \tag{4.70}$$

As one example, the electrostatic double-layer force was calculated for a sphere of $R_1 = 3\,\mu m$ radius interacting with a flat surface ($R_2 = \infty$). The surface charge was adjusted by $\sigma_{1/2} = \varepsilon\varepsilon_0\kappa\psi_{1/2}$ so that at large distances both lead to the same potential.

So far all surfaces considered have a homogeneous charge distribution. In many applications, charges are, however, heterogeneously distributed. The relevant length scale of whether a charge distribution is homogeneous or not is determined by the Debye length. If the distance between surface charges is much smaller than the Debye length, we take it as homogeneous. Heterogeneous charge distributions can lead to various effects, depending on how the charges are distributed [435, 436]. Even net neutral surfaces can interact. The force can be either attractive or repulsive depending on whether the regions of like and unlike charges are in opposition [437].

4.6
The DLVO Theory

It has been known for more than 100 years that many aqueous dispersions precipitate upon addition of salt [438]. Schulze and Hardy observed that most dispersions precipitate at concentrations of 25–150 mM of monovalent counterions [439, 440]. For divalent ions, they found far smaller precipitation concentrations of 0.5–2 mM. Trivalent counterions lead to precipitation at even lower concentrations of 0.01–0.1 mM. For example, gold colloids are stable in NaCl solution, as long as the NaCl concentration does not exceed 24 mM. If the solution contains more NaCl, then the gold particles aggregate and precipitate. The appropriate concentrations for KNO_3, $CaCl_2$, and $BaCl_2$ are 23, 0.41, and 0.35 mM [441], respectively.

This coagulation can be understood as follows. The gold particles are negatively charged and repel each other. With increasing salt concentration, the electrostatic repulsion decreases. The particles, which move around thermally, have a higher chance of approaching each other to a few Ångströms. Then, the van der Waals attraction causes them to aggregate. Since divalent counterions weaken the electro-static repulsion more effectively than monovalent counterions, only small concen-trations of $CaCl_2$ and $BaCl_2$ are necessary for coagulation.

About seven decades ago, Derjaguin, Landau, Verwey, and Overbeek developed a theory to explain the aggregation of aqueous dispersions quantitatively [413, 442, 443]. This theory is called DLVO theory. This theory explains coagulation of dispersed particles by the interplay between two forces: the attractive van der Waals force and the repulsive electrostatic double-layer force. These forces are sometimes referred to as DLVO forces. Van der Waals forces promote coagulation while the double-layer force stabilizes dispersions. Taking into account both components, we can approx-imate the energy per unit area between two infinitely extended solids that are separated by a gap x:

$$V^A(x) = 64c_0 k_B T \lambda_D \alpha^2 \cdot e^{-x/\lambda_D} - \frac{A_H}{12\pi x^2}, \tag{4.71}$$

with α defined in expression (4.23).

Figure 4.8 shows the interaction energy between two identical spherical particles calculated with the DLVO theory. In general, it can be described by a very weak attraction at large distances (secondary energy minimum), an electrostatic repulsion at intermediate distances, and a strong attraction at short distances (primary energy minimum). At different salt concentrations, the three regimes are more or less pronounced and sometimes even completely missing. At low and intermediate salt concentrations, the repulsive electrostatic barrier prevents the particles from aggre-gating. With increasing salt concentration, the repulsive energy barrier decreases. At low salt concentration, the energy barrier is so high that particles in a dispersion have practically no chance of gaining enough thermal energy to overcome it. At high salt concentration, this energy barrier is drastically reduced and the van der Waals attraction dominates. This leads to precipitation. In addition, the surface potential usually decreases with increasing salt concentration thus lowering the energy barrier even more (this effect was not taken into account in Figure 4.9).

For small distances DLVO theory predicts that the van der Waals attraction always dominates. Please remember, the van der Waals force between identical media is always attractive irrespective of the medium in the gap. Thus thermodynamically, or after long periods of time, we expect all dispersions to precipitate. Precipitation might, however, be so slow that it exceeds the life span of any human observer. Once in contact, particles should not separate again, unless they are strongly hit by a third object and gain a lot of energy.

A closer look at the interaction at large distances shows the weak attractive energy. This secondary energy minimum can lead to a weak, reversible coagulation without leading to direct molecular contact between the particles.

For many systems, this is indeed observed and many aqueous dispersions and emulsions are efficiently stabilized by electrostatic double-layer forces. Electrostatic stabilization is, however, sensitive to salts. If the salt concentration increases, both the range and the amplitude of electrostatic double-layer forces decrease. The dispersion

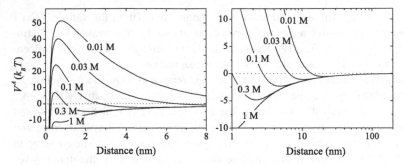

Figure 4.9 Gibbs interaction energy (in units of $k_B T$) versus distance for two identical spherical particles of $R_p = 100\,nm$ radius in water, containing different concentrations of monovalent salt. The calculation is based on DLVO theory using Eqs. (4.71) and (2.67). The Hamaker constant was $A_H = 7 \times 10^{-21}\,J$, the surface potential was set to $\psi_0 = 30\,mV$. Both figures show the same curves, only at different scales.

might become unstable and flocculate. Some aqueous dispersions are stable even at high salt concentrations. Then, short range hydration forces more than compensate the van der Waals attraction. Hydration forces are discussed in Section 10.2. If one cannot rely on hydration repulsion, one way to prevent coagulation is to coat particles with polyelectrolytes. This leads to an electrosteric repulsion (see Section 11.3).

Experimentally, electrostatic double-layer forces versus distance were first quantitatively measured in foam films [444–446]. Aqueous foam films with adsorbed charged surfactant at air–liquid interfaces are stabilized by double-layer forces, at least for some time. Voropaeva *et al.* measured the height of the repulsive barrier between two platinum wires at different applied potentials and in different electrolyte solutions [447]. Usui *et al.* [448] observed that the coalescence of two mercury drops in aqueous electrolyte depends on the applied potential and the salt concentration. Accurate measurements between solid–liquid interfaces were first carried out between rubber and glass with a special setup [449]. In the late 1970s, DLVO force could be studied systematically with the surface forces apparatus [424, 450, 451]. With the introduction of the atomic force microscope, DLVO forces between dissimilar surfaces could be measured [198, 199, 452, 453].

For example, Figure 4.10 shows force-versus-distance curves between two mica cylinders in aqueous electrolyte [424]. Results measured at different concentrations of the monovalent salt KNO_3 are plotted. Under the conditions of the experiment, the mica surfaces are negatively charged. This leads to an exponentially decaying double-layer repulsion. On the logarithmic scale, the straight lines indicate that for distances larger than the respective Debye length a single exponential is sufficient to describe force versus distance. The measured decay lengths agreed with the calculated Debye lengths within the accuracy of the experiment.

In DLVO theory van der Waals attraction and electrostatic forces are added. One may question whether this superposition is justified. Could it not be that electrodynamic effects influence the electrostatic force or vice versa and that electric charges change the van der Waals interaction? In a strict sense, van der Waals and electrostatic

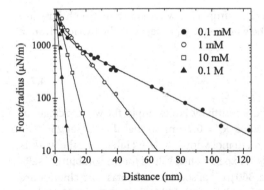

Figure 4.10 Force versus distance between two curved mica cylinders measured with the surface forces apparatus [424]. The force is divided by the radius of curvature of the cylinders. The experiment was carried out in aqueous electrolyte containing different concentrations of KNO_3. Continuous lines are to guide the eye.

forces indeed cannot be completely decoupled [454]. Practically, the superposition is valid if we consider distances above a few atoms. Thus, for distances where continuum theory can be applied, the superposition of double-layer and van der Waals forces is also valid.

4.7
Electrostatic Forces in Nonpolar Media

"Early in 1954 a large tank in Shell's refinery at Pernis exploded 40 minutes after the start of a blending operation in which a tops-naphtha mixture was being pumped into straight-run naphtha. The fire was quickly brought under control and the salvaged contents transferred to another tank. On the following day a second attempt was made to blend these materials and again an explosion occurred 40 minutes after starting the pumps. This striking and unusual coincidence could only be explained by the assumption that both explosions had been caused by static electricity." This description is quoted from the report of Klinkenberg and van der Minne, in which the effect of electrostatic charging in a nonpolar liquid is vividly described [455].

The charging of surfaces in nonpolar liquids is relevant not only when oil is transported but also to understand surface forces and the stabilization of dispersions in nonpolar liquids. In this section, we describe electrostatic forces in media with low dielectric constant (reviewed in Ref. [456]).

A major difference between nonpolar liquids and water is that ions do not easily dissociate because the work required to separate two charges is higher. Let us, for example, take the dissociation of the carboxylic group on benzoic acid: $C_6H_5COOH \rightleftharpoons C_6H_5COO^- + H^+$. The pK of the dissociation is 4.24 in water with $\varepsilon = 78.4$ at 25 °C. In a 1 : 1 water–ethanol mixture with $\varepsilon = 50.4$, it increases to $pK = 5.87$ [457]. To estimate the electrostatic work required to separate two charges, we apply the Coulomb[10] equation. For simplicity, we assume that both ions are initially separated by a distance d. The work required to separate the two ions from a distance d to infinity is

$$U = \frac{e^2}{4\pi\varepsilon\varepsilon_0 d}.$$ (4.72)

It decreases with increasing dielectric permittivity. For example, the work to create an ion pair with both ions having a radius $R_i = 0.2$ nm so that $d = 2R_i$ at 25 °C is 7.4×10^{-21} J $= 1.8 k_B T$ in water. In methanol with $\varepsilon = 32.6$ (Table 4.1), the work is $4.4\, k_B T$. In acetone, it requires $6.9\, k_B T$. Experimental values for the solubility of salts confirm this prediction. For example, $360\, \mathrm{g\,l^{-1}}$ of sodium and potassium chloride are soluble of both salts in water. In methanol, the solubilities are 30.2 and $53\, \mathrm{g\,l^{-1}}$, in acetone the values are 0.00042 and $0.00091\, \mathrm{g\,l^{-1}}$, respectively. In less polar media, the solubility decreases further and below $\varepsilon \leq 5$, it is negligible. Ion dissociation in nonpolar liquids has been extensively studied [458, 459, 460, 461], in particular by conductivity measurements.

Equation (4.72) not only shows how important the dielectric permittivity for ion dissociation is but also tells us that ion dissociation requires less energy if the ions are large. If we take the distance of closest approach d to be the sum of the two ionic radii, the dissociation energy decreases for large ions.

Practically, it is useful to distinguish three regimes [456]:

- For $\varepsilon \geq 11$, in the semipolar regime, dispersions can be charge stabilized as in aqueous media.

Table 4.1 Dielectric permittivity ε of various liquids.

Liquid	ε	Liquid	ε
Water (0 °C)	87.9	*n*-Hexane	1.88
Water (10 °C)	84.0	*n*-Octane	1.94
Water (20 °C)	80.2	*n*-Decane	1.98
Water	78.4	*n*-Dodecane	2.02
Water (30 °C)	76.6	Methanol	32.6
Water (40 °C)	73.2	Ethanol	24.3
Benzene	2.27	1-Propanol	20.5
Toluene	2.37	1-Butanol	17.8
Dichloromethane (CH_2Cl_2)	8.93	1-Pentanol	15.1
Chloroform ($CHCl_3$)	4.71	1-Hexanol	12.5
1,4-Dioxane ($C_4H_8O_2$)	2.21	1-Octanol	9.86
Tetrahydrofuran	7.43	1,2-Propanediol ($C_3H_8O_2$)	28.6
Acetone	20.5	Glycerol	45.6
Dimethyl sulfoxide	46.8	Diethylformamide	37.2
Methyl ethyl ketone	18.5	Methylformamide	181.6

If not otherwise mentioned the temperature is 25 °C.

10) Charles Augustin de Coulomb, 1736–1806. French physicist.

- For $5 \leq \varepsilon \leq 11$, in the low-polar regime, charge stabilization is still possible provided some dissociated electrolyte is present.
- For $\varepsilon \leq 5$, in the apolar regime, the electric conductivity of the pure liquid is very low. Screening by free ions is negligible and electrostatic stabilization of dispersions is problematic.

Nonpolar liquids in their pure form have low electric conductivities. Typical specific conductivities in the semipolar regime range from 10^{-5} to $5 \times 10^{-4}\,S\,m^{-1}$ and 10^{-7} to $10^{-5}\,S\,m^{-1}$ in the low-polar regime. In the apolar regime, conductivities are even lower. For example, typical conductivities of light distillates range from 0.01 to $10\,pS\,m^{-1}$. Electric conductivities in particular in the low-polar regime also can not be increased by the addition of simple salts due to the poor dissociation.

In their report, Klinkenberg and van der Minne drew the conclusion that conductivity is determined by some trace compounds present in the oil [455]. In particular, ionic surfactants such as metal salts of carboxylic acids were found to significantly increase the conductivity of nonpolar liquids [462]. Ions remain dissociated only in low dielectric media if they are large or contained in some large structure (reviewed in Ref. [461]). Ionic surfactants dissociate to a minor extent. As inverse micelles, they are large enough to keep that charge or to solubilize ions in their interior. Micelles also get charged by collisions with other micelles [463, 464]. One surfactant used frequently is sodium di-2-ethylhexylsulfosuccinate (sodium aerosol-OT or NaAOT). It dissociates into $C_8H_{17}COOCH_2CH(SO_3^-)OOC_8H_{17}$ and Na^+. When adding NaAOT to liquid alkanes, the conductivity first increases due to charged monomers [465]. At a certain concentration, the critical micelle concentration (CMC), surfactants start to form inverse micelles. In an inverse micelle, the hydrophobic tails of the surfactant molecules point outside while the hydrophilic head groups are inside. The interior of inverse micelles also tends to accommodate some water. When reaching the CMC, the reverse micelles dominate as charge carriers (Figure 4.11, top).

Charging of surfaces in nonpolar liquids not only occurs when the liquid is pumped through a tube. Particles often charge spontaneously. This was demonstrated by measuring the mobility of charged particles in electric fields. Traditionally, the zeta potential is obtained from such electrophoresis experiments [466, 467]. Recently, the sensitivity could be increased to such a degree that very few and even single charges were detected [469, 468].

In nonpolar media, less charge is needed to reach a certain surface potential. Quantitatively, this is deduced from the Grahame equation (4.32). At low potentials, the surface potential is given by Eq. (4.33):

$$\psi_0 = \frac{\sigma}{\varepsilon\varepsilon_0\varkappa} = \frac{\sigma}{e}\sqrt{\frac{k_BT}{2c_0\varepsilon\varepsilon_0}}. \tag{4.73}$$

With decreasing ε the surface potential for a given surface charge density σ increases. Since only few charges are needed to reach high potentials, surface potentials of 50–100 mV are common.

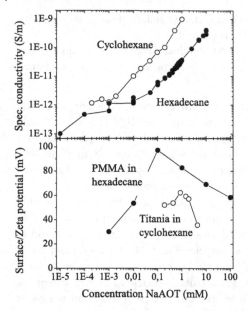

Figure 4.11 Specific electric conductivity (top) and zeta potential versus concentration of NaAOT in hexadecane (•) [464] and cyclohexane (○) [467]. Zeta potentials were measured on titania [467] and surface potentials on spherical poly(methyl methacrylate) (PMMA) particles [464].

Although it is not completely clear how surfaces are charged in nonpolar media, two processes seem to dominate [470]: surfactant adsorption and proton exchange. If surfactant is present in the solution, it will strongly influence the surface charge of particles due to adsorption to the surface [469]. The effect of surfactant is often not simple. Very different dependencies of the zeta potential on the concentration have been observed [471]. Two examples are plotted in Figure 4.11 (bottom). One shows the zeta potential of titania particles in cyclohexane and the other the surface potential of PMMA particles in hexadecane, both at different concentrations of NaAOT.

In the absence of surfactants, the dissociation of surface groups prevails surface charging. The ion that is almost exclusively responsible for the charge transfer between liquid and surface is the proton. If we represent the solid by SH and the solution by HB, the sign of the surface charge due to proton acceptance or donation is governed by

$$SH_2^+ + B^- \rightleftharpoons SH + HB \rightleftharpoons S^- + H_2B^+$$

This implies that the charge of surfaces depends on the proton accepting/donating properties of the liquid. It was indeed observed that the charge of particles depends on the liquid. For example, titanium dioxide particles were observed to be negative in *n*-butylamine and positive in *n*-butanol [472]. α-Fe_2O_3 particles are negative in cyclohexanone and positive in isopropanol [473]. Labib and Williams related the charge of particles to the electron donacity scale of the solvent [474].

A practically relevant question is if electrostatic forces are sufficient to stabilize dispersions in nonpolar liquids. In principle, we can apply DLVO theory also to

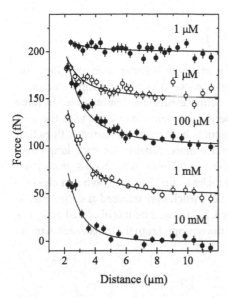

Figure 4.12 Electrostatic force between charged PMMA particles of 1.2 m diameter in hexadecane for different concentrations of NaAOT [464]. Forces for successive concentrations have been offset by 50 fN. Solid lines are fits with the screened Coulomb potential (Eq. (4.74)). (Force curves were provided by E. Dufresne.)

nonaqueous dispersions [456]. We only have to insert the correct Hamaker constants and the correct Debye length. In fact, some dispersion can be stabilized by electrostatic forces [456, 472, 475, 476]. Practically, it is, however, rarely used and particles in nonpolar media are usually sterically stabilized (see Chapter 11). For a reliable electrostatic stabilization of dispersions, two conditions have to be fulfilled [456, 477]. First, the particles should carry sufficient surface charge. Second, the Debye length should not be too short so that the potential decays steeply enough. Otherwise, the repulsive force is not sufficiently strong to prevent the particles to come close. The first condition can be fulfilled, although surfactant is required for the particles to acquire sufficient charge. In addition, trace amounts of water can drastically change the charging mechanism [476, 467].

The second condition is more difficult to fulfill. As seen from Eq. (4.8), the Debye length is predicted to decrease with decreasing ε. In the semipolar regime, this is indeed observed. However, for the low-polar or apolar regime, the low dissociation becomes the dominant effect. The low ion concentration leads to long Debye lengths. Debye lengths in low-polar or apolar liquids can be much larger than the size of dispersed particles. Then, we explicitly have to take into account the spherical shape of the particles. The electrostatic double-layer repulsion is then better described by a screened Coulomb potential [478]

$$V(d_{cc}) = \frac{(e\psi_0 R_p)^2}{k_B T \lambda_D} \cdot \frac{e^{-\frac{d_{cc}-2R_p}{\lambda_D}}}{d_{cc}}, \tag{4.74}$$

$$F(d_{cc}) = \frac{(e\psi_0 R_p)^2}{k_B T \lambda_D^2} \cdot \frac{e^{-\frac{d_{cc}-2R_p}{\lambda_D}}}{d_{cc}} \cdot \left(\frac{\lambda_D}{d_{cc}} + 1\right).$$

Here, V is the Gibbs free energy of interaction between two particles and $d_{cc} = D + 2R_p$ is the center-to-center distance between two interacting particles.

The long range of double-layer forces in nonpolar media is demonstrated by force measurements. The first direct force experiments were carried out with the SFA [479]. More recently, forces in nonpolar liquids have been measured with optical techniques [464, 478]. Optical techniques are more suitable for such long-range interaction potentials. One example is shown in Figure 4.12. Sainis *et al.* [464] measured the force between two PMMA particles by first bringing both particles into close proximity by optical tweezers. Then, the particles are released and they start to diffuse apart, driven by the electrostatic force. Their movement is hindered by Stokes friction. Their position is tracked by optical microscopy. From the trajectories of many events, the force versus distance was calculated.

4.8
Summary

- Most solid surfaces in water are charged. Reason: Due to the high dielectric permittivity of water, ions are easily dissolved. The resulting electric double layer consists of an inner Stern or Helmholtz layer, which is in close contact with the solid surface, and a diffuse layer, also called the Gouy–Chapman layer.
- The electric potential in the diffuse layer of a planar surface decays exponentially

$$\psi = \psi_0 \cdot e^{-x/\lambda_D},$$

  provided the potential does not exceed 50–80 mV.
- For a monovalent salt at 25 °C, the Debye length is given by $\lambda_D = 3.04/\sqrt{c_0}$ Å, with the concentration c_0 in mol L^{-1}.
- In an aqueous medium, the electrostatic double-layer force is present. For distances x larger than the Debye length λ_D, it decays roughly exponentially: $F \propto \exp\left(-x/\lambda_D\right)$.
- The stability of dispersions in aqueous media can often be described by the DLVO theory, which contains the double-layer repulsion and the van der Waals attraction. Specific force versus distance curves depend on the Hamaker constant, the surface charges, and the salt concentration. In general, DLVO theory predicts an electrostatic repulsion at intermediate distance and a van der Waals dominated attraction at short distance.
- Charging and electrostatic forces in nonpolar liquids are less understood than in water and polar liquids. Charging is mainly due to protonation/deprotonation of surface groups or due to the adsorption of surfactants. Debye lengths even in the presence of surfactants are usually so large that an effective stabilization of dispersions is problematic.

4.9

Exercises

4.1. Compare the Debye length of 0.1 mM NaCl solution of water and ethanol ($\varepsilon = 24.3$).

4.2. Plot the potential versus distance curves for surface potentials of 60, 100, and 140 mV using the solution of the linearized and the full Poisson–Boltzmann equation for an aqueous solution with 2 mM KCl.

4.3. Silicon oxide has a typical surface potential in an aqueous medium of -70 mV in 50 mM NaCl at pH 9. Which concentration of cations do you expect to be close to the surface? What is the average distance between two adjacent cations? What is the local pH at the surface?

4.4. Follow Debye and Hückel [380] and calculate the potential distribution around a spherical particle with a low surface potential ψ_0. Use polar coordinates. For this geometry, only the dependency in r remains and the Poisson equation reads $\frac{1}{r^2}\frac{d}{dr}\left(r^2\frac{d\psi}{dr}\right) = -\frac{\varrho_e}{\varepsilon\varepsilon_0}$. The general solution of the linearized Poisson–Boltzmann equation is $\psi = Ae^{-\kappa r}/r + Be^{\kappa r}/r$, where the constants A and B are determined by the boundary conditions. Verify that the total charge of the particle is given by $Q = 4\pi\varepsilon\varepsilon_0 R_p\psi_0\left(1 + R_p/\lambda_D\right)$.

4.5. What is the Gibbs energy of the electrostatic double layer around a sphere of radius R_P that has a surface charge density σ? Derive an equation where the Gibbs energy of one sphere is given as a function of the total charge (see Exercise 4.4). You can assume a low surface potential.

4.6. In an aqueous electrolyte, we have spherical silicon oxide particles. The dispersion is assumed to be monodisperse with a particle radius of 1 μm. Please estimate the concentration of monovalent salt at which aggregation sets in. Use the DLVO theory and assume that aggregation starts, when the energy barrier decreases below $10\,k_B T$. The surface potential is assumed to be independent of the salt concentration at -20 mV. Use a Hamaker constant of 0.4×10^{-20} J.

5
Capillary Forces

Everyone who has been at a beach knows that dry sand flows easily. It is impossible to shape dry sand. To build a sand castle, the sand has to be wet. Wet sand can be shaped because particles adhere to each other. The strong adhesion is caused by liquid menisci, which form around the contact areas of two neighboring particles. The force caused by such a liquid meniscus is called "capillary force," also called meniscus force. In the mid-1920, Fisher [480] and Haines [481] were the first to realize the significance of capillary forces. Capillary forces must be taken into account in studies of powders, soils, and granular materials [482–490], the adhesion between particles or particles to surfaces [491–493], adhesion of insects [494], friction [495, 496], and sintering of ceramic and metallic particles [497]. Capillary forces are also important in technological interfaces such as heads on storage disks where they cause stiction [498–501].

Capillary forces are not only caused by liquid menisci in gaseous environment but can also be caused by capillary bridges of one liquid in another immiscible liquid [502, 503]. Bloomquist and Shutt, for example, observed that particles suspended in organic liquids tend to aggregate when trace amounts of water are dissolved in the organic liquid [504]. When the particles happen to come close to each other by Brownian motion, a phase separation into water and organic liquid can occur in the gap. The water forms a liquid bridge and the particles are attracted by capillary action.

In this chapter, the term "surface forces" has an additional meaning. Surface forces are forces that act between interfaces. Capillary forces not only act between surfaces but are also mediated via interfaces.

In this chapter, we first get to know some of the fundamental equations describing liquid surfaces: the equation of Young and Laplace, the Kelvin equation, and Young's equation. Each equation describes a physical effect. The equation of Young and Laplace describes the shape of a liquid surface. The Kelvin equation relates the vapor pressure of a liquid to the curvature of its surface. Young's equation describes wetting of a liquid on a solid surface. Then, we describe how the capillary force between two perfectly smooth spheres can be calculated. Capillary forces for objects with other geometries and the influence of roughness are discussed in the next sections. All calculations assume equilibrium. If the contact between the two interacting solid surfaces is formed or ruptured fast, such as for a

Surface and Interfacial Forces. Hans-Jürgen Butt and Michael Kappl
Copyright © 2010 Wiley-VCH Verlag GmbH & Co. KGaA
ISBN: 978-3-527-40849-8

flowing powder or if two rough surfaces slide over each other, we have to take the kinetics into account. This is discussed in the next section. Then, capillary forces involving liquid–liquid interfaces are described. Finally, we introduce lateral capillary forces that act between two particles floating at a surface or being immersed in a thin liquid film.

5.1
Equation of Young and Laplace

The equation of Young[1] and Laplace[2] describes one of the fundamental laws in interface science: If an interface between two fluids is curved, there is a pressure difference across it provided the system is in equilibrium. The Young–Laplace equation relates the pressure difference between the two phases ΔP and the curvature of the interface. In the absence of gravitation, or if the objects are so small that gravitation is negligible, the Young–Laplace equation is

$$\Delta P = \gamma_L \cdot \left(\frac{1}{r_1} + \frac{1}{r_2} \right). \tag{5.1}$$

Here, r_1 and r_2 are the two principal radii of curvature. ΔP is also called Laplace pressure. Equation (5.1) is also referred to as the Laplace equation.

The factor $1/r_1 + 1/r_2$ describes the local curvature of an interface at a point on an arbitrarily curved surface. The curvature is obtained as follows. At the point of interest, we draw a normal through the surface. Then, we pass a plane through this line and the intersection of this line with the surface. One angle of orientation of this plane is not defined and can be chosen conveniently. The line of intersection will, in general, be curved at the point of interest. The radius of curvature r_1 is the radius of a circle inscribed to the intersection at the point of interest. The second radius of curvature is obtained by passing a second plane through the normal line but perpendicular to the first plane. This gives the second intersection and leads to the second radius of curvature r_2. So, the planes defining the radii of curvature must be perpendicular to each other and contain the surface normal. Otherwise, their orientation is arbitrary. A law of differential geometry says that the value $1/r_1 + 1/r_2$ for an arbitrary interface is invariant and does not depend on the orientation as long as the radii are determined in perpendicular directions.

Let us illustrate the curvature for two examples. For a sphere with radius r, we have $r_1 = r_2 = r$ and the curvature is $1/r + 1/r = 2/r$ (Figure 5.1a). For a cylinder of radius r_c, a convenient choice is $r_1 = r_c$ and $r_2 = \infty$ so that the curvature is $1/r_c + 1/\infty = 1/r_c$ (Figure 5.1b).

---

1) Thomas Young, 1773–1829. English physician and physicist, professor in Cambridge.
2) Pierre-Simon Laplace, Marquis de Laplace, 1749–1827. French natural scientist.

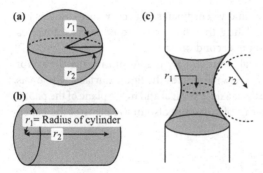

Figure 5.1 Illustration of the curvature of a sphere, a cylinder, and a drop placed between the flat ends of two cylinders.

■ **Example 5.1**

How high is the pressure in a spherical bubble with a radius r_b of 1 mm and a bubble of 10 nm radius in pure water, compared to the pressure outside? For a bubble, the two radii of curvature are identical to that of a sphere: $r_1 = r_2 = r_b$. Therefore,

$$\Delta P = \frac{2\gamma_L}{r_b}. \tag{5.2}$$

With $r_b = 1$ mm, we get

$$\Delta P = 0.072 \, \mathrm{Jm^{-2}} \times \frac{2}{10^{-3} \, \mathrm{m}} = 144 \, \mathrm{Pa}.$$

With $r_b = 10$ nm, the pressure is $\Delta P = 0.072 \, \mathrm{Jm^{-2}} \times 2/10^{-8} \, \mathrm{m} = 1.44 \times 10^7 \, \mathrm{Pa} = 144$ bar. The pressures inside the bubbles are, therefore, 144 and 1.44×10^7 Pa, respectively, higher than the outside pressure.

The Young–Laplace equation has several fundamental implications:

- If we know the shape of a liquid surface, we know its curvature and we can calculate the pressure difference.
- In the absence of external fields (e.g., gravity), the pressure is the same everywhere in the liquid; otherwise, there would be a flow of liquid to regions of low pressure. Thus, ΔP is constant and the Young–Laplace equation tells us that the surface of the liquid has the same curvature everywhere.
- With the help of the Young–Laplace equation, it is possible to calculate the equilibrium shape of a liquid surface. If we know the pressure difference and some boundary conditions (such as the volume of the liquid and its contact line), we can calculate the geometry of the liquid surface.

In practice, it is usually not trivial to calculate the geometry of a liquid surface with Eq. (5.1) and numerical procedures have to be applied. The shape of the liquid surface can be mathematically described by a function $z = z(x, y)$. The z-coordinate of the surface is given as a function of its x- and y-coordinates. The curvature involves the second derivative. As a result, calculating the shape of a liquid surface involves solving a partial differential equation of second order.

In many cases, we deal with rotational symmetric shapes. Assuming that the axis of symmetry is identical to the z-axis, $z = z(x)$ describes the liquid surface, where x is the radial coordinate. Let us assume the z-axis is vertical and in the plane of the paper. Then, it is convenient to put one radius of curvature also in the plane of the paper. This radius is given by

$$\frac{1}{r_2} = \frac{z'}{x\sqrt{1+z'^2}}. \tag{5.3}$$

The other principal radius of curvature is in a plane perpendicular to the plane of the paper and oriented parallel to r_2. It is given by

$$\frac{1}{r_1} = \frac{z''}{\sqrt{(1+z'^2)^3}}, \tag{5.4}$$

where z' and z'' are the first and second derivatives with respect to x.

When applying the equation of Young and Laplace to simple geometries, it is usually obvious at which side the pressure is higher. For example, both inside a bubble and inside a drop, the pressure is higher than outside. In other cases, this is not so obvious because the curvature can have an opposite sign. One example is a drop placed between the planar ends of two cylinders (Figure 5.1c). Then, the two principal curvatures, defined by

$$C_1 = \frac{1}{r_1} \quad \text{and} \quad C_2 = \frac{1}{r_2}, \tag{5.5}$$

can have a different sign. We count it positive if the interface is curved concave with respect to the liquid. The pressure difference is defined as $\Delta P = P_{\text{liquid}} - P_{\text{gas}}$.

■ **Example 5.2**

For a drop in a gaseous environment, the two principal curvatures are positive and given by $C_1 = C_2 = 1/r_d$. The pressure difference is positive, which implies that the pressure inside the liquid is higher than outside.

For a bubble in a liquid environment, the two principal curvatures are negative: $C_1 = C_2 = -1/r_b$. The pressure difference is negative and the pressure inside the liquid is lower than inside the bubble.

For a drop hanging between the ends of two cylinders (Figure 5.1c) in a gaseous environment, one curvature is conveniently chosen to be $C_1 = 1/r_1$. The other curvature is negative, $C_2 = -1/r_2$. The pressure difference depends on the specific values of r_1 and r_2.

The shape of a liquid surface is determined by the Young–Laplace equation. In large structures, we have also to consider the hydrostatic pressure. Then, the equation of Young and Laplace becomes

Table 5.1 Saturation vapor pressure P_0, surface tension γ_L, density ϱ, capillary constant $\varkappa_c$, and Kelvin length λ_K of liquids at 25 °C.

Substance	P_0 (Pa)	γ_L (mNm^{-1})	ϱ (kgm^{-3})	$\varkappa_c$ (mm)	λ_K (nm)
Water	3169	71.99	1000	2.71	0.52
Toluene	3790	27.93	866	1.81	1.20
Diiodomethane (CH_2CI_2)	172	49.90	3308	1.24	1.63
Chloroform ($CHCI_3$)	26200	26.67	1465	1.36	0.88
n-Pentane	68300	15.49	626	1.59	0.72
n-Hexane	20200	17.89	655	1.67	0.95
n-Octane	1860	21.14	699	1.76	1.39
Benzene	12700	28.22	879	1.81	1.01
Methanol	16900	22.07	787	1.67	0.36
Ethanol	7870	21.97	789	1.68	0.52
1-Propanol	2760	23.32	803	1.72	0.70
1-Butanol	860	24.93	810	1.77	0.92
1-Pentanol	259	25.36	811	1.79	1.11
1-Hexanol	110	25.81	814	1.80	1.31
1-Octanol	10	27.10	817	1.84	1.74
Acetone	30800	23.46	785	1.75	0.70
1,2-Propanediol ($C_3H_8O_2$)	20	40.10	1036	1.99	1.19
Methyl ethyl ketone (MEK)	12.000	28.00	810	1.74	0.86
Mercury	1.6×10^{-4}	485.48	13534	1.91	2.90

$$\Delta P = \gamma_L \cdot \left(\frac{1}{r_1} + \frac{1}{r_2} \right) + \varrho g h. \tag{5.6}$$

Here, g is the acceleration of free fall and h is the height coordinate.

What is a large and what is a small structure? In practice, this is a relevant question because for small structures we can neglect $\varrho g h$ and use the simpler equation. It is convenient to define the capillary constant

$$\varkappa_c = \sqrt{\frac{\gamma_L}{\varrho g}}. \tag{5.7}$$

For liquid structures whose curvature is much smaller than the capillary constant, the influence of gravitation can be neglected. Please note that sometimes $\sqrt{2\gamma_L/\varrho g}$ is also defined as the capillary constant. Capillary constants for some common liquids are listed in Table 5.1.

5.2
Kelvin Equation and Capillary Condensation

The subject of the Kelvin[3] equation is the vapor pressure of a liquid. Tables of vapor pressures for various liquids and different temperatures can be found in common

---

3) William Thomson, later Lord Kelvin, 1824–1907. Physics professor at the University of Glasgow.

textbooks or handbooks of physical chemistry. These vapor pressures are reported for vapors that are in thermodynamic equilibrium with liquids having *planar* surfaces. When the liquid surface is curved, the vapor pressure changes. The vapor pressure of a drop is higher than that of a flat, planar surface. In a bubble, the vapor pressure is reduced. The Kelvin equation tells us how the vapor pressure depends on the curvature of the liquid. Like the Young–Laplace equation, it is based on thermodynamic principles and does not refer to a special material or special conditions. It is valid only in equilibrium.

The cause for this change in vapor pressure is the Laplace pressure. The raised Laplace pressure in a drop causes the molecules to evaporate more easily. In the liquid, which surrounds a bubble, the pressure with respect to the inner part of the bubble is reduced. This makes it more difficult for molecules to evaporate. Quantitatively, the change of vapor pressure for curved liquid surfaces is described by the Kelvin equation:

$$RT \cdot \ln \frac{P}{P_0} = \gamma_L V_m \cdot \left(\frac{1}{r_1} + \frac{1}{r_2} \right), \tag{5.8}$$

where P is the vapor pressure of the vapor equilibrium with the liquid of curvature $1/r_1 + r_2$, P_0 is the vapor pressure of a vapor in equilibrium with the liquid having a planar surface, and V_m is the molar volume of the liquid. Please keep in mind that in equilibrium the curvature of a liquid surface is constant everywhere.

The Kelvin equation is derived by equating the chemical potential of the liquid molecules in the meniscus, $\mu_0 + V_m \Delta P = \mu_0 + \gamma_L V_m (1/r_1 + 1/r_2)$, with that of molecules in the vapor phase, $\mu_0 + RT \ln (P/P_0)$. Here, μ_0 is the standard chemical potential of the liquid with a planar surface. We assumed that the gas behaves like an ideal gas.

When applying the Kelvin equation, it is instructive to distinguish two cases: a drop in its vapor (or more generally, a positively curved liquid surface) and a bubble in liquid (a negatively curved liquid surface).

Drop in its Vapor   For a spherical drop of radius r_d, the Kelvin equation can be simplified:

$$RT \cdot \ln \frac{P}{P_0} = \frac{2\gamma_L V_m}{r_d} \quad \text{or} \quad P = P_0 \cdot e^{\frac{2\gamma_L V_m}{RT r_d}}. \tag{5.9}$$

The constant $2\gamma_L V_m / RT$ characterizes the curvature for which the vapor pressure changes by a factor e. For convenience, we call

$$\lambda_K = \frac{\gamma_L V_m}{RT} \tag{5.10}$$

"Kelvin length". Kelvin lengths for several liquids are given in Table 5.1.

The vapor pressure of a drop is higher than that of a liquid with a planar surface. One consequence is that an aerosol of drops (fog) should be unstable. To see this, let us assume that we have a box filled with many drops in a gaseous environment. Some drops are larger than others. The small drops have a higher vapor pressure than the large drops. Hence, more liquid evaporates from their surface. This tends to condense into large drops. Within a population of drops of different sizes, the bigger drops will grow at the expense of the smaller ones – a process called Ostwald ripening.[4] These drops will sink down and, finally, bulk liquid fills the bottom of the box.

For a given vapor pressure, there is a critical drop size. Every drop bigger than this size will grow. Drops at a smaller size will evaporate. If a vapor is cooled to reach oversaturation, it cannot condense (because every drop would instantly evaporate again), unless nucleation sites are present. In that way, it is possible to explain the existence of oversaturated vapors and also the undeniable existence of fog.

Bubble in a Liquid In case of a bubble, the curvature is negative. As a result, we get

$$RT \cdot \ln \frac{P}{P_0} = -\frac{2\gamma_L V_m}{r_b}.$$ (5.11)

Here, r_b is the radius of the bubble. The vapor pressure inside a bubble is therefore reduced. This explains why it is possible to overheat liquids: When the temperature is increased above the boiling point occasionally, tiny bubbles are formed. Inside the bubble, the vapor pressure is reduced, the vapor condenses, and the bubble collapses. Only if a bubble larger than a certain critical size is formed is it more likely to increase in size rather than to collapse.

■ **Example 5.3**

The relative vapor pressure of a spherical water drop of 1 μm radius is

$$\frac{P}{P_0} = \exp\left(\frac{2\gamma_L V_m}{RT r_d}\right) = \exp\left(\frac{2 \cdot 0.072 \ \mathrm{Nm^{-1}} \cdot 18 \times 10^{-6} \ \mathrm{m^3 \ mol^{-1}}}{8.31 \ \mathrm{J \ mol^{-1} \ K^{-1}} \cdot 298 \ \mathrm{K} \cdot 10^{-6} \ \mathrm{m}}\right) = 1.001.$$

For a drop of 10 nm radius, it increases to $P/P_0 = 1.114$. For a bubble of 1 μm radius, the curvature is negative and thus $P/P_0 = 0.999$. In a bubble radius of 10 nm, the relative vapor pressure decreases to 0.901.

At this point, it is instructive to discuss one question that sometimes causes confusion: How does the presence of an additional background gas change the properties of a vapor? For example, does pure water vapor behave differently from water vapor at the same partial pressure in air (in the presence of nitrogen and oxygen)? Answer: To a first approximation, there is no difference as long as the system is in thermodynamic equilibrium. "First approximation" means as long as interactions between the vapor molecules and the molecules of the background gas are negligible. However, processes and kinetic phenomena such as diffusion can be completely different and certainly depend on the background gas. This is, for instance, the reason why drying in a vacuum is much faster than drying in air.

---

4) In general, Ostwald ripening is the growth of large objects at the expense of smaller ones. Friedrich Wilhelm Ostwald, 1853–1932. German physicochemist, professor in Leipzig, Nobel Prize for chemistry, 1909.

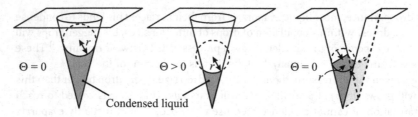

Figure 5.2 Capillary condensation into a narrow conical pore and a slit pore at a given vapor pressure P. For the conical pore, the perfectly wetting case ($\Theta = 0$) and the case of finite contact angle ($\Theta > 0$) are shown.

5.2.1
Capillary Condensation

Capillary condensation is the condensation of vapor into capillaries or fine pores even at vapor pressures below P_0. Lord Kelvin was the one who realized that the vapor pressure of a liquid depends on the curvature of its surface. In his words, this explains why "moisture is retained by vegetable substances, such as cotton cloth or oatmeal, or wheat-flour biscuits, at temperatures far above the dew point of the surrounding atmosphere" [505].

Capillary condensation can be illustrated by the model of a conical pore with a totally wetting surface (Figure 5.2, left). At a given vapor pressure P, liquid will condense in the tip of the pore. Condensation continues until the bending radius of the liquid has reached the value given by the Kelvin equation. Since the liquid surface is that of a spherical cap, the situation is analogous to that of a bubble and we can write

$$RT \cdot \ln \frac{P}{P_0} = -\frac{2\gamma_L V_m}{r}. \tag{5.12}$$

Here, r corresponds to the capillary radius at the point where the meniscus is in equilibrium. The curvature of the liquid surface is $-2/r$.

Many surfaces are not totally wetted, but they form a certain contact angle Θ with the liquid. In this case, less liquid condenses into the pore (Figure 5.2, middle).

Attention has to be paid as to which radius is inserted into the Kelvin equation. In general, there is no rotational symmetric geometry. For example, in a fissure or crack (Figure 5.2, right) one radius of curvature is infinitely large, $r_1 = \infty$, while the other is the bending radius vertical to the fissure direction, $r_2 = -r$, so that $1/r_1 + 1/r_2 = -1/r$.

■ **Example 5.4**

We have a porous solid with cylindrical pores of all dimensions. It is in water vapor at 20 °C. The humidity is 90%. What is the size of the pores, which fill up with water? The solid is supposed to be hydrophilic with $\Theta = 0$.

$$r_C = -\frac{2\gamma_L V_m}{RT \cdot \ln 0.9} = -\frac{2 \cdot 0.072 \, \text{J m}^{-2} \cdot 18 \times 10^{-6} \, \text{m}^3}{8.31 \, \text{J K}^{-1} \cdot 293 \, \text{K} \cdot \ln 0.9} = 10 \, \text{nm}.$$

Capillary condensation has been studied by various methods, and the validity of the previous description has been confirmed for several liquids and radii of curvature down to a few nanometers [506, 507].

5.3
The Young's Equation

To describe the stability of thin films, we first need to introduce the Young's equation. On solid hydrophobic surfaces, water forms a high contact angle. To appreciate the significance of this observation, we first need to introduce Young's equation. Young's equation is the basis for a quantitative description of wetting phenomena. If a drop of a liquid is placed on a solid surface, there are two possibilities: the liquid completely spreads on the surface (contact angle $\Theta = 0°$) or a finite contact angle is established. In the second case, a three-phase contact line – also called wetting line – is formed. At this line, three phases are in contact: the solid, the liquid, and the vapor (Figure 5.3). The Young's equation relates the contact angle to the interfacial tensions of the solid–vapor, liquid–vapor, and solid–liquid interfaces γ_S, γ_L, and γ_{SL}, respectively [508, 509]:

$$\gamma_L \cdot \cos \Theta = \gamma_S - \gamma_{SL}. \tag{5.13}$$

If the interfacial tension of the bare solid surface is higher than that of the solid–liquid interface ($\gamma_S > \gamma_{SL}$), the right-hand side of the Young's equation is positive. Then, $\cos \Theta$ has to be positive and the contact angle is smaller than 90°; the liquid partially wets the solid. If the solid–liquid interface is energetically less favorable than the bare solid surface ($\gamma_S < \gamma_{SL}$), the contact angle will exceed 90° because $\cos \Theta$ has to be negative. If the contact angle is greater than 90°, the liquid is said not to wet the solid.

There is another aspect to the fact that water forms a high contact angle on hydrophobic surfaces: one cannot put a thin layer of water on a hydrophobic surface. Thin films are not stable and will spontaneously rupture (see Section 7.6.2).

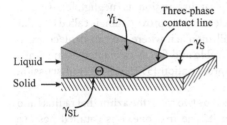

Figure 5.3 Rim of a liquid drop with a contact angle Θ on a solid surface.

Practically, this is a pain when trying to coat hydrophobic surfaces with an aqueous solution.

5.4
Capillary Forces Calculated with the Circular Approximation

Two particles in contact necessarily create a narrow slit around the contact area. If the surfaces are lyophilic with respect to a surrounding vapor, some vapor will condense and form a meniscus [510]. The meniscus causes an attractive force. This is for two reasons [480, 481, 511, 512]. First, the direct action of the surface tension of the liquids around the periphery of the meniscus pulls the particles together. Second, the curved surface of the liquid causes a Laplace pressure, which is negative with respect to the outer pressure. This negative pressure acts over the cross-sectional area of the meniscus and attracts the particles toward each other.

We start this chapter by discussing the capillary force between a perfectly smooth sphere and a plane. Spheres are the first approximation for real particles. The calculation for a sphere and a plane is instructive and the results are simple. Then, we generalize the treatment to two spheres of arbitrary radius and contact angle. Though spheres are a model for the interaction between particles, some fundamental properties are unique for spheres. For example, the capillary force between perfectly spherical surfaces does not (or only weakly) depend on the vapor pressure. Therefore, we also discus other geometries and in particular the influence of roughness.

5.4.1
Capillary Force Between a Sphere and a Plane

A hard, smooth sphere of radius R_1 is supposed to be a distance D away from a hard, smooth plane (Figure 5.4). The liquid is assumed to form a contact angle Θ_1 with the sphere and Θ_2 with the plane. To calculate the shape of the liquid meniscus, we would need to solve the Laplace equation with appropriate boundary conditions. The Young–Laplace equation is a second-order, nonlinear partial differential equation, which is not trivial to solve. Solutions of the Laplace equation for a radial symmetric geometry lead to a class of curves, which are called nodoids [513]. Numerical procedures to calculate the shape of the meniscus have been reported [514–516]. For small menisci, where the extension of the liquid meniscus is much smaller than the capillary constant so that gravitation is negligible, we can approximate both radii of curvature by circles. This approximation is called circular or toroidal approximation [480]. For small menisci, where we can neglect gravitation, the circular approximation is applicable and forces calculated numerically agree with the results of the circular approximation [513, 517–519] (discussion below).

The curvature of the liquid is characterized by two radii: the azimuthal radius l and the meridional radius r, perpendicular to it. The first one, l, is counted positive because it is concave with respect to the liquid. The other, r, is counted negative. The

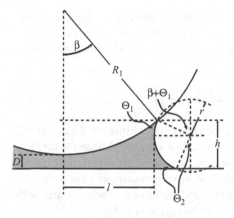

Figure 5.4 Spherical particle of radius R_1 a distance D away from a planar surface. The angle β describes the position of the three-phase contact line on the particle surface. The height of the liquid meniscus is h.

total radius of curvature of the liquid surface $1/r_1 + 1/r_2$ is $1/l - 1/r$. The pressure is, therefore, $\Delta P = \gamma_L(1/l - 1/r)$. Since $l > r$, the curvature is negative, that is, the pressure in the liquid is lower than in the outer vapor phase. It acts upon a cross-sectional area πl^2 leading to an attractive force of $-\pi l^2 \Delta P$. The total capillary force is

$$F = 2\pi l\gamma_L - \pi l^2 \Delta P. \tag{5.14}$$

Here, $2\pi l\gamma_L$ is the force caused directly by the surface tension of the liquid. Since capillary forces are always attractive, it is convenient to apply a positive sign to attraction, while in the other chapters repulsive forces are counted positive.

To calculate the capillary force, it is convenient to introduce the parameter

$$c = \frac{1}{2}[\cos(\Theta_1 + \beta) + \cos\Theta_2]. \tag{5.15}$$

Later, we make the use of the fact that for small menisci, where $R_1 \ll l$, the angle β is small and can be neglected. Then c is the mean cosine of the contact angles: $c \approx (\cos\Theta_1 + \cos\Theta_2)/2$.

We express the two radii of curvature by angle β. With the given geometry, we get

$$h = 2rc \quad \text{and} \quad h = R_1(1 - \cos\beta) + D. \tag{5.16}$$

Setting both expressions equal leads to

$$r = \frac{R_1(1 - \cos\beta) + D}{2c} \tag{5.17}$$

and

$$l = R_1\sin\beta - r[1 - \cos(\Theta_1 + \beta)]. \tag{5.18}$$

We could use Eq. (5.14) and insert r and l as given by Eqs. (5.17) and (5.18) to derive the capillary force. In this case, the narrowest part of the meniscus is used for the calculation. It is, however, more convenient to calculate the force at the three-phase contact line, at which the surface tension component is not directed vertically and only a component $\sin(\Theta + \beta)$ contributes to the force. The capillary force is [483, 484, 512]

$$F = \pi \gamma R_1 \sin\beta \left[2 \sin(\Theta_1 + \beta) + R_1 \sin\beta \cdot \left(\frac{1}{r} - \frac{1}{l} \right) \right]. \tag{5.19}$$

When trying to calculate the capillary force for a given vapor pressure, the reader will realize that Eq. (5.19) is not an explicit equation. One cannot calculate the capillary force by inserting the vapor pressure on the right-hand side. To calculate the capillary force for a given particle size, contact angle, and vapor pressure, we have to vary the parameter β and calculate r and l. Then, the capillary force can be calculated with Eq. (5.19) and the vapor pressure with the Kelvin Equation (5.8):

$$\frac{P}{P_0} = \exp\left[-\lambda_K \left(\frac{1}{r} - \frac{1}{l} \right) \right]. \tag{5.20}$$

Finally, one has to correlate the vapor pressure with the capillary force.

This is a rather awkward procedure and it would be helpful to have an explicit expression. Fortunately, an approximate expression can be derived for $R_1 \gg l \gg r, D$. Therefore, we first simplify Eq. (5.19). The first term in square brackets is caused by the direct action of the surface tension. It is usually only a small percentage of the second term, which is caused by the Laplace pressure. For $R_1 \gg l \gg r$, the factor $R_1(1/r - 1/l)$ is much larger than unity. Unless $\beta \approx 1$ and $\Theta_1 \gg 0$, the first term is negligible and the second term completely dominates. In addition, we neglect $1/l$ compared to $1/r$. This leads to

$$F \approx \pi \gamma_L R_1^2 \sin^2\beta \cdot \left(\frac{1}{r} - \frac{1}{l} \right) \approx \pi \gamma_L R_1^2 \sin^2\beta \cdot \frac{1}{r}. \tag{5.21}$$

To get an expression for $\sin^2\beta$, we rearrange Eq. (5.17):

$$\cos\beta = 1 - \frac{2rc - D}{R_1}.$$

With $\cos^2\beta = 1 - \sin^2\beta$, we get

$$1 - \sin^2\beta = \left[1 - \frac{2rc - D}{R_1} \right]^2 \approx 1 - \frac{4rc - 2D}{R_1} \Rightarrow \sin^2\beta = \frac{4rc - 2D}{R_1}. \tag{5.22}$$

In the second step, we used $(1-x)^2 \approx 1 - 2x$ for $x \ll 1$. Inserting Eq. (5.22) into Eq. (5.21) we finally get the analytical expression for the capillary force [84, 520]:

$$F = 2\pi \gamma_L R_1 \left(2c - \frac{D}{r} \right) \approx 2\pi \gamma R_1 \left(\cos\Theta_1 + \cos\Theta_2 - \frac{D}{r} \right). \tag{5.23}$$

At constant vapor pressure and in equilibrium, the radius r is constant and can easily be calculated with Kelvin's equation. The capillary force decays linearly with distance and the slope is given by the vapor pressure via

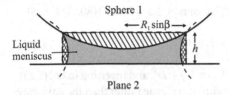

Figure 5.5 Schematic of the contributions to calculate the volume between a sphere and a plane.

$$r = -\frac{\lambda_K}{\ln(P/P_0)}. \tag{5.24}$$

Equation (5.23) is quite useful in situations where the liquid meniscus is in equilibrium with its vapor. Then, the radius r is fixed and given by Kelvin's equation. If the separation of the two particles is faster than the time required to establish equilibrium by condensation or evaporation, the radius r might deviate from its equilibrium value given by Kelvin's equation (5.24). In the extreme case of a very fast separation or a slowly evaporating/condensing liquid, we are more likely to have a constant volume of the meniscus.

Therefore, we now calculate the capillary force at constant volume. As a first step, we need to calculate the volume of the meniscus, also called pendular ring (Figure 5.5). We consider the volume of a cylinder of radius $R_1 \sin \beta$ and height h and subtract the volume of the spherical cap (diagonal lines):

$$V = \pi h R_1^2 \sin^2\beta - \frac{\pi}{6} R_1 (1-\cos\beta)\left[3R_1^2\sin^2\beta + R_1^2(1-\cos\beta)^2\right]. \tag{5.25}$$

We neglect the volume of the rim (cross hatched). For $R_1 \ll r$, the volume of the rim turns out to be only a small percentage of the total volume.

For small angle β, we can substitute $1-\cos\beta = \sin^2\beta/2$:

$$V = \pi h R_1^2 \sin^2\beta - \frac{\pi}{12} R_1 \sin^2\beta \left[3R_1^2\sin^2\beta + R_1^2\frac{\sin^4\beta}{4}\right].$$

Simplifying and inserting h from Eq. (5.16):

$$V = \pi R_1^2 [R_1(1-\cos\beta) + D]\sin^2\beta - \frac{\pi}{4} R_1^3 \sin^4\beta \left(1 + \frac{\sin^2\beta}{12}\right)$$

$$\approx \frac{\pi}{2} R_1^2 \left[R_1\sin^2\beta + 2D\right]\sin^2\beta - \frac{\pi}{4} R_1^3 \sin^4\beta.$$

We neglected the last term in the left brackets since β is small and $1 \gg \sin^2\beta/12$. Simplification leads to

$$V = \frac{\pi}{2} R_1^2 \sin^2\beta \cdot \left(2D + \frac{R_1}{2}\sin^2\beta\right).$$

We use expression (5.22) and replace $\sin^2\beta$ to obtain the volume [480, 524, 525]:

$$V = \pi R_1(2rc-D)\cdot[2D+(2rc-D)] = \pi R_1(4r^2c^2-D^2).\tag{5.26}$$

Solving for r, which leads to $r = 1/(2c)\cdot\sqrt{V/(\pi R_1)+D^2}$, and inserting in Eq. (5.23), we get an expression for the capillary force with the volume rather than the curvature $1/r$ as a parameter [520, 521]:

$$F = 4\pi\gamma_L cR_1\left(1-\frac{D}{\sqrt{D^2+\frac{V}{\pi R_1}}}\right)\tag{5.27}$$

$$\approx 2\pi\gamma_L R_1(\cos\Theta_1+\cos\Theta_2)\left(1-\frac{D}{\sqrt{D^2+\frac{V}{\pi R_1}}}\right).\tag{5.28}$$

■ **Example 5.5**

What is the capillary force for a sphere of 1 mm radius lying on a planar surface in a 90% saturated vapor of octanol? The contact angle of both materials with octanol is supposed to be $\Theta_1 = \Theta_2 = 20°$. Calculate the curvature and the volume of the meniscus.
At contact, the force is

$$F = 4\pi\gamma_L R_1\cos\Theta = 4\pi\cdot0.0271\,\mathrm{Nm}^{-1}\cdot10^{-3}\,\mathrm{m}\cdot0.940 = 3.20\times10^{-4}\,\mathrm{N}.\tag{5.29}$$

With Eq. (5.24) and $\lambda_K = 1.74$ nm, we get a curvature

$$r = -\frac{\lambda_K}{\ln(P/P_0)} = -\frac{1.74\,\mathrm{nm}}{\ln 0.9} = 16.5\,\mathrm{nm}.$$

According to Eq. (5.26) and with $c \approx 2\cos 20° = 1.88$, the volume of the meniscus is

$$V = \pi\cdot10^{-3}\,\mathrm{m}[4\cdot(16.5\times10^{-9}\,\mathrm{m})^2\cdot1.88^2] = 3.02\,\mu\mathrm{m}^3$$

5.4.2
Two Different Spheres

The results obtained above can be directly applied to calculate the force between two different spheres. In the general case, the two interacting spheres can have different contact angles, Θ_1 and Θ_2, and different radii, R_1 and R_2. The height coordinate h describes in this case the distance between the surface of sphere 1 and sphere 2. Rather than considering the shape of each sphere explicitly, we transform the geometry and consider the equivalent case of plane interacting with a sphere of effective radius

$$R^* = \frac{R_1 R_2}{R_1 + R_2}. \tag{5.30}$$

For two identical spheres of radius R_s, the effective radius is $R^* = R_s/2$. For a sphere of radius R_1 interacting with a plane, we have $R^* = R_1$.

The force is simply derived by replacing R_1 in Eqs. (5.23) and (5.28) [84, 520]:

$$F = 2\pi\gamma_L R^* \left(\cos \Theta_1 + \cos \Theta_2 - \frac{D}{r} \right), \tag{5.31}$$

$$F = 2\pi\gamma_L R^* (\cos \Theta_1 + \cos \Theta_2) \left(1 - \frac{D}{\sqrt{\frac{V}{\pi R^*} + D^2}} \right). \tag{5.32}$$

Both equations are valid at the same time. Depending on the boundary condition, one or the other is more suitable to use. If the vapor phase is in equilibrium with the condensed liquid all the time, the radius r is constant and Eq. (5.31) is convenient to use. For nonvolatile liquids and fast processes, the volume is likely to be constant and Eq. (5.32) is appropriate. The volume of the liquid meniscus is [484, 522, 523]

$$V = \pi R^* [4r^2 (\cos \Theta_1 + \cos \Theta_2)^2 - D^2]. \tag{5.33}$$

The adhesion force, that is, the force required to separate two spherical surfaces from each other, is in both cases [511]

$$F_{\text{adh}} = 2\pi\gamma_L R^* (\cos \Theta_1 + \cos \Theta_2). \tag{5.34}$$

It only depends on the radii of the particles and the surface tension of the liquid. Neither does it depend on the actual radius of curvature of the meniscus nor does it depend on the vapor pressure. This at first sight surprising result is due to the fact that with increasing vapor pressure the cross section of the meniscus l increases. At the same time, the capillary pressure decreases because r increases. The product of cross-sectional area and pressure difference, $\pi l^2 \Delta P$, remains constant and both effects compensate each other.

As one example, we calculated the capillary force versus distance between two spheres of $5\,\mu m$ radius (Figure 5.6). Two different boundary conditions were considered. First, we assumed that the meniscus is formed by condensing water at a humidity of 80%. The meniscus is assumed to be in equilibrium with the surrounding vapor at all times. The force decays linearly until a separation of $3.1\,nm$ is reached. Then, the meniscus becomes unstable and the force drops to zero. Calculations were carried out with Eqs. (5.19) and (5.23). For comparison also, the results of the approximation Eqs. (5.31) and (5.32) are shown. The slight difference between the two graphs is caused by neglecting the influence of the curvature $1/l$.

Second, we assumed that the volume of the water meniscus is constant. The volume of the liquid was set to $9.2 \times 10^{-23}\,m^3$, which is equal to the volume condensing into the meniscus at zero distance for a humidity of 80%. At constant

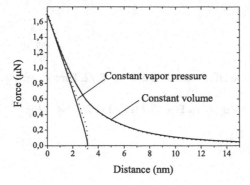

Figure 5.6 Capillary force versus distance for two similar spheres ($R_P = 5\,\mu m$, contact angle $\Theta = 40°$ with respect to water). Curves were calculated with two different boundary conditions: for constant relative humidity ($P/P_0 = 0.8$) and for constant volume ($V = 9.2 \times 10^{-23}\,m^3$). Exact solutions are plotted in continuous lines. Approximations are dotted.

volume, the capillary force is longer ranged and it does not decrease linearly anymore. The exact solution using the exact volume and Eq. (5.19) is almost indistinguishable from the approximate solution given by Eq. (5.28).

Several aspects of the above equations have been experimentally verified. McFarlane and Tabor measured the adhesion force required to pull smooth glass spheres of 0.25–1.0 mm radius from a flat glass surface [526]. In saturated water vapor, the adhesion force indeed increased linearly with the radius. Also, the predicted proportionality with the surface tension was observed when other liquids were used. Cross and Picknett also measured the force between glass surfaces [511]. They used an involatile liquid and studied the capillary force at constant volume rather than constant vapor pressure (and thus curvature). By modifying the solid surfaces they could vary the contact angle and verify the dependence of the capillary forces on the contact angle. Particularly appealing are images of liquid menisci, which have been taken optically [489, 490, 527] and by environmental scanning electron microscopy [528, 529].

An illustrative example of a force versus distance measurement at constant volume is shown in Figure 5.7. In this case, the forces between two spheres of 1.5 cm radius and between a sphere and a plane are plotted. The spheres were immersed in aqueous medium. Between the spheres, a drop of an organic liquid, a mixture of di-n-butyl phthalate and a liquid paraffin, was placed. The organic liquid is immiscible with water so that its volume is constant. A mixture was chosen to adjust the density of the organic liquid to that of water. Therefore, effects due to gravitation could be neglected. The surfaces of the spheres and the plane were treated in such a way that the contact angle with respect to the organic liquid in water was zero. The results verify that the absolute force for sphere–plane compared to sphere–sphere is a factor of 2 higher. The range of the force increases with the volume of the meniscus. Experimental curves agree with calculated curves reasonably well. The only systematic deviation is at large ranges: at a certain range the bridge formed by the organic liquid breaks and the capillary force decreases to zero. This instability is not considered in Eqs. (5.31) and (5.32).

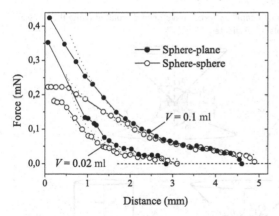

Figure 5.7 Capillary force versus distance for a sphere of 1.5 cm radius interacting with another sphere of similar radius (open circles) and with a plane (filled circles). The spheres are in aqueous medium and the menisci are formed by a mixture of di-*n*-butyl phthalate and a liquid paraffin. The organic liquid perfectly wets the surfaces ($\Theta_1 = \Theta_2 = 0$). Its volume was fixed to either 0.02 or 0.1 mL. The dotted lines are calculated with Eqs. (5.28) and (5.32) using an interfacial tension of 0.032 N m^{-1}. (Results redrawn from Ref. [533].)

5.4.3
Other Geometries

Capillary forces for different geometries have been calculated using the circular approximation (Table 5.2). When considering geometries other than spheres, the capillary force can change fundamentally. First, not only the strength but also the distance dependency changes. Second, the dependence of the capillary force on the vapor pressure changes. Although for spherical contacts the capillary force is almost independent of the vapor pressure, for other geometries it can drastically change.

■ **Example 5.6**

As an example, Figure 5.8 shows the force versus distance and the adhesion force versus humidity for a sphere of $R_1 = 3\,\mu m$ radius and a plane. It is compared to the force between a cylinder of $r_c = 200$ nm radius with a conical end (opening angle $\Phi = 88°$) and a plane. Both surfaces are assumed to be perfectly wetted ($\Theta_1 = \Theta_2 = 0$). Force versus distance was calculated at a relative vapor pressure of water of $P/P_0 = 0.9$ leading to the radius of curvature of $r = -0.52\,nm/\ln 0.9 = 5.0$ nm (Eq. (5.24)). The humidity dependence is plotted for contact ($D = 0$) and is thus equal to the adhesion force.

For the sphere, the force decreases linearly with increasing distance until the meniscus breaks at $D = 8.5$ nm. The force is almost independent of humidity, except at very high humidity where it decreases and eventually goes to zero for $P/P_0 \to 1$. For $P/P_0 \leq 0.9$, it is well described by Eq. (5.29): $F = 4\pi\gamma_L R_1 = 2.7\,\mu N$ (dashed).

Table 5.2 Different geometries for which capillary forces have been calculated using the circular approximation [531, 532]. The symbols are indicated

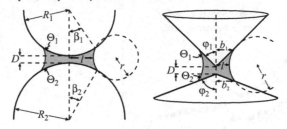

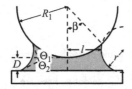 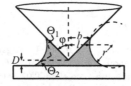 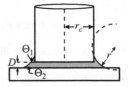

Sphere–sphere	$F = \pi\gamma_L R_1 \sin\beta_1 \left[2\sin(\Theta_1+\beta_1) + R_1 \sin\beta_1 \cdot \left(\frac{1}{r}-\frac{1}{l}\right)\right]$
[483, 484, 523]	$r = \frac{R_1(1-\cos\beta_1) + R_2(1-\cos\beta_2) + D}{\cos(\Theta_1+\beta_1) + \cos(\Theta_2+\beta_2)}$
	$l = R_1 \sin\beta_1 - r[1-\sin(\Theta_1+\beta_1)]$
	$l = R_2 \sin\beta_2 - r[1-\sin(\Theta_2+\beta_2)]$
Cone–cone	$F = \pi\gamma_L b_1 \left[2\cos(\phi_1-\Theta_1) + b_1\left(\frac{1}{r}-\frac{1}{l}\right)\right]$
	$r = \frac{b_1/\tan\phi_1 + b_2/\tan\phi_2 + D}{\sin(\phi_1-\Theta_1) + \sin(\phi_2-\Theta_2)}$
	$l = b_1 - r[1-\cos(\phi_1-\Theta_1)] = b_2 - r[1-\cos(\phi_2-\Theta_2)]$
Sphere–plane	$F = \pi\gamma_L R_1 \sin\beta \left[2\sin(\Theta_1+\beta) + R_1 \sin\beta \cdot \left(\frac{1}{r}-\frac{1}{l}\right)\right]$
[511, 513]	$r = \frac{R_1(1-\cos\beta) + D}{\cos(\Theta_1+\beta) + \cos\Theta_2}$
	$l = R_1 \sin\beta - r[1-\sin(\Theta_1+\beta)]$
Cone–plane	$F = \pi\gamma_L b \left[2\cos(\phi-\Theta_1) + b\left(\frac{1}{r}-\frac{1}{l}\right)\right]$
[519]	$r = \frac{b/\tan\phi + D}{\sin(\phi-\Theta_1) + \cos\Theta_2}$
	$l = b - r[1-\cos(\phi-\Theta_1)]$
Plane–plane	$F = 2\pi r_c\gamma_L \cos\Theta_1 + \pi r_c^2\gamma_L \cdot \left(\frac{1}{r}-\frac{1}{l}\right)$
[494, 530]	for $D \leq rc$ and $F = 0$ for $D > rc$ with $c = \cos\Theta_1 + \cos\Theta_2$

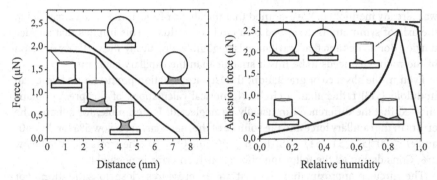

Figure 5.8 Capillary force versus distance and adhesion force versus humidity for a sphere and a plane and for a cylinder with a conical end and a plane calculated with the respective equations in Table 5.2.

For the cylinder, the force for distances below 2.7 nm is given by (Table 5.2, bottom)

$$F \approx 2\pi r_c \gamma_L \frac{\pi r_c^2 \gamma_L}{r}$$

$$= 2\pi \cdot 2.0 \times 10^{-7} \cdot 0.072\,\text{N} + \frac{\pi \cdot (2.0 \times 10^{-7})^2 \cdot 0.072}{5.0 \times 10^{-9}}\,\text{N} = 1.90\,\mu\text{N}.$$

The first term (0.091 µN) is caused by the direct action of the surface tension. The second term (1.81 µN) is due to the Laplace pressure acting over an area πr_c^2. For distances above 2.7 nm, the meniscus does not extend to the full circumference of the cylinder anymore and the conical part dominates. Then, the equations in the row "cone–plane" of Table 5.2 were applied. For a cone, the capillary force decays more steeply until at $D = 7.4$ nm it is zero.

The humidity dependence is also different from the humidity dependence between spheres. The capillary force at contact increases with humidity until it reaches a maximum at $P/P_0 = 0.86$. In this part of the curve, the meniscus only extends over the conical part. With increasing humidity, the meniscus increases in size and thus in circumference and cross-sectional area. This increase is more significant than the effect of the decrease in the Laplace pressure (in contrast to the sphere where both effects compensate each other). After the maximum at $P/P_0 = 0.86$, the capillary force steeply decreases. Here, the meniscus has reached the cylindrical part. Its circumference and the cross-sectional area remain constant but the curvature $1/r$ decreases. As a result, the force decreases steeply to a value of $2\pi \gamma_L r_c = 0.09\,\mu\text{N}$ for $P/P_0 \to 1$.

5.4.4
Assumptions and Limits

Applying the circular approximation allowed us to calculate capillary forces between axisymmetric objects analytically. What are the limits and errors involved? One limit was already mentioned: We assumed that the shape of the liquid surface parallel to the axis of symmetry is described by a circle of radius r. This implies that surface tension dominates and that gravitation is negligible. Practically, the vertical extension of the meniscus needs to be much smaller than the capillary constant.

Even in the absence of gravitation, the Laplace equation predicts a nodoid or an unduloid [513] rather than a circle. Numerical calculations of the precise shape showed that the difference can usually be neglected. For example, for spheres the errors in the capillary force and the volume of the meniscus are below 5% for $\beta \leq 40°$ and $\Theta \leq 90°$ [513, 515, 517, 518]. For $\beta \leq 20°$ and $\Theta \leq 40°$, the errors are even below 2%. Considering all the following effects, such an error is negligible.

The circular approximation fails at vapor pressures close to saturation. For $P/P_0 \rightarrow 0$, the total curvature and thus the Laplace pressure goes to zero. In addition, the circumference of the meniscus might become large. As a result, the contribution of the capillary pressure term in the total force might become insignificant and the direct action of the surface tension dominates. For two equal spheres in contact, this leads to an adhesion [584] of

$$F_{adh} = \frac{4}{3}\pi\gamma_L R_1 \cos\Theta \quad \text{for} \quad \frac{P}{P_0} \rightarrow 0 \tag{5.35}$$

in the limit of saturation.

Several aspects are often neglected in the calculation and discussion of capillary forces, although they are intensely debated with respect to wetting or adsorption [535, 536, 805]:

- **Surface heterogeneity:** Solid surfaces are usually not perfectly homogeneous. Different crystal surfaces are exposed, and defects and variations in the chemical composition lead to local changes in the contact angle. This can influence the capillary force.

- **Line tension:** The extension or shrinkage of a liquid meniscus is usually accompanied by a change in the length of the wetting line, also called three-phase contact line. The energy required to extend the wetting line per unit length is called line tension. Typical line tensions are in the order of 10^{-10} N, but in some cases significantly higher *effective* line tensions have been determined [537]. Since liquid menisci around contacts between particles are relatively small, line tension might play a significant role.

- **Microscopic contact angle:** In all calculations, we used the macroscopic contact angle and did not take the effect of surface forces into account. On the sub-10 nm scale, surface forces between the solid–liquid and the liquid–vapor interfaces can change the contact angle. For contact angle phenomena, the distinction between macroscopic and microscopic contact angle was introduced more than 20 years ago [538]. Since then, it has been extensively analyzed. For capillary forces, a thorough analysis is still lacking. Surface forces, for example, cause the vapor to form an adsorbed film. This can change the capillary force [539]. Calculations of the line tension, which are based on the analysis of surface forces, are reviewed in Ref. [540].

- **Surface deformation:** For soft surfaces, the meniscus deforms the solid surfaces, which can change the effective contact angle [541].

Kelvin lengths are typically twice the diameter of the molecules in a liquid (Table 5.1). It is questionable if at such length scales the liquid behaves like a continuum. Experiments with the SFA showed that the discrete molecular nature of the liquid does not seem to play a crucial role down to dimensions of 0.8 nm for hexane and 1.4 nm for water, or even lower [506, 507, 534]. Molecular dynamics simulations of two silica surfaces, interacting across a water bridge agreed with predictions using Kelvin's equation [542]. Monte Carlo simulations of the interaction between a sphere and a flat surface in a vapor showed that either the adhesion force increases with humidity or the force versus humidity curve shows a maximum [543, 544]. Such simulations are, however, limited to sphere sizes of the order of at most few 10 molecular diameters. They complement continuum theory, which is applicable only for larger particle radii.

5.5
Influence of Roughness

Until now, we have only considered perfectly smooth surfaces. Several observations cannot be explained with such an assumption. One such observation is the dependence of adhesion between hydrophilic surfaces at different humidities. The flow behavior of powders can depend critically on the content of moisture [483, 485, 486]. The adhesion force between hydrophilic particles either increases continuously or shows an increase, a maximum, and a decrease [491–493, 545–547]. This includes particles, which are used as carriers for pharmaceutical substances [492, 548, 549]. Similarly, the force between AFM tips and hydrophilic surfaces depends on humidity [545, 550–555]. Two typical results of adhesion force versus humidity experiments are shown in Figure 5.9. The adhesion force was measured between a hydrophilic glass sphere interacting with a hydrophilic silicon wafer surface. It increases monotonically with relative humidity. In other cases, adhesion force versus humidity curves

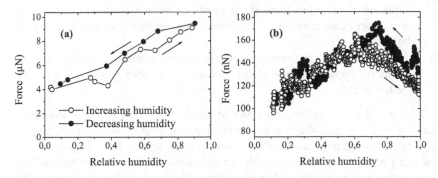

Figure 5.9 Adhesion force versus relative humidity curves. (a) A hydrophilic glass sphere of 20 μm radius interacting with a naturally oxidized silicon wafer as measured by AFM [493]. (b) Force between a microfabricated silicon nitride AFM tip and a silicon wafer [553].

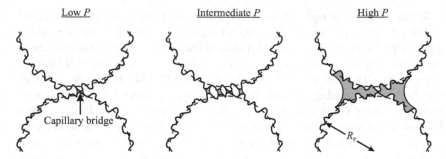

Figure 5.10 Schematic of two particles in contact at different vapor pressures of a condensing liquid. Macroscopically, both particles are assumed to be spherical and described by the apparent radius R_P. On the nanometer scale, they are rough.

show a maximum. One example, which was measured with a silicon nitride AFM tip on a silicon wafer, is plotted in Figure 5.9b.

The relevant length scale of surface structures that determine capillary forces is given by the Kelvin length. Typically, the Kelvin length is of the order of 0.5–2 nm (Table 5.1). This is particularly relevant when considering surface roughness. Even on seemingly smooth surfaces, roughness is usually a few nanometers. As a result, when two solid surfaces get into contact, real contact is only established at asperities (Figure 5.10). This can drastically change the meniscus force. It is instructive to discriminate three regimes [556]:

1) For low vapor pressure, only one capillary bridge is formed at the outmost asperity. To calculate the capillary force, the asperity is usually assumed to have a spherical shape and the formalism developed in the previous section is applied. Since the asperity has a small radius of curvature, the capillary force will be low. For the remaining part, the asperity creates an effective gap between the two surfaces, which prevents liquid from condensing unless a certain minimal vapor pressure is reached (see Exercise 5.2) [484, 553, 557].
2) At intermediate vapor pressure, more and more capillary bridges are formed. To quantify the capillary force, the force of one asperity is either multiplied by the number of asperities or a distribution of asperities is assumed [496, 498, 500, 549].
3) At high vapor pressure, the menisci merge into one continuous capillary bridge. The mean, apparent shape of the interacting surfaces dominates the total capillary forces.

Whether all three regimes exist and at which vapor pressure they dominate depends on the precise shape of the surfaces on the length scale of Kelvin length. Qualitatively, roughness has the effect of lowering the capillary force, in particular at low vapor pressure. As an example, the effect of surface roughness on the capillary force between two spheres of 10 μm radius is illustrated in Figure 5.11. In this particular case, roughness was described by a rectangular asperity distribution on each particle with an effective maximal asperity height δ_0 of 0.5, 1, and 1.5 nm. Thus,

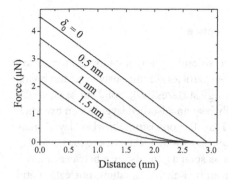
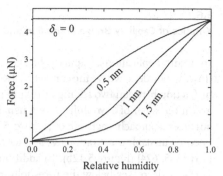

Figure 5.11 Capillary force caused by condensing water for two rough spheres of 10 μm apparent radius and a contact angle of 0°. Roughness is characterized by the maximal asperity height δ_0. On the left, the capillary force was calculated against distance at 70% humidity. On the right, the capillary force is plotted versus humidity at contact. See Ref. [558] for details.

the real height on each particle is distributed between $\pm\delta_0/2$ around the apparent surfaces. For comparison, results for the perfectly smooth sphere ($\delta_0 = 0$) are also plotted. For the smooth sphere, the capillary force decreases linearly with distance. This changes as soon as the surfaces show even a low degree of roughness. Surface roughness decreases the capillary force. In addition, the linear decrease does not continue for all distances. At high distances, the slope of the force versus distance curve decreases. Even more drastic is the effect of roughness on force versus humidity curves. Rather than the constant force predicted for smooth spheres, the capillary force starts at zero for low humidity. Then, it increases to reach its maximum. The maximal capillary force is slightly lower than the prediction for smooth spheres of $2\pi\gamma R_1 \cos\Theta$.

Surface roughness explains the monotonic increase of adhesion forces with humidity often observed between hydrophilic particles (e.g., Figure 5.9, left). For a microfabricated AFM tip and a silicon wafer (Figure 5.9, right), the adhesion force first increases, reaches a maximum at 70% relative humidity, and then decreases again. The increase at low humidity in both cases is an indication of roughness. It demonstrates that surface features on the 1 nm scale determine capillary forces. One consequence is that capillary forces are difficult to predict because it is practically impossible to determine the topology of surfaces with a precision on the order of 1 nm.

Another problem are changes of the surface in the contact region, in particular at high humidity and with oxides. In many experiments, a difference between the adhesion forces measured during increasing and decreasing humidity is observed. To a certain degree, this is also evident in Figure 5.9. This difference may be caused by wear in the contact regime [559]. Features on the 1 nm scale change during contact. Considering the extremely high stress in the contact regime and the presence of condensed water, this is hardly surprising.

5.6
Kinetics of Capillary Bridge Formation and Rupture

In many applications, capillary bridges form and rupture rapidly. Examples are flowing powders, where the contacts between particles continuously form and break, or friction between two sliding or rolling rough surfaces [560]. Solid surfaces, which are in the vapor of a lyophilic liquid, usually bear an adsorbed layer. When two such surfaces approach each other (Figure 5.12a), the opposing absorbed layers may become unstable due to attractive surface forces, grow in thickness, and form a bridge [561–565, 624] (Figure 5.12b). In addition, as soon as the distance between the two surfaces decreases below the threshold given by Kelvin's equation, nucleation and capillary condensation are possible [560, 566, 578]. With decreasing gap width, nucleation becomes more and more likely and eventually a liquid bridge is formed (Figure 5.12c).

After contact (Figure 5.12d) upon retraction the liquid will form a bridging meniscus (Figure 5.12e). The meniscus breaks at a distance where the condition set by the Kelvin equation cannot be fulfilled anymore [565, 567] (Figure 5.12f). Even for constant volume, the meniscus becomes unstable at a certain distance [527] because no solutions of the Laplace equation exist anymore [515, 568, 569]. The meniscus also breaks if the retraction velocity is so high and the liquid is so viscous that it becomes kinetically unstable [520, 570]. For layers of nonvolatile liquids, for example, for lubrication layers on magnetic hard disks, capillary condensation and evaporation are not an option. The liquid in the bridging meniscus has to be provided by direct flow from the adsorbed layer.

Here, we first consider menisci, which are formed by capillary condensation. Condensation is limited by diffusion of vapor molecules toward the growing meniscus. Vapor molecules diffuse toward the meniscus because directly at the liquid surface the local vapor pressure is determined by the curvature via Kelvin's equation. This vapor pressure is lower than the vapor pressure far away from the surface. Kohonen, Maeda, and Christenson [571] derived an approximate expression for the change of the radius of curvature of a liquid meniscus between a sphere and a flat surface for zero contact angle. Their expression can be generalized to two spheres of different radii:

$$\frac{dr}{dt} = \frac{D_d M_W P_0}{\varrho R T R^*} \left(\frac{P}{P_0} - e^{-\lambda_K/r} \right). \tag{5.36}$$

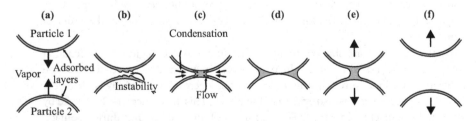

Figure 5.12 Schematic of the approach and retraction of two particles with an adsorbed liquid layer.

Here, D_d is the diffusion coefficient of vapor molecules in the surrounding gas in $m^2 s^{-1}$ and M_W is the molar mass of the molecules considered. Please note that P is the vapor pressure in the surrounding atmosphere far away from the meniscus. Close to the meniscus, the local vapor pressure can be different.

We assumed that diffusion in the gap between the two particles is the same as in the bulk. This assumption is valid as long as the gap is much larger than the mean free path for collision between gas molecules. At room temperature and normal pressure, typical free path lengths are of the order of 100 nm. In some cases, the gap width can be significantly below 100 nm. Then collisions with the walls become more likely than collision with other gas molecules. This is called Knudsen[5] flow. It can significantly slow down the process [571, 572].

■ **Example 5.7**

A silica sphere of 5 μm radius is placed on a glass plate at 80% humidity. Both surfaces are perfectly wetted by water. At normal pressure and at 25 °C, the diffusion coefficient of water molecules in air is $D_d = 2.4 \times 10^{-5} m^2 s^{-1}$ and the saturation vapor pressure is $P_0 = 3169$ Pa. Thus, condensation and evaporation is described by

$$\frac{dr}{dt} = \frac{2.4 \times 10^{-5} \, m^2 \, s^{-1} \cdot 0.018 \, kg \, mol^{-1} \cdot 3169 \, Nm^{-2}}{1000 \, kg \, m^{-3} \cdot 8.31 \cdot 298 \, J \, mol^{-1} \cdot 5 \times 10^{-6} \, m} \cdot \left(0.8 - e^{-0.52 \, nm/r}\right)$$

$$= 1.105 \times 10^{-4} \, ms^{-1} \cdot \left(0.8 - e^{-0.52 \, nm/r}\right).$$

(5.37)

By solving this differential equation numerically and using $l^2 \approx 4rR_P$, the radius of the meniscus l can be calculated (Figure 5.13). The meniscus first increases in size and after a characteristic time saturates to the value given by Kelvin's equation.

To obtain a characteristic condensation or evaporation time we take dr/dt at $t = 0$ and extrapolate it linearly until the equilibrium value $r = -\lambda_K / \ln(P/P_0)$ is reached. The linear extrapolation is described by

$$r = \frac{D_d M_W P}{\varrho R T R^*} \cdot t.$$

(5.38)

Setting these two expressions equal leads to a characteristic condensation time of

$$\tau = -\frac{\varrho R T R^* \lambda_K}{D_d M_W P \ln(P/P_0)} = -\frac{R^* \gamma_L}{D_d P \ln(P/P_0)}.$$

(5.39)

5) Martin Hans Christian Knudsen, 1871–1949. Danish physicist and oceanographer.

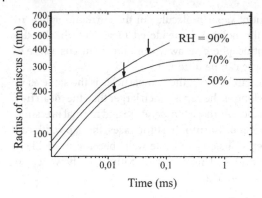

Figure 5.13 Radius of a meniscus l versus time (in milliseconds) calculated with Eq. (5.36). Water vapor in air at normal pressure is condensing into the gap between a sphere of $R_1 = 5\,\mu m$ and a flat surface, both being perfectly wetted by water. The arrows indicate the respective characteristic time constants calculated with Eq. (5.39).

It increases with the radius of the particle because the equilibrium volume of the meniscus increases. For the $5\,\mu m$ sized particle on a flat plate (Figure 5.13), the characteristic time constants are 13, 19, and $50\,\mu s$ for a relative humidity of 0.5, 0.7, and 0.9, respectively.

For nonvolatile liquids, meniscus formation is different [499, 501]. The liquid layer is not formed by spontaneous adsorption from the vapor but by adding a certain amount of liquid to the system. For example, a lubricant is added to the surface of a hard disk. Assuming that the liquid spreads spontaneously, the layer thickness is simply the total amount added divided by the total surface area available. The fact that the liquid spreads spontaneously implies that we have a repulsive disjoining pressure between the solid–liquid and the liquid–gas interfaces. For layers bound by van der Waals forces, this pressure is given by

$$f = \frac{A_H}{6\pi h^3}.$$

(5.40)

The pressure depends on film thickness h. For thin films, the liquid–gas interface is pushed away from the solid–liquid interface more strongly than for thick films. Thus, liquid tends to flow from thick regions into thin regions, forming a homogeneous layer.

When a particle or an asperity gets in contact with the liquid, a meniscus is initialized and a negative Laplace pressure is set up in the film around the contact because of the concave shape of the meniscus. The liquid is thus drawn toward the meniscus. The flow is continuous until the meniscus has reached its equilibrium volume. In equilibrium, the disjoining pressure is equal to the Laplace pressure in the meniscus. Using film hydrodynamics, the formation of the meniscus can be predicted [499, 572, 573]. Assuming continuum mechanics, a constant viscosity η of the liquid in the film, and radial symmetry of the contact, Gao and Bhushan [501] calculated the volume of the meniscus to increase as

$$V(t) = V_\infty \left(1 - e^{-\frac{C_0 h^3 t}{V_\infty \eta}}\right). \tag{5.41}$$

Here, C_0 is a constant with the dimension of pressure and V_∞ is the volume of the final meniscus for $t \to \infty$. For typical situations such as a head on a magnetic storage disk with film thicknesses of the order of 1 nm and low viscosity, typical time constants are of the order of 1 min to hours [501, 573].

5.7
Capillary Forces in Immiscible liquid Mixtures and Other Systems

Capillary forces not only occur in gaseous environment by capillary condensation of a liquid. For example, in a slurry of hydrophobic particles bubbles can form between hydrophobic particles [514]. In this case, either gas molecules dissolved in the water nucleate and form a continuous gas phase between two adjacent particles or the gas phase is formed by water vapor alone [574–576] (Figure 5.14, left). When two glass surfaces in mercury approach each other, a cavity is formed and forms a bridge [577]. Menisci can also be formed by one liquid B in another liquid A (Figure 5.14, right). The two liquids can be completely immiscible. For example, if hydrophobic particles, water, and a small amount of oil are vigorously shaken, the particles aggregate. Oil tends to form menisci between the particles and cause a strong attraction. The two liquids might also be partially miscible. For example, toluene can be dissolved in water up to a concentration of 5.8 mM at room temperature. When two hydrophobic surfaces are brought together in water with dissolved toluene, a liquid toluene phase will form a meniscus and attract the surfaces. Capillary forces can be considered in a general way in which one fluid phase forms a meniscus in another fluid phase [579, 580], and at least one fluid phase is a liquid. An interesting case are, for example, liquid crystals. When an isotropic liquid crystal is confined between two surfaces, it can undergo a phase transition to the nematic or smectic phase [581]. The ordered phase then forms a meniscus, which leads to a capillary force [582].

To illustrate phase separation in liquid mixtures, we consider the phase diagram of methanol–n-hexane, a typical binary liquid (Figure 5.15). Above the critical solution temperature of 34.5 °C, the two liquids mix at all volume ratios. Below 34.5 °C they only mix at low volume ratios of either methanol or hexane. If we go along an

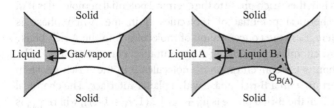

Figure 5.14 Menisci of gas and vapor between two lyophobic surfaces (a) and a liquid phase B in a liquid phase A (b).

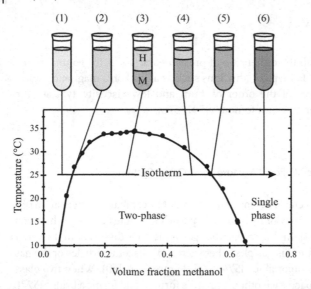

Figure 5.15 Phase diagram of a binary mixture of methanol and *n*-hexane with an upper critical solution temperature of 34.5 °C. Methanol (M) has a higher density than hexane (H) and will therefore form the bottom phase.

isotherm at 25 °C (arrow in Figure 5.15), a homogeneous mixture is found below a volume fraction of methanol of 9%, (1). A few methanol molecules are dissolved and we have one phase, which is mostly hexane. At a volume fraction of 9% the saturation concentration of methanol in hexane is reached (2). Further addition of methanol (3) leads to a phase separation into a hexane-rich phase (with saturated concentration of methanol) and a methanol-rich phase (with saturated concentration of hexane). At some volume fraction, methanol becomes the majority phase (4). Once the concentration of methanol reaches the saturation concentration of hexane in methanol (5), the hexane-rich phase disappears. We reach again a one-phase region, in which few molecules of hexane are dissolved in methanol (6).

Let us turn to the general case of meniscus formation. We consider two surfaces that approach each other in an A-rich phase with a little dissolved B. We further assume that the surfaces attract B more than A. As a result, the contact angle of B on the surfaces in A, $\Theta_{B(A)}$, is lower than 90°. In this case, B will tend to form a meniscus of a B-rich phase between the two surfaces. To derive volume equation, we equate the chemical potential of molecules B in the A-rich phase to the chemical potential of molecules B in the meniscus. The chemical potential of molecules B in the A-rich phase is $\mu_B^0 + RT \ln(c_B/c_B^0)$. Here, c_B is the concentration of molecules B in the A-rich phase and μ_B^0 is the standard chemical potential of B at saturation concentration c_B^0. The saturation concentration is the concentration of molecules B in the A-rich phase in contact and in equilibrium with a B-rich phase with a planar interface. The chemical potential of molecules B in the B-rich phase is $\mu_B^0 + \gamma_{AB} V_m^B (1/r_1 + 1/r_2)$, where γ_{AB} is the interfacial tension between the A-rich phase and the B-rich phase and V_m^B is the molar volume of molecules B in the B-rich phase. r_1 and r_2 are the two principal radii of curvature of the liquid. This leads to the generalized Kelvin equation:

$$RT \cdot \ln \frac{c_B}{c_B^0} = \gamma_{AB} V_m^B \cdot \left(\frac{1}{r_1} + \frac{1}{r_2} \right). \tag{5.42}$$

Capillary forces can be calculated as described above once the curvature of the liquid interface and thus its shape is known. Since interfacial tensions between two liquid phases are usually lower than typical liquid–vapor interfacial tensions, the Kelvin length $\lambda_K = \gamma_{AB} V_m^B / RT$ is longer and the capillary force can be more long range [583]. Force measurements can even be used to measure very low interfacial tensions [584], which are difficult to measure with conventional techniques.

5.8
Lateral Forces Between Particles at a Fluid Interface

Let us place a particle at the interface between a liquid A and a fluid B. For simplicity, we assume it is a spherical particle. Typically, liquid A will be water and fluid B is air. The particle will usually remain at the interface unless it is completely wetted by one of the two phases and the contact angle is zero [585, 586]. In the absence of gravitation and in equilibrium, the interface will assume a planar shape (Figure 5.16a). Practically, gravitation effects can usually be neglected for particles with diameters below 10 μm. If the particle is large and its density is high, gravitation sets in. The particle sinks into the liquid (Figure 5.16b) and deforms the liquid surface [587] until the normal capillary force caused by the curved surface is equal to the gravitational force minus buoyancy.

If we place a second particle on the interface, it will notice the deformed liquid surface (Figure 5.16c) and vice versa [588, 589]. As a result, the three-phase contact line is not radially symmetric anymore. The asymmetry of the contact line leads to a lateral force between the two floating particles. It is called lateral flotation force (reviewed in Ref. [590]).

Lateral forces between particles floating at a fluid interface or immersed in a thin liquid film are observed and utilized in some extraction and separation flotation processes [591]. They are important in stabilizing emulsions and foams, for making porous materials from emulsions and foams [593, 594], and for preparing colloidal crystals [595, 596]. The presence of lateral capillary forces has

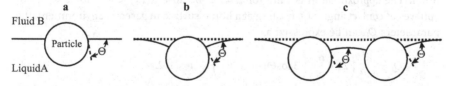

Figure 5.16 A spherical particle with contact angle Θ at a liquid–fluid interface without (a) and including gravitation (b). Two particles at constant contact angle in the liquid–fluid interface (c). The density of the particles is assumed to be higher than that of liquid A.

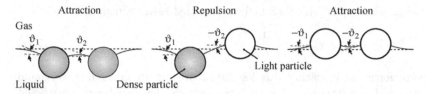

Figure 5.17 Capillary flotation forces between two particles at a liquid surface can be attractive or repulsive depending on the meniscus slopes ϑ_1 and ϑ_2.

long been known, for example, for two parallel plates placed vertically through a liquid surface [597]. For colloidal particles at liquid surfaces, lateral capillary forces have been calculated [588–590, 598, 599] and basic features have been verified by measurements [600, 601].

Two similar particles always attract each other. This attraction appears because the liquid meniscus deforms in such a way that the gravitational potential energy of the two particles decreases as they approach each other. In terms of forces, they attract each other because the horizontal component of the surface tensional force is stronger in the direction toward the neighboring particles than in the opposite direction.

For dissimilar particles, the lateral force can also be repulsive. The sign of the interaction is determined by the meniscus slope angles ϑ_1 and ϑ_2 (Figure 5.17). It is attractive for $\sin\vartheta_1 \sin\vartheta_2 > 0$. It is repulsive for $\sin\vartheta_1 \sin\vartheta_2 < 0$. No lateral capillary force is present for $\sin\vartheta_1 \sin\vartheta_2 = 0$. Nonspherical, regular objects can even form complex two-dimensional arrays on liquid surfaces [602].

The flotation force for spherical particles can be quantified by applying the superposition principle. In this case, the deformation caused by two particles is assumed to be equal to the sum of the deformations caused by the isolated individual particles. For two spheres of radius R_1 and R_2 with slope angles ϑ_1 and ϑ_2, the lateral force is [590, 598, 599]

$$F = -2\pi\gamma_L \frac{Q_1 Q_2}{\varkappa_c} \cdot K_1(L/\varkappa_c) \approx -2\pi\gamma_L \frac{Q_1 Q_2}{L} \tag{5.43}$$

with $Q_1 = R_1 \sin\vartheta_1$ and $Q_2 = R_2 \sin\vartheta_2$. The minus sign indicates that the force is attractive if $\sin\vartheta_1$ and $\sin\vartheta_2$ are of equal sign. L is the center-to-center distance between the particles. K_1 is the modified Bessel function of the second kind and first order. The approximation is valid for $L \ll \varkappa_c$ because $K_1(x) \approx 1/x$ for $x \ll 1$. For spheres of contact angles Θ_i floating at a liquid surface in gaseous environment, the parameters Q_i can be expressed as

$$Q_i = \frac{R_i^3}{6\varkappa_c^2}\left(2-4\frac{\varrho_i}{\varrho_L} + 3\cos\Theta_i - \cos^3\Theta_i\right), \quad i = 1, 2. \tag{5.44}$$

Here, ϱ_i and ϱ_L are the densities of the particles and liquid, respectively. With $\varkappa_c = \sqrt{\gamma_L/\varrho_L g}$, we see that the lateral flotation force is proportional to $R_1^3 R_2^3/\gamma_L$.

■ **Example 5.8**

Calculate the flotation force between two hydrophobized silica particles with a contact angle of $80°$, $R_p = 20\,\mu m$, and $\varrho_P = 2500\,kg\,m^{-3}$ in a water surface for a distance of $D = 20\,\mu m$.

Since $L = D + 2R_P \ll \varkappa_c$, we can use the approximation. Inserting Eq. (5.44) into Eq. (5.43) leads to

$$F = -\frac{\pi}{18}\frac{\gamma_L R_P^6}{\varkappa_c^4 L}\left(2-4\frac{\varrho_P}{\varrho_L}+3\cos\Theta-\cos^3\Theta\right)^2. \qquad (5.45)$$

With $\varkappa_c = 2.71\,mm$ (Table 5.1) and $\varrho_L = 1000\,kg\,m^{-3}$, we get $F = -2.49\times 10^{-16}\,N\cdot(-7.48)^2 = -1.39\times 10^{-14}\,N$.

In general, lateral capillary forces are caused by a deformation of a liquid surface, which is flat in the absence of particles. The larger the deformation, the lager the interaction. In the case of floating particles, the cause of deformation is the weight of the particles. What about small particles where gravitation is negligible? Small particles also interact (reviewed in Ref. [603]). The effect of lateral interaction is observed when imaging small particles at the air–water or oil-water interface with an optical microscope [604, 605]. The interaction is, however, not dominated by the weight-induced capillary force. The lateral force between small particles depends on the surface charge and the salt concentration. Weakly charged particles or particles in a subphase with high salt concentration tend to form aggregates due to short-range van der Waals attraction [605]. Highly charged particles floating on water with little salt tend to repel each other. The reason is a repulsive dipole–dipole interaction [604, 606]. Surface charges plus the counterions in the diffuse layer cause an electric dipole moment. Since the dipoles of two floating particles are oriented parallel, they repel each other. For the simple case of two point dipoles and in the framework of linearized Poisson–Boltzmann theory for the electric double layer, the potential energy between two floating particles is [607, 608]

$$U = \frac{Q_1 Q_2}{2\pi\varepsilon\varepsilon_0 L}\left(\frac{\varepsilon^2}{\varepsilon^2-1}e^{-L/\lambda_D}+\frac{\lambda_D^2}{\varepsilon L^2}\right). \qquad (5.46)$$

Since the particles are immersed in two media in which the electric field distribution is very different, the force also contains two components with different distance dependencies. The first part, which decays exponentially, is caused by the screened electric fields in the aqueous medium. The second term, in which the distance decays like L^{-3}, is caused by the electric field in air.

Experimentally, a weak, long-range lateral attraction has been observed in addition to the shorter ranged dipole–dipole repulsion [609], even for particles smaller than $10\,\mu m$. In some cases, two-dimensional ordered aggregates, foams, voids or chains, called mesostructures, were observed [610–612]. The origin, range, and strength of this attraction are still under debate. It has been suggested that an irregular contact line, caused by either a heterogeneous surface or an irregular

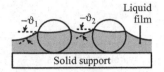 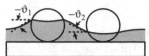 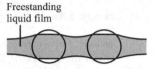

Figure 5.18 Capillary immersion forces between particles in a liquid film on a solid support can be attractive (left) or repulsive (center) depending on ϑ_1 and ϑ_2. In freestanding films (right), immersion forces are always attractive.

topography of the particles, deforms the liquid interface and causes a capillary attraction [613, 614]. In fact, an irregular particle at a liquid interface will cause a deformation of the interface that should result in a capillary force [615]. The electric field of the dipole moments of the charged particles might deform the liquid–fluid interface, which again could cause a capillary attraction [616]. Calculations, however, showed that this effect should not be strong enough to overcome the dipole–dipole repulsion between the particles [617]. A third suggestion attributes the attraction to fluctuations. Any fluid interface fluctuates thermally. These fluctuations are modified by the particles, which can lead to an effective attraction [618, 619]. Finally, mesostructures were explained by the presence of contaminations at the water interface [620].

Lateral capillary forces also occur when the particles are partially immersed in a thin liquid film on a solid support (Figure 5.18) [590, 595]. In this case, they are called immersion forces. Immersion forces on solid surfaces occur always when suspensions of solid particles are dried. In the last state of evaporation, the particles are only partially immersed in the thinning liquid film and attractive immersion forces lead to an aggregation of particles (Figure 5.19). Immersion forces can be used to self-assemble particles in two-dimensional arrays [595, 596]. The deformation of the liquid surface is related to the wetting properties of the particles, that is, to the position of the contact line and the contact angle rather than the weight. For this reason, also very small particles such as proteins are affected.

Immersion forces can also be calculated with Eq. (5.43). They can be attractive or repulsive, depending on the meniscus angles ϑ_1 and ϑ_2. The same rule applies as for flotation forces: attraction for $\sin\vartheta_1 \sin\vartheta_2 > 0$ and repulsion for $\sin\vartheta_1 \sin\vartheta_2 < 0$. Usually, the particles attract each other since the contact angles are low; if the particles were not wetted by the liquid it would be difficult to keep them dispersed in solution. One example of a force versus distance measurement for two immersed spheres is shown in Figure 5.20.

A third kind of lateral capillary attraction acts on particles in freestanding liquid films (Figure 5.18) [586]. Particles in freestanding films are, for example, relevant for the stabilization of foams [593, 594]. Membrane proteins in biological membranes may be considered as nanoparticles in a liquid film [596, 621–623]. The lateral capillary force in freestanding films is also called immersion force. It is determined by the wetting properties and therefore it can be significant even for nanoparticles. In freestanding films, the lateral capillary force is always attractive, unlike the flotation force and the immersion force on solid supported films.

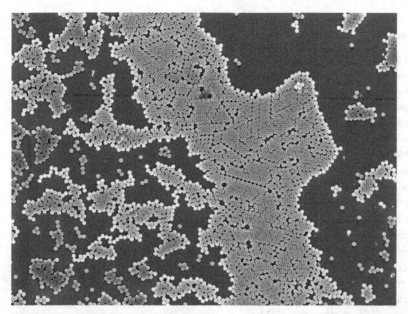

Figure 5.19 Polystyrene (PS) particles of 0.64 μm radius dried from aqueous suspension on a solid surface and imaged by scanning electron microscopy. Due to immersion forces, the particles tend to aggregate rather than being isolated as in solution. Bare PS particles are hydrophobic and would aggregate in aqueous solution. Therefore, these PS have covalently bound carboxyl groups at their surface. At some places, the particles even formed an ordered hexagonal array.

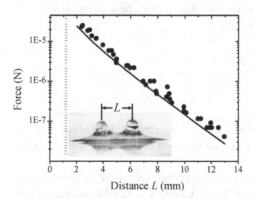

Figure 5.20 Lateral immersion force versus center-to-center distance for two glass spheres of $R_1 = 0.6$ mm radius measured with a torsion balance [601]. The two spheres are partially immersed in aqueous surfactant solution with $\gamma_L = 0.0368$ Nm^{-1} and $\varkappa_c = 1.94$ mm. Each particle is kept at the liquid surface by a support from below, which plays the role of the solid substrate. With $\Theta \approx 0$, we have contact line radii of $r \approx R_1$ and slope angles of $\vartheta \approx 90°$. The continuous line was calculated with Eq. (5.43). The dotted line indicates the position of closest contact between the two spheres.

5.9
Summary

- The vapor pressure of a liquid depends on the curvature of its surface. For drops, it is increased compared to the vapor pressure of a planar surface. For bubbles, it is reduced. Quantitatively, this is described by the Kelvin equation (5.8).
- Capillary condensation is the spontaneous condensation of liquids into pores and capillaries with lyophilic surfaces. It can cause the adhesion of particles. The condensing liquid forms a meniscus around the contact region between two particles, which leads to capillary attraction.
- Capillary forces for axial symmetric menisci, which are much smaller than the capillary constant, can be well calculated with the circular approximation. In the circular approximation, the two radii of curvature are approximated by circles.
- For two spherical particles, the capillary force is given by $F = 2\pi\gamma R^* \cos\Theta$. Two spheres are a unique case. Unlike other contact geometries, the capillary force does not (or only weakly) depend on the vapor pressure. In general, the capillary force depends sensitively on the vapor pressure.
- At constant vapor pressure, the capillary force between two spheres decreases linearly with distance (Eq. (5.31)). If the volume of the meniscus is constant, the decrease is less steep (Eq. (5.32)).
- The relevant length scale for the contact geometry and for force versus distance curves is given by $-2\lambda_K / \ln(P/P_0)$ (Eq. (5.24)). It is of the order of the Kelvin length and increases with the relative vapor pressure. Since Kelvin lengths are of the order of 1 nm, surface roughness plays a significant role.
- The characteristic time of evaporation increases with the radius of interacting particles and the surface tension. It decreases with increasing diffusion coefficient of vapor molecules in the gas phase and the vapor pressure (Eq. (5.39)).
- Lateral capillary forces act between particles floating on a liquid surface or which are partially immersed in a liquid film.

5.10
Exercises

5.1. Verify that the capillary force between a perfectly smooth sphere of radius R_P in contact with a plane in a vapor of a liquid that wets the surfaces is $F = 4\pi\gamma_L R_P$ from first principles.

5.2. Calculate the capillary force versus humidity for two spheres of 5 μm radius and 40° contact angle with respect to water. A simple way to take roughness into account is to assume that an asperity keeps the particles a certain distance H apart. Asperities create an effective gap between the two particles. Consider the case of $H = 1$ and 2 nm.

5.3. Stiction of a head on a magnetic storage disk. The head of a hard disk rests on the disk. A lubricant layer of 1 nm thickness and with a surface tension of 25 m Nm^{-1} is homogeneously distributed on the disk. The Hamaker constant for the organic lubricant layer–disk–air system is $A_H = 2 \times 10^{-20}$ J. We assume that the lubricant wets the head and the head can be described as a sphere of radius of curvature of 1 μm. What is the volume of the meniscus in equilibrium? What is the capillary force? How do volume and capillary force change if lubricant is added and the film grows to 1.5 nm thickness?

5.4. Discuss how the characteristic condensation time given by Eq. (5.39) depends on the relative vapor pressure.

5.5. Liquid mixtures. Calculate the capillary force versus distance for a sphere of 5 μm radius interacting with a planar surface in water containing toluene at a concentration of 5.0 and 5.6 mM ($c_B^0 = 5.8$ mM). The interfacial tension between water and toluene at room temperature is 31 mN m^{-1}. The contact angle of toluene in water on the particular surfaces is assumed to be 85°. $\varrho_{tol} = 866$ kg m^{-3}, $M_W = 92.13$ g mol^{-1}.

6
Hydrodynamic Forces

Many applications in mineral processing, food science, paper making, and so on involve the vigorous stirring of colloidal dispersions or emulsions. The interacting interfaces are not stationary but move relative to each other. Then, hydrodynamic forces can be significant, sometimes even dominant.

In this chapter, we discuss the principles of how to calculate fluid flow. As we shall see, hydrodynamics is governed by a partial differential equation, the Navier[1]–Stokes[2] equation. It can be solved analytically only for a few simple cases. A systematic introduction into hydrodynamics is beyond the scope of this book. For an instructive introduction, we recommend Refs [625, 626]. New methods for the calculation of hydrodynamic interactions in dispersions are described in Ref. [627]. As one important example, we derive the hydrodynamic force between a rigid sphere and a plane in an incompressible liquid. Finally, hydrodynamic interactions between fluid boundaries are discussed.

6.1
Fundamentals of Hydrodynamics

6.1.1
The Navier–Stokes Equation

If we slide one plate over another, parallel plate and the interstitial space is filled with a fluid (Figure 6.1), experiments showed us that the shear force required is

$$\frac{F}{A} = \eta \frac{\Delta v}{\Delta z}. \tag{6.1}$$

Here, A is the area of the plates, Δv is the velocity difference tangential to the orientation of the plates, and Δz is the distance between the plates. In fact, Eq. (6.1) is strictly only valid for laminar flow. Laminar flow dominates for small distances and

1) Claude Louis Marie Henri Navier, 1785–1836. French engineer, professor in Paris.
2) Sir George Gabriel Stokes, 1819–1903. English mathematician and physicist, professor in Cambridge.

Surface and Interfacial Forces. Hans-Jürgen Butt and Michael Kappl
Copyright ©2010 Wiley-VCH Verlag GmbH & Co. KGaA
ISBN: 978-3-527-40849-8

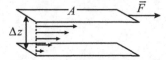

Figure 6.1 The viscous force required to slide a plane of area A over another, parallel plane at a distance Δz across a fluid of viscosity η is $F = \eta A \Delta v / \Delta z$.

low velocity. How "small" and how "low" will be discussed in Section 6.1.2. F/A is proportional to the difference in velocity divided by the distance. It is also proportional to the constant η, called dynamic viscosity or just viscosity.[3] The viscosity of air at 25 °C is 18.4 μ Pas. Viscosities of various liquids are listed in Table 6.1. The velocity gradient $\Delta v / \Delta z$ is called shear rate. We denote it by $\dot{\gamma}$. It is in units of s^{-1}. If η does not change with the shear rate, the fluid is called a Newtonian[4] fluid.

In continuum mechanics, the flow of a Newtonian fluid is described by the Navier–Stokes equation. In order to make the equation plausible, we consider an infinitesimal quantity of the liquid having a volume $dV = dx \cdot dy \cdot dz$ and a mass dm. If we want to write Newton's equation of motion for this volume element, we have to consider different forces.

The first contribution is a viscous force, caused by gradients in the shear stress of the fluid. The flow velocity is denoted by $\vec{v} = (v_x, v_y, v_z)$. To quantify the viscous

Table 6.1 Viscosities η of various liquids in mPa s or 10^{-3}Pa s.

Liquid	η	Liquid	η
Water (0 °C)	1.79	n-Hexane	0.30
Water (20 °C)	1.00	n-Octane	0.51
Water	0.89	n-Decane	0.84
Water (50 °C)	0.55	n-Dodecane	1.38
Water + 20% glycerol	1.54	n-Tetradecane	2.13
Water + 40% glycerol	3.18	n-Hexadecane	3.03
Water + 60% glycerol	8.82	Methanol	0.54
Water + 80% glycerol	45.9	Ethanol	1.07
Glycerol	934	1-Propanol	1.95
Water + 20% sucrose	1.92	1-Butanol	2.54
Water + 40% sucrose	5.98	1-Pentanol	3.62
Benzene	0.60	1-Hexanol	4.58
Toluene	0.56	1-Octanol	7.29
Dichloromethane (CH_2Cl_2)	0.41	Tetrahydrofuran	0.46
Chloroform ($CHCl_3$)	0.54	Acetone	0.31
1,4-Dioxane ($C_4H_8O_2$)	1.18	Dimethyl sulfoxide	1.99
Dimethylformamide	0.79	Methyl ethyl ketone	0.68
Methylformamide	1.68		

If not otherwise mentioned, the temperature is 25 °C. Solutions are given in mass percent.

3) Sometimes, the kinematic viscosity defined by $\eta_k = \eta / \varrho$ in units of m^2/s is also used.
4) Sir Isaac Newton, 1643–1727. British mathematician, physicist, and astronomer. Founder of mechanics and geometric optics.

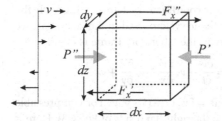

Figure 6.2 Fluid volume element in a flow field. In this case, the flow is supposed to be in x-direction, with a gradient in z-direction. As a result, a shear force is acting between the upper and the lower surface elements. The pressure acting on the right surface, P', and on the left surface, P'', can be different.

force, let us suppose that the fluid is flowing in x-direction (Figure 6.2). Also, suppose that there is a velocity gradient in the flow in z-direction: $\partial v_x/\partial z \neq 0$. This implies that a tangential force acting on the upper face of the volume element in x-direction is F_x''. It is different from the tangential force on the lower face, F_x'. The net force acting in x-direction is $F_x'' - F_x'$. Since the volume element is infinitesimally small, the force difference is also infinitesimally small and we write $dF_x = F_x'' - F_x'$. Applying Eq. (6.1) to our volume element, we can express the force difference by

$$dF_x = \eta \, dx \, dy \frac{\partial v_x}{\partial z} = \eta \frac{\partial^2 v_x}{\partial z^2} dxdydz.$$

Since $\partial v_x/\partial z$ is a shear rate, $\partial^2 v_x/\partial z^2$ is the gradient in the shear rate. In three dimensions, the viscous force, caused by gradients in the shear rate of the fluid, is a vector with components in all directions. It is given by $\eta \nabla^2 \vec{v} \cdot dV$. Here, ∇^2 is the Laplace operator and $\nabla^2 \vec{v}$ is a vector, which is written in full in Eq. (6.10).

The second force acting on dV is caused by a pressure gradient. The pressure acting on the left area, P'', is different from the pressure P' on the right side of the volume element. The resulting force in x-direction is

$$dF_x = (P'' - P')dydz = -\frac{\partial P}{\partial x} dxdydz.$$

In three dimensions, a pressure gradient causes the force $-(\nabla P)dV$. Here, ∇ is the Nabla operator and ∇P is the vector $(\partial P/\partial x, \partial P/\partial y, \partial P/\partial z)$.

In addition, external forces might influence the movement of the volume element. For example, gravity causes a hydrostatic pressure. An electrostatic force, caused by the action of an electric field on the ions in solution, is $\varrho_e \vec{E} \, dV$, where ϱ_e is the ion density and $\vec{E}$ is the field strength. Electrostatic forces are relevant for electrokinetic effects.

According to Newton's law, the sum of these forces is equal to the mass dm times its acceleration:

$$dm\frac{d\vec{v}}{dt} = (\eta \nabla^2 \vec{v} - \nabla P)dV. \tag{6.2}$$

With $\varrho = dm/dV$ and adding the external force density $\vec{f}$ in $N\,m^{-3}$, we get

$$\varrho\frac{d\vec{v}}{dt} = \eta \nabla^2 \vec{v} - \nabla P + \vec{f}. \tag{6.3}$$

Equation (6.3) was implicitly derived in a coordinate system moving with the volume element. If we transform to a coordinate system fixed in the laboratory, we have to consider that the change in velocity in the laboratory frame is $\partial\vec{v}/\partial t$ and

$$\frac{d\vec{v}}{dt} = \frac{\partial\vec{v}}{\partial t} + \frac{\partial\vec{v}}{\partial x}\frac{\partial x}{\partial t} + \frac{\partial\vec{v}}{\partial y}\frac{\partial y}{\partial t} + \frac{\partial\vec{v}}{\partial z}\frac{\partial z}{\partial t} = \frac{\partial\vec{v}}{\partial t} + v_x\frac{\partial\vec{v}}{\partial x} + v_y\frac{\partial\vec{v}}{\partial y} + v_z\frac{\partial\vec{v}}{\partial z}.$$

In a compact form, this is written as $d\vec{v}/dt = \partial\vec{v}/\partial t + (\vec{v}\cdot\nabla)\vec{v}$. For compressible fluids, we have to add $(\varsigma + \eta/3)\cdot\nabla(\nabla\cdot\vec{v})$ on the right-hand side of Eq. (6.3). Here, ς is the so-called bulk or second viscosity. This additional term is due to energy dissipation when the volume element is compressed. Except for dense gases, it can usually be neglected. Since we are interested in liquids, we ignore it. This finally brings us to the Navier–Stokes equation:

$$\varrho\left[\frac{\partial\vec{v}}{\partial t} + (\vec{v}\cdot\nabla)\vec{v}\right] = \eta\nabla^2\vec{v} - \nabla P + \vec{f}. \tag{6.4}$$

We derive another important equation. Starting point is the fact that mass is not destroyed or created. This can be expressed in an equation for mass conservation:

$$\frac{\partial\varrho}{\partial t} + \nabla\cdot(\varrho\vec{v}) = \left(\frac{\partial}{\partial t} + \vec{v}\cdot\nabla\right)\varrho + \varrho\,\nabla\cdot\vec{v} = 0. \tag{6.5}$$

Many liquids are practically incompressible. Their mass density ϱ is constant over space and in time. Then, we can neglect the left term and obtain the continuity equation:

$$\nabla\cdot\vec{v} = \frac{\partial v_x}{\partial x} + \frac{\partial v_y}{\partial y} + \frac{\partial v_z}{\partial z} = 0. \tag{6.6}$$

Equation (6.6) basically tells us that any liquid flowing into a volume element has to be compensated by the same amount of liquid flowing out of that volume element. The Navier–Stokes and the continuity equations are the basic equations describing the flow of an incompressible liquid.

6.1.2
Laminar and Turbulent Flow

In all following calculations, we assume that the flow is laminar and not turbulent. In laminar flow, the different layers do not mix due to hydrodynamic flow. Mixing is possible only by diffusion. Laminar flow dominates if the inertial components of the flow are low compared to frictional effects. Inertial forces become apparent due to the transformation from the comoving to the laboratory reference frame. They are represented by the term $\varrho(\vec{v}\cdot\nabla)\vec{v}$, which is a force density in $N\,m^{-3}$. If we denote a typical velocity by $\tilde{v}$ and L is the length scale over which the velocity changes, we can approximate $\nabla\vec{v} \approx \tilde{v}/L$ and $\varrho(\vec{v}\cdot\nabla)\vec{v} \approx \varrho\tilde{v}^2/L$.

Frictional forces are due to the viscous term $\eta\nabla^2\vec{v}$. It can be approximated by $\eta\nabla^2\vec{v} \approx \eta\tilde{v}/L^2$. The ratio between both force densities is Reynolds[5] number:

---

5) Osborne Reynolds, 1842–1912. Irish physicist, professor in Manchester.

$$Re = \frac{\varrho \bar{v} L}{\eta}.$$ (6.7)

Reynolds number allows us to predict what kind of flow we expect. As a rule of thumb, turbulences occur for large Reynolds numbers, $Re \gg 1$. Laminar flow dominates for $Re \ll 1$.

■ **Example 6.1**

In a microfluidic device, water flows through channels of 50 µm width at a central velocity 100 µm/s. Would you expect laminar flow or turbulences to occur? Compare it to normal household water pipe 1 cm diameter in which water flows with a central speed of $0.2 \, \mathrm{m \, s^{-1}}$. The viscosity of water at 20 °C is $\eta = 0.001 \, \mathrm{Pa \, s}$.

In the microfluidic device, the Reynolds number can be estimated to be

$$Re = \frac{1000 \, \mathrm{kg \, m^{-3}} \cdot 10^{-4} \, \mathrm{m \, s^{-1}} \cdot 5 \times 10^{-5} \, \mathrm{m}}{0.001 \, \mathrm{kg \, m^{-1} \, s^{-1}}} = 0.005.$$

For the pipe, we find $Re = 2000$. In the microfluidic device we certainly have laminar flow, while the flow in the water pipe is expect to be turbulent.

In the microscopic domain, Reynolds numbers are typically low and we can often neglect the inertial term in the Navier–Stokes equation. The Navier–Stokes equation simplifies to

$$\varrho \frac{\partial \vec{v}}{\partial t} = \eta \, \nabla^2 \vec{v} - \nabla P + \vec{f}.$$ (6.8)

6.1.3
Creeping Flow

In many cases, we can safely make another approximation. We neglect the explicit time dependence of $\vec{v}$ compared to the viscous term and the influence of pressure differences. This corresponds to a steady-state flow, in which the velocity is constant ($\partial \vec{v}/\partial t = 0$). It is referred to as creeping flow. For creeping flow and in the absence of external forces, the Navier–Stokes equation reduces to

$$\eta \nabla^2 \vec{v} - \nabla P = 0.$$ (6.9)

Written in full and in Cartesian coordinates, the three components of this vector equation read

$$\eta \left(\frac{\partial^2 v_x}{\partial x^2} + \frac{\partial^2 v_x}{\partial y^2} + \frac{\partial^2 v_x}{\partial z^2} \right) = \frac{\partial P}{\partial x},$$

$$\eta \left(\frac{\partial^2 v_y}{\partial x^2} + \frac{\partial^2 v_y}{\partial y^2} + \frac{\partial^2 v_y}{\partial z^2} \right) = \frac{\partial P}{\partial y},$$ (6.10)

$$\eta \left(\frac{\partial^2 v_z}{\partial x^2} + \frac{\partial^2 v_z}{\partial y^2} + \frac{\partial^2 v_z}{\partial z^2} \right) = \frac{\partial P}{\partial z}.$$

6.2

Hydrodynamic Force between a Solid Sphere and a Plate

A solid spherical particle of radius R_p moving with velocity v_p in a fluid has to overcome a hydrodynamic drag force

$$\vec{F} = -6\pi\eta R_p \vec{v}_p. \tag{6.11}$$

Equation (6.11) is referred to as Stokes' law. The minus sign indicates that the force is directed opposite to the direction of the drift velocity. Equation (6.11) can also be read in a different way: If a force $\vec{F}$ is acting on a spherical particle, then it will drift with a velocity $\vec{v}_p$.

> ■ **Example 6.2**
>
> A water drop of 1 μm diameter is falling in still air. Its velocity is calculated by inserting gravitation as the driving force:
>
> $$F = \frac{4}{3}\pi R_p^3 \varrho g = 6\pi\eta v_p R_p \Rightarrow v_p = \frac{2R_p^2 \varrho g}{9\eta}. \tag{6.12}$$
>
> Inserting $\varrho = 1000 \, \text{kg m}^{-3}$ and $\eta = 18.4 \, \text{Pa s}$, we find (6.13)
>
> $$v_p = \frac{2 \cdot (0.5 \times 10^{-6} \, \text{m})^2 \cdot 1000 \, \text{kg m}^{-3} \cdot 9.81 \, \text{m s}^{-2}}{9 \cdot 18.5 \times 10^{-6} \, \text{Pa s}} = 29.5 \, \mu\text{m s}^{-1}. \tag{6.13}$$

When a particle moves toward a planar surface, the hydrodynamic force increases and it depends on the distance. The hydrodynamic force increases because fluid has to be removed out of the closing gap.

6.2.1

Force in Normal Direction

The hydrodynamic interaction of a solid sphere moving in normal direction toward a solid plane has been calculated by Happel and Brenner [325, 327, p. 330] and later by Cox [628]. We follow a derivation of Chan and Horn [629], which is valid for small distances, that is, for $D \ll R_p$. The shape of the sphere, which confines the liquid film between the sphere and the plane, is described by the function $h(x, y)$ (Figure 6.3).

When a particle approaches the planar surface, liquid is squeezed out of the gap in between. In particular for small distances, the flow of the liquid is primarily in radial direction. Thus, for $D \ll R_p$, we neglect the vertical component of the flow and set $v_z = 0$. One consequence of the last line in Eq. (6.10) is that in such a case the pressure depends only on x, y and not on the vertical z-component since for $v_z = 0$ also $\partial P/\partial z = 0$. We further assume that variations in the flow velocity v_x, v_y in x- and y-directions are small compared to variations of the velocity v_x, v_y in z-direction. This is known as the lubrication approximation first applied by Reynolds [630]. Then, $\partial^2 v_x/\partial x^2, \partial^2 v_x/\partial y^2, \partial^2 v_y/\partial x^2$, and $\partial^2 v_y/\partial y^2$ in the first two lines of Eq. (6.10) can be

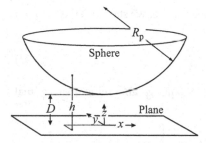

Figure 6.3 Symbols and variables used to calculate the hydrodynamic force between a sphere and a plane.

neglected. Still remaining are

$$\eta \cdot \frac{\partial^2 v_x}{\partial z^2} = \frac{\partial P}{\partial x},$$ (6.14)

$$\eta \cdot \frac{\partial^2 v_y}{\partial z^2} = \frac{\partial P}{\partial y}.$$ (6.15)

Integration of Eq. (6.14) with respect to z leads to

$$\eta \frac{\partial v_x}{\partial z} + C_1 = z \frac{\partial P}{\partial x}.$$ (6.16)

C_1 is the first integration constant. The integration is easy because $\partial P/\partial x$ does not depend on z, since P does not depend on z. With the second integration, we obtain

$$\eta v_x + C_1 z + C_2 = \frac{z^2}{2} \frac{\partial P}{\partial x}.$$ (6.17)

To determine the two integration constants C_1 and C_2, we apply the so-called no-slip boundary conditions. No-slip boundary conditions require that the liquid molecules directly in contact with the surfaces are stationary:

$$v_x = 0 \quad \text{for} \quad z = 0 \quad \text{and} \quad v_x = 0 \quad \text{for} \quad z = h(x, y).$$ (6.18)

Inserting the first condition in Eq. (6.17) directly leads to $C_2 = 0$. Applying the second condition to Eq. (6.17):

$$C_1 = \frac{h}{2} \frac{\partial P}{\partial x}.$$ (6.19)

Please keep in mind that C_1 is constant with respect to z, but it might still depend on x and y. Inserting this expression for C_1 in Eq. (6.17):

$$\eta v_x + z \frac{h}{2} \frac{\partial P}{\partial x} = \frac{z^2}{2} \frac{\partial P}{\partial x} \Rightarrow v_x = \frac{z}{2\eta} \frac{\partial P}{\partial x} (z - h).$$ (6.20)

For y-direction, we can derive a similar equation in the same way:

$$v_y = \frac{z}{2\eta} \frac{\partial P}{\partial y} (z - h).$$ (6.21)

In the second step, we derive an expression for the pressure distribution in relation to the approaching speed of the sphere $v_\perp = dD/dt$. The index $\perp$ indicates that we consider the velocity of the particle normal to the surface. We start with the equation of continuity:

$$\frac{\partial v_z}{\partial z} = -\frac{\partial v_x}{\partial x} - \frac{\partial v_y}{\partial y}. \tag{6.22}$$

Inserting expressions (6.20) and (6.21) for v_x and v_y and remembering that $h = h(x, y)$:

$$\frac{\partial v_z}{\partial z} = \frac{z}{2\eta} \left[\frac{\partial}{\partial x} \left(\frac{\partial P}{\partial x} (h-z) \right) + \frac{\partial}{\partial y} \left(\frac{\partial P}{\partial y} (h-z) \right) \right]$$

$$= \frac{z}{2\eta} \left[\frac{\partial h}{\partial x} \frac{\partial P}{\partial x} + \frac{\partial h}{\partial y} \frac{\partial P}{\partial y} + (h-z) \left(\frac{\partial^2 P}{\partial x^2} + \frac{\partial^2 P}{\partial y^2} \right) \right]. \tag{6.23}$$

We integrate in z taking into account that neither P nor h nor any of their derivatives depend on z:

$$v_z + C_3 = \frac{z^2}{4\eta} \left[\frac{\partial h}{\partial x} \frac{\partial P}{\partial x} + \frac{\partial h}{\partial y} \frac{\partial P}{\partial y} + \left(h - \frac{2z}{3} \right) \left(\frac{\partial^2 P}{\partial x^2} + \frac{\partial^2 P}{\partial y^2} \right) \right]. \tag{6.24}$$

The boundary condition $v_z = 0$ at $z = 0$ tells us that $C_3 = 0$. The no-slip boundary condition at the approaching surface reads $v_z = v_\perp$ for $z = h(x, y)$. Inserting leads to

$$v_\perp = \frac{h^2}{4\eta} \left(\frac{\partial h}{\partial x} \frac{\partial P}{\partial x} + \frac{\partial h}{\partial y} \frac{\partial P}{\partial y} \right) + \frac{h^3}{12\eta} \left(\frac{\partial^2 P}{\partial x^2} + \frac{\partial^2 P}{\partial y^2} \right). \tag{6.25}$$

At this point, we have to specify the shape of the approaching particle. The spherical shape can be described by

$$h(x, y) = D + R_p - \sqrt{R_p^2 - r^2} \approx D + \frac{x^2 + y^2}{2R_p} = D + \frac{r^2}{2R_p}. \tag{6.26}$$

Here, r is the radial coordinate in cylindrical coordinates. The approximation is called parabolic approximation because we approximate the sphere by a parabola of the same curvature. Since $r = \sqrt{x^2 + y^2}$, it follows that

$$\frac{\partial r}{\partial x} = \frac{x}{r}, \tag{6.27}$$

$$\frac{\partial P}{\partial x} = \frac{dP}{dr} \frac{\partial r}{\partial x} = \frac{dP}{dr} \frac{x}{r}, \tag{6.28}$$

$$\frac{\partial^2 P}{\partial x^2} = \frac{d}{dr} \left(\frac{\partial P}{\partial x} \right) \frac{\partial r}{\partial x} = \frac{d}{dr} \left(\frac{dP}{dr} \frac{x}{r} \right) \frac{x}{r} = \frac{x^2}{r^2} \frac{d^2 P}{dr^2} + \frac{y^2}{r^3} \frac{dP}{dr}. \tag{6.29}$$

The last step involved differentiation and rearrangements. Inserting this and the corresponding expression for y into Eq. (6.25) leads to

$$v_\perp = \frac{1}{12\eta} \left[\left(\frac{3rh^2}{R_p} + \frac{h^3}{r} \right) \frac{dP}{dr} + h^3 \frac{d^2 P}{dr^2} \right], \tag{6.30}$$

which can also be written as

$$v_\perp = \frac{1}{12\eta r}\frac{d}{dr}\left(rh^3\frac{dP}{dr}\right).$$ (6.31)

For the derivatives with respect to r, we do not need to distinguish partial and full derivatives because at the end r is the only parameter that determines the lateral properties. To obtain the pressure distribution, we integrate $12\eta r v_\perp$ from a certain r to ∞:[6]

$$6\eta r^2 v_\perp = rh^3\frac{dP}{dr} + C_4.$$ (6.32)

The integration constant C_4 is zero because at $r = 0$, the pressure gradient dP/dr is also zero. Integration of

$$\frac{dP}{dr} = \frac{6\eta r v_\perp}{h^3}$$ (6.33)

leads to the pressure distribution in the gap:

$$P(r) = P(\infty) - \frac{3\eta R_p v_\perp}{h^2}.$$ (6.34)

For the integration we used the fact that

$$\frac{d}{dr}\frac{1}{h^n} = -\frac{n}{h^{n+1}}\frac{dh}{dr} = -\frac{n}{h^{n+1}}\frac{r}{R}.$$ (6.35)

In the last step, we use Eq. (6.34) to calculate the force. The hydrodynamic force is obtained by integrating the pressure over the whole cross-sectional area:

$$F = 2\pi\int_0^\infty [P(r) - P(\infty)]r\,dr = -6\pi\eta v_\perp R_p\int_0^\infty \frac{r}{h^2}\,dr.$$ (6.36)

With Eq. (6.35), the integration can be carried out and we finally arrive at [628, 629, 631]

$$F = -\frac{6\pi\eta v_\perp R_p^2}{D}.$$ (6.37)

This expression differs from Stokes equation (6.11) by an additional factor R_p/D. The closer the particle gets to the surface the stronger the hydrodynamic drag becomes.

■ Example 6.3

To drag a microsphere of 10 μm diameter at a speed of $100\,\mu m\,s^{-1}$ through water at 20 °C, a force

---

6) One might object that for a sphere the integration to ∞ is impossible because the largest possible value for r is the particle radius. Since we have replaced the sphere by a paraboloid, r can indeed extend to ∞. The difference is not important because those parts of the sphere, which are closest to the planar surface, contribute most.

$$F = 6\pi \cdot 0.001 \, \text{Pa} \, \text{s} \cdot 5 \times 10^{-6} \, \text{m} \cdot 10^{-4} \, \text{m} \, \text{s}^{-1} = 9.4 \, \text{pN}$$

is necessary. When the particle approaches a planar surface and is a distance $0.2 \, \mu\text{m}$ away, the force to keep up the same speed would be

$$F = 9.4 \times 10^{-12} \, \text{N} \cdot \frac{5 \, \mu\text{m}}{0.2 \, \mu\text{m}} = 236 \, \text{pN}.$$

For a particle of radius R_1 moving toward another stationary particle of radius R_2 with a velocity v_p, we can generalize

$$F = -\frac{6\pi\eta v R^{*2}}{D} \quad \text{with} \quad R^* = \frac{R_1 R_2}{R_1 + R_2}. \tag{6.38}$$

As for the Stokes force in Eq. (6.11), on the right-hand side of Eq. (6.37), we have a minus sign. For an approaching particle, $dD/dt = v_\perp$ is negative because D is decreasing. The hydrodynamic force increasingly resists further approach toward the surface. When the particle is moved away from the surface ($dD/dt > 0$), the hydrodynamic drag is directed toward the surface. It keeps the particle from separating from the surfaces.

In Eq. (6.37), the force diverges to infinity for $D \to 0$. This implies that we would never be able to bring a submerged sphere and a flat surface into contact, which contradicts daily experience. The reason for this obvious false prediction of hydrodynamics is that it is a continuum theory based on local thermodynamic equilibrium. At molecular dimensions, it breaks down. Practically, we have to define a minimal molecular distance of the order of $0.2 \, \text{nm}$ that defines contact.

In the derivation of Eq. (6.37), we used the assumption of close distances, $D \ll R_p$. Cox [628] points out that this is only the first term in a series. For larger distances, we need to take more terms into account. The full expression is [325]

$$F = -6\pi\eta v_\perp R_p f^*, \tag{6.39}$$

$$f^* = \frac{4}{3} \sinh \alpha \cdot \sum_{n=1}^{\infty} \left\{ \frac{n(n+1)}{(2n-1)(2n+3)} \cdot \left[\frac{2 \sinh\left[(2n+1)\alpha\right] + (2n+1)\sinh 2\alpha}{4 \sinh^2\left[(n+0.5)\alpha\right] - (2n+1)^2 \sinh^2 \alpha} - 1 \right] \right\},$$

with

$$\alpha = \frac{1}{\cosh \tilde{D}} \quad \text{and} \quad \tilde{D} = \frac{D + R_p}{R_p}. \tag{6.40}$$

Here, $\tilde{D}$ is a normalized, relative distance of the particle. For distances smaller than $\approx 0.1 R_p$, the approximation (6.37) leads to similar results as the full expression (6.39). For larger distance, it is obvious that Eq. (6.37) predicts the wrong hydrodynamic force because for $D \to \infty$, the force decreases to zero while it should approach Eq. (6.11). One can easily extend approximation (6.37) to larger distances by adding the distance-dependent hydrodynamic drag given by Eq. (6.37) to Stokes force equation (6.11) [632]:

$$F = -6\pi\eta v_\perp R_p \left(1 + \frac{R_p}{D}\right). \tag{6.41}$$

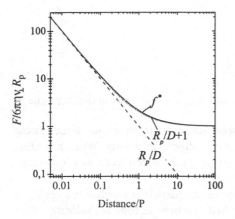

Figure 6.4 Hydrodynamic force on a sphere approaching a planar surface versus the distance calculated with approximation (6.37) and accurately with Eq. (6.39). The force is scaled by division with $6\pi\eta R_p v_\perp$ so that for Eq. (6.37) R_p/D is plotted while for (6.39) the function f^* is shown. For comparison, we also plotted $1 + R_p/D$ (Eq. (6.41)). The distance is scaled by division with the radius of the particle.

Figure 6.4 shows that Eq. (6.41) is a good approximation for the force between a sphere and a stationary wall.

Equation (6.39) has been verified by a number of experiments. In classical experiments, a sphere falling in gravity toward a planar surface has been observed [633–635]; see also Exercise 6.1. A typical experimental force versus distance curve, which is dominated by hydrodynamic drag, is shown in Figure 6.5. In this experiment a borosilicate glass sphere of 18 μm diameter was moved toward a silicon wafer in aqueous electrolyte with 0.2 M KCl at 25 °C. Both surfaces are hydrophilic. The sphere is attached to an atomic force microscope cantilever of spring constant 0.26 N m^{-1}. The cantilever was moved with a constant velocity of 40 μm s^{-1} toward

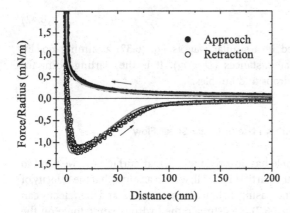

Figure 6.5 Hydrodynamic force versus distance curves between a microsphere of $R_p = 9$ μm and a silicon wafer in aqueous electrolyte measured with an AFM. The sphere is attached to a cantilever that was driven at a speed of 40 μm s^{-1}. Curves simulated with Eq. (6.37) are plotted in gray. (The results were provided by E. Bonaccurso.)

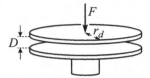

Figure 6.6 Circular disk approaching another parallel, fixed disk in a fluid in vertical direction.

and away from the planar surface. On approach, water is driven out of the closing gap, which leads to the long-range repulsive hydrodynamic force. When retracting the sphere, water has to flow into the opening gap, which leads to a long-range attractive force.

In measurements of hydrodynamic forces, the particle is attached to a spring or cantilever. It is the cantilever or spring that is moved at constant velocity, rather than the particle. This has several consequences. First, it is impossible to write down an analytical expression for force versus distance curves because $v_\perp$ changes during the approach. Force curves have to be simulated from a differential equation with the variable D, dD/dt, and the deflection of the spring [629, 636]. Second, the properties of the cantilever can influence the measurement. In AFM experiments, therefore, relatively stiff springs should be used [637]. Third, care should be taken that the cantilever does not influence the result. Typically, its hydrodynamic drag is subtracted and taken to be constant. For small spheres, however, the drag on the cantilever might depend on the distance. Therefore, the microspheres used in the AFM for hydrodynamic experiments should be larger than $10\,\mu m$ in diameter [638].

Without derivation, we also report the force between two parallel circular disks of radius r_d. One surface is assumed to be fixed while the other is moving in normal direction with a velocity dD/dt (Figure 6.6). The force required to move the disk was already measured by Stefan [639] and calculated by Reynolds [630]:

$$F = -\frac{3\pi\eta r_d^4}{2D^3}\frac{dD}{dt}.$$ (6.42)

Equation (6.42) can be derived in a similar way as Eq. (6.37) assuming no-slip boundary conditions and small distances ($D \ll r_d$). It is the starting point for calculating the force between drops and bubbles.

6.2.2
Force on a Sphere in Contact with a Plate in Linear Shear Flow

Let us assume a spherical particle has adsorbed to a planar surface and we want to remove it by applying a flow. If the particle is small, we can assume that the velocity of the fluid changes linearly with increasing distance from the plane and the velocity can be described by $v_x = \dot\gamma_0 z$ (Figure 6.7). To estimate the hydrodynamic force on the particle we apply Stokes equation (6.11) and insert the mean velocity of the fluid around the particle. The mean velocity of the fluid is $\dot\gamma_0 R_p$, which leads to an estimated force of $6\pi\eta\dot\gamma_0 R_p^2$. Exact calculations show that this expression has to be corrected by

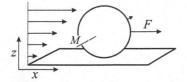

Figure 6.7 Spherical particle in contact with a planar surface with a linearly increasing fluid flow.

an additional factor [640, 642]. The hydrodynamic force on the particle tangential to the plane is

$$F = 1.701 \cdot 6\pi\eta\dot{\gamma}_0 R_p^2. \tag{6.43}$$

In addition, the fluid applies a torque to the sphere because the flow at the top side is stronger than that on the bottom side. The torque with respect to an axis going through the center of the sphere is

$$M = 0.472 \cdot 8\pi\eta\dot{\gamma}_0 R_p^3. \tag{6.44}$$

■ **Example 6.4**

Water at 25 °C is flowing through a circular tube of radius $r_c = 50\ \mu m$. The total flow rate, that is, the volume transported per unit time, is $dV/dt = 5 \times 10^{-12}\ m^3$. A latex particle of $R_p = 100\ nm$ sticks to the surface of the tube. What is the tangential hydrodynamic force on the particle?

Since $r_c \gg R_p$, the wall of the tube for the particle is like a flat surface. To calculate the shear rate, we recall that the flow profile in a cylinder in laminar flow is parabolic: $v = v_0(1 - r^2/r_c^2)$. Here, v_0 is the flow velocity in the center of the tube along the tubes axis and v is the flow velocity at a radial position r. Note that at the walls $r = r_c$ and $v = 0$. The flow rate through the tube is $dV/dt = \pi r_c^2 v_0/2$. Thus, $v_0 = 1.27$ mm s^{-1}. With $Re = 0.14$, we can safely assume laminar flow. The shear rate at the walls is

$$\dot{\gamma}_0 = -\frac{dv}{dr}(r = r_c) = \frac{2v_0}{r_c} = 50.8\ \text{s}^{-1}.$$

The hydrodynamic force tangential to the wall is

$$F = 1.701 \cdot 6\pi \cdot 0.89 \times 10^{-3}\ \text{Pa s} \cdot 50.8\ \text{s}^{-1} \cdot (5 \times 10^{-5}\ \text{m})^2 = 3.62\ \text{nN}. \tag{6.45}$$

6.2.3
Motion of a Sphere Parallel to a Wall

The Navier–Stokes equation for laminar flow (Eq. (6.8)) is a linear differential equation. For this reason, an arbitrary motion of a sphere relative to a planar wall can be separated into motion perpendicular and parallel to the boundary. The motion perpendicular was treated in Section 6.2.1. Here, we consider the case of a sphere moving parallel to a plane wall in a quiescent fluid (Figure 6.8). The velocity is slow so that we can assume creeping flow. The situation is, however, more complicated

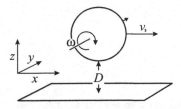

Figure 6.8 Spherical particle at a distance D moving parallel to a planar wall with velocity $v_\parallel$ and rotating with an angular velocity ω in a stationary fluid. We consider a force $\vec{F}$ and a torque $\vec{M}$ directed parallel to the surface.

than for the motion in normal direction because we have to take into account a torque acting at the sphere. The shear on the side facing the wall is stronger than the shear on the opposite side. The result is a torque. This leads to a rotation of spheres moving parallel to walls [641].

The force F and the torque M acting on a sphere moving parallel to a planar wall can be described by [326]

$$F = 6\pi\eta R_\mathrm{p}(v_\parallel F^\mathrm{t} + \omega R_\mathrm{p} F^\mathrm{r}), \tag{6.46}$$

$$M = 8\pi\eta R_\mathrm{p}^2(v_\parallel M^\mathrm{t} + \omega R_\mathrm{p} M^\mathrm{r}). \tag{6.47}$$

Here, F^t, F^r, M^t, and M^r are dimensionless functions, which depend on the normalized distance D/R_p. The index "t" stands for translation, "r" for rotation. The exact solutions for functions were calculated numerically [643]. Analytical approximations for close distances ($D \ll R_\mathrm{p}$) are [326]

$$F^\mathrm{t} = \frac{8}{15}\ln\left(\frac{D}{R_\mathrm{p}}\right) - 0.9588, \quad F^\mathrm{r} = -\frac{2}{15}\ln\left(\frac{D}{R_\mathrm{p}}\right) - 0.2526, \tag{6.48}$$

$$M^\mathrm{t} = -\frac{1}{10}\ln\left(\frac{D}{R_\mathrm{p}}\right) - 0.1895, \quad M^\mathrm{r} = \frac{2}{5}\ln\left(\frac{D}{R_\mathrm{p}}\right) - 0.3817. \tag{6.49}$$

Please note that $\ln(D/R_\mathrm{p})$ is negative because D/R_p was assumed to be much smaller than unity.

For large distances, analytical approximations have been reported [325, 326]:

$$F^\mathrm{t} = -\left(1 - \frac{9}{16\,\tilde{D}} + \frac{1}{8\,\tilde{D}^3} - \frac{45}{256\,\tilde{D}^4} - \frac{1}{16\,\tilde{D}^5}\right)^{-1}, \tag{6.50}$$

$$F^\mathrm{r} = \frac{1}{8\,\tilde{D}^4}\left(1 - \frac{3}{8\,\tilde{D}}\right), \tag{6.51}$$

$$M^\mathrm{t} = \frac{3}{32\,\tilde{D}^4}\left(1 - \frac{3}{8\,\tilde{D}}\right), \tag{6.52}$$

$$M^\mathrm{r} = -\left(1 + \frac{5}{16\,\tilde{D}^3}\right). \tag{6.53}$$

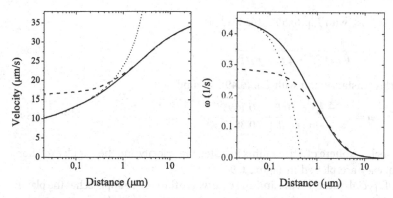

Figure 6.9 Velocity and angular rotation frequency versus distance for a silica sphere of 3 μm radius falling in water at 25 °C parallel to a planar wall. With a density of the silica of 2650 kg m^{-3}, we have $\Delta\varrho = 1650$ kg m^{-3} and $F = 1.83 \times 10^{-12}$ N. In the absence of a wall, the sphere would sink with a velocity of 36.4 μm s^{-1}. Results were calculated with Eq. (6.57). The continuous line was calculated with the full numerical solution for F^t, F^τ, M^t, and M^τ from Refs. [326, 643]. For large distances, Eqs. (6.54) and (6.55) provide a good approximation (dashed line). For close distances, Eqs. (6.48) and (6.49) can be used (dotted).

Here, $\tilde{D}$ is again the normalized, relative distance $\tilde{D} = (D + R_p)/R_p$.

We suggest to use the following equations, which approximate the exact solution as well:

$$F^t = -\left(1 + \frac{5}{8\,\tilde{D}} + \frac{5}{8\,\tilde{D}^3}\right), \quad F^\tau = \frac{1}{8\,\tilde{D}^4}, \tag{6.54}$$

$$M^t = \frac{3}{32\,\tilde{D}^4}, \quad M^\tau = -\left(1 + \frac{1}{8\,\tilde{D}^3} + \frac{5}{8\,\tilde{D}^5}\right). \tag{6.55}$$

Equations (6.54) and (6.55) are simpler than the ones reported, they show the same limiting behavior for $D \to \infty$, and they fit the exact solutions better at small distances (Figure 6.9).

■ **Example 6.5**

As an example, we consider a sphere falling in a fluid parallel to a vertical wall. The force on the sphere is given by

$$F = \frac{4}{3}\pi R_p^3 g \Delta\varrho. \tag{6.56}$$

Here, $\Delta\varrho$ is the density difference between the material of the sphere and that of the surrounding fluid. If the sphere is made of a material that is denser than the fluid, the sphere will sink. If the fluid is denser, the sphere will rise driven by buoyancy. Since the sphere is free to rotate, the torque is zero. With $M = 0$, we obtain the angular frequency of the sphere:

$$\omega = -\frac{v_\parallel}{R_p}\frac{M^t}{M^\tau}. \tag{6.57}$$

For $D \geq R_p$ with Eq. (6.55), this leads to

$$\omega = \frac{v_{\parallel}}{R_p} \frac{3}{32\, \tilde{D}^4 + 4\, \tilde{D} + 20/\tilde{D}}. \tag{6.58}$$

At close distance and with Eq. (6.49), we find

$$\omega = \frac{v_{\parallel}}{R_p} \frac{-\frac{1}{10}\ln(D/R_p) - 0.1895}{-\frac{2}{5}\ln(D/R_p) + 0.3817}. \tag{6.59}$$

For a silica sphere of 3 μm radius in water, the velocity and the angular rotation frequency are plotted in Figure 6.9.

Of special interest is the limiting case where the sphere approaches the plane. For $D \to 0$, we have $-\ln(D/R_p) \gg 1$ and we can neglect the constant terms. Then,

$$\omega = \frac{v_{\parallel}}{4 R_p}. \tag{6.60}$$

Since in the case of no slip we expect $\omega R_p = v_{\parallel}$, Eq. (6.60) shows that a sphere falling parallel to a wall or "rolling" down an inclined plane must slip. Thus, the sphere cannot be in physical contact with the wall and D cannot be zero.

6.3
Hydrodynamic Boundary Condition

While the Navier–Stokes equation is a fundamental, general law, the boundary conditions are not at all clear. In fluid mechanics, one usually relies on the assumption that when liquid flows over a solid surface, the liquid molecules adjacent to the solid are stationary relative to the solid and that the viscosity is equal to the bulk viscosity. We applied this no-slip boundary condition in Eq. (6.18). Although this might be a good assumption for macroscopic systems, it is questionable at molecular dimensions. Measurements with the SFA [644–647] and computer simulations [648–650] showed that the viscosity of simple liquids can increase many orders of magnitude or even undergo a liquid to solid transition when confined between solid walls separated by only few molecular diameters; water seems to be an exception [651, 652]. Several experiments indicated that isolated solid surfaces also induce a layering in an adjacent liquid and that the mechanical properties of the first molecular layers are different from the bulk properties [653–655]. An increase in the viscosity can be characterized by the position of the plane of shear. Simple liquids often show a shear plane that is typically 3–6 molecular diameters away from the solid–liquid interface [629, 644, 656–658].

A tenet of textbook continuum fluid dynamics is the "no-slip" boundary condition, which means that the ensemble average of the velocity of fluid molecules directly at the surface of a solid is the same as the velocity of the solid. A possible slip was discussed only in the mainstream literature for complex liquids, for example, polymer melts [659, 660]. Recent experiments, however, indicated that simple liquids might also slip past smooth surfaces [661–666].

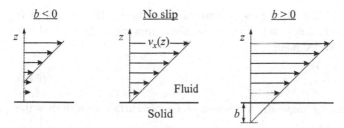

Figure 6.10 Schematic of different hydrodynamic boundary conditions at a solid–fluid interface. Left: The layer adjacent to the solid surface is bound. Middle: No-slip boundary condition. Right: Slip characterized by the slip length b.

To quantify the slip, the slip length has been introduced. The slip length b is the distance behind the interface at which the liquid velocity extrapolates to zero. It is defined by [667]

$$b = \frac{v_x(z=0)}{\partial v_x / \partial z(z=0)}. \qquad (6.61)$$

Here, $v_x(z=0)$ is the fluid velocity directly at the surface, the so-called slip velocity and $\partial v_x / \partial z(z=0)$ is its gradient (Figure 6.10).

An important microscopic parameter is the interaction between the liquid molecules and the solid wall. For weak liquid–wall interaction, the liquid molecules interact more strongly with each other than with the solid wall. Experimentally, weak liquid–wall interaction implies that the contact angle Θ of the liquid on the solid surfaces is higher than $90°$. For strong liquid–wall interaction, the contact angle is low ($\Theta < 90°$) or the liquid even wets the solid completely ($\Theta = 0$). Computer simulations [668, 669, 670, 671, 672] showed that for low fluid–wall interactions, slippage might occur.

Both interfacial effects discussed—slip and a change in viscosity—are experimentally related. The same effect as real surface slip, where the liquid molecules adjacent to the solid wall are actually moving along the wall, can be caused by a change in the viscosity close to the solid wall. If the viscosity of the near-to-wall layer is characterized by a viscosity η_S, the effective slip length is

$$b = \delta \left(\frac{\eta}{\eta_S} - 1 \right), \qquad (6.62)$$

where δ is the thickness of the surface layer. For example, if we assume that the viscosity of a 1 nm thick layer is reduced by a factor of 2, the slip length is $b = 0.5$ nm. For most liquids, however, the viscosity at solid surfaces is increased rather than decreased.

Today, it is not yet clear which boundary condition for which liquid and for which shear rate is correct. Most experiments have been done with aqueous liquids. Experimental results obtained with different methods agree that we have no slip on hydrophilic surfaces [637]. Slip might be present on hydropic surfaces [661, 673, 674] and it seems to increase with the shear rate [675, 676]. Other reports find no slip, even for water on hydrophobic surfaces [673, 677]. Recent reviews are Refs [678, 679].

In the derivation of Eq. (6.37), the no-slip boundary condition (6.18) was applied. If we allow for slip, the hydrodynamic force for a sphere moving normal to a planar surface has to be modified [680]:

$$F = -\frac{6\pi\eta v_\perp R_p^2 g^*}{D}.$$ (6.63)

The correction function g^* depends on the distance. For two surfaces, which show the same slip, it is given by

$$g^* = \frac{D}{3b}\left[\left(1 + \frac{D}{6b}\right)\ln\left(1 + \frac{D}{6b}\right) - 1\right].$$ (6.64)

For gaseous media, Goren [681] derived an equation in which the hydrodynamic force depends on the mean free path of the gas molecules and slip is characterized by an interaction parameter between the gas molecules and the solid surfaces.

6.4
Gibbs Adsorption Isotherm

In the next section, we discuss hydrodynamic forces between fluid interfaces. The interaction between fluid interfaces is strongly influenced by surfactants and contaminants at the interfaces. Therefore, we first need to introduce a fundamental relation between the amount of substance adsorbed at a fluid interfaces and the surface tension. This is quantitatively expressed in the Gibbs[7] adsorption isotherm. We only introduce the Gibbs adsorption isotherm for a two-component system, that is, a liquid and one dissolved substance. It is

$$\Gamma = -\frac{a}{RT}\frac{d\gamma_L}{da}\Big|_T.$$ (6.65)

Here, Γ is the surface excess of the solute and a is its activity. The change in surface tension of the liquid is at constant temperature; that is the reason why Eq. (6.65) is called *isotherm*. Equation (6.65) tells us that when a solute is enriched at the interface ($\Gamma > 0$), the surface tension decreases when the solution concentration is increased. Such solutes are said to be surface active and they are called surfactants or surface-active agents. Often, the term amphiphilic molecule or simply amphiphile is used. When a solute avoids the interface ($\Gamma < 0$), the surface tension increases by adding the substance. Experimentally, Eq. (6.65) can be used to determine the surface excess by measuring the surface tension versus the bulk concentration. If a decrease in the surface tension is observed, the solute is enriched in the interface. If the surface tension increases upon addition of solute, then the solute is depleted in the interface.

---

7) Josiah Gibbs, 1839–1903. American physicist, professor at Yale.

■ **Example 6.6**

We add 0.5 mM SDS (sodium dodecylsulfate, $NaSO_4(CH_2)_{11}CH_3$) to pure water at 25 °C. This leads to a decrease in the surface tension from 71.99 to 69.09 mJ m^{-2}. What is the surface excess of SDS?

At such low activities and as an approximation, we replace the activity a by the concentration c and get

$$\frac{d\gamma_L}{da} \approx \frac{\Delta\gamma_L}{\Delta c} = \frac{(0.06909-0.07199)\text{ N m}^{-1}}{(0.0005-0)\text{ mol l}^{-1}} = -5.80 \text{ N l mol}^{-1}\text{ m}^{-1}. \quad (6.66)$$

In this case, for SDS dissociates we have to take into account the counterion. This brings a factor of 2. It follows that

$$\Gamma = -\frac{a}{2RT}\frac{d\gamma_L}{da} = \frac{0.0005 \text{ mol l}^{-1}}{2 \cdot 8.31 \cdot 298 \text{ J mol}^{-1}} \cdot 5.80 \text{ N l mol}^{-1}\text{ m}^{-1}$$

$$= 5.86 \times 10^{-7} \text{ mol m}^{-2}. \quad (6.67)$$

Every molecule occupies an average surface area of 2.8 nm².

Plots of surface tension versus concentration for n-pentanol [682], LiCl (based on Ref. [683]), and SDS in an aqueous medium at room temperature are shown in Figure 6.11. The three curves are typical for three different types of adsorptions. The SDS adsorption isotherm is typical for amphiphilic substances. In many cases, above a certain critical concentration defined aggregates called micelles are formed. This concentration is called the critical micellar concentration (CMC). In the case of SDS at 25 °C, this is at 8.9 mM. Above the CMC, the surface tension does not change significantly any further because any added substance goes into micelles not to the liquid–gas interface.

The adsorption isotherm for pentanol is typical for lyophobic substances, that is, substances that do not like to stay in solution, and for weakly amphiphilic substances. They become enriched in the interface and decrease the surface tension. If

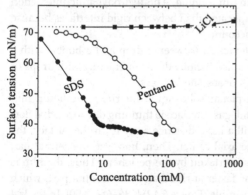

Figure 6.11 Plots of surface tension versus concentration for n-pentanol [682], LiCl (based on Ref. 683), and SDS in an aqueous medium at room temperature.

water is the solvent, most organic substances show such a behavior. The LiCl adsorption isotherm is characteristic of lyophilic substances. Many ions in water show such behavior.

Gibbs adsorption isotherm is valid in thermodynamic equilibrium. In equilibrium, the surface tension should not depend on the total surface area; if new surface is produced, surfactant from the bulk diffuses to the surface and the same surface tension is established as before. If, however, the system is not given enough time to equilibrate, the local surface tension changes with an expansion or shrinkage of the geometric surface area A. This is characterized with the surface elasticity E, also called surface dilatational modulus [684]. The surface elasticity is defined as

$$E = -A\frac{d\gamma}{dA} = -\frac{d\gamma_L}{d\ln A}.$$ (6.68)

E quantifies how much the surface tension changes upon a change in area. E is zero if the surface is in rapid equilibrium with a large body of bulk solution and the exchange of surfactant at the surface and in the bulk phase is fast. The surface elasticity depends on the specific process of surface extension.

6.5
Hydrodynamic Forces Between Fluid Boundaries

The situation is more complicated if the sphere approaching a planar surface is deformable, such as a bubble in a liquid or an oil drop in water. We can also think of the inverse situation: a solid particle interacting with liquid surface such as a bubble or drop. A particle approaching a liquid–fluid interface will lead to a deformation of the interface. Then, there are three possibilities: The particle is repelled by the interface and remains in the first liquid, it goes into the interface and forms a stable three-phase contact, or it crosses the interface and enters the fluid phase completely. The second fluid can be a gas. An example, is interaction of particles with bubbles in a liquid [685]. This interaction is of fundamental importance for flotation [591]. Another example is bubble interacting in a liquid. The hydrodynamic interaction between fluid interfaces is more complicated than between rigid interfaces because we have to take a deformation into account.

In this section, we discuss the interaction between a drop and a bubble with a planar solid surface. In Chapter 7, we also include drops interacting with drops or bubbles and particles interacting with drops and bubbles.

When moving a bubble toward a planar solid–liquid interface, the liquid film between the solid–liquid and the liquid–gas interface is thinning due to a radial flow of the liquid. When the bubble is still a large distance away from the surface, the distance of closest approach is in the axial center. Then, however, the central part approaches slower and slower and is overtaken by the periphery. Thus, the film is most often not plane parallel. It thins faster at rim than in the central part, which causes the formation of the so-called dimple (Figure 6.12) [686–688, 690]. In the last stage, the central part also approaches the solid surface until it has reached the

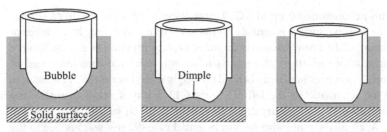

Figure 6.12 Bubble at the end of a capillary approaching a planar rigid surface. A dimple is formed at the intermediate stage.

equilibrium distance. The equilibrium film thickness is reached when the repulsive interfacial forces, also called disjoining pressure Π, equals the Laplace pressure inside the bubble: $\Pi = 2\gamma_L/r_b$. The concept of the disjoining pressure will be introduced in the next chapter.

The drainage of liquid films between a solid–liquid and a liquid–fluid interface has been studied experimentally and theoretically [690–695, 702]. The formation of the dimple and the close distance at the rim hinders the liquid in the center to flow out. As a consequence, surface forces indirectly influence the drainage time. A strong repulsion leads to a large film thickness in the rim and a fast drainage. Weak repulsion leads to a closely approaching rim and a slow drainage [697].

■ **Example 6.7**

To analyze hydrodynamic interaction, Connor and Horn [698] drove a mica plate toward a mercury drop immersed in aqueous electrolyte (Figure 6.13). The

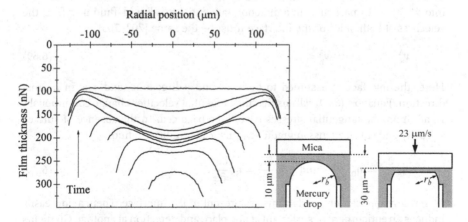

Figure 6.13 Experimental results of aqueous film thickness plotted as a function of radial position r at discrete times, showing dimple formation [698]. Note the difference in the vertical and horizontal scales. From bottom to top, the curves correspond to times $t = -0.02$, 0.02, 0.10, 0.18, 0.34, 0.62, 1.22, 2.62, 5.62, and 13.62 s. (Results were provided by R. Horn.)

electrolyte contained 0.1 mM KCl. Under these conditions, the mica surface bears a negative surface potential of -100 mV. The mercury drop had a radius of curvature of 1.92 mm leading to an inner Laplace pressure of 443 Pa. From a initial distance of 10 μm the mica surface was driven at constant speed of 23 μm s^{-1} to a position 20 μm behind the original mercury surface. Please note that the schematic on the left of Figure 6.13 is out of scale. In reality, the deformation of the mercury drop of ≈ 20 μm is much smaller than the drop size. At all times, a negative surface potential of -492 mV was applied to the mercury. This led to a strong electrostatic double-layer repulsion between the mica and the mercury surfaces, which stabilized the aqueous film at eventually a thickness of 90 nm. To measure the shape of the mercury–water interface, optical interferometry was used.

As it turns out, dimpling is a relatively general phenomenon also observed when two bubbles or two drops approach each other [689, 691, 699] or when a solid sphere approaches a liquid–liquid interface [686]. Roberts [449] observed dimpling even between a soft rubber and glass in aqueous medium. Dimpling can be rather complex. In the initial phase of the approach, sometimes a secondary minimum in the thickness is observed [686, 700]. Due to the shape, it was termed wimple [700]. A periodic dimpling was observed in aqueous films between two oil drops in the presence of surfactants [701]. This cyclic dimpling is caused by a redistribution of surfactant between the oil and the aqueous phases. Dimple formation is important because drainage of the liquid film is limited by the close approach of the rim. We return to the drainage of liquid films after having discussed the hydrodynamic boundary conditions.

To understand hydrodynamic interactions, the boundary conditions need to be known. In contrast to solid surfaces, liquid–fluid interfaces are mobile. For a mobile interface between two fluid phase A and B, the no-slip boundary condition translates into $\vec{v}^A = \vec{v}^B$. In particular in a direction tangential to a liquid–fluid interface, the velocities at both sides of the interface must be the same [702–705]:

$$v_x^A = v_x^B \quad \text{and} \quad v_y^A = v_y^B. \tag{6.69}$$

Here, the interface is assumed to be in x- and y-directions, and z is in normal direction. Equation (6.69) tells us that the tangential velocities change continuously. In addition, the tangential stresses have to be balanced. In the absence of surface viscosity and surface tension gradients, this leads to the condition

$$\eta_A \frac{\partial v_x}{\partial z} = \eta_B \frac{\partial v_x}{\partial z} \quad \text{and} \quad \eta_A \frac{\partial v_y}{\partial z} = \eta_B \frac{\partial v_y}{\partial z}. \tag{6.70}$$

In many applications, surfactants are present in the interface. Then, a flow easily induces an enrichment of surfactant at one place and depletion at another. Gradients in the surface tension are one result. Such gradients cause a tangential stress: $(\partial \gamma_{AB}/\partial x, \partial \gamma_{AB}/\partial y)$. This gradient has to be added in Eq. (6.70). If we take the example of a liquid of viscosity η and a bubble with zero viscosity, we obtain

$$\eta \frac{\partial v_x}{\partial z} = \frac{\partial \gamma_L}{\partial x} \quad \text{and} \quad \eta \frac{\partial v_y}{\partial z} = \frac{\partial \gamma_L}{\partial y}. \tag{6.71}$$

Equation (6.71) also explains the Marangoni[8] effect. In 1871, Marangoni observed the vigorous spread of a small drop of oil in a pond of water [706]. Even earlier, Thomson described "that if, in the middle of the surface of a glass of water, a small quantity of alcohol or strong spirituous liquor be gently introduced, a rapid rushing of the surface is found to occur outward from the place where the spirit is introduced" [707]. In both cases, a gradient in the surface tension caused the flow of a liquid.

Equation (6.70) tells us that the drag on a drop or bubble in a fluid depends on the viscosities of the two media. For example, the hydrodynamic drag on a bubble should be lower than the hydrodynamic drag on solid sphere of the same size. Hadamard [708] and Rybczynski [709] calculated the velocity of a spherical bubble and a falling or rising spherical drop in a liquid. Depending on viscosities of the surrounding fluid, η, and the fluid in the spherical bubble or drop, η_s, they get a different rise/fall velocity:

$$v = \frac{2gr_s^2 \Delta \varrho}{9\eta} \frac{\eta_s + \eta}{\eta_s + 2\eta/3}. \tag{6.72}$$

Here, r_s is the radius of the bubble or drop and $\Delta \varrho$ is the difference in density of the two fluids. If the fluid in the drop is less dense than the surrounding fluid, the sphere will rise. If it is denser, the sphere will sink. For a bubble with $\eta_s \rightarrow 0$, we get [710]

$$v = \frac{2gr_s^2 \Delta \varrho}{3\eta}. \tag{6.73}$$

For a rigid sphere, for which we can take $\eta_s \rightarrow \infty$, the velocity is $v = 2gr_s^2 \Delta \varrho/(9\eta)$. Thus, from a careful measurement of the rising velocity of a bubble one should be able to determine the boundary condition.

Indeed, the rise velocity of small bubbles in ultrapure water and electrolyte solutions agrees with Eq. (6.73) [711, 712]. When, however, adding even trace amounts of surfactant or surface-active contaminants, a bubble rises slower than the one in purified water [713]. This phenomenon is explained by the Marangoni effect caused by surfactant adsorption to the bubble surface [704, 710, 714–716]. While the bubble is rising, there exists a surface concentration profile of surfactant along the bubble surface. Adsorbed surfactant is swept off from the front of the bubble and accumulates at the rear. In steady state, this surfactant gradient stops the flow of liquid at the liquid–gas interface, which leads to an effective no-slip boundary condition. For a fast exchange of surfactant between bulk and interfaces and thus a low surface elasticity, the effect is lower than that for a slow exchange [717].

This effect was even earlier described for the drainage of foam and emulsion films. Drainage of such liquid films could in many cases be described by effectively using a no-slip boundary condition [718]. Such a no-slip boundary condition is attributed to gradients in surfactant concentration caused by the flow, which resisted further flow at the surfaces [719–721].

8) Carlo Marangoni, 1840–1925. Italian physicist, professor at a Lyceum in Florence.

Force measurements with fluid interfaces lead to a similar picture. Manor *et al.* [722] attached a bubble to the end of an AFM cantilever and measured dynamic forces across the nanometer thick intervening aqueous film on mica. The hydrodynamic response of the air–water interface ranged from a fully immobile, no-slip surface in the presence of added surfactants to a partially mobile interface without added surfactants. A simple Marangoni model, explicitly taking Eq. (6.71) and surfactant distributions at the interface into account, gave even a better fit to force curves than the phenomenologically no-slip boundary condition [723]. The same was theoretically found for colliding drops [724]. AFM measurements of hydrodynamic forces between deformable decane and tetradecane drops stabilized by surfactants in water [270, 727] and also between oil drops and solid particles [696] indicate that an effective no-slip hydrodynamic boundary condition applies at the oil–water interface in the presence of surfactants. An exception are results obtained with aqueous films between a deformable mercury drop and an approaching and receding mica plate as described above. Direct observations [698, 700] and modeling [697, 728] indicate the applicability of the no-slip boundary condition at the water–mercury interface even in the absence of surfactants.

6.6
Summary

- The flow of a liquid is described by the Navier–Stokes and the continuity equations.
- For Reynolds numbers $Re = \varrho \bar{v} d / \eta$ below unity, usually laminar flow prevails. For $Re \gg 1$, we expect turbulent flow. As a result, in the microscopic regime liquids usually flow laminar rather than turbulent.
- The hydrodynamic drag force of a sphere approaching a planar surface and assuming no-slip boundary conditions is $6\pi\eta v_\perp R_p(1 + R_p/D)$ (Eq. (6.41)).
- For particles moving parallel to a wall, one has to take into account a possible rotation. The hydrodynamic drag increases with decreasing distance.
- One of the unsolved questions is still the hydrodynamic boundary condition on solid surfaces. It is not clear to which extent slip occurs and which conditions and parameters it depends on. For fluid interfaces, the tangential velocities change continuously and the tangential stresses are balanced. Trace amounts of surfactants lead to an effective no-slip boundary condition due to a Marangoni effect.

6.7
Exercises

6.1. A sphere of radius R_p is falling in a liquid toward a horizontal planar surface. Derive an equation for the time Δt it needs to fall from a height D_1 to a height D_2. The density difference between sphere and liquid is $\Delta \varrho$. Assume that the sphere is already close to the surface so that you can apply Eq. (6.37) for the hydrodynamic force.

6.2. A glass sphere of density $2500\,\text{kg}\,\text{m}^{-3}$ and radius $R_p = 15\,\mu\text{m}$ is falling in tetradecane (density $760\,\text{kg}\,\text{m}^{-3}$, $\eta = 0.00213\,\text{Pa s}$) from a large height toward a planar surface. Plot distance versus time and velocity versus distance for the last $30\,\mu\text{m}$ of the fall. Take the distance of closest approach to be $0.2\,\text{nm}$.

6.3. Calculate the shear rate $\partial v_x / \partial z$ in the fluid between a flat surface and an approaching sphere versus the radial distance r at the surface of the plane. Therefore, choose the radial distance to be along the x-axis and calculate $\partial v_x / \partial z$ at $z = 0$. Derive an equation for the maximal shear. Plot the shear rate versus radial distance for a sphere of $10\,\mu\text{m}$ radius approaching a plane with a velocity of $1\,\mu\text{m/s}$ at distances of 10, 20, and $40\,\text{nm}$.

6.4. Calculate the diffusion coefficient for a sphere at a distance D from a planar wall for the component parallel and normal to the wall in a quiescent fluid. Assume creeping flow and no-slip boundary conditions. Plot both diffusion coefficients for a sphere of $50\,\text{nm}$ radius in water at $25\,^\circ\text{C}$ ($\eta = 0.89\times 10^{-3}$ Pa s) versus distance. The diffusion coefficient and the force are related by the generalized Stokes–Einstein equation $D_d = \mu_d k_B T$. Here, μ_d is the mobility of the particle defined as v_p/F, where F is the driving force causing a drift with velocity v_p.

7
Interfacial Forces Between Fluid Interfaces and Across Thin Films

7.1
Overview

This chapter deals with interfacial forces across fluid interfaces. Fluid interfaces are involved in many natural and technical processes, and a variety of methods have been developed to analyze them. One example are drops of liquid A interacting in a continuous phase of an immiscible liquid B (Figure 7.1a). Liquid A can for, example, be water, B might be oil or vice versa. This is an essential process to understand the structure and dynamics of emulsions [729]. At large distance, two drops will interact via hydrodynamic forces. For small drops at large distance $(D \ll R_1, R_2)$, this force is not much different from the force between two solid particles, we only have to consider the right boundary condition (see Section 6.5) [705, 730]. At closer distance, however, the drops will deform. The liquid of the continuous phase drains out of the closing gap. Depending on long-range interfacial forces, the viscosities of the two liquids, and the flow situation, the film of liquid A ruptures and the drops coalesce [691, 693], or a stable film of liquid B between A is formed. We call the film of one liquid between two other liquids an emulsion film.

The interaction between drops has been studied experimentally with different techniques [731]. The most important technique for studying interfacial forces across thin films is the thin film balance [699, 732]. In fact, a whole class of thin film balances have been developed. They are described in detail in Sections 7.4 and 7.6. In another technique, two drops at the end of two capillaries are moved toward each other [733] or one drop is moved toward a planar interface [734]. The process is observed by optical microscopy. Direct force measurements have become possible with the atomic force microscope by attaching tiny oil drops to AFM cantilevers [270, 696]. Recently, such drops have also been produced in microfluidic channels [735].

A rather similar process occurs with bubbles (Figure 7.1b). In beer or champagne, for example, small gas bubbles are produced. While rising to the top, they interact with other bubbles, some form an intervening liquid film – we call it foam

Surface and Interfacial Forces. Hans-Jürgen Butt and Michael Kappl
Copyright ©2010 Wiley-VCH Verlag GmbH & Co. KGaA
ISBN: 978-3-527-40849-8

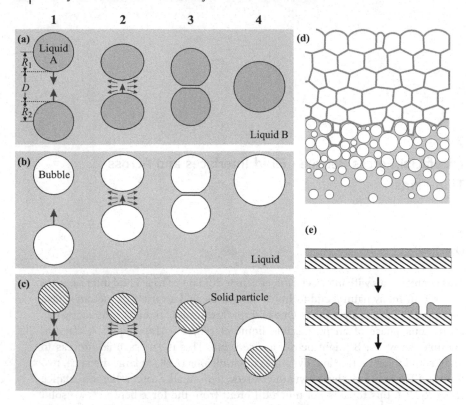

Figure 7.1 Different systems in which fluid interfaces interact. (a) Two drops of liquid A interacting in a continuous, immiscible liquid B. Different stages of the interaction are the approach (1), drainage of the intervening film (2), leading to a thin film (3), and rupture of the film with coalescence (4). (b) The interaction between a rising bubble and a fixed bubble in a liquid. (c) A particle interacting with a bubble (or a drop). Interacting fluid interfaces are also important for the structure and stability of foams (d) and for liquid films on solid surfaces (e) that can either form stable continuous films or rupture and disintegrate into small drops.

film – and some eventually coalesce [736]. At the top, they interact with the planar air–liquid interface. One might consider this as a special case of a small bubble interacting with an infinitely large one. Such a configuration allows to observe the process in detail by mounting a microscope and a camera on top and observe the approaching bubble with high time resolution [689]. As with drops, the interaction between two bubbles in a liquid has also been directly observed [91]. A difference between gas bubbles and liquid drops in an immiscible liquid is that gravity is often not important in emulsions whereas for gas bubbles gravity is essential. In addition, the viscosity of the gas in a bubble is negligible.

If surfactants or other additives are present, the bubbles might form a foam. The structure and stability of foams is another important example where surface forces play an essential role (Figure 7.1d) [729, 738]. Liquid drains out of the lamella into the plateau border and continues to flow to the bottom. This destabilizes the foam. On the other hand, repulsive surface forces between the two air–liquid interfaces stabilize the foam.

In addition to two fluid interfaces interacting with each other, the interaction of fluid interfaces with solid–liquid interfaces is important [686, 695]. One example is the interaction of particles with bubbles or drops (Figure 7.1c). The interaction of particles with bubbles in aqueous liquid is the key process in flotation [591]. The interaction with drops is essential in oil recovery. Direct particle bubble force measurements have been carried out with the AFM [739—741, 1216] (reviewed in Ref. [742]). Also, the force between individual particles and oil drops in water has been measured by atomic force microscopy [743, 744].

For the last stage of bubble–particle interaction, a liquid film is formed on the surface of the solid particle. Liquid films on solid surfaces have a much wider significance, for example, for coating of surfaces. Therefore, thin films on solid surfaces are explicitly considered in Section 7.6 (Figure 7.1e). The films either can be stable and wet the surface or can be meta- or unstable and eventually rupture and form individual drops on the solid surface.

Until now we did not mention the interaction between liquid drops in a gas. In principle, such drops interact like solid surfaces. In the absence of electrostatic charging, this interaction is dominated by van der Waals attraction. We just have to take into account a possible deformation of the surfaces. Therefore, we do not discuss it here. We would, however, like to mention one effect, which is typical for the interaction of fluid interfaces: very often, the systems are not in equilibrium and the interaction between fluid interfaces is influenced or even dominated by nonequilibrium effects [733, 745]. One is unique for drops of volatile liquids. If the liquid is not in a saturated atmosphere of its vapor, it will evaporate. The flow of the vapor emanating from the liquid surface can lead to an effective repulsive force. Such a repulsion was indeed noticed by Prokhorov, who measured the interaction between two water drops [746]. He observed a repulsion that increased with decreasing relative humidity.

It is beyond the scope of this book to describe the interactions between all the mentioned phenomena in detail. We can, however, distinguish different stages when drops, bubble, and particles interact [591, 592, 695]. For large distances, drops and bubbles interact via hydrodynamic forces much like solid particles. The only differences are their inertia and the respective hydrodynamic boundary conditions. At shorter distance, drainage of liquid films sets in. Depending on the special situation, liquid drains out between a solid and a fluid phase, between two liquids, or between two gas phases. Interfacial force acting between two interfaces is another central issue. The process of rupture of the intervening film is a third general topic.

Drainage, interstitial forces, and rupture of thin liquid films are the topic of this chapter. We start by introducing the concept of the disjoining pressure. Then, we describe the drainage of liquid in thin films. Here, we distinguish between vertical films, in which gravitation dictates the direction of the flow, and horizontal films. After introducing thin film balance as the main device to measure forces across liquid films, we discuss interstitial forces across foam and emulsion films. Finally, thin liquid films on solid supports are described.

7.2
The Disjoining Pressure

The disjoining pressure is a useful concept to describe surface forces in general. It is particularly suitable for thin liquid films. Therefore, we introduce it here.

We consider a medium between two planar, parallel surfaces separated by a distance x. The medium is supposed to be in contact with a large reservoir and we neglect edge effects. In the interfacial zones close to the surfaces, the properties of the medium can be different from its bulk properties. For example, surface charges change the ion concentration of water close to a charged surface. Van der Waals forces can impose a dielectric property different from the bulk.

When the two surfaces are far apart, the two interfacial zones do not overlap. Decreasing the separation x does not require work, unless energy is dissipated by viscous or friction forces. The situation changes when the separation becomes so small that the interfacial zones start to overlap. Now, the separation can be changed only by doing positive or negative work. This implies that the pressure in the interfacial layer differs from the pressure in the bulk reservoir. In 1936, Derjaguin called this additional pressure the "disjoining pressure" [747]. The disjoining pressure Π is equal to the difference between the pressure in a film between two surfaces (more precise, the normal component of the pressure tensor) and the pressure in the bulk reservoir provided the system is in equilibrium (Figure 7.2). It is defined as the change in Gibbs energy with distance and per unit area at constant cross-sectional area, temperature, and external pressure:

$$\Pi = -\frac{1}{A} \cdot \frac{\partial G}{\partial x}\Big|_{A,T,P.}$$ (7.1)

The sign of Π is positive for repulsive surface forces. It is negative for attractive surface forces.

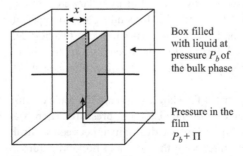

Box filled with liquid at pressure P_b of the bulk phase

Pressure in the film
$P_b + \Pi$

Figure 7.2 Schematic of the disjoining pressure between two parallel plates.

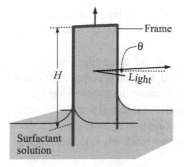

Figure 7.3 Schematic of how a liquid film can be formed by carefully lifting a frame from a surfactant solution.

7.3
Drainage

7.3.1
Vertical Foam Films

Pure liquid films are highly unstable. Surfactants are added to increase their stability. Traditionally, the main surfactants used were soaps. Soaps are salts of fatty acids, which contain at least eight carbon atoms. Therefore, such films are sometimes called soap films. The more general term is foam film. Still, one should keep in mind that liquid films are always metastable at best and will eventually rupture. Here, we focus on aqueous films.

Vertical foam films have a preferred direction due to gravitation. We focus on freestanding aqueous films since most of the experimental work has been done in that area to better understand the stability of foams.

Vertical foam films have been extensively studied and for a long time. Hooke[1] [748], Newton[2] [749], Fusinieri[3] [750], Dewar[4] [751], and Lawrence [752] studied the drainage and structure of vertical foam films (for a review on the history, see Ref. [718]). Vertical foam films can be produced by carefully lifting a frame vertically out of a surfactant solution (Figure 7.3). Practically, sometimes the liquid surface is lowered and the frame is kept stable. To measure the thickness of a liquid film, one can observe interference of light that is reflected from the film. The reflected light is composed of two small fractions of the original beam. One is reflected from the front side of the film (water to air). The other is reflected from the backside (air to water). Because of the opposite sign of the change in refractive index, the light coming from the backside is phase shifted by 180°. Constructive interference is obtained if the light

1) Robert Hooke, 1635–1703. English physicist, mathematician, biologist, and inventor.
2) Sir Isaac Newton, 1643–1727. English mathematician, physicist, and astronomer. Founder of mechanics and geometrical optics.
3) Ambrogio Fusinieri, 1775–1852. Italian scientist.
4) James Dewar, 1842–1923. Scottish physicist and chemist.

path in the liquid film is 1/2, 3/2, 5/2, ... times the wavelength of light λ. By analyzing the intensity of the reflected light, the thickness can be obtained (see Section 7.4).

One case, that of a black film, is particularly important. Soap films, which are thinner than $\approx 30\,\text{nm}$ appear black [718, 749, 750, 752]. Newton was the first to have reported on black films. Very thin films appear black because the light reflected from the backside is phase shifted by $\lambda/2$ and the path length in the film is negligibly small. For this reason, the phase difference between the two rays is $\approx \lambda/2$ and they interfere destructively. No light is reflected and the film appears black when viewed against a black background.

Two types of black films can be distinguished [754, 755]: common black films, which are stabilized by electrostatic double-layer repulsion, and Newton black films, which are stabilized by short-range forces. Common black films are 6–30 nm thick. Newton black films basically consist of a bilayer of surfactant. They have a defined thickness of 4–5 nm. For lipids, it is slightly larger [756]. If we consider a Newton black film formed from an oil phase in water, we are left with a lipid bilayer that still contains some oil molecules. Such films are extremely good electric insulators and are used in electrophysiological studies of membrane proteins [757].

Even cursory observation of the drainage of vertical films formed from a variety of solutions shows that one can distinguish between slow and fast draining films [718, 758]. Slowly draining films, also called rigid films, are formed when the surfactant causes an effective no-slip boundary condition rather than the mobile boundary condition. Rigid films are, for example, observed with water-containing sodium dodecyl sulfate below the CMC and some added 1-dodecanol [718]. In fast-draining films, turbulences are visible, in particular at the borders [718, 751, 799]. In both types of films, areas of black films can be observed.

To calculate the drainage of rigid thin films, we assume creeping flow. We start with Eq. (6.10). Let z be the vertical coordinate starting at the level of the liquid reservoir. The y-direction is horizontal and parallel to the film surfaces and x is normal to its surface (Figure 7.4). The zero point of x is supposed to be in the center of the film. In y-direction, the film is infinitely extended. Practically, this is realized by using a cylindrical film with a radius much larger than the capillary length. A cylinder has no borders at the side. Let H denote the total height of the film. The hydrostatic pressure in the film is $P = \varrho g(H-z)$. It leads to $\partial P/\partial z = -\varrho g$. We only have to consider the last line in Eq. (6.10) and can neglect flow components in x- and y-directions. Due to symmetry, v_z can not depend on y and $\partial^2 v_z/\partial y^2 = 0$. Also, the

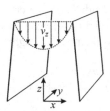

Figure 7.4 Schematic cross section of a draining liquid.

change in v_z in x-direction is much stronger than in z-direction and we neglect $\partial^2 v_z/\partial z^2$ compared to $\partial^2 v_z/\partial x^2$. We are left with

$$\eta \frac{\partial^2 v_z}{\partial x^2} = -\varrho g. \tag{7.2}$$

Integration in x leads to

$$\eta \frac{\partial v_z}{\partial x} = -\varrho g x. \tag{7.3}$$

We used $\partial v_z/\partial x = 0$ at $x = 0$ due to symmetry. Equation (7.3) tells us that in steady state the weight of the interior liquid is balanced by viscous forces acting across the lamellae. We integrate a second time:

$$\eta v_z = -\frac{\varrho g x^2}{2} + C. \tag{7.4}$$

The constant C is determined by the boundary conditions. Now comes a decisive step: we assume effectively no-slip boundary condition. As we had seen in Section 6.5, in the presence of surfactants the no-slip boundary condition usually describes hydrodynamic flow of fluid boundaries correctly. No slip means $v_z = 0$ at $x = \pm h/2$. Here, $h(z)$ is the film thickness at a vertical position z. With the no-slip boundary condition, we get

$$v_z = \frac{\varrho g}{2\eta} \left[\left(\frac{h}{2} \right)^2 - x^2 \right]. \tag{7.5}$$

Please note that we count v_z as positive if it is directed downward. Equation (7.5) tells us that the flow profile is parabolic.

In the next step, we calculate the total flow through a horizontal cross section at a vertical position z. It can be obtained by integrating the velocity:

$$Q = \int_{-h/2}^{h/2} v_z \, dx = \frac{\varrho g h^3}{12\eta}. \tag{7.6}$$

In fact, Q is the flow (transported volume per second) per unit length in y-direction. The viscosity strongly influences the flow. The higher the viscosity, the slower the films drain. Since the film thickness h depends on z, the flow also changes with the vertical position. Let us consider a certain segment of the film at height z, of thickness h, width in y-direction l and small vertical extension Δz. If the flow at the top side of that segment is not equal to the flow at the bottom side, the segment will shrink or expand in size, depending on whether the influx is higher or lower than the outflux. Any increase in the segment can be realized only by an increase in h. In steady state and assuming that the liquid is incompressible and no liquid is lost due to evaporation, the change of the flux is related to the change in film thickness:

$$\frac{\partial h}{\partial t} = \frac{\partial Q}{\partial z}. \tag{7.7}$$

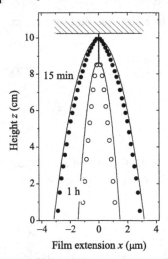

Figure 7.5 Drainage of a slowly draining water film stabilized by surfactant. The film is initially slightly higher than 10 cm. After 15 min, the film has drained to a thickness of 5.8 μm at its bottom. The experimental points (●) show the successive interference fringes obtained with red light [718]. With increasing height z, the film thickness decreases. The top 1–2 mm of the film has thinned to a black film. It is plotted as a vertical line. In addition to the experimental results, the film thickness calculated with Eq. (7.9) is plotted. After 60 min, the film has further drained and the Newton black film has extended to $z = 8.6$ cm. Experimental results (○) show a slightly thinner film than the one calculated with Eq. (7.9). This small difference can be due to various factors such as residual evaporation or imperfections in the film.

Inserting Eq. (7.6) leads to

$$\frac{\partial h}{\partial t} = \frac{\varrho g h^2}{4\eta} \frac{\partial h}{\partial z}. \tag{7.8}$$

If we further assume that the film is not supplied by liquid from the top, Eq. (7.8) is solved by

$$\frac{H-z}{t} = \frac{\varrho g h^2}{4\eta}. \tag{7.9}$$

Starting with an infinitely extending film (h large for $0 < z < H$), the film soon develops a parabolic flow profile. With time, the width of the parabola decreases. One example of the drainage of an aqueous film is shown in Figure 7.5.

7.3.2
Horizontal Foam Films

Let us consider a horizontal foam film between two bubbles or an emulsion film between two drops of another immiscible liquid. We neglect gravitation and assume an axisymmetric symmetry. When two bubbles or drops approach each other, the liquid in between is squeezed out. Thus, drainage and hydrodynamic forces are intimately related. The force between the two bubbles (or drops) influences the rate of drainage and drainage limits their further approach. As a first try, we can take the film to be plane parallel. Using this assumption, drainage can be calculated based on Reynolds equation (6.42). In the following, we describe the equation for the interaction between two bubbles. The results can be immediately applied to drops by replacing the surface tension of the liquid, γ_L, by the interfacial tension between two immiscible liquids, γ_{AB}.

For fluid interfaces, it is appropriate to replace the force in Eq. (6.42) by the pressure $\Delta P = \pi r_f^2$:

$$-\frac{dh}{dt} = \frac{2h^3}{3\eta r_f^2}\Delta P. \tag{7.10}$$

Here, r_f is the radius of the film. We replaced the distance D, conveniently used to describe the separation between rigid objects, with the film thickness h. The pressure $\Delta P = P_c - \Pi$ contains two contributions: the capillary pressure, $P_c = 2\gamma_L/r_b$, caused by the curvature of the bubble and the disjoining pressure caused by interfacial forces, $\Pi(h)$.

For a bubble interacting with a planar solid surface, Scheludko[5] replaced the no-slip boundary conditions used to derive Eq. (6.42) with a boundary condition for an ideal mobile surface. As a result, he obtained a higher drainage rate [686, 759]:

$$-\frac{dh}{dt} = \frac{16h^3}{3\eta r_f^2}\Delta P. \tag{7.11}$$

One might assume that for a foam film we should apply the mobile boundary condition to both surfaces. In that case, any film would immediately rupture. In fact, however, as discussed in Section 6.5 gradients in the surface concentration of surfactants usually lead to an effective no-slip boundary condition [714, 721, 724, 760].

Experiments, however, showed a systematic deviation from the predictions of Eqs. (7.10) and (7.11) [726, 753]. For example, the speed of thinning, $-dh/dt$, was observed to be proportional to $1/r_f^{0.8}$ rather than $1/r_f^2$. In addition, usually a faster drainage than predicted by Eqs. (7.10) and (7.11) is observed. There are several reasons for these discrepancies:

- Films are usually not plane parallel, but dimpling occurs. When a dimple forms, most of the resistance to flow is in the rim barrier ring. The effect of dimpling on the drainage has been calculated by various authors [703, 761, 762]. The presence of a repulsive disjoining pressure can largely increase the flow because it keeps the ring open.
- As already mentioned, gradients in the radial distribution of surfactants change the interaction [693, 720, 725, 760, 763, 764]. In the simplest case, this effect changes the hydrodynamic boundary condition from mobile to effectively no slip.

---

5) Alexei Scheludko, 1920–1995. Bulgarian physicist, professor in Sofia.

- Surface tension-driven flow can lead to hydrodynamic instabilities. Such instabilities in the films and in the rim of a dimple lead to the formation of heterogeneous, channel-like structures that facilitate drainage [765]. Nonhomogeneous thickness distributions have indeed been observed and can explain an increased drainage [766].

Empirically, it turned out that for films with $h < 100\,\text{nm}$ film thinning follows an exponential drainage law [766, 767]:

$$h = h_0 \cdot e^{-\alpha t}. \tag{7.12}$$

Here, h_0 is some initial thickness from which the time is started to count and α is an experimentally determined drainage coefficient. Since most of the drainage time is required for the last $100\,\text{nm}$, Eq. (7.12) is often used to compare drainage rates obtained from different experiments.

7.4
Thin Film Balance

The direct measurement of the disjoining pressure as a function of film thickness can be achieved by the thin film balance introduced by Derjaguin *et al.* [91] and Scheludko and Exerowa [768]. The principle of this so-called "Scheludko cell" is shown in Figure 7.6. The film is formed within a ring with 2–4 mm diameter. A small hole at the side of the ring allows to draw liquid from the film and thus adjust and measure the capillary pressure applied. The film thickness is deduced via optical interferometry. Almost exclusively aqueous solutions are studied. Originally, the thin film balance was constructed to measure the force across foam films. Modified versions are also used to measure forces across emulsion films [732, 734].

This early design had two main restrictions. The first is the limited range of capillary pressures that could be applied. The maximum value of capillary pressure is given by the entrance pressure of the small hole. It has the value $2\gamma_L/r_c$, where r_c is the radius of the supplying capillary. For capillary pressures higher than this value, air would start to enter the hole. The second limitation is the uneven drainage of the film during thinning.

To overcome these limitations, the porous plate technique, also called "Mysels cell", was introduced by Mysels [446, 769]. It was further improved by Exerowa and

Scheludko cell   Mysels cell

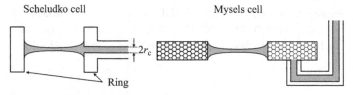

Figure 7.6 Schematic of thin film balance cells.

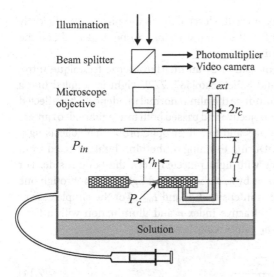

Figure 7.7 Schematic of a thin film balance. The gas pressure P_{in} inside the sealed chamber can be adjusted via the syringe relative to the external reference pressure P_{ext} to enforce thinning of the film and balance the disjoining pressure. Film thickness is monitored via microinterferometry and film drainage can be observed by video microscopy.

Scheludko [770] (Figure 7.6). The thin film is formed inside a hole that was drilled into a glass frit. A glass capillary tube is welded to this frit to measure the capillary pressure. In such a configuration, the maximum pressure is still given by the same equation, but the radius r_c has to be taken as the maximum pore size of the glass frit, which can be made much smaller than the drilled hole. Furthermore, drainage will now occur radially in all directions.

A complete setup of a thin film balance is shown in Figure 7.7. The holder with the film is placed inside a sealed chamber, which contains the same solution as the film for equilibration at the bottom. The holder has been saturated with the solution of interest by immersing it before mounting for several hours. The free end of the capillary tube is exposed to a reference pressure, which in the simplest case is just ambient pressure. The capillary pressure in the film spanning the hole can be controlled by varying the pressure inside the sealed chamber. A simple way to do this with high precision is to connect to the chamber a syringe pump that is driven by a micrometer screw and measure the chamber pressure with a pressure transducer.

In equilibrium, the disjoining pressure Π in the flat part of the film is equal to the capillary pressure P_c at the rim [771]: $\Pi = P_c$. Balancing all forces for the setup and assuming complete wetting of the glass frit by the solution, one obtains

$$\Pi = P_{in} - P_{ext} + \frac{2\gamma_L \cos\Theta}{r_c} - \varrho g H. \tag{7.13}$$

Here, P_{ext} is the external reference pressure, r_c is the radius of the glass capillary, Θ is the contact angle formed between the liquid and the wall of the capillary, and H is the height of the liquid in the capillary above the level of the liquid film. For sufficiently high applied pressures ($P_{in} > 1\,kPa$), the two last terms can be neglected and the pressure difference $P_{in} - P_{ext}$ dominates.

The film thickness is usually measured by an interferometric technique introduced by Derjaguin, Kussakov, and Scheludko [687, 772]. Light is coupled into a reflection optical microscope to illuminate the film at normal incidence. The reflected light is collected by the microscope objective and passed both to a digital video camera to allow observation of film dynamics and to a fiber optic probe that collects light from the central part of the film. During thinning of the film, light reflected from the top side of the film interferes with light reflected from the bottom side. For each change of optical film thickness by $\lambda/4$, the reflected light will go through one minimum and maximum reflection intensity, I_{min} and I_{max}. For the simple case of a homogeneous water film with refractive index n and illumination with a fixed wavelength λ, the film thickness h_L can be calculated as

$$h_L = \frac{\lambda}{2\pi n}\arcsin\sqrt{\frac{\Delta}{1 + \frac{4R}{(1-R)^2}(1-\Delta)}},\qquad(7.14)$$

with

$$\Delta = \frac{I - I_{min}}{I_{max} - I_{min}} \quad \text{and} \quad R = \frac{(n-1)^2}{(n+1)^2}.$$

Here, I is the observed reflected intensity at a certain film thickness. By measuring the maximum and minimum intensities during thinning and correctly counting the number of multiples of $\lambda/4$ still left – the final thin film of interest will be less than $\lambda/4$ thick – one obtains the film thickness.

For real systems, we do not have a homogeneous water film but surfactant layers at both interfaces. These layers can be accounted for by correcting h with a three-layer model derived by Duyvis [773]. The total film thickness is calculated from

$$h = h_L - 2\left(\frac{n_s^2 - n^2}{n^2 - 1}\right)x_s.\qquad(7.15)$$

Here, n_s and x_s are refractive index and thickness of one molecular layer of the surfactant, respectively. These values depend on the adsorption of the surfactant to the interface and may lead to a slight uncertainty in the determination of h. Other approaches to determine film thickness were X-ray reflectometry [775] and ellipsometry [774]. The latter is applicable to the study of wetting films on surfaces, where the surface can be attached to the glass frit from below.

The whole setup is usually enclosed in a temperature-controlled chamber and mounted on a vibration isolation system. Special care must be taken when attempting to apply small pressures to the system. Bergeron and Radke described an optimized setup [779] that was used to measure the stepwise thinning of multilayered surfactant films down to pressures of 10 Pa. The reference pressure P_{in} was set by connecting a 1 m^3 stainless steel chamber with saturated humidity to the capillary to avoid fluctuations. The minimal pressure that can be applied is determined by the capillary pressure in the meniscus P_c. It is fixed by the dimensions of the hole used to hold the film. For hole dimension where the diameter is smaller than the frit thickness at the inner rim and for zero contact angle, the minimum capillary pressure is determined by the hole radius r_h being equal to $P_c = 2\gamma_L/r_h$. However, if the glass frit thickness gets smaller than the hole radius, the film cannot form a zero contact angle at the rim, which leads to a larger radius of the meniscus and thus to a smaller value of P_c. Bergeron and Radke used a strong tapering of the rim to achieve a frit thickness as small as 0.1 mm while keeping hole diameter at 2 mm.

One disadvantage of the Mysels cell with the bulky glass frit is its large surface area within the that can give rise to unwanted desorption of surface-bound contamination. To avoid frit contamination, a new holder is therefore used for each experiment with a different surfactant. For the study of substances that are available only in very small quantities, this can be problematic, since the whole frit has to be loaded with the substance. In case of very high molecular weight substance, diffusion time through the frit may also become an issue. Therefore, Velev *et al.* [776] have fabricated a Scheludko cell with a capillary diameter of 280 μm and a cross section of 50×90 μm^2 using laser drilling. While extending the range of accessible pressures and allowing much faster film thinning compared to classical Scheludko cells, film drainage is still inhomogeneous. A recent design called bike-wheel microcell [777] combines miniaturization of a Scheludko cell with the use of 24 radially distributed access channels that facilitate uniform film drainage.

7.5
Interfacial Forces Across Foam and Emulsion Films

7.5.1
Shape of a Liquid Film

After drainage, foam and emulsion films can remain in a metastable state if the two interfaces are kept from getting into contact by repulsive forces. What is the shape of such a film? For the interaction between two bubbles, the shape of the bubble far away from the film is that of a sphere of radius r_b. Inside the bubble, the pressure is higher with respect to the outside pressure by $\Delta P = 2\gamma/r_b$ (Eq. (5.2)). In the region far away from the contact region, this is balanced by the curvature of the liquid–air interface and the capillary pressure $P_c = 2\gamma/r_b$. In the region of the thin film, the liquid surface is planar. The capillary pressure is zero. The internal pressure is balanced by surface forces according to $\Pi = \Delta P$. What about the transition region? Since a sharp edge in the liquid interface is not allowed (it would lead to a high curvature and thus high capillary pressure), there must be a gradual transition. In this transition region, the curvature of the liquid surface gradually increases from zero to $2/r_b$.

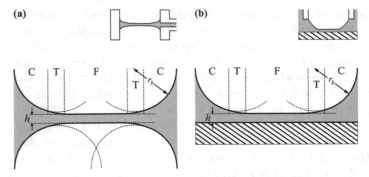

Figure 7.8 Schematic of a foam film (a) and a film on a solid surface (b). The region of the meniscus (C), the transition regions (T), and the planar films (F) are indicated.

The situation of the thin film balance is shown in Figure 7.8. We denote the radius of the meniscus by r_b as a reminiscence on the bubble–bubble interaction. The general equation describing the shape of the interfaces is a generalized Laplace equation [759]:

$$\gamma_L \left(\frac{1}{r_1} + \frac{1}{r_2} \right) = P_c - \Pi(h). \tag{7.16}$$

The left term is the local capillary pressure caused by the local curvature $(r_1^{-1} + r_2^{-1})$. This expression describes the actual shape of the fluid interfaces. P_c is the capillary pressure caused by the curvature of the meniscus far away from the film. It is given by the Laplace equation and determined by the curvature of the meniscus (indicated by r_b). In the meniscus far away from the film where h is so large that Π is zero, the curvature is equal to the curvature of the meniscus. In the transition zone, interfacial forces come into play and $\Pi \neq 0$. Since P_c is constant, the local curvature decreases below that in the meniscus. In the film region, the local curvature is zero, $(r_1^{-1} + r_2^{-1}) = 0$, and the inner pressure of the gas phase is fully balanced by the disjoining pressure in the film.

7.5.2
Quasiequilibrium

Different components contribute to the force across a foam or emulsion film. We superimpose all contributions:

$$\Pi = \Pi_{vdW} + \Pi_e + \Pi_{st} + \cdots . \tag{7.17}$$

Here, Π_{vdW} is the van der Waals force, Π_e is the electrostatic double-layer force, and Π_{st} contains steric and short-range structural forces. In addition, other forces might contribute.

Van der Waals forces across liquid films are always attractive because the two media interacting across the liquid film are identical. This is different from thin liquid films on solid surfaces, where van der Waals forces can be attractive or repulsive. For short distances, the van der Waals attraction is described by

$$\Pi_{vdW} = -\frac{A_H}{6\pi h^3}. \tag{7.18}$$

In the range from 10 to 100 nm, retardation effects become relevant and the distance dependency changes to $\Pi_{vdW} \propto h^{-4}$ [445].

Van der Waals attraction destabilize liquid foam films. As a result, if we do not take care to add a repulsive force component, liquid films become highly unstable and rupture immediately. The first and most common approach is to add charged surfactants such as sodium dodecyl sulfate (SDS). With dissociated surfactants present, the air–water interface changes. This leads to an electrostatic double-layer repulsion between the two interfaces. Since the system is symmetric with respect to the two interfaces, electrostatic forces are repulsive. The double-layer repulsion for low surface potentials and thick films ($h > \lambda_D$) can be expressed as (Eq. (4.57))

$$\Pi_e(h) = \frac{2\varepsilon\varepsilon_0}{\lambda_D^2} \cdot \psi_0^2 \cdot e^{-h/\lambda_D}. \tag{7.19}$$

For high surface potentials and thinner films, we have to specify the electrostatic boundary condition. For constant surface potential, we apply Eq. (4.60). For constant surface charge, Eq. (4.61) is suitable.

That electrostatic double-layer forces play a dominant role in the stabilization of soap films can be demonstrated by measuring the change in film thickness with a variation of the salt concentration. With increasing ionic strength, the Debye length decreases and we expect the film to shrink in thickness. Such experiments were

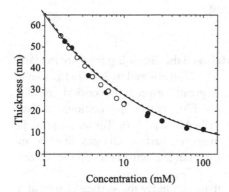

Figure 7.9 Thickness of two aqueous foam films versus the concentration of added salt. In one classical experiment (○), carried out by Scheludko and Exerowa [444], a thin film balance was used to study an aqueous solution containing saponin plus different concentrations of KCl. In the second experiment (●), a vertical film of SDS solution at various concentrations of LiCl was analyzed [445]. The curves were calculated with Eq. (7.20) with $\psi_0 = 65$ mV and $P_{ext} = 73$ Pa (continuous line) and $\psi_0 = 40$ mV and $P_{ext} = 25$ Pa (dashed).

indeed carried out [446, 780–782]. The results belong to the most powerful verifications of DLVO theory. Two examples are shown in Figure 7.9. In one case, a thin film balance was used to study a water film stabilized by saponin at different concentrations of the background salt KCl [444]; saponin is a commercial mixture of surfactants. An external pressure of 73 Pa was applied. In the other case, a vertical rectangular frame was raised slowly from a submerged position in an aqueous SDS solution [445]. SDS was used at a low concentration ($\ll$CMC). LiCl was added as a salt. The thickness of the film was measured interferometrically at a height corresponding to an hydrostatic pressure of 25 Pa.

For a rough estimation of the film thickness, we can assume that the film thickness is determined only by the electrostatic double-layer repulsion. We apply Eq. (7.19) and set the disjoining pressure equal to the external or hydrostatic pressure: $P_{\text{ext}} = \Pi_{e}(h)$. This leads to a film thickness of

$$h = -\lambda_D \cdot \ln\left(\frac{P_{\text{ext}}\lambda_D^2}{2\varepsilon\varepsilon_0\psi_0^2}\right). \tag{7.20}$$

The film thickness of foam films can usually be well described by Eq. (7.20) (reviewed in Ref. [783]). In particular for low salt concentration and thick films, the agreement is quantitative. Figure 7.9, however, also shows that for film thicknesses at or below 30–50 nm, films are systematically thinner than predicted. One reason is the presence of van der Waals forces. Van der Waals attraction tends to decrease the film thickness.

When dealing with foam films, the surface charges play an essential role. Foam films are stable, at least some time, only if surfactants are present. For ionic surfactants, the charge comes usually from the dissociation of groups, such as with SDS

$$C_{12}H_{25}NaO_4S \rightarrow C_{12}H_{25}O_4S^- + Na^+$$

or cetyltrimethylammonium bromide (CTAB)

$$C_{19}H_{42}BrN \rightarrow C_{19}H_{42}N^+ + Br^-$$

DLVO theory is usually sufficient to understand the disjoining pressure in foam films stabilized by ionic surfactants [446, 780, 782] (reviewed in Ref. [783]) as long as the surfactant concentration does not greatly exceed the critical micelle concentration (CMC). Also, the thickness of films containing nonionic surfactants [781, 784–786] can usually be well described by Eq. (7.20). The assumption is that even nonionic surfactants lead to negative surface charges due to an enrichment of OH^- ions [781, 785]. Thus, long-range electrostatic force stabilizes the film.

In this context, it is interesting to note that even today the surface charge of a surfactant-free air–water or water–oil interface is not clear. In the presence of simple inorganic salts, usually the surfaces are negatively charged due to the adsorption of anions. Anions adsorb preferentially since they have a higher polarizability than the smaller cations. In the absence of salts, classical experiments show that bubbles or oil drops tend to be negatively charged in water, presumably by an enrichment of OH^-

ions (overview in Ref. [787]). In contrast, simulations (overviews in Refs. [788, 789]) and resonant second harmonic generation experiments [790] indicate an enrichment of H_3O^+ ions.

If the surfactant concentration is much higher than the CMC, a stepwise thinning of foam films is observed (for a history, see Refs. [752, 779]). This stratification is caused by individual layers of micelles being expelled from the thinning film. In equilibrium, an oscillatory force is observed [779]. Such structural forces will be discussed in Section 11.6.

In many applications involving foams, a polymer is added. Polymers can affect both the drainage and the interfacial forces. This often has a dramatic effect on the stability of foams. One effect is that polymer slows down drainage by increasing the viscosity of the liquid solution (see Eqs. (7.9)(7.11)). If polymers are surface active, they can change the rheological properties of the surface. If in addition surfactants are present, the polymer might interact with the surfactant. Therefore, the picture soon becomes complex since we are dealing with many parameters such as the chemical nature of the polymer and surfactant, the size of the polymer, and the charge of the polymer and surfactant – both depend on the concentration of added salts and pH. With polyelectrolytes, a layered structure and stratification has been observed [783, 794–796]. It is beyond the scope of this book to discuss the influence of polymers in detail.

As one example, Figure 7.10 shows a foam film with added polyelectrolyte [796]. A polyelectrolyte is a polymer that in water is charged due to dissociation of specific groups. To illustrate the stepwise thinning, photographs of the film at different thicknesses are presented. The circular area in the middle of each photo is the plane parallel part of the film surrounded by the plateau border. The intensity reflected from the film is homogeneous and bright before the first jump (Figure 7.10a). The film presents an interferometer as described in Section 7.4. In the present case, the reflected intensity decreases with decreasing film thickness. During the transition, the new film thickness spreads out in the form of darker spots (Figure 7.10b) that unify to a new homogeneous film thickness. When the external pressure is increased further, jumps occur (Figure 7.10c and d). The spots around the newly formed black film (Figure 7.10d) are drops and are thicker than the rest of the film. They are caused by hydrodynamic instabilities. The rate of extension of the block film is faster than the rate at which the liquid can be pressed out of the film.

7.5.3
Rupture

Though foam and emulsion films might exist for a long time, on some timescale they will collapse. The rupture of foam and emulsion films has been studied by various methods both experimentally [797] and theoretically [798]. It is obvious that the stability of foam films is influenced by surface forces. For example, in 1924 Bartsch reported that electrolytes decrease the life time of certain foams [799], presumably by decreasing electrostatic stabilization. Surface forces alone, however, do not determine the life time of a soap film.

(a)

(b)

(c)

(d)

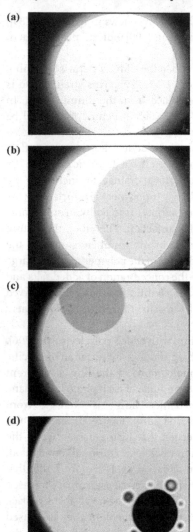

Figure 7.10 Photo of an aqueous foam film in a thin film balance with a hole diameter of 1 mm. The film was stabilized with a mixture of commercial alkylpolyglycosides (APG) with alkyl chain lengths of 12 and 14 carbon atoms at a concentration four times lower than the CMC. In addition, a positively charged polymer poly (diallyl dimethyl ammonium chloride) (DADMAC) at a concentration of 8 mM (with respect to the monomer concentration) was added. The polymer forms a layered structure in the film, which leads to stratification when the film is thin. Thinning is induced by increasing the external pressure. (The picture was provided by R. von Klitzing [796].)

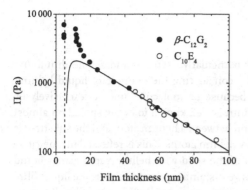

Figure 7.11 Disjoining pressure versus film thickness of thin aqueous foam films stabilized by β-C$_{12}$G$_2$ and C$_{10}$E$_4$. The solutions contain 0.1 mM NaCl. The data are fitted with the DLVO theory from which the surface charge density is calculated to be σ = 1.1 mC m^{-2}. The continuous line was calculated using Eqs. (7.17)(7.19) with λ$_D$ = 28 nm and A$_H$ = 10^{-20} J. The vertical dashed line indicates the Newton black film of ≈ 5 nm thickness.

■ **Example 7.1**

That factors other than surface forces influence the life time of foam films was demonstrated by Stubenrauch *et al.* [800]. They measured disjoining pressure versus film thickness and the rupture pressures of foam films stabilized by two nonionic surfactants, namely, *n*-dodecyl-β- D-maltoside (β-C$_{12}$G$_2$) and tetraethyleneglycol-monodecyl ether (C$_{10}$E$_4$). In Figure 7.11, the Π(h) curves of an aqueous solution with 68.5 μM β-C$_{12}$G$_2$ and a 110 μM solution C$_{10}$E$_4$ solution are shown. Both concentrations are below the respective CMCs, which are 0.86 mM for C$_{10}$E$_4$ and 0.17 mM for β-C$_{12}$G$_2$. With respect to surface forces, no significant difference is detected between the two curves. In both cases, the surface charge density was calculated to be 1.1 × 10^{-3} C m^{-2}. However, the stabilities of the corresponding foam film are very different. While the β-C$_{12}$G$_2$ film is stable up to 9000 Pa, the C$_{10}$E$_4$ film ruptures at pressures around 800 Pa. The β-C$_{12}$G$_2$ film is more stable than the corresponding C$_{10}$E$_4$ film, although both systems have the same surface charge density.

Several experiments indicate that the life time of a foam film is correlated with the surface elasticity [738, 786, 800]. One explanation is that high surface elasticities dampen fluctuations in the film [786, 791]. Fluctuations are one possible reason for film rupture. For the same reason, surface viscosity influences the stability of films [792, 800]. In particular for large surface-active molecules such as proteins, this has been analyzed for emulsion films due to the importance in food science [721, 793]. The rupture of thin films has been extensively studied for liquid films on solid surfaces. Therefore, we describe it in more detail in Section 7.6.3.

7.6
Thin Wetting Films

The concept of disjoining pressure is particularly useful to understand thin liquid films on solid surfaces. Let us start by considering the example of liquid helium. Liquid helium is particularly suited because its molecules interact exclusively by van der Waals interaction. Liquid helium is well known to avidly spread on almost any solid surface because the interaction between a helium atom and the substrate is usually much stronger than with fellow helium atoms. This is reflected in a negative Hamaker constant for the interaction of the solid with helium vapor across liquid helium, which corresponds to a repulsive disjoining pressure across the liquid film. If helium is placed in a beaker, it rapidly climbs up the walls. When the film climbs up a smooth wall, the gain in van der Waals energy is at the expense of gravitational energy. In equilibrium, the disjoining pressure at a given height is balanced by the hydrostatic pressure in the film. As a result, the thickness of the film h will decrease with height z according to

$$\Pi(h) = -\frac{A_H}{6\pi h^3} = \varrho g z \Rightarrow h = \left(-\frac{A_H}{6\pi\varrho g z}\right)^{1/3}. \tag{7.21}$$

By measuring the thickness of a liquid helium film at 1.38 K on smooth single-crystal SrF$_2$ and BaF$_2$ surfaces, Anderson and Sabisky [801] verified the theory of van der Waals forces (Figure 7.12). For films up to 10 nm thickness, the predicted

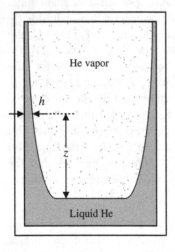
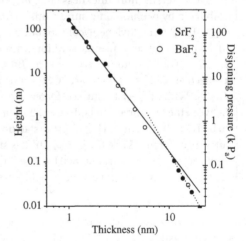

Figure 7.12 left: Schematic of liquid helium climbing up the wall of a container due to the repulsive van der Waals force between the solid–liquid and liquid–vapor interfaces. right: Film thickness of liquid helium at 1.35 K on single-crystal SrF$_2$ and BaF$_2$ surfaces (redrawn from Ref. [801]). For heights of up to 10 cm, the thickness was determined at saturated vapor pressure at different height. For thinner films, the vapor pressure was varied and converted to a corresponding height (see Exercise 7.1). On the right, the height scale is converted to a disjoining pressure by $\Pi = \varrho g z$ (Eq. (7.21)) with $\varrho = 125$ kg m^{-3}. The dashed line is a fit for $\Pi \propto h^{-4}$.

$\Pi \propto z \propto h^{-3}$ dependence was indeed observed. It is plotted as a continuous straight line in Figure 7.12. For thicker films, retardation leads to a steeper decrease in the disjoining pressure and $\Pi \propto z \propto h^{-4}$ for $h \geq 10$ nm.

There is, however, a practical limit. To obtain very thin films, z needs to be high. For example, with a Hamaker constant for SrF_2/liquid helium/vapor of -5.9×10^{-21} J and a density of liquid helium of $\varrho = 125$ kg m^{-3}, the film is expected to be 63 nm thick at a height of 1 mm. For $z = 1$ m, it is still 6.3 nm thick. To reach a thickness of 1 nm, the wall would need to be 255 m high. Not only that, this would require a unique laboratory building, and preparing clean and homogeneous surfaces of such dimensions would be demanding.

Fortunately, we have an alternative: we can change the film thickness by changing the vapor pressure. When we reduce the vapor pressure, the film shrinks and vice versa. To obtain a quantitative relationship between the relative vapor pressure, P/P_0, and the film thickness, we calculate the change in chemical potential in the vapor and in the liquid film. When the vapor pressure is reduced from saturation pressure P_0 to a pressure P, the chemical potential in the vapor phase changes by

$$\Delta\mu_v = RT \ln \frac{P}{P_0}. \tag{7.22}$$

Here, we assumed that the vapor behaves like an ideal gas. Please note that $\ln(P/P_0)$ is negative because $P/P_0 \leq 1$. Let us turn to the liquid film. At constant temperature, the chemical potential, that is, the Gibbs energy per mole, is the product of the molar volume and the change in pressure, $d\mu_l = V_m dP_{ext}$. Here, V_m is the molar volume of the liquid and P_{ext} is the external pressure compressing the liquid. In our case, the role of the external pressure is taken by the disjoining pressure and we replace dP_{ext} by $-\Pi(h)$. The negative sign takes into account that a negative film pressure corresponds to a compression of the liquid. The change in chemical potential of the liquid film with respect to the thick film at saturated vapor pressure is $\Delta\mu_l = -V_m[\Pi(h)-\Pi_{sat}]$. Here, Π_{sat} is the film pressure at saturated vapor pressure. For $P \to P_0$, we get condensation, the film thickness grows to infinity, $h \to \infty$, and the film pressure approaches zero ($\Pi_{sat} = 0$). Thus [759],

$$\Delta\mu_l = -V_m \Pi(h). \tag{7.23}$$

In equilibrium, the chemical potential of molecules in the vapor phase is equal to the chemical potential of molecules in the liquid film. Setting $\Delta\mu_v = \Delta\mu_l$, we get

$$\Pi(h) = -\frac{RT}{V_m} \ln \frac{P}{P_0} = -\frac{\varrho RT}{M_w} \ln \frac{P}{P_0}. \tag{7.24}$$

By varying the relative vapor pressure of the substance studied and measuring the film thickness, the disjoining pressure can be determined.

In this chapter, we described two methods to measure the disjoining pressure across thin liquid films on solid surfaces: the thickness versus height method and the thickness versus vapor pressure method. To illustrate that both methods are equivalent, we reconsider the situation shown in Figure 7.12. In equilibrium, the temperature in the closed vessel with a reservoir of liquid at the bottom is the same everywhere. Directly at the liquid surface, the vapor pressure is equal to P_0. If we go up in the vapor phase, the vapor pressure, however, decreases. According to the barometric distribution law, it decreases according to $P/P_0 = \exp(-M_w g h/RT)$. Inserting this expression into Eq. (7.24) leads to $\Pi(h) = \varrho g z$, which is identical to Eq. (7.21).

Although thickness versus height measurements are suitable for very thick films thickness versus vapor pressure measurements are also applicable to thinner films. h versus P/P_0 measurements on the other hand are difficult at high vapor pressures because adjusting precisely the vapor pressure above 99% is demanding and the error can be considerable. Thus, both methods complement each other. In fact, results for films below a film thickness of 10 nm in Figure 7.12 were determined by thickness versus P/P_0 measurements. One should also keep in mind that the whole concept of having a continuous film on a surface is not applicable for films of only molecular thickness.

Equation (7.24) is of fundamental importance in surface science. It relates interfacial forces with the adsorption of a vapor to a solid surface. The concept goes back to Michael Polanyi[6] and it is known as the potential theory of adsorption [802]. The basic idea is that vapor molecules close to a surface feel a potential–similar to the gravitation field of the earth. The potential isothermally compresses the gas close to the surface. Once the pressure becomes higher than the equilibrium vapor pressure, it condenses and forms a liquid film.

One cause for this potential is the van der Waals attraction. When inserting the van der Waals interaction into Eq. (7.24), we arrive at

$$h = \left[\frac{A_H M_w}{6\pi\varrho RT\ln(P/P_0)}\right]^{1/3}. \tag{7.25}$$

In a more general form, the equation is frequently referred to as the Frenkel–Halsey–Hill (FHH) equation [803–805]:

$$\left(\frac{h}{h_{mon}}\right)^r = \frac{C}{\ln(P/P_0)}. \tag{7.26}$$

Here, h_{mon} is the thickness of a monolayer and C and r are constants. Empirically, often values of $r = 2.5-3$ are found.

For $P/P_0 \geq 0.2$, Eqs. (7.25) and (7.26) fit adsorption isotherms of noble gases and small, nonpolar molecules such as nitrogen or short-chain alkanes reasonably well [806]. Two examples, helium adsorbing to single-crystal SrF_2 and BaF_2 at $T = 1.35$ K and hydrogen adsorbing to gold at 14 K, are shown in Figure 7.13. Typically, however, other effects such as surface roughness and thermal fluctuations have to be taken into account, in particular when analyzing adsorption isotherms recorded at different temperatures [807].

---

6) Michael Polanyi, 1891–1976. Hungarian physicist who worked in Berlin and Manchester. His son, John Charles Polanyi, received the Nobel Prize in chemistry in 1986.

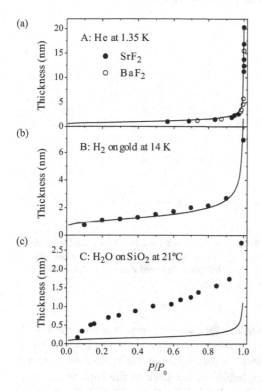

Figure 7.13 Adsorption isotherms for He adsorbing to SrF$_2$ and BaF$_2$ at 1.35 K (a), H$_2$ adsorbing to gold at 14 K (b), and H$_2$O adsorbing to SiO$_2$ at 21 °C (c). The experimental results were redrawn from Refs. [801, 807, 808]. Continuous lines were calculated with Eq. (7.25) with $A_H = -0.6 \times 10^{-20}$ J, $M_W = 0.004$ kgm^{-3}, $\varrho = 125$ kgm^{-3} for (a), $A_H = -14 \times 10^{-20}$ J, $M_W = 0.00201$ kgm^{-3}, $\varrho = 70.8$ kgm^{-3} for (b), and $A_H = -10^{-20}$ J, $M_W = 0.018$ kgm^{-3}, $\varrho = 1000$ kgm^{-3} for (c). In (a), the same results are plotted as in Figure 7.12.

For polar molecules or for large, nonspherical molecules, other interactions contribute significantly to the force. Water adsorbing to silicon oxide is one example (Figure 7.13c). The first up to three monolayers form an ice-like network that is dominated by hydrogen bonds and in which the rotation and translation is strongly reduced. Only at a relative vapor pressure above 30% does liquid water start to appear. A real liquid layer is present only for $P/P_0 \geq 60\%$ [808]. Van der Waals forces contribute only partially to the total disjoining pressure.

A third method to measure the disjoining pressure across thin films is schematically shown in Figure 7.14. It is sometimes called "modified thin film balance" and was developed and improved by Derjaguin and Kussakov [687], Platikanov [688], Princen and Mason [809], Schelduko *et al.* [810], Kitchener *et al.* [811], and others (reviewed in Ref. [812]). In a container filled with the liquid investigated, a bubble of air or nitrogen is pressed through a capillary onto the planar surfaces. Assuming that a stable film is formed and thus the solid–liquid and liquid–gas interfaces repel each other, the bubble is flattened as soon as it gets close to the solid surface. In the center,

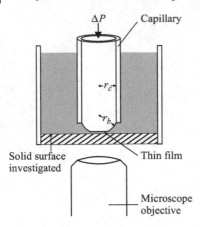

Figure 7.14 Schematic of a thin film balance. A capillary of radius r_c is moved close to the planar surface. Gas is pressed through the capillary with a pressure ΔP. Between the end of the capillary and the plane a bubble is formed with a radius $r_b = 2\gamma_L/\Delta P$. At the end, the bubble is flattened due to repulsive surface forces.

the disjoining pressure just compensates the Laplace pressure inside the bubble ($\Pi = \Delta P$). The film thickness is measured interferometrically with a microscope. In a typical experiment, the film thickness is measured versus the applied pressure ΔP. Since the radius of curvature of the bubble r_b is related to the Laplace pressure, $\Delta P = 2\gamma_L/r_b$, the maximal pressure is limited by the radius of the capillary $\Delta P \leq 2\gamma_L/r_c$. Typical values are $r_c = 0.2\text{–}2$ mm leading to a maximal pressure of 50–500 Pa.

In addition to measuring the equilibrium thickness of thin films, the method is widely used to analyze film stability and drainage [810, 811, 813]. In many practical applications, a system is far away from equilibrium and highly dynamic. One example is a flotation cell in which particles and bubbles are mixed. The attachment of a particle to a bubble is limited by the hydrodynamic interaction rather than equilibrium surface forces [695]. When a particle and a bubble approach each other, the liquid in between needs to have time to flow out of the closing gap [728]. This process of film drainage is also studied with the thin film balance.

7.6.1
Stability of Thin Films

The stability of thin liquid films on solid surfaces is a major topic both in fundamental and in applied science. In flotation, the aqueous film formed between a particle and an approaching bubble determines the interaction and thus the efficiency of the process [695, 753, 813, 814]. Polymer films preserve, isolate, or decorate materials [815–820]. Thin metal films in microelectronics serve as electrodes [821–823]. When such films dewet, a solid surface complex pattern is formed (Figure 7.15) [824].

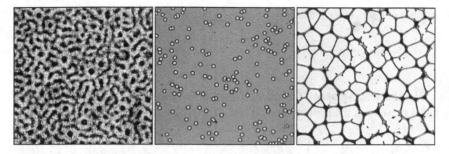

Figure 7.15 Morphologies of different polymer films after dewetting. Left: A poly (dimethyl siloxane) film under water on a silicon wafer with a layer of grafted identical molecules. Middle: Early stage of a dewetting polystyrene film on a silicon wafer. Right: Late stage of the same dewetting polystyrene film. Images are $200\,\mu m \times 200\,\mu m$. (Images provided by G. Reiter [817].)

Even this short list of examples – aqueous, polymer, and metal films – shows that we deal with very different situations. Polymer and metal films are usually liquid only above room temperature when the material is heated above the melting temperature. In addition, the polymer or metal is usually not in contact with a reservoir. The thermodynamic condition for equilibrium is therefore not a constant chemical potential but a constant amount of material. We now focus on the case that the liquid is practically nonvolatile and its amount is supposed to be constant.

Let us start with macroscopic films. In macroscopic films, which are thicker than $\approx 100\,nm$, interfacial forces between the solid–liquid and the liquid–gas interfaces are usually negligible. A liquid film is stable, if the free energy of creating a solid–liquid and a liquid–gas interface is less than the free energy of the bare solid surface: $\gamma_L + \gamma_{LS} \leq \gamma_S$. Here, γ_L, γ_{LS}, and γ_S are the surface tensions of the liquid–gas, the solid–liquid, and the solid–gas interfaces, respectively. If the spreading coefficient, defined by $S = \gamma_S - \gamma_L - \gamma_{LS}$, is positive, the film is stable.

For $S < 0$, we talk about partial wetting. The contact angle of such a liquid on the solid is finite and given by Young's equation (Figure 7.16). If we take a small amount of such a liquid and place it on the solid, it will form a drop that is shaped like a spherical cap. If we increase the amount of liquid, the drop will grow and change its shape from the ideal spherical cap to a more flattened cap. Eventually, it will form a pancake-like structure with a constant height and rounded rims. In equilibrium, that is, in coexistence of bare solid surface with liquid films of height

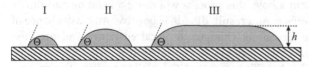

Figure 7.16 Schematic of a drop, which is so small that gravitation can be neglected (I), a drop, which is already slightly deformed by gravitation (II), and a continuous pancake-like film of height h.

h_0, an energy balance leads to

$$\gamma_S = \gamma_{SL} + \gamma_L - \frac{\varrho g h_0^2}{2}. \tag{7.27}$$

The last term represents the gravitational work required to form a film of thickness h_0. It represents the hydrostatic pressure. The height is

$$h_0^2 = \left(1 - \frac{\gamma_S - \gamma_{SL}}{\gamma_L}\right) \frac{2\gamma_L}{\varrho g}. \tag{7.28}$$

With the capillary constant $\varkappa_c = \sqrt{\gamma_L / \varrho g}$ (Eq. (5.7)) and applying Young's equation, we can write

$$h_0^2 = (1 - \cos\Theta) \frac{2}{\varkappa_c^2}. \tag{7.29}$$

With the geometric relation $1 - \cos\Theta = 2\sin^2(\Theta/2)$, we get [825]

$$h_0 = \frac{2}{\varkappa_c} \sin\left(\frac{\Theta}{2}\right). \tag{7.30}$$

A liquid film of actual thickness h is stable if $h \geq h_0$. If the film prepared is thinner, $h < h_0$, it will eventually form holes and dry patches until the patches have a thickness $h = h_0$ given by Eq. (7.30).

For films thinner than ≈ 100 nm, interfacial force between the solid–liquid and the liquid–gas interfaces might change the situation. Repulsive forces are required for a stable film. Attractive forces lead to unstable films. One might even say that film rupture is the manifestation of attractive interfacial forces. The condition for a stable film is [759]

$$\frac{d\Pi}{dh} < \frac{dP_c}{dh}. \tag{7.31}$$

Here, P_c is the local capillary pressure (Section 5.1). The left-hand side of Eq. (7.31) is uniquely defined for a given set of materials. It is the change in disjoining pressure upon changing the separation of two parallel interfaces. The right-hand side is more problematic because it depends on the way the interface is deformed. It is the change in capillary pressure upon a certain deformation that at some point leads to a minimal separation h. For a film with a perfectly planar surface, $P_c = 0$. If we deform the film to form a hole, this process will cause a bending and thus a curvature of the fluid interface. As a result, dP_c/dh is positive. Any disturbance of the film, preceding hole formation, changes the local curvature and thus the capillary pressure. The gradient of the attractive surface forces, $d\Pi/dh$, needs to be lower than the change in local capillary pressure. Calculating a critical thickness is not a simple task because relation (7.30) inherently involves the dynamics of the rupturing process.

7.6.2
Rupture of Thin Films

There are different approaches to describe rupture of liquid films. The first was proposed by Derjaguin and Gutop [826]. They postulated that film rupture is initiated by nucleation. Owing to thermal fluctuations, there will always be tiny holes in a film. They are, however, unstable unless they have reached a certain critical size. Once they have exceeded that size, they are likely to grow further and form stable holes. Practically, some disturbances or small nuclei create defects in the film. When the defect exceeds a critical size, it tends to grow in size and the film permanently ruptures [827]. Growth of a hole in a liquid film has been analyzed extensively [825, 828, 829]. First, a rim is formed that increases in size to a mature stage. Then, the hole grows with a rim keeping its cross-sectional shape [829]. Depending on the liquid and the hydrodynamic boundary condition, holes increase in size with different growth rates. For a liquid on a solid support with no-slip boundary conditions, the radius of a hole, r_h, increases with constant velocity: $dr_h/dt = v_d$ [828]. For polymer films, other dependencies have also been observed, which might indicate slippage [829].

A second mechanism was proposed by Vrij [830] and Scheludko [759]. They assumed that thermal fluctuating waves on fluid interfaces grow under the influence of attractive forces. It is called spinodal dewetting. The fundamental difference between nucleation and spinodal dewetting is that in the first case the film is metastable. The system has to overcome an energy barrier before nucleation and rupture occur. Spinodal dewetting occurs spontaneously and no energy barrier exists. The process might, however, take a very long.

To describe spinodal dewetting, we recall that in any liquid film of mean thickness h_0 surface corrugations occur due to thermal fluctuations. As a result, at any given moment some parts of the film are thinner than h_0. Surface tension tends to smoothen the film again. If, however, attractive surface forces exist, the system can gain free energy by thinning the already thinner part even further. This balance of the two effects is quantified in Eq. (7.31). Surface forces also act on those parts of the film that are thicker than h_0. Since the surface forces decrease with distance, the effect on the thicker parts is less than that on the thinner parts. This leads to hole formation and dewetting. In the theory, surface fluctuations are described by a Fourier spectrum of waves. The film is unstable with respect to fluctuations with wavelengths larger than a critical wavelength λ_c [830]:

$$\lambda_c = \sqrt{\frac{2\pi^2 \gamma_L}{d\Pi/dh}}. \qquad (7.32)$$

Here, the gradient of the force $d\Pi/dh$ at a thickness h_0 has to be inserted.

To calculate the time of film rupture, one needs to consider the hydrodynamic flow in the film. For large wavelengths, flow takes much longer than for short wavelength. For this reason, film rupture will be dominated by fluctuations with a wavelength only slightly larger than the critical wavelength. The dominating length scale of spinodal dewetting is $\lambda_s = \sqrt{2}\,\lambda_c$ [830]. For a film of constant viscosity and assuming no slip at the solid–liquid but a mobile interface at the liquid–gas interface, the mean time for dewetting can be estimated [831]:

$$t_s = \frac{12\gamma_L\eta}{h_0^3}\frac{1}{(d\Pi/dx)^2}. \tag{7.33}$$

Vrij and others used a linear theory with respect to deviations in the film thickness. Including nonlinear effects, Williams and Davis found that rupture times are typically 10 times shorter [832, 833]. Simulations demonstrated that the morphology of structures formed by spinodal dewetting depends sensitively on the precise interfacial forces and the film thickness. But the characteristic length scale is correctly predicted by the theory [834]. For very thin films, also the fact that the viscosity of the film is not constant but changes with the thickness and the specific height position in the film needs to be taken into account [835].

■ **Example 7.2**

An 8 nm thick polystyrene film is spin coated on an oxidized silicon wafer [820]. Immediately after spin coating, the film is in a nonequilibrium state. However, since polystyrene is a solid material at room temperature, it is practically stable. When heating, the film becomes a viscous liquid and dewets the surface. What is the characteristic size of structures formed? Assume that attractive van der Waals forces dominate and the Hamaker constant is $A_H = 2.2 \times 10^{-20}$ J. The surface tension of the heated polystyrene is $31\,\text{mNm}^{-1}$.

The van der Waals force is described by $\Pi = -A_H/6\pi h^3$. With Eq. (7.32) and $\lambda_s = \sqrt{2}\,\lambda_c$, we get

$$\lambda_s = h_0^2 \sqrt{\frac{8\pi^3\gamma_L}{A_H}}$$

$$= (8 \times 10^{-9}\,\text{m})^2 \cdot \sqrt{\frac{8\pi^3 \cdot 0.031\,\text{Nm}^{-1}}{2.2 \times 10^{-20}\,\text{J}}} = 1.2\,\mu\text{m} \tag{7.34}$$

for the characteristic length scale of dewetting structures.

Whether nucleation or spinodal dewetting dominates the rupture of a liquid film depends on the specific system. The $\lambda_s \propto h_0^2$ dependence predicted by Eq. (7.34) has been observed in polymer [815, 816, 820] and metal [821] films. For polymer films in many cases, nucleation is faster [818, 820]. For aqueous films on hydrophobic surfaces, small bubbles are generally accepted to initiate nucleation and nucleation dominates [813, 814].

The same formalism was also applied to describe the rupture of foam and emulsion films [791, 798, 831, 836].

7.7
Summary

- We distinguish different types of liquid films. Foam films are liquid lamellae between two gas phases. They form, for example, between two bubbles. Emulsion films are films of liquid A between two immiscible liquids B. This can, for example, be a water film between two oil drops. When a solid surface is involved, we have a liquid film on top of a solid substrate. Such films form, for example, when a particle collides with a bubble.
- Foam and emulsion films are never thermodynamically stable and will eventually rupture. Solid-supported liquid films can be stable.
- When dealing with the interaction between bubbles or drops, it is useful to distinguish different phases: (1) long-range hydrodynamic forces, (2) drainage, leading either to the formation of a thin film or (3) to the rupture of the liquid film. While the first two steps are dominated by hydrodynamics, the thickness of a possible metastable film is given by the disjoining pressure between the two liquid–fluid interfaces.
- Interfacial forces across foam and emulsion films are measured with the thin film balance. Surface forces across thin liquid films on solid surfaces can also be analyzed by measuring the thickness versus the vapor pressure or the thickness versus height.
- Films rupture either by nucleation or by a spinodal process.

7.8
Exercises

7.1. Calculate the relative vapor pressure of liquid helium ($M_W = 4.00\,\mathrm{g\,mol^{-1}}$) 1 m above the reservoir at 1.35 K. Which height corresponds to a relative vapor pressure of 0.5? Do the same for water vapor at 20 °C ($M_W = 18.0\,\mathrm{g\,mol^{-1}}$).

7.2. A smooth silica plate is fixed 1 m above a pool of liquid hexane. The hexane adsorbs to the silica surface and forms a thin supported film. Calculate the thickness of the film at 20 °C assuming that only van der Waals forces are acting ($A_H = -10^{-20}\,\mathrm{J}$ for silica/hexane/gas) and assuming that thermodynamic equilibrium has been established.

7.3. A bubble of 0.2 mm radius rises in water containing 1 mM KNO_3 until it hits a horizontal glass window. The liquid–gas interface and the glass surface are assumed to have a surface potential of $-20\,\mathrm{mV}$. The Hamaker constant of glass–water–air is $A_H = -10^{-20}\,\mathrm{J}$. Calculate the thickness and the area of the thin film region between the bubble neglecting rim effects. Use the fact that the disjoining pressure multiplied with the area has to compensate the buoyancy.

7.4. Derive an equation for the rupture time of a liquid film of constant viscosity for an exponentially decaying interfacial force $\Pi = -Ce^{-x/\lambda}$ assuming spinodal dewetting. Estimate the time for rupture of a 5 nm thick water film ($\eta = 0.89 \times 10^{-3}\,\mathrm{Pa\,s}$) on a hydrophobic surface assuming that the force is exponentially decaying with $C = 400\,\mathrm{Pa}$ and $\lambda = 12\,\mathrm{nm}$.

8
Contact Mechanics and Adhesion

Adhesion between surfaces is obviously of high practical relevance in science and technology as well as in everyday life. Bonding of materials with glues is clearly a case where one wants to have either maximum adhesion or reversible removal and reattachment. Other applications such as self-cleaning or nonsticking surfaces need to avoid adhesion completely. In fact, both types of systems also occur in nature: animals such as flies and geckos have found ways to reversibly stick to surfaces and leafs of plants, show remarkable self-cleaning properties.

When we bring two bodies in mechanical contact, attractive surface forces will lead to adhesion. The adhesion force (i.e., the maximum force necessary to separate the two bodies again) will depend on the strength of the attractive interaction, the contact area between the bodies, and the minimum distance between them. If we know the shape of the bodies and the interaction energy $V^A(X)$ per unit area between the materials, we can in principle use the Derjaguin approximation to calculate the adhesion force. Bradley [10] and Derjaguin [84] calculated in this manner the adhesion force F_A between a rigid sphere and a rigid planar surface to be

$$F_{adh} = -2\pi w_{adh} R_p, \tag{8.1}$$

with w_A being the adhesive energy per surface area, that is, the work necessary to separate a unit area of the two surfaces from contact. As a first approximation, one can identify

$$w_{adh} = V^A(D_0), \tag{8.2}$$

where $w(D)$ is the interaction potential between the surfaces and D_0 is the distance of closest approach (interatomic distance).

The adhesion force between two objects can arise from a combination of different contributions such as the van der Waals force, electrostatic force, chemical bonding, and hydrogen bonding forces, capillary forces, and others (e.g., bridging or steric forces on polymer-coated surfaces):

$$F_{adh} = F_{vdW} + F_{el} + F_{chem} + F_H + F_{cap} + \cdots \tag{8.3}$$

The adhesion force between two materials may therefore depend not only on the materials themselves but also on the ambient conditions. For micro- and

Surface and Interfacial Forces. Hans-Jürgen Butt and Michael Kappl
Copyright © 2010 Wiley-VCH Verlag GmbH & Co. KGaA
ISBN: 978-3-527-40849-8

nanocontacts, capillary condensation and thus the relative humidity may have a strong influence on the adhesion force. In such cases, a description with a simple value of interaction energy per unit area will fail, and an explicit calculation of forces for a given contact geometry will be mandatory.

Furthermore, the assumption of rigid bodies is in most cases a too strong simplification to correctly describe adhesion between solids. Bodies in contact will deform due to either external or surface forces. This means for understanding the phenomenon of adhesion, we need to know not only the adhesion energy of the materials but also the deformations. In this chapter, we will first discuss the surface energy of solids and its connection to adhesion. Then, we will give an overview of the classical theories of contact mechanics and finally discuss important parameters influencing adhesion.

8.1
Surface Energy of Solids

For liquids, the energy dW necessary to increase the surface area by an increment dA is

$$dW = \gamma_L dA. \tag{8.4}$$

This means that for liquids, surface energy and surface tension are identical. The increase in surface area in a liquid is directly related to an increase in the number of molecules at the surface while keeping the area per molecule σ_A constant. When this concept is applied to solids, a problem arises. The surface area of a solid can be increased by two possible routes. The first is to increase the number of molecules N at the surface just as in the case of the liquid, the second one is an elastic stretching of the surface, which increase the value of σ_A without changing the number of surface molecules. This means the surface energy γ^S will have both elastic and plastic contributions. With each surface molecule, an excess energy E_S is associated. The surface energy is then given as E_S/σ_A. The work done to increase the surface area by dA is

$$dW = \gamma^S dA = d(E_S N) = E_S \frac{\partial N}{\partial A} dA + N \frac{\partial E_S}{\partial A} dA, \tag{8.5}$$

which is equivalent to

$$\gamma^S = E_S \frac{\partial N}{\partial A} + N \frac{\partial E_S}{\partial A}. \tag{8.6}$$

When considering a purely plastic change in surface area, σ_A stays constant. Thus, the change in surface area is $dA = \sigma_A dN$ and we obtain

$$\left(\frac{\partial E_S}{\partial A}\right)_{\text{plastic}} = \left.\frac{\partial E_S}{\partial (N\sigma_A)}\right|_{\sigma_A} = \frac{1}{\sigma_A} \frac{\partial E_S}{\partial N} = 0, \tag{8.7}$$

which is equal to zero since the excess energy per molecule does not depend on the total number of molecules. For the surface intensive parameter, we obtain

$$\gamma_{\text{plastic}}^S = \left(E_S \frac{\partial N}{\partial A} \right)_{\text{plastic}} = E_S \frac{\partial A/\sigma_A}{\partial A} \bigg|_{\sigma_A} = \frac{E_S}{\sigma_A} \equiv \gamma_S. \tag{8.8}$$

In the case of a plastic deformation, the change in surface energy is similar to that of a liquid. We call it *surface tension* and use the symbol γ_S if we have to deal with several phases and want to explicitly denote the surface tension of the solid.

For a purely elastic deformation, N is constant and any change in surface area is due to a change in the area per molecule: $dA = N \, d\sigma_A$. Since $\partial N/\partial A = 0$, we write Eq. (8.6) as

$$\gamma_{\text{elastic}}^S = \left(N \frac{\partial E_S}{\partial A} \right)_{\text{elastic}} = \frac{N}{N} \frac{\partial E_S}{\partial \sigma_A} \bigg|_N = \frac{\partial E_S}{\partial \sigma_A}. \tag{8.9}$$

Inserting the result of Eq. (8.8) into Eq. (8.9) leads to

$$\gamma_{\text{elastic}}^S = \frac{\partial(\gamma_S \sigma_A)}{\partial \sigma_A} = \gamma + \sigma_A \frac{\partial \gamma_S}{\partial \sigma_A}, \tag{8.10}$$

which is the so-called Shuttleworth equation [837].

For a purely elastic increase in surface area, the elastic strain $\varepsilon_{\text{elastic}}$ is identical to $dA/A = d\sigma_A/\sigma_A$ and we get

$$\gamma_{\text{elastic}}^S = \gamma_S + A \frac{\partial \gamma_S}{\partial A} = \gamma_S + \frac{\partial \gamma_S}{\varepsilon_{\text{elastic}}}. \tag{8.11}$$

The change in surface energy equals the sum of the surface tension γ and its derivative with respect to the elastic surface strain $\partial\gamma/\partial\varepsilon_{\text{elastic}}$ or

$$\gamma_{\text{elastic}}^S \equiv \Upsilon = \gamma_S + \frac{\partial \gamma_S}{\partial \varepsilon_{\text{elastic}}}. \tag{8.12}$$

The quantity Υ is also called *surface stress*. With this formulation of the Shuttleworth equation, the surface stress is given as both the sum of surface tension and the change of surface tension γ with the elastic strain $\varepsilon_{\text{elastic}}$.

In general, we have plastic and elastic creation of new surface area concurrently. By replacing E_S with $\gamma_S \sigma_A$ in Eq. (8.6), we obtain

$$\gamma^S = \gamma_S \sigma_A \frac{\partial N}{\partial A} + N \frac{\partial(\gamma_S \sigma_A)}{\partial A} = \gamma_S \sigma_A \frac{\partial N}{\partial A} + N\gamma_S \frac{\partial \sigma_A}{\partial A} + N\sigma_A \frac{\partial \gamma_S}{\partial A}. \tag{8.13}$$

With $d(N\sigma_A) = N d\sigma_A + \sigma_A dN$ and $A = N\sigma_A$, we get

$$\gamma^S = \gamma_S \frac{\partial(N\sigma_A)}{\partial A} + N\sigma_A \frac{\partial \gamma_S}{\partial A} = \gamma_S + \frac{\partial \gamma_S}{\partial \varepsilon_{\text{tot}}}. \tag{8.14}$$

To highlight plastic and elastic contributions, we rewrite Eq. (8.14), using $\partial A/\partial A = \partial(A_{\text{elastic}} + A_{\text{plastic}})/\partial A$ and $d\varepsilon_{\text{tot}} = dA/A$:

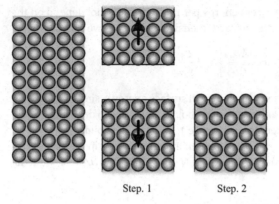

Step. 1   Step. 2

Figure 8.1 Cleavage of a crystal: thought experiment for the calculation of the surface energy.

$$
\begin{aligned}
\gamma^S &= \gamma_S \frac{\partial(A_{\text{elastic}} + A_{\text{plastic}})}{\partial A} + A \frac{\partial \gamma_S}{\partial A_{\text{elastic}}} \frac{\partial A_{\text{elastic}}}{\partial A} \\
&= \gamma_S \frac{\partial A_{\text{plastic}}}{\partial A} + \left(\gamma_S + A \frac{\partial \gamma_S}{\partial A_{\text{elastic}}} \right) \frac{\partial A_{\text{elastic}}}{\partial A} \\
&= \gamma_S \cdot \frac{d\varepsilon_{\text{plastic}}}{d\varepsilon_{\text{tot}}} + \Upsilon \cdot \frac{d\varepsilon_{\text{elastic}}}{d\varepsilon_{\text{tot}}}.
\end{aligned}
\tag{8.15}
$$

The change in the Gibbs energy is given by the reversible work $\gamma^S dA$ required to expand the surface against the surface tension γ_s and the surface stress Υ.

8.1.1
Relation Between Surface Energy and Adhesion Energy

Let us carry out the following thought experiment: we take a crystalline block of material with cross-sectional area A and cleave it in the middle (step 1 in Figure 8.1). This corresponds to breaking all bonds in the cleavage plane to create a new surface area of $2A$.

The surface energy of the crystal would then be given by

$$
\gamma^S = N_B E_B / 2A,
\tag{8.16}
$$

where N_B is the number of bonds within the cross-sectional area A and E_B is the bond energy.

■ **Example 8.1**

The energy per bond between two carbon atoms in diamond is 376 kJ mol^{-1}. If we split diamond at the (111) face, then 1.83×10^{19} bonds per m^2 break. Thus,

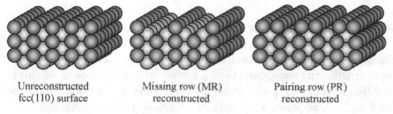

| Unreconstructed | Missing row (MR) | Pairing row (PR) |
| fcc(110) surface | reconstructed | reconstructed |

Figure 8.2 Typical reconstructions of face-centered cubic (110) surfaces.

$$\gamma^S \approx \frac{376\ \text{kJ mol}^{-1} \times 1.83 \times 10^{19}\ \text{m}^{-2}}{(2 \times 6.02 \times 10^{23}\ \text{mol}^{-1})} = 5.7\ \text{Nm}^{-1}.$$

This simple calculation demonstrates that the surface energies of solids can be much higher than the corresponding values of liquids.

However, only considering step 1 in Figure 8.1 to estimate the surface energy of a solid will lead to wrong results. Usually, the structure of crystal surfaces will not be that of a simple half of a bulk crystal. The first few molecular layers will usually undergo a structural change called surface relaxation, as indicated in step 2 of Figure 8.1 or a so-called surface reconstruction as (Figure 8.2).

The corrected surface energy taking surface relaxation and reconstruction account should then be equal to half the adhesion energy between the two halves of the material, which is called the *cohesive energy* of the material.

If we separate two blocks of different materials 1 and 2, the energy difference that equals the adhesive energy (also called Dupré[1] work of adhesion) will be given by

$$w_{adh} = \gamma_1^S + \gamma_2^S - \gamma_{12}, \tag{8.17}$$

where γ_{12} is the interfacial energy for materials 1 and 2 in contact and $\gamma_{1,2}^S$ are their respective surface energies.

In principle, the relation between adhesion energy w_{adh} given by Eq. (8.17) is reasonably justified from a thermodynamical point of view. However, in practice experimental adhesion forces may be significantly lower than expected from surface energies determined by other independent methods. This discrepancy is in most cases due to the fact that real surface have a certain surface roughness and often are more or less contaminated by adsorbates. Under ambient conditions, capillary forces may also contribute significantly and in such a situation, surface energies may contribute only indirectly via the respective contact angles (see chapter 5). Therefore, in many practical situations, the adhesion energy w_{adh} should be seen as a parameter to characterize the adhesion strength of a given system rather than a material constant.

---

1) Athanase Dupré, 1808–1869, French mathematician, professor at the University of Rennes.

8.1.2
Determination of Surface Energies of Solids

Determining surface energy parameters such as the surface tension, the surface stress, the internal surface energy, and so on is a difficult task. This is partly due to the fact that different methods probe different parameters and results cannot be compared directly. In addition, solid surfaces are usually not in equilibrium.

Bonds in *covalent solids* are dominated by short-range interactions. The internal surface energy is simply calculated as half the energy necessary to split the bonds that pass through a certain cross-sectional area. This is called the *nearest-neighbor broken bond model*. The Gibbs free surface energy is not much different from this value because, at room temperature, the entropic contribution is usually negligible.

Noble gas crystals are held together only by van der Waals forces. To calculate their surface energy, we proceed as indicated in Figure 8.1. In a thought experiment, we split the crystal and calculate the required work. In the second step, we allow the molecular positions to rearrange according to the new situation. The final value of the surface energy will, therefore, be lower than that obtained from the cleavage step alone.

Examples of the surface energies for noble gas crystals at 0 K are given in Table 8.1. The fact that a range of values is given is due to different crystal planes that lead to a variation in the surface energies γ^s.

A similar calculation can be carried out for *ionic crystals* (Table 8.2). In this case, the Coulomb interaction is taken into account, in addition to the van der Waals attraction and the Pauli repulsion. Although the van der Waals attraction contributes little to the three-dimensional lattice energy, its contribution to the surface energy is significant and typically 20–40% [838]. The calculated surface energy is sensitive to the particular choice of the interatomic potential.

For *metals*, there are two methods used to calculate their surface energy. (1) As in the case of noble gases and ionic crystals, the surface energy is calculated from the interaction potential between atoms. (2) Alternatively, we can use the model of free electrons in a box, whose walls are the surfaces of the metal.

There are only a few and often indirect methods of measuring surface energy parameters of solids. The problems are that surfaces contaminate and that they are usually inhomogeneous. For a review, see Ref. [840].

- For low energy solids, such as many polymers, the surface tension can be obtained from contact angle measurements.
- From the value of the liquid: We measure the surface tension of the melt and rely on the fact that, close to the melting point the free surface enthalpy of the solid is 10–20% larger.

Table 8.1 Internal surface energies u^σ of noble gas crystals.

Ne	19.7–21.2 mN m^{-1}
Ar	43.2–46.8 mN m^{-1}
Kr	52.8–57.2 mN m^{-1}
Xe	62.1–67.3 mN m^{-1}

A range is given because of different crystal orientations that lead to different surface energies.

Table 8.2 Calculated surface tensions γ and surface stresses ϒ of ionic crystals for different surface orientations compared to experimental results [839, 840].

Crystal	Orientation	γ_{calc}	γ_{exp}	Υ_{calc}	Υ_{exp}
LiF	(100) plane	480	340[a]	1530	
	(110) plane	1047		407	
CaF_2	(111) plane		450[a]		
NaCl	(100) plane	212	300[a], 190[b]	415	375
	(110) plane	425		256	
KCl	(100) plane	170	110[a], 173[b]	295	320
	(110) plane	350		401	

All values are given in $mN\,m^{-1}$.
a) From cleavage experiments.
b) Extrapolated from liquid.

- From the decrease in the lattice constant in small crystals caused by the compression due to the surface stress: The lattice spacings can be measured with the help of X-ray diffraction or by LEED (low-energy electron diffraction) experiments.
- From the work necessary for cleavage: We measure the work required to split a solid. The problem is that often mechanical deformations consume most of the energy and that the surfaces can reconstruct after cleaving.
- From adsorption studies with the help of inverse gas chromatography.
- From mechanical measurement of the surface tension: If the surface tension of a solid changes, the surface tends to shrink (if ϒ increases) or expand (if ϒ decreases). This leads, for instance, to the deflection of bimetallic cantilevers or a contraction of ribbons.
- From the heat generated upon immersion: Material in the form of a powder with known overall surface is immersed in a liquid. The free surface enthalpy of the solid is set free as heat and can be measured with precise calorimeters.

8.2
Contact Mechanics

At the beginning of this section, we just want to recall some definitions and terms relevant in mechanics of elastic solids. The stress acting on a material is defined as the force per cross-sectional area:

$$\sigma = \frac{F}{A}. \tag{8.18}$$

It has the units of $Nm^{-2} = Pa$. The strain is the amount of relative elongation of a material

$$\varepsilon_x = \frac{\Delta L_x}{L_x}, \tag{8.19}$$

where L_x is the total length of the object in x-direction and ΔL_x is its change in length in x-direction. Compression of a material corresponds to a negative strain. For an isotropic elastic material, applying a stress σ_x acting only in x-direction result in a strain of magnitude ε_x in x-direction, where

$$\sigma_x = E\varepsilon_x \tag{8.20}$$

with the Young's modulus E. Typical values of the Young's modulus are some 10 MPa for rubbers, some GPa for polymers, around 200 GPa for steel, and more than 1200 GPa for diamond. Note that Eq. (8.20) is valid only within the elastic limit of the materials, which means only for very small strains of $\varepsilon < 1\%$. This also implies that applying stresses larger than $\sim 1\%$ of the Young's modulus leads to plastic deformation or even failure of the material.

Most materials show the Poisson effect, which means when they are stretched in one direction they will contract in the directions perpendicular to that direction. The amount of contraction is characterized by the Poisson's ratio ν: for a material stretched in x-direction, the contractions in y- and z-directions are given by

$$\varepsilon_y = \varepsilon_z = -\nu\varepsilon_x. \tag{8.21}$$

The value of Poisson's ratio for a perfectly incompressible material is $\nu = 0.5$, which is also the maximum value that ν can take. Typical values of the Poisson's ratio are in the range 0.2–0.3. Rubbers have a value close to 0.5, and there exist even some special materials with negative values of ν (the theoretical limits are $-1 \leq \nu \leq 0.5$). For isotropic materials, the general form of Eq. (8.21) (i.e., describing nonuniaxial deformations) is given by

$$\begin{aligned}
\varepsilon_x &= \frac{1}{e}[\sigma_x - \nu(\sigma_y + \sigma_z)] \\
\varepsilon_y &= \frac{1}{e}[\sigma_y - \nu(\sigma_x + \sigma_z)] \\
\varepsilon_z &= \frac{1}{e}[\sigma_z - \nu(\sigma_x + \sigma_y)].
\end{aligned} \tag{8.22}$$

A quantity often used in contact mechanics is the so-called reduced Young's modulus E^*. For the contact between two materials 1 and 2, with Young's moduli E_1 and E_2 and Poisson's ratios of ν_1 and ν_2, it is defined as

$$E^* = \left(\frac{1-\nu_1^2}{E_1} + \frac{1-\nu_2^2}{E_2}\right)^{-1}. \tag{8.23}$$

In the case of an elastic material 1 and a rigid material 2, it simplifies to

$$E^* = \frac{E_1}{1-\nu_1^2}. \tag{8.24}$$

We will now give a short overview of the most important theories of contact mechanics for linear elastic solids. For a more detailed discussion on these topics, we refer to the comprehensive textbooks [841, 843] and reviews [842, 844, 845].

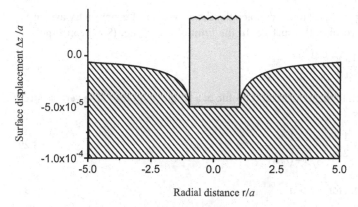

Figure 8.3 Indentation of an elastic half-space by a flat cylindrical punch with contact radius a.

8.2.1
Elastic Contact for a Flat Punch

The most basic configuration for an elastic contact is the indentation of an elastic half-space by a rigid axisymmetric frictionless punch. Frictionless means that we assume that no shear stress can develop between the punch and the half-space. While historically the first solution of such a problem was given by Hertz[2] for the case of a spherical indenter [846], we will start with a flat rigid cylindrical punch (Figure 8.3) that was first worked out by Boussinesq in 1885 [847] and solved in all details by Sneddon in 1946 [848].

For a cylindrical flat punch indenting an elastic half-space with the flat side parallel to the plane of the half-space, the contact area is simply given by πa^2, where a is the radius of the cylinder. For an applied load F_L, the indentation δ (i.e., the change in distance of the center points beyond the first contact of the bodies) of the punch becomes

$$\delta = \frac{1-\nu^2}{E}\frac{F_L}{a} = \frac{F_L}{2aE^*}. \tag{8.25}$$

Here, E and ν are the Young's modulus and Poisson's ratio of the half space material. The indentation of a flat punch on an elastic half-space is proportional to the applied load. The contact area stays constant independent of load. We obtain a linear relation between load and displacement with an effective spring constant of $E \cdot a/(1 - \nu^2) = E^* a$. The vertical surface displacement $\Delta z(r)$ of the elastic half-space is given by

$$\Delta z(r) = \frac{1-\nu^2}{\pi E}\frac{F_L}{r}\arcsin\left(\frac{a}{r}\right) = \frac{2\delta}{\pi}\arcsin\left(\frac{a}{r}\right) \quad \text{for} \quad r \geq a \tag{8.26}$$

and

$$\Delta z(r) = \delta \quad \text{for} \quad r \leq a, \tag{8.27}$$

2) Heinrich Hertz, 1857–1984 German physicist, professor in Karlruhe and Bonn.

as shown in Figure 8.3. Here r denotes the distance from the symmetry axis of the punch that is vertical to the surface. In the limit of $r \gg a$, Eq. (8.27) converges to

$$\Delta z(r) = \frac{1-\nu^2}{\pi E} \frac{F_L}{r},$$

(8.28)

which is identical to the deformation profile of a single-point load. The vertical stress $\sigma_z(r)$ is given as

$$\sigma_z(r) = -\frac{1}{2} \frac{F_L}{\pi a^2} \frac{1}{\sqrt{1-(r/a)^2}} \quad \text{for} \quad r < a,$$

(8.29)

$$\sigma_z(r) = 0 \quad \text{for} \quad r > a,$$

(8.30)

The negative sign is to indicate that for a positive applied load, the stress is compressive. The vertical stress reaches half the applied pressure in the center of the punch and diverges at the rim of the punch (Figure 8.4).

Since we have assumed frictionless contact, the surface of the elastic half-space is free to slide below the indenting punch. The radial displacement of the surface within the contact area is given by

$$\Delta r(r) = -\frac{(1-2\nu)(1+\nu)}{2\pi E} \frac{F_L}{r} \left[1 - \sqrt{1-\left(\frac{r}{a}\right)^2} \right]$$

(8.31)

$$= -\frac{\delta}{\pi} \frac{1-2\nu}{1-\nu} \frac{a}{r} \left[1 - \sqrt{1-\left(\frac{r}{a}\right)^2} \right] \quad \text{for} \quad r < a.$$

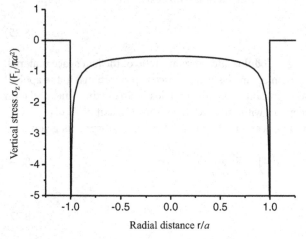

Figure 8.4 Vertical stress σ_z in units of the applied pressure $F_L/\pi a^2$ for flat cylindrical punch with contact radius a.

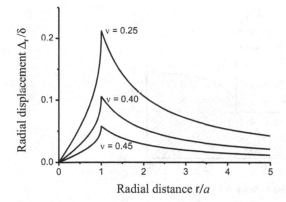

Figure 8.5 Radial displacements $-\Delta r/\delta$ for different values of the Poisson's ratio ν for indentation of an elastic half-space by a flat cylindrical punch with contact radius a.

The radial displacement outside the punch area is given by

$$\Delta r(r) = -\frac{(1-2\nu)(1+\nu)}{2\pi E}\frac{F_L}{r} \quad \text{for} \quad r > a. \tag{8.32}$$

The radial displacements Δr at the surface are plotted for different values of the Poisson's ratio in Figure 8.5. For a value of the Poisson's ratio of 0.5, the radial displacements vanish and the surface is deformed only vertically.

For the more general case of an elastic punch with Young's modulus E_1 and Poisson's ratio ν_1, indenting a half-space with Young's modulus E_2 and Poisson's ratio ν_2, the above equations simply have to be modified by replacing the reduced Young's modulus $E/(1-\nu^2)$ for the single material by the reduced Young's modulus for two materials as defined by Eq. (8.23)

8.2.2
Adherence of a Flat Punch (Kendall)

If we introduce an adhesion between the surfaces of the flat punch and the elastic half-space, all equations of Section 8.2.1 remain valid and can also be applied for the situation where the load F_L becomes negative, that is, a traction force (Figure 8.6).

The equations will be valid only up to the point where the punch detaches from the deformed surface of the elastic half-space. This will occur at some critical negative load, which just equals the adhesion force F_{adh}. The value of F_{adh} can be obtained from simple thermodynamic arguments as first shown by Kendall [849]. The adherence will become instable, when the so-called energy release rate $\mathcal{G}$ equals the adhesive energy per unit area w_{adh}.

$$\mathcal{G} = \left[\frac{\partial U_E}{\partial A} + \frac{\partial U_P}{\partial A}\right]_{F_L} = w_{adh}, \tag{8.33}$$

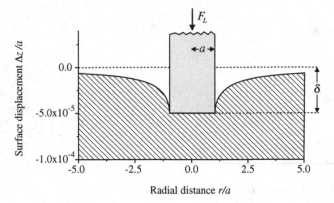

Figure 8.6 Adherence of a flat cylindrical punch with contact radius a to an elastic half-space.

where U_E and U_P are the elastic and potential energies and A is the contact area. Equation (8.33) is known as Griffith's criterion in fracture mechanics. For a given applied load F_L, the potential energy is applied

$$U_P = -F_L\delta = -\frac{F_L^2}{2E^*} \tag{8.34}$$

and the stored elastic energy is

$$U_E = \int F_L(\delta)d\delta = \frac{1}{2}F_L\delta = \frac{F_L^2}{4E^*}. \tag{8.35}$$

Using $A = \pi a^2$, we obtain the energy release rate as

$$\mathcal{G} = \frac{1}{2\pi a}\frac{\partial}{\partial a}\left[-\frac{F_L^2}{4aE^*}\right] = \frac{F_L^2}{8\pi a^3 E^*} = \frac{E^*\delta^2}{2\pi a}. \tag{8.36}$$

Combining Eqs. (8.33) and (8.36), we obtain

$$F_{adh} = -\sqrt{8\pi a^3 E^* w_{adh}} \tag{8.37}$$

as the adhesion force of a flat punch to an elastic half-space. Note that the adhesion force depends non-linearly on the contact area and on the adhesion energy. For the general case of an arbitrary axisymmetric punch shape, the energy release rate is given by

$$\mathcal{G} = \frac{1}{E^*}\frac{(F_1-F_L)^2}{8\pi a^3}, \tag{8.38}$$

where F_L is the load in the presence of adhesion that leads to a contact radius a and F_1 is the load that would be necessary to obtain the same contact radius in the absence of adhesion ($w_{adh} = 0$).

8.2.3
Elastic Contact of Spheres: Hertz Model

The problem of the elastic contact of a flat cylindrical punch is less complex than that of a sphere due to the fact that for the flat punch, the contact radius is just given by the radius of the cylinder and thus known a priori. The problem of the elastic contact between a sphere and a planar surface and between two spheres was solved by Hertz in 1882 [846]. Under the assumption that the contact radius a is small compared to the sphere radii, that the contact is frictionless and no tensile stress exists within the area of contact, Hertz derived an equation for the contact radius a between the spheres:

$$a^3 = \frac{3F_L R^*}{4E^*},$$ (8.39)

where E^* again is the reduced Young's modulus (Eq. (8.23)) and R^* is the so-called reduced radius defined as

$$\frac{1}{R^*} = \frac{1}{R_1} + \frac{1}{R_2}.$$ (8.40)

The penetration δ is given by

$$\delta = \frac{a^2}{R^*} = \left(\frac{9F_L^2}{16E^{*2}R^*}\right)^{1/3}.$$ (8.41)

Correspondingly, the force to achieve a certain penetration δ is given as

$$F_L = \frac{4}{3}E^*\sqrt{R^*}\delta^{3/2}.$$ (8.42)

For a rigid sphere indenting an elastic half-space, the vertical displacement of the half-space outside the contact area is given by

$$\Delta z(r) = \frac{a^2}{\pi R_P}\left[\sqrt{\frac{r^2}{a^2}-1} + \left(2-\frac{r^2}{a^2}\right)\arcsin\frac{a}{r}\right] \quad \text{for} \quad r > a,$$ (8.43)

as shown in Figure 8.7.

The vertical stress distribution in the contact follows an elliptical shape:

$$\sigma_z = \frac{3}{2}\frac{F_L}{\pi a^2}\sqrt{1-\frac{r^2}{a^2}},$$ (8.44)

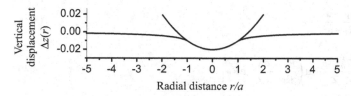

Figure 8.7 Contact geometry between a rigid sphere and an elastic half-space with contact radius a as derived from the Hertz theory.

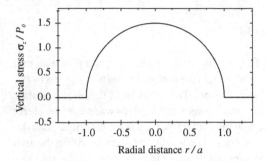

Figure 8.8 Vertical stress (contact pressure) between a sphere and an elastic half-space forming a contact with radius a.

which corresponds to a maximum contact pressure in the center of the contact of

$$p_{max} = \frac{3}{2}\frac{F_L}{\pi a^2} = \frac{3}{2}P_0 \tag{8.45}$$

which is 1.5 times the value of the average contact pressure P_0 (Figure 8.8).

The indentation force increases with a power of 3/2 with indentation depth. The contact area πa^2 increases as $F_L^{2/3}$ and the mean contact pressure $F_L/\pi a^2$ increases with applied load as $F_L^{1/3}$. In contrast to the flat punch case, the contact does no longer act as a linear spring since F_L is proportional to $\delta^{3/2}$ due to the fact that the contact area changes with load. The effective spring constant or contact stiffness can be defined as the slope of the curve $F_L(\delta)$:

$$\frac{dF_L}{d\delta} = \frac{d}{d\delta}\frac{4}{3}E^*\sqrt{R^*}\delta^{3/2} = 2E^*\sqrt{R^*\delta} = 2E^*a. \tag{8.46}$$

This is twice the value of that for a flat cylindrical punch with the same contact radius a. The elastic energy U_E stored for a given indentation δ can be calculated from

$$U_E = \int F_L(\delta)d\delta = \int \frac{4}{3}E^*\sqrt{R^*}\delta^{3/2}d\delta = \frac{2}{5}\frac{4}{3}E^*\sqrt{R^*}\delta^{5/2} = \frac{2}{5}F_L\delta. \tag{8.47}$$

Deriving of the above equations, the Hertz theory assumes that $a \ll R$. This will in most practical cases be fulfilled since for larger indentations materials will often be no longer within their elastic limit and the Hertz theory does not apply. Indentation of an elastic half-space by a spherical indenter with radius R_P was studied by Sneddon [850] and Ting [851] who derived equations without the approximation $a \ll R_P$:

$$\delta = \frac{a}{2}\ln\frac{R_P + a}{R_P - a} \tag{8.48}$$

$$F_L = E^*\left[\frac{R_P^2 + a^2}{2}\ln\left(\frac{R_P + a}{R_P - a}\right) - aR_P\right]. \tag{8.49}$$

A comparison of their results with those of Hertz shows that for values of a/R_P of up to 0.4, the Hertz theory is an excellent approximation.

8.2.4
Adhesion of Spheres: JKR Theory

The Hertz theory allows to calculate the contact shape and forces between spheres under the influence of an external force. It does not include any surface force and therefore does not lead to an expression for the adhesion force. When moved apart, the bodies separate at the point where $\delta = 0$ and $a = 0$ without any adhesion force. A first model to include adhesive forces based on the Hertz theory was introduced by Derjaguin in 1934 [84]. He assumed that the contact shape is that given by the Hertz theory and that the total energy of the system is the elastic energy as given by the Hertz model minus the energy due to the formation of the contact area πa^2. In Derjaguin's model, the force to achieve a certain indentation δ is reduced by

$$F(\delta) = F_L(\delta) - \pi R^* w_{adh}, \tag{8.50}$$

where F_L is the corresponding value of the load in the Hertz model as defined in Eq. (8.42). Separation of the surfaces therefore occurs at a pull-off or adhesion force of

$$F_{adh} = -\pi R w_{adh} \tag{8.51}$$

and a contact radius of $a = 0$. Note that this force is smaller by a factor of 2 than the value calculated by Bradley as well as that resulting from the Derjaguin approximation, which was introduced in the same paper [84].

An extension of the Hertz theory taking adhesive interactions and their influence on the contact shape into account was introduced in 1971 by Johnson, Kendall and Roberts [852] and it has become well known as the JKR theory. Their basic assumption was to take into account adhesive interaction only within the contact zone and neglect any interactions outside the contact zone.

To calculate the penetration δ for adhesive interactions, they suggested the thought experiment sketched in Figure 8.9. Let us assume that the spheres form a contact with radius a in the presence of adhesion and an external load F_L. We define F_1 as the loading force necessary to obtain the same contact radius a for a Hertzian contact

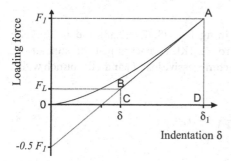

Figure 8.9 Loading (0–A) and partial unloading (A–B) process for deriving indentation and elastic energy in the JKR model.

without adhesion. According to Eq. (8.42) and Eq. (8.41),

$$F_1 = \frac{4}{3} E^* \sqrt{R^*} \delta^{3/2} = \frac{4}{3} \frac{a^3 E^*}{R^*} \tag{8.52}$$

When loading the contact up to the force F_1 in the absence of adhesion ($w_A = 0$), we follow the Hertzian $P(\delta)$ curve from 0 to A. At A, we have the contact radius a and the indentation $\delta_1 = a^2/R^*$ and the stored elastic energy is (Eq. (8.47))

$$U_E^{(A)} = \frac{2}{5} \frac{F_1 a^2}{R^*} = \frac{2}{5} F_1 \delta_1. \tag{8.53}$$

We now start to decrease the load down to a value F_L while simultaneously increasing the adhesion energy to its value w_A so that the contact area stays constant. Reducing the load at constant a corresponds to the unloading of a flat punch and follows the straight line A–B. The change in indentation along this line is given by Eq. (8.25):

$$\Delta\delta = \frac{F_1 - F_L}{2aE^*} \tag{8.54}$$

and the change in elastic energy is given by the area of the trapezium ABCD:

$$\Delta U_E = \Delta\delta \frac{F_1 + F_L}{2} = \frac{1}{4aE^*}(F_1^2 - F_L^2). \tag{8.55}$$

The indentation δ in the presence of adhesion is given by

$$\delta = \delta_1 - \Delta\delta = \frac{a^2}{3R^*} + \frac{F_L}{2aE^*} \tag{8.56}$$

The elastic energy at point B is

$$U_E = U_E^{(A)} - \Delta U_E = \frac{F_L^2}{4aE^*} + \frac{4a^5 E^*}{45 R^{*2}}. \tag{8.57}$$

The last result was obtained by combining Eqs. (8.52), (8.53) and (8.55). The distribution of the normal stress σ_z across a JKR contact is just the sum of the distributions for a Hertzian contact with compressive load F_1 and a flat punch with a tensile load $F_1 - F_L$ (Figure 8.10):

$$\sigma_z(r) = \frac{F_1 - F_L}{2\pi a^2} \frac{1}{\sqrt{1 - (r/a)^2}} - \frac{3}{2} \frac{F_1}{\pi a^2} \sqrt{1 - (r/a)^2}. \tag{8.58}$$

The inner circle of the contact area will be under compressive stress, whereas outer annular zone will be under tensile stress. The radius of the inner compressed zone

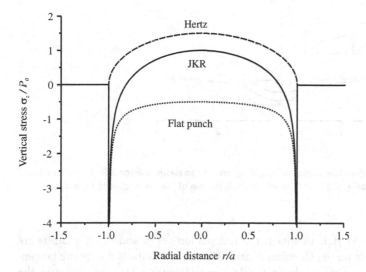

Figure 8.10 Vertical stress (contact pressure) between a sphere and an elastic half-space forming a contact with radius a. The stress distribution according to the JKR theory (solid line) is the sum of a Hertzian contact pressure (dashed line, compressive stress) and a flat punch contact stress (dotted line, tensile stress).

is given by

$$r' = a\sqrt{1 - \frac{F_1 - F_L}{3F_1}}. \qquad (8.59)$$

The tensile stresses go to infinity at the rim of the contact, but the total stress integrated over the whole contact area remains finite. Outside the contact area, the vertical stress is zero since the JKR model assumes that no surface forces act outside the contact area. The infinite stresses at the edge of the contact are physically not possible. Obviously, the description of that region down to molecular scales by a continuum model as the JKR theory cannot be realistic. As soon as one assumes realistic interaction potentials between the molecules at the rim of the contact, these singularities will disappear. However, such detailed models will usually be too complicated to allow their routine use in contact mechanics.

The vertical displacement outside the contact area is the sum of the displacement of a Hertzian contact under load F_1 and a flat punch under a tensile force of $F_1 - F_L$.

$$\Delta z = \frac{1}{E^*}\frac{F_1 - F_L}{\pi a}\arcsin\left(\frac{a}{r}\right) + \frac{a^2}{\pi R^*}\left[\sqrt{\left(\frac{r}{a}\right)^2 - 1} + \left(2 - \left(\frac{r}{a}\right)^2\right)\arcsin\left(\left(\frac{a}{r}\right)^2\right)\right]. \qquad (8.60)$$

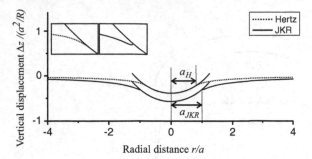

Figure 8.11 Profiles of the contact of a rigid sphere with an elastic half-space for the Hertz (dotted line) and JKR model (solid line). The insets show the rim of the contact zones in detail.

Compared to a Hertzian contact, the indentation depth and contact radius are increased. While for the Hertzian contact, the surface contacts the sphere tangentially, it forms a right angle in the JKR case (Figure 8.11). This leads to the characteristic shape of the JKR contact with a small neck at the edge of the contact zone (Figure 8.11).

The energy release rate is given by Eq. (8.38) as

$$\mathcal{G} = \frac{1}{E^*} \frac{(F_1 - F_L)^2}{8\pi a^3} \tag{8.61}$$

$$= \frac{(4a^3 E^*/3R^* - F_L)^2}{8\pi a^3 E^*} \tag{8.62}$$

$$= \frac{E^*}{2\pi a} \left(\delta - \frac{a^2}{R^*} \right)^2. \tag{8.63}$$

Combining Griffith's criterion (8.33) with Eq. (8.61)

$$(F_1 - F_L)^2 = 8\pi a^3 E^* w_{adh}. \tag{8.64}$$

Expressing F_1 and F_L as dimensionless quantities $F_1/3\pi w_A R$ and $F_L/3\pi w_A R$ and using Eq. (8.52), we obtain

$$\left(\frac{F_1}{3\pi w_{adh} R^*} - \frac{F_L}{3\pi w_{adh} R^*} \right)^2 = \frac{8\pi a^3 E^* w_{adh}}{(3\pi w_{adh} R^*)^2} = \frac{8a^3 E^*}{9\pi w_{adh} R^{*2}} = \frac{2F_1}{3\pi R^* w_{adh}}. \tag{8.65}$$

This quadratic equation is solved by

$$F_1 = F_L + 3\pi w_{adh} R^* \pm \sqrt{6\pi w_{adh} R^* F_L + (3\pi w_{adh} R^*)^2}. \tag{8.66}$$

From the stability condition that $\partial \mathcal{G}/\partial A > 0$, it can be shown that only the positive determinant in Eq. (8.66) has to be taken into account. By using again Eq. (8.52), to replace F_1 with a^3, we obtain the JKR equation for the contact radius

$$a^3 = \frac{3R^*}{4E^*}\left(F_L + 3\pi w_{adh} R^* + \sqrt{6\pi w_{adh} R^* F_L + (3\pi w_{adh} R^*)^2}\right).$$ (8.67)

The first summand in this equation is identical to the Hertzian contact radius defined in Eq. (8.39), the second and the third one are due to the adhesive interaction and lead to an increased contact radius compared to the adhesionless case. By using again Griffith's criterion (8.33) in combination with Eq. (8.63), we obtain the equation for the JKR indentation

$$\delta = \frac{a^2}{R^*} - \sqrt{\frac{2\pi a w_{adh}}{E^*}}.$$ (8.68)

The relation between load and contact radius is given by

$$F_L(a) = \frac{4E^* a^3}{3R^*} - 2\sqrt{2\pi E^* w_{adh} a^3}.$$ (8.69)

For zero external load $F_L = 0$, we obtain the contact radius from Eq. (8.62) with $\mathcal{G} = w_{adh}$:

$$a_0 = \left(\frac{9\pi w_{adh} R^{*2}}{2E^*}\right)^{1/3}$$ (8.70)

and the indentation

$$\delta_0 = \frac{a_0^2}{3R^*} = \left(\frac{\pi^2 w_{adh}^2 R^*}{E^{*2}}\right)^{1/3}.$$ (8.71)

The pull-off occurs at a negative loading force $F_L = -F_1 = F_{adh}$. Using Eq. (8.65) we obtain the adhesion force:

$$F_{adh} = -\frac{3}{2}\pi w_{adh} R^*.$$ (8.72)

At pull-off, the contact radius is given as

$$a_{min} = \left(\frac{9}{8}\frac{\pi w_{adh} R^{*2}}{E^*}\right)^{1/3} = 0.63 a_0$$ (8.73)

and the height of the neck (=negative indentation) as

$$\delta_{min} = -\left(\frac{\pi^2 w_{adh}^2 R^*}{\frac{64}{3} E^{*2}}\right)^{1/3}.$$ (8.74)

While the contact radius (Eq. 8.73) and neck height (Eq. 8.74) depend on the reduced Young's modulus E^*, the expression for the adhesion force (Eq. 8.72) does not depend on the elastic properties of the materials. This result is counter-intuitive, since for a soft material, a larger deformation will occur and the contact area will be larger and one might expect a higher value of F_{adh}. However, when pulling the two bodies out of contact, the stored elastic energy will be recovered. This balance between adhesive and elastic energy leads to the independence from the Young's modulus. The adhesion force given by Eq. (8.72) is smaller than the value calculated by Bradley for rigid materials (Eq. (8.1)). Since Eq. (8.72) does not depend on the value of E^*, it should also hold for the limit of rigid bodies. This clear contradiction gave rise to strong discussions about the validity of the JKR theory. We will come back to that issue at the end of the next section.

8.2.5
Adhesion of Spheres: DMT Theory

Soon after the introduction of the JKR theory, Derjaguin, Muller, and Toporov in 1975 [854] came up with an alternative contact model that became known as the DMT theory. As in the Derjaguin model from 1934 [84], they assumed that within the contact, the stresses and deformations are given by the Hertz theory but that attractive surface forces acting in an annular zone around the contact (Figure 8.12). The additional load induced by the surface forces with that so-called cohesive zone can be calculated by using the Derjaguin approximation. This force is then simply added as an additional load to obtain the correct indentation and contact area from the Hertz theory. It is assumed that the surface forces do not deform the surfaces outside the contact area and thus the surface profile is the same as for a Hertzian contact (see Figure 8.7).

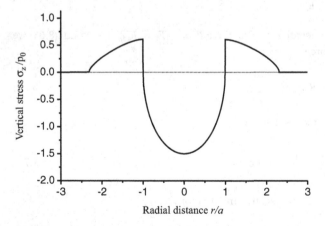

Figure 8.12 Stress distribution for the DMT model. In an annular zone outside the contact area, the so-called cohesive zone, surface forces lead to a tensile stress. Within the contact zone, the stress distribution is that of a Hertzian contact.

The pull-off in the DMT theory occurs at zero contact radius and the adhesion force is given by

$$F_{adh} = -2\pi w_{adh} R^*, \tag{8.75}$$

In their original paper, DMT used a thermodynamic approach to calculate the contribution of the adhesion energy. They found that the surface forces give an additional contribution of $2\pi w_{adh} R^*$ to the normal load at zero indentation that decreases to $\pi w_{adh} R^*$ with increasing indentation δ. In a later paper [855], the authors used a force balance calculation and found that the additional load *increases* with indentation, starting from a value of $2\pi w_{adh} R^*$ for zero indentation. In his calculations on the JKR–DMT transition, Maugis (see Section 8.2.6) used a fixed that value of $2\pi w_{adh} R^*$. Most references to the DMT theory actually refer to this formulation of the DMT theory that was introduced by Maugis.

The radius of contact at zero external load ($F_L = 0$) for that variant of the DMT theory is

$$a_0 = \left(\frac{3\pi w_{adh} R^{*2}}{2E^*} \right)^{1/3}. \tag{8.76}$$

Note that the value of the adhesion force in the DMT theory is identical to the value obtained by Bradley and in contradiction to the JKR result. This discrepancy between JKR and DMT theories caused some debate. Both theories have obvious limitations. The JKR theory does not account for surface forces outside the contact zone but allows deviations from the Hertzian contact shape. The DMT model takes into account the surface forces outside the contact area but not the deformations due to these forces. Therefore, one might expect that the JKR model is more appropriate for the case of large soft spheres and high surface energies, whereas the DMT model should be better suited for the case of small rigid spheres and low surface energies, where deformations close to the contact zone should remain small. This was put on a more quantitative basis by Tabor in 1977 [856], who recognized that a consequence of the neck formation in the JKR theory is that the gap width increases steeply outside the contact, rendering surface forces outside the contact area ineffective. He concluded that the JKR theory should apply if the neck height is of order or larger than the range of the action of the surface forces. He introduced the so-called Tabor parameter

$$\mu_T = \left(\frac{w_{adh}^2 R^*}{E^{*2} D_0^3} \right)^{1/3}, \tag{8.77}$$

which is the ratio between the neck height at separation (Eq. (8.74)) and the typical range of the surface forces, which was set equal to the equilibrium atomic distance z_0. For values of $\mu_T \ll 1$, the DMT theory should be valid and for $\mu_T \gg 1$, the JKR theory should be applicable. In 1980, Muller *et al.* [853] showed by numerical calculations using a Lennard–Jones interaction potential between the molecules that there is indeed a transition between the two theories and that is governed by the value of the parameter μ_T.

■ **Example 8.2**

A silicon oxide sphere with $20\,\mu m$ diameter sits on a silicon oxide surface. Estimate the contact radius at zero external force and calculate the adhesion force. $E = 5.4 \times 10^{10}\,Pa$, $\nu = 0.17$, $\gamma_S = 50\,mN\,m^{-1}$, density $\varrho = 2600\,kg\,m^{-3}$. With $R_1 = 10\,\mu m$ and $R_2 = \infty$, the effective particle radius is $R^* = R_1 = 10\,m$.

$$\frac{1}{E^*} = 2 \cdot \frac{1-0.17^2}{5.4 \times 10^{10}\,Pa} \Rightarrow E^* = 2.8 \times 10^{10}\,Pa.$$

The contact radius is estimated using the JKR model. Without external forces, we have

$$a^3 = \frac{3}{4} \cdot \frac{10^{-5}\,m}{2.8 \times 10^{10}\,Pa} \cdot (6\pi \cdot 2 \cdot 0.05\,N\,m^{-1} \cdot 10^{-5}\,m)$$

$$= 5.05 \times 10^{-21}\,m^3 \Rightarrow a = 1.71 \times 10^{-7}\,m.$$

The neck height is

$$h_n \approx \left(\frac{0.05^2\,N^2\,m^{-2} \cdot 10^{-5}\,m}{(2.8 \times 10^{10}\,Pa)^2}\right)^{1/3} = 3.2 \times 10^{-10}\,m.$$

The neck height is about as large as an atomic diameter. Therefore, the DMT model is suitable and the adhesion is

$$F_{adh} = -4\pi \cdot 0.05\,N\,m^{-1} \cdot 10^{-5}\,m = -6.3\,\mu N$$

8.2.6
Adhesion of Spheres: Maugis Theory

Another approach to the problem of the JKR–DMT transition was introduced by Maugis in 1992 [857]. By using a Dugdale potential (Figure 8.13) Maugis could derive a set of analytical equations that describe the transition between the JKR and DMT limits. The Dugdale potential simply has a hard wall response at distance z_0 and a

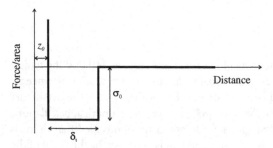

Figure 8.13 Dugdale potential used in the Maugis theory.

constant attractive interaction σ_0 over a certain distance δ_t; it is zero for distances greater than $z_0 + \delta_t$. The values of σ_0 and δ_t are chosen such that $\sigma_0\delta_t = w_{adh}$ and that σ_0 matches the minimum adhesive stress of a Lennard–Jones potential with equilibrium distance z_0, which implies that $\delta_t = 0.97z_0$.

The transition in the Maugis model is governed by the parameter μ_M defined as

$$\mu_M = \frac{2\sigma_0}{(16\pi w_{adh} E^{*2}/9R^*)^{1/3}}. \tag{8.78}$$

Since $\sigma_0\delta_t = w_A$ and $\delta_t = 0.97z_0$, the transition parameter λ is almost identical to the one defined by Tabor:

$$\mu_M = 1.1570\,\mu_T \tag{8.79}$$

Maugis introduced the dimensionless variables $\bar{a}$ for the contact radius $\bar{F}_L$ for the load, and $\bar{\delta}$ for the indentation defined as

$$\bar{a} = \frac{a}{(3\pi w_{adh} R^*/4E^*)^{1/3}}, \tag{8.80}$$

$$\bar{F}_L = \frac{F_L}{\pi w_{adh} R^*}, \tag{8.81}$$

$$\bar{\delta} = \frac{\delta}{16\pi^2 w_{adh}^2 R^*/9E^{*2}}. \tag{8.82}$$

For spheres that are soft and have a large radius, μ_M will become large and this should correspond to the JKR limit. For small, hard spheres, $\mu_M \rightarrow 0$ corresponding to the DMT limit. Using this set of variables, the Hertz equations reduce to

$$\bar{a} = \bar{F}_L^{1/3}, \tag{8.83}$$

$$\bar{\delta} = \bar{a}^2. \tag{8.84}$$

The JKR equations are

$$\bar{a}^3 = \bar{F}_L + \bar{a}\sqrt{6\bar{a}}, \tag{8.85}$$

$$\bar{\delta} = \frac{\bar{a}^2 + 2\bar{F}_L}{3\bar{a}} = \bar{a}^2 - \frac{2}{3}\sqrt{6\bar{a}}. \tag{8.86}$$

The governing equations based on the Dugdale potential were found by Maugis as

$$\bar{F}_L = \bar{a}^3 - \mu_M\bar{a}^2\left(\sqrt{m^2-1} + m^2\arctan\sqrt{m^2-1}\right), \tag{8.87}$$

$$\bar{\delta} = \bar{a}^2 - \frac{4}{3}\bar{a}\mu_M\sqrt{m^2-1}, \tag{8.88}$$

$$\frac{\mu_M \bar{a}^2}{2}\left[\sqrt{m^2-1}+(m^2-2)\arctan\left(\sqrt{m^2-1}\right)\right]$$
$$+\frac{4\mu_M^2 \bar{a}}{3}\left[\sqrt{m^2-1}\arctan\left(\sqrt{m^2-1}\right)-m+1\right]=1, \qquad (8.89)$$

where m is the ratio of the radius $c = a + \delta_t$ of the adhesive zone and the contact radius a:

$$m = c/a. \qquad (8.90)$$

The implicit character of the Maugis equations makes a somewhat evolved procedure necessary to plot the relations between load, contact radius, and indentation. For a given value of the parameter μ_M and contact radius a, one can use Eq. (8.89) to obtain the corresponding value of m. This value is then plugged into Eqs. (8.87) and (8.88) to calculate the desired quantities. In Figures 8.14–8.16, the relations between different quantities for several values of μ_M are plotted together with the limiting cases of the Hertz, JKR, and DMT theories. Note that the curves in these figures correspond to the full set of values obtained form these equations. Some parts of the curves, however, will be of no physical relevance, since the surfaces will separate before, as described in the figure captions.

From the shapes of the curves in Figures 8.14–8.16, we can conclude that the JKR model is already a good approximation for values of $\mu_M > 3$. To justify the use of the DMT model, the value of μ_M should be at least lower than 0.1 but better even lower than 0.01. Especially in Figure 8.16, it becomes obvious that prediction of the DMT model that the rupture of the contact occurs at zero contact radius will be fulfilled only in the limit of $\mu_M \rightarrow 0$, which corresponds to the hard sphere limit of Bradley.

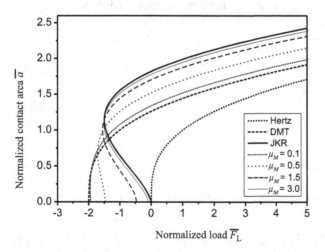

Figure 8.14 Relation between normalized contact radius and normalized load calculated from the Maugis theory for different values of μ_M. In the limit of the DMT theory, pull-off occurs at a value of −2. For the other curves, pull-off occurs at the point where the tangent to the curves becomes vertical.

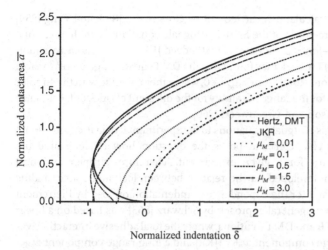

Figure 8.15 Relation between normalized contact radius and normalized indentation calculated from the Maugis theory for different values of μ_M. In the case of the DMT and Hertz model, separation occurs at a value of zero contact radius at zero indentation. In other cases, rupture of the adhesive contact occurs at finite contact radius and at negative indentation (point where the tangent to the curve becomes vertical).

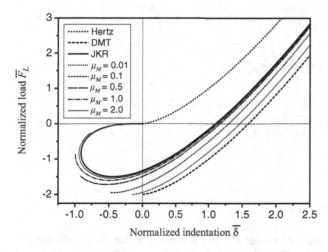

Figure 8.16 Relation between normalized applied load and normalized indentation calculated from the Maugis theory for different values of μ_M. For the Hertz case, detachment occurs at zero load and indentation. In the DMT case, detachement occurs at a normalized load of −2 and zero indentation. For all other cases, detachment under a constant applied force occurs at negative load between −2 and −1.5 and at negative indentation in the point where the tangent to the curve becomes horizontal.

Detailed numerical calculations carried out by Greenwood [858] and Feng [859] showed reasonable agreement of the Maugis–Dugdale equations with their calculations using a Lennard–Jones potential and that indeed JKR is a good approximation for $\mu_M > 3$. For the DMT model, they found that the DMT equations give a reasonable description of the data only for the limit $\mu_M \rightarrow 0$, where they coincide with the simpler Bradley model. A map for the range of validity for the different elastic contact models was proposed by Johnson and Greenwood [860].

The use of the Maugis–Dugdale equations to fit experimental data is cumbersome due to their implicit character and poses the question how to determine the appropriate value of μ_M for a given experiment. Therefore, simpler analytical expressions have been developed for the relation between load and contact radius by Carpick et al. [861] and for the dependence of indentation on load by Pietrement and Troyon [862]. A more general approach by Schwarz [863] was based on a linear superposition of the JKR and DMT solutions, where the total adhesive interaction was split up in a short-range component w_{adh1} (JKR) and a long-range component w_{adh2} (DMT) with $w_{adh} = w_{adh1} + w_{adh2}$. In this model, no additional transition parameter is introduced, but the value of the adhesion force depends on the relative contributions of adhesive interaction:

$$F_{adh} = -\frac{3}{2}\pi R^* w_{adh1} - 2\pi R^* w_{adh2}. \tag{8.91}$$

The relations for contact area, indentation, and vertical stress distribution were found to be

$$a = \left(\frac{3R^*}{4E^*}\right)^{1/3}\left(\sqrt{3F_{adh} + 6\pi w_{adh}R^*} + \sqrt{F_L - F_{adh}}\right)^{2/3}, \tag{8.92}$$

$$\delta = \frac{a^2}{R^*} - 4\sqrt{\pi a}4E^*\left(\frac{F_{adh}}{\pi R^*} + 2w_{adh}\right), \tag{8.93}$$

$$\sigma_z(r) = \frac{4E^*a}{2\pi R^*}\sqrt{1-\left(\frac{r}{a}\right)^2} - \sqrt{\frac{4E^*}{\pi a}\left(\frac{F_{adh}}{\pi R} - 2w_{adh}\right)}\left(1-\left(\frac{r}{a}\right)^2\right)^{-1/2}. \tag{8.94}$$

8.3
Influence of Surface Roughness

All the equations derived for contact models discussed so far were valid for the assumption of perfectly smooth surfaces. Surface roughness will have a pronounced influence on adhesion. First, adhesion limits the average distance of closest approach between two bodies and thus reduces the interaction. Second, bringing the two rough bodies in close contact involves the deformation of the highest surface asperities. The elastic energy stored during this process will be released again during separation of the bodies. A combination of both effects has the consequence that adhesion between rough solids with a high Young's modulus will almost disappear [865]. The extension of contact mechanic models for single spheres to rough surfaces can be accomplished by a statistical description of the surface asperities that are modeled as spherical protrusions with a certain height distribution. A well-known example is that of Greenwood and Williamson [864] that will be discussed in Section 9.1.1 in the context of friction forces between rough surfaces. Based on this concept, Fuller and Tabor [865] developed their model of adhesive contact of rough surfaces by assuming a JKR-type interaction for each of the contacting asperities. A central quantity in their model was the so-called adhesion index, which is the ratio of adhesion force to the elastic force needed to push the contacting sphere by a depth σ into the surface. The quantity σ is the standard deviation of the assumed Gaussian height distribution of the surface asperities. Limitations of this model are that it is valid only for situations where the real contact area (i.e., the area where the surfaces make real contact with the deformed asperities) is a small fraction of the apparent (projected) contact area and that it assumes that contact breaks uniformly across the apparent contact area. In reality, the latter assumption will usually not be fulfilled since rupture will typically occur by crack propagation.

Another example of models that use the approximation of real surface profiles by spherical asperities is that of Rumpf [866], where the adhesion between a sphere with radius R_P and a rough surface with asperities of radius r_a due to van der Waals interaction is given as

$$F_{adh} = -\frac{A_H}{6}\left[\frac{r_a}{D_0^2} + \frac{R_P}{(r_a + D)^2}\right],\tag{8.95}$$

where D_0 is the atomic distance for closest possible approach between surfaces. The description of surface roughness by a single asperity radius is very simplistic. Rabinovich et al. [867] characterized surface roughness by two parameters, root mean square roughness r_{rms} and average lateral distance d between asperities (with $d \gg r_{rms}$), and found that

$$F_{adh} = -\frac{A_H R_P}{6D_0^2}\left[\frac{1}{1 + \frac{32cR_Pr_{rms}}{d^2}} + \frac{1}{\left(1 + \frac{cr_{rms}}{D_0}\right)^2}\right],\tag{8.96}$$

where c is proportionality factor ($c = 1.817$). This model was found to lead to a more realistic description of adhesion forces for their AFM experiments [868].

A different approach to the characterization of surface roughness uses fractals. The basic idea was introduced by Archard [869] even before the term fractals was coined. Majumdar and Bhushan [870] noticed that surfaces often exhibit roughness on different length scales and could best be described as self-affine. Corresponding theories that link the fractal dimension of the surface with the dependence of true contact area on applied load have been developed recently by the groups of Persson

[871] and Robbins [872]. They predict an exponential increase of load with increasing indentation:

$$F_L \propto \exp(-\delta/\delta_0), \tag{8.97}$$

where δ_0 depends on the nature of the surface roughness (being of the order of the rms roughness) and is independent of F_L. Such an exponential relation between repulsive force and distance for randomly rough surfaces was confirmed experimentally with the SFA (but not for surfaces with regular patterns) [872] and for the contact between elastomers with an asphalt surface [873].

8.4
Adhesion Force Measurements

Due to the high practical relevance of adhesion forces in industrial and everyday applications, a broad spectrum of experimental methods to measure adhesion forces has been established and there are, for example, standardized procedures such as peel tests for adhesive tapes or tack tests for pressure-sensitive adhesives. We will focus here on some representative examples of experimental work targeted toward a fundamental understanding of contact mechanics and adhesion phenomena.

The adhesion of spherical particles to surfaces has, for instance, been measured by a centrifuge for more than 40 years (Figure 8.17). A significant part of our knowledge about the behavior of powders stems from such experiments [875]. The centrifugal force, which is required to detach particles from a planar surface, is measured [875–878] by mounting the surfaces on an ultracentrifuge and rotating them at a defined speed. Detachment force of particles is determined from pictures of the surfaces taken after each run with increasing speed of rotation. Particle size and centrifugal force needed to remove the particle are obtained by digital image analysis. Usually, the detachment forces of many particles are measured in a single series of runs, allowing statistical evaluation of the data. This is especially useful in the case of irregularly shaped particles where the contact area and adhesion force will depend on the random orientation of the particles relative to the surface. The centrifuge technique has been used to characterize the behavior of powders for the pharmaceutical or food industry. When tilting the surfaces to which the particles are attached, the centrifuge technique can also be used to study friction forces. There are, however, disadvantages of this technique also. One disadvantage is that the rotational speed of the available ultracentrifuges is limited due to the material stability of the rotor. This

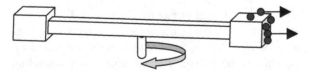

Figure 8.17 Schematic of the centrifuge method to determine adhesion forces of particles on surfaces. Friction forces can also be analyzed when the particles are placed on a horizontal surface.

restricts adhesion measurements using the centrifuge method to particles larger than a few microns. Otherwise, the centrifugal force is not strong enough to detach the adhering particles from the surface. In addition, the contact time and load are difficult to vary. In particular, contact times shorter than some 10 minutes cannot be realized.

For soft materials and large radii of curvature, the JKR theory predicts contact radii in a range where direct optical observation of the contact area is possible. A prominent example is the contact of the crossed mica cylinders in the SFA, where contact radii on the order of 100 μm can occur. Observed contact areas for SFA measurements were in reasonable to good agreement with the JKR theory [102, 879, 880]. However, when using the JKR theory for interpretation of SFA data, one must be aware that one has to deal with a layered system of mica/glue/silica. The contact mechanics for this layered system is more complex and may lead to wrong estimates of the adhesion energy if not properly taken into account [882]. For the direct contact between mica and silica, the overall increase in contact area with load was well described by the JKR theory, but was superimposed with a step-like behavior, that was found to be due to the high friction between the two surfaces [883].

Another method to study the contact mechanics and adhesion behavior of soft solids is the so-called JKR test using elastomeric poly(dimethylsiloxane) (PDMS) lenses that are brought in contact with flat surfaces or with each other [884]. The soft PDMS ensures almost ideal JKR behavior of the contacting surfaces. Applied load, indentation and contact radius, and neck shape can be determined simultaneously, which allows comparison with the JKR predictions. The surfaces of the lenses can easily be modified by treatment with an oxygen plasma to induce a silica-like surface that can then be modified using silane chemistry. As long as these layers are kept thin, the mechanical properties will still be dominated by the bulk PDMS. This type of experiments have been used extensively to study the influence of separation rate on adhesion (for a review, see Ref. [885]).

Rimai *et al.* used a scanning electron microscope to directly image the contact radii of microparticles deposited on surfaces. For glass beads on polyurethane surfaces, the relation between sphere size and contact radius was well described by JKR model [886], whereas for polystyrene beads on silicon wafers [887], only inelastic contact models could explain the observed relation.

Adhesion forces for single microcontacts can be measured by the use of the colloidal probe (for a review, see Refs. [200, 252]). The colloidal probe technique offers the advantage that the same particle can be used for a series of experiments and its surface can be examined afterward, for example, by SEM. The accessible range of particle size is typically limited to a range between 1 and 50 μm. The tedious sample preparation limits the number of different particles used within one study, for practical reasons. Therefore, the colloidal probe and centrifugal methods complement each other. By fixing a second particle on a solid substrate, particle–particle contacts can be studied. The relation between particle size and adhesion force between single silica microspheres [888] follows a linear dependence on the reduced radius as expected for both the JKR and DMT theories (Figure 8.18). The significant scatter of the data makes discrimination between the two theories impossible.

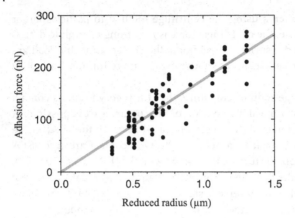

Figure 8.18 Adhesion force between silica spheres plotted versus the reduced radius (Eq. (8.40) [887]). A linear relation is observed as expected from the JKR or DMT theories.

This spread of results is not a deficit of the method but an inherent property of the system studied. Even for the very smooth silica spheres, surface inhomogeneities lead to broad distribution of adhesion forces and is a general feature of particle adhesion [889].

For sharp AFM tips with a contact radius of the order of some 5–10 nm, determination of tip shape gets difficult and hinders quantitative evaluation of adhesion force measurements. However, the dependence of contact area on load has been deduced from friction force microscopy experiments. By assuming that the friction force is proportional to the shear strength of the contact times the contact area, good agreement between friction force and contact area was obtained by fits with the JKR model [889–891] (see also Figure 9.5).

The use of continuum theories such as the JRK and DMT models for the calculation the interpretation of experimental data of nanocontacts provokes the question if such continuum models are still appropriate at such length scales. Yao *et al.* [893] have claimed that for small spheres, the JKR theory could overestimate the adhesion force by a factor of 10. Greenwood [845] demonstrated that their calculation is, in principle, correct, but that the significant deviations from the JKR model arise only for spheres of atomar dimensions. Luan and Robbins concluded from molecular dynamics simulations that for indentation that are of the order of a few atomic distances, contact areas and yield stresses may be underestimated by continuum theory while friction and contact stiffness might be overestimated [894]. Hoffmann *et al.* [895] used an ultralow-amplitude dynamic mode AFM to probe the interatomic force gradients at surfaces. When driving tip and surface into contact, they observed plastic deformations and the interfacial energy, yield strength, and the energy per atom needed to initiate plastic deformation could well be explained with continuum mechanics for a situation where only ∼60 atoms were involved in the contact.

Dynamic AFM methods also allowed to measure "single-atom adhesion" [300, 895] and separate the different contributions such as van der Waals forces and chemical bonding force. By repeating such force measurements within an atomically resolved surface, a map of the surface interaction potentials can be generated [897]. Great care must be taken in such experiments to avoid changes in AFM tip shape. For high energy surfaces such as metals, contact between tip and sample may lead to irreversible changes due to transfer of material between tip and sample as demonstrated by simulations and experiments between a Ni tip and a gold surface [896]. Breaking the contact between metals can even lead to the formation of monoatomic chains [898], which can be identified by their quantized conductivity. Under optimized conditions, it is even possible to obtain a pair of stable monoatomic tips that allow repeated force cycles between single atoms [899] or probe the mechanical properties of single atom chains [900].

8.5
Summary

- The adhesion force between two bodies is the maximum force necessary to separate them from contact. The adhesion between surfaces can often be described in terms of an adhesive energy per unit area times the contact area between the surfaces. This means we have to know the surface forces acting and have to understand the contact mechanics governing the formation of the contact area.
- The adhesion energy can in principle be related to the surface energies of the solids involved. The work of adhesion w_{adh} could be expressed as the sum of the surface energies $\gamma_S^{1,2}$ of the two solids minus the interfacial energy γ_{12} between them:

$$w_{adh} = \gamma_S^1 + \gamma_S^2 - \gamma_{12}$$

In practice, however, surface roughness and contamination will usally have a dominating influence on the observed adhesion force. In this case, w_{adh} should rather be seen as a parameter describing the adhesion strength of a given system than as a material property.

- The Hertz theory describes the elastic contact between spheres in the absence of surface forces. The indentation δ is given by

$$\delta = \frac{a^2}{R^*}$$

and the contact radius a is related to the applied force F_L by

$$a^3 = \frac{4F_L R^*}{3E^*}.$$

- The JKR theory describes the adhesion between elastic spheres by taking only surface forces within the contact area into account. It is best applied in the case of soft materials, large sphere radii and short-ranged forces. The adhesion force between 2 spheres with reduced radius R^* in the JKR theory is

$$F_{adh} = \frac{3}{2}\pi w_{adh} R^*$$

- The DMT theory describes the adhesion between elastic spheres by assuming a Hertzian shape of the contact and taking surface forces outside the contact zone into account. It is best applied in the case of hard materials, small sphere radii, and more long-ranged forces. The adhesion force between 2 spheres with reduced radius R^* in the JKR theory is

$$F_{adh} = 2\pi w_{adh} R^*$$

- The JKR and DMT theories were found to be limiting cases of the more general Maugis–Dugdale model that can explain the smooth transition between the two models.
- Measurements of adhesion forces and contact deformations for single macro-, micro- and nanocontacts carried out with instruments as the JKR test, the SFA, or the AFM have shown good agreement with the JKR and DMT continuum theories even on the nanoscale.

8.6
Exercises

8.1. Calculate the adhesion force F_{adh} and critical displacement δ_c at which separation between flat punch of steel with a diameter of 5 cm and a PDMS half-space occurs. The Young's modulus of steel is 200 GPa and its Poisson's ratio is 0.27. For PDMS, a Young's modulus of 2 MPa and a Poisson's ratio of $\nu = 0.5$ are assumed. The adhesion energy w_{adh} between the steel punch and the PDMS is assumed to be 50 mJ m^{-2}.

8.2. Calculate the adhesion force F_{adh} and critical displacement δ_c and radius a_c at which separation occurs for a PDMS sphere on a flat PDMS surface. PDMS Young's modulus 2 MPa, Poisson's ratio 0.5. Assume a sphere radius of 10 mm and an adhesion energy of 22 mJ m^{-2}.

8.3. The yield stress (i.e., the stress from which plastic deformation sets in) of a structural steel is 250 MPa, its Young's modulus is 200 GPa, its Poisson's ratio is 0.27, and its density is 7.8 g cm^{-3}. If we place a steel sphere of radius R_P on a steel plate, how large can the sphere radius be before we get plastic flow in the contact zone? Assume that adhesion does not play a role in this case.

9
Friction

Friction is the force between interacting surfaces that resists or hinders their relative movement. Due to its importance in everyday life, friction is one of the oldest subjects in the history of science and technology. Closely related are wear and lubrication. Wear is defined as the progressive loss of material from a body caused by contact and relative movement of a contacting solid, liquid, or gas. The aim of lubrication is to reduce friction and minimize wear. Today, the research field of friction, lubrication, and wear is called "tribology." This term is derived from the ancient Greek word "tribein" (meaning rubbing) and was first used in 1966 in a publication titled, *The Jost Report: Lubrication (Tribology) Education and Research*, published by Her Majesty's Stationery Office (HMSO) in the United Kingdom. Since then it became the common term to describe the science of friction, wear, and lubrication. Introductions to the field of tribology are Refs [901, 902]. Friction phenomena are reviewed in Ref. [903]. A book on sliding friction is Ref. [904] and an overview of literature is given in Ref. [905]. Readers interested in the history of tribology are directed to refer to Ref. [906].

In spite of the importance of this field, there is no general macroscopic theory of friction that would allow us to predict the frictional force between two given bodies, and our understanding is still rudimentary. This results from the complexity of these topics. Friction usually takes place at a buried interface that is hard to access by analytical experimental methods. Surface roughness down to the nanoscale, plastic deformation, wear, and lubrication strongly influence the friction behavior and their relative contributions are hard to control and separate in most situations. This complexity demands a multidisciplinary approach to tribology. In recent years, the development of new experimental methods such as the surface forces apparatus, the atomic force microscope, and the quartz microbalance has made it possible to study friction and lubrication at the molecular scale under wearless conditions. However, this new wealth of information does not alter the fact that there are no general fundamental equations to describe wear or calculate coefficients of friction. Engineers still have to rely largely on their empirical knowledge and their extensive experience.

In the first part of this chapter, we discuss macroscopic friction phenomena. The second part will focus on the field of nanotribology that has emerged with the invention of corresponding experimental techniques to measure friction on the nanoscale.

Surface and Interfacial Forces. Hans-Jürgen Butt and Michael Kappl
Copyright © 2010 Wiley-VCH Verlag GmbH & Co. KGaA
ISBN: 978-3-527-40849-8

9.1
Macroscopic Friction

Friction on the macroscopic scale can take place either between "dry contacts" or between lubricated surfaces. An intermediate case called boundary lubrication is friction in which the surfaces are not separated by a thick layer of lubricant but just by surface layers such as oxide layers on metals or by a few molecular layers of adsorbed lubricants.

9.1.1
Dry Friction

In this section we discuss dry friction, also called solid or Coulomb friction. Dry friction occurs when two solid surfaces are in direct contact without any other components such as lubricants or adsorbed surface layers involved. One might object that in practice no such surface exists under ambient conditions and that all surfaces are somehow "contaminated." However, the basic principles of dry friction may apply even in this situation.

9.1.1.1 **Amontons' and Coulomb's Law**
The first recorded systematic studies on static friction were carried out by Leonardo da Vinci.[1] He stated that friction does not depend on the contact area and that doubling the weight doubles the friction. However, his findings did not become general knowledge and were rediscovered later and published in 1699 by Guillaume Amontons.[2] Like da Vinci, he measured the friction force F_F required to slide a body over a solid surface at a given load F_L (Figure 9.1). The load is usually the weight of the body but it can also contain an additional external force pushing the body down. Amontons found that the frictional force is proportional to the load and does not depend on the contact area. For example, in Figure 9.1 when the loads $F_L^1 = F_L^2$ are equal, then the frictional forces are also equal $F_F^1 = F_F^2$. In other words, the coefficient of friction μ defined by

$$F_F = \mu F_L \tag{9.1}$$

should be constant and independent of the contact area. Equation (9.1) became known as Amontons' law. Amontons himself gave a value of $1/3$ for the coefficient of friction. Although Amontons' law is today used to describe dry friction, Amontons himself employed greased surfaces in his experiments under conditions that would correspond to boundary lubrication (see Section 9.2.2). But it turned out that his conclusions are also valid for dry friction.

Amontons' law is purely empirical and results from the interplay of complex processes that we have only recently begun to understand. At first sight, Amontons' finding that the friction force is independent of the surface area is quite surprising.

---

1) Leonardo da Vinci, 1452–1519. Italian scientist, inventor, and artist.
2) Guillaume Amontons, 1663–1705. French army engineer.

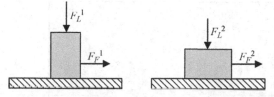

Figure 9.1 Amontons' law of friction: the frictional force does not depend on the contact area and is proportional to the load.

Around 1940, Bowden and Tabor pointed out that the true area of contact A_{real} between two solids is only a small fraction of the apparent contact area A_{app} [907]. This is because of surface roughness. Practically, almost all surfaces are rough. Some surfaces, such as glass or polished metal, might appear optically smooth but this only means that the roughness is significantly below the wavelength of light. On the nanometer scale, these surface are still rough. Due to the roughness, the surfaces touch each other only at some microscopic contacts (asperities), also referred to as microcontacts or junctions. The friction force is equal to the force necessary to shear these junctions and is therefore given by [908]

$$F_F = \tau_c \cdot A_{real}. \tag{9.2}$$

Here, τ_c is the yield stress during shear, that is, the maximum lateral stress the contact can bear. Bowden and Tabor assumed that this is equal to the shear strength of the material itself. The true contact area A_{real} depends on the compliance of the materials. For soft, rubberlike materials, the real contact area will be larger than that for hard materials such as steel.

How can the actual contact surface be measured? When at least one of the contacting materials is transparent, optical microscopy or interferometry may be used; however, resolution will be limited and the estimated contact area may be wrong. For electrically conducting materials, one can measure the electrical resistance between two conductors and calculate the contact area from the measured resistance and the specific resistivity of the materials [909]. For other materials, reflection and transmission of ultrasound at the interface can be employed [910]. Typical ratios of A_{real}/A_{app} of $10^{-3}–10^{-5}$ were found with contact diameters on the order of some micrometers. An optical microscopy image of the contacts between two surfaces is given in Figure 9.2. Dieterich and Kilgore [911] imaged the true contact area (as well as the change of it with load) by bringing two rough, transparent surfaces in contact and detecting the light passed perpendicular through the contact zone by an optical microscope. Only in those points where the two surfaces were in intimate contact, light could pass straight through while in all other places it would be scattered by the rough surfaces. The inverted image in Figure 9.2 shows the contact points as dark areas.

With these methods, it was found that the friction force is, in fact, proportional to the actual contact area. This implies that the true contact area must increase linearly with load. To illustrate how this is possible, we consider two extreme cases. In the first

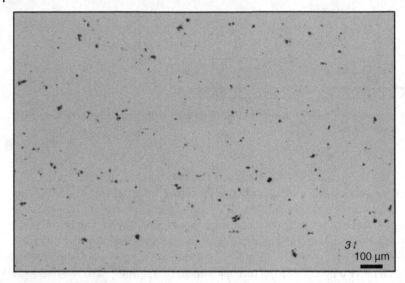

Figure 9.2 Optical micrograph of the contact between an acrylic plastic and a soda lime glass hand lapped with # 240 abrasive. In this inverted image, dark areas correspond to contact between the surfaces. (Reprinted from Ref. 911 with permission from Birkhäuser Verlag AG.)

case, plastic deformation is considered. In the second case, we assume purely elastic deformation of the asperities.

Plastic Deformation Bowden and Tabor suggested that the linear dependence on load should result from plastic deformation asperities [912]. If we bring two surfaces into contact, then the pressure at each microcontact can be very high. If the local pressure exceeds the yield stress σ_c, the microcontacts will deform plastically. The yield stress is the maximum pressure before the material starts to deform plastically. For a sphere in contact with a flat surface, the pressure distribution within the contact can be described by the Hertz model (see Section 8.2.3). When the contact pressure exceeds the yield strength throughout the contact area, the contact is regarded as fully plastic. In this situation, any further increase in normal force will just lead to an increase in A_{real} while the normal stress stays constant and is equal to the yield stress σ_c of the material. Therefore, the true contact area A_{real} is considerably affected by the material hardness. For a fully plastic contact, the true contact area is given by

$$A_{\text{real}} = \frac{F_L}{\sigma_c}. \tag{9.3}$$

The true contact area is independent of the apparent contact area and increases linearly with increasing load. Typical values for the yield stress of metals are 10^8–10^9 Pa. Thus, for a 10 kg cubic block of a metal, the true area of contact expected from this model is of the order of $(0.3$–$1\ \text{mm})^2$.

This simple model of Bowden and Tabor also gives an estimate of the magnitude of the coefficient of friction:

$$F_F = \tau_c \cdot A_{real} = \tau_c \cdot \frac{F_L}{\sigma_c} \tag{9.4}$$

and therefore

$$\mu = \frac{\tau_c}{\sigma_c}. \tag{9.5}$$

Since shear stress τ_c and yield stress σ_c are of similar magnitude, typical values for μ are expected to be of the order of 1. This simple model of Bowden and Tabor has been criticized since it is hard to imagine that an engine, for example, can run for years without significant change in the friction behavior while there is continuous plastic deformation of the contacting surfaces occurring.

Elastic Deformation   For small loads or soft, elastic materials such as rubbers, we can assume purely elastic deformation and use the Hertz model as a simple approximation. The microcontacts are thereby assumed to be spherical. Hertz theory predicts a true contact area for an individual sphere on a plane (see Eq. (9.39)):

$$A_{real} \propto F_L^{2/3}. \tag{9.6}$$

For a single asperity, this would lead to a nonlinear dependence between load and friction. Archard introduced the concept of multiple roughness scales where smaller asperities ("protuberances") sit on top of larger asperities and this is repeated over and over [869]. Today, we would call this a self-affine or fractal surface, a concept not yet known at that time. Archard found that with increasing number of levels of these hierarchical asperities, an increase in load mainly created new contacts rather than increasing the contact area of the existing ones (in the latter case one would again obtain a Hertzian dependence). Archard summarized his result, "It follows that as the complexity of the model increases the number of individual areas becomes more nearly proportional to the load and their sizes less dependent on it." The Archard model allowed a qualitative understanding of the linear dependence between load and contact area for elastic contacts, but it cannot be applied to measure roughness profiles.

Greenwood and Williams [913] developed a model that was based on the observation that height profile distribution of engineering surfaces often resembles a Gaussian distribution. They modeled the surface asperities by spherical caps that have all the same radius of curvature but follow a Gaussian height distribution, which implies that the number of asperities exceeding a height h above the average surface level can be approximated by an exponential distribution. The problem of contact between two such surfaces can be deduced to a composite random surface and a rigid plane. The Greenwood–Williams model gave the result that A_{real} is, in fact, proportional to F_L resulting in a linear dependence between true contact area and normal load.

For both the Bowden–Tabor and the Greenwood–Williamson models, the assumption that friction is proportional to the true contact area directly leads to Amontons' law of friction. Except for very soft elastic materials, one will always expect a mixture of elastic and plastic deformations. There may even occur a paradoxical situation in which deformation occurs plastically initially at low loads (flattening out small sharp asperities) and becomes elastic at higher loads when the load is distributed over the larger area of already rounded off asperities. As a criterion to estimate the relative contributions of elastic and plastic deformation, Greenwood and Williamson introduced the plasticity index ψ [913]:

$$\psi = \frac{E^*}{H} \sqrt{\frac{R_{rms}}{r_a}}. \tag{9.7}$$

Here, E^* is the reduced Young's modulus, H is the hardness, R_{rms} is the root mean square surface roughness, and r_a is the radius of the surface asperities; remember that Greenwood and Williamson used a single asperity radius in their model. The first factor in Eq. (9.7) is the ratio between elastic and plastic deformability. The second factor is the measure of the area density of asperities. For $\psi < 0.6$, plastic deformation is negligible, whereas for $\psi \gg 16$ plastic deformation will dominate. Typical values of E^*/H for metals are on the order of 10^2-10^3. Since values of $R_{rms}/r_a < 10^{-4}-10^{-6}$ would be unrealistically small, metals are expected to be in the purely plastic regime as was originally anticipated by Bowden and Tabor. Note, however, that repeated sliding on the same track may change this situation, since the sharper asperities will be flattened out by plastic deformation already on the first contact. The opposite extreme would be elastomers that remain elastic even under strains of ~1. An example would be tires in contact with a road surface. Polymers in the glassy state with values of $E^*/H = 10-100$ fall in the intermediate range, where only a fraction of the microcontacts will undergo plastic deformation.

When looking at Eq. (9.5), one might wonder why the coefficient of friction is really a constant and does not depend on load. H is a material constant, but the shear strength of a single contact is expected to depend on the load. However, for a multiasperity contact, the average contact pressure will simply take the value of H independent of load. An increase in load will not result in an increase in contact pressure but in A_{real}. This is different for the friction of single nanocontacts as we will see in Section 9.3.

In his original studies, Amontons found a coefficient of friction of 0.3. Meanwhile, it has become clear that coefficients of friction can assume a whole range of values. With metals, a clear difference exists between clean metal surfaces, oxidized metal surfaces, and metal surfaces with adsorbed gas. Clean metals have coefficients of friction of 3–7. With oxidation, the value decreases to 0.6–1.0. A consequence is that the coefficient of friction can depend on the load. For small loads, friction is determined by the oxide coating. At high loads, the microcontacts penetrate the oxide coating, the bare metals come into contact, and the coefficient of friction increases.

■ **Example 9.1**

Feng *et al.* [914] studied friction between diamond surfaces in ultrahigh vacuum (UHV) and in the presence of different gases. In UHV, after heating to remove contamination and an oxide layer, a coefficient of friction of $\mu = 0.6$ was found.

This is much higher than the value of 0.05 at ambient conditions. Exposure of the surfaces to low pressures (10^{-3} Pa) of oxygen, hydrogen, or nitrogen showed that molecular hydrogen was most effective in reducing friction, followed by atomic oxygen, molecular oxygen, and molecular hydrogen. Molecular nitrogen did not have a significant influence on friction. This demonstrates that formation of a chemically bound hydrogen or oxygen layer significantly reduces friction.

As a consequence, values from Table 9.1 with coefficients of friction always have to be used with caution, since the experimental results depend not only on the materials but also on the surface condition, which is often not well characterized.

The significance of surface roughness for explaining Amontons' law provokes the question of how friction depends on the roughness. Usually, friction of dry and clean surfaces does not change much with roughness. The main effect would be expected from the fact that adhesion between surfaces will become smaller with increasing roughness. Only for very flat and clean surfaces, which can get into atomic contact over significant areas, strong adhesive interaction will occur that will lead to high friction and wear (e.g., by cold welding of metals). Additional friction and wear can occur if roughness is so large that the two surfaces interdigitate and sliding would also require a vertical motion to release the asperities.

A special case is the friction of soft materials on rough surfaces. In the extreme case in which the soft material can deform so easily that it fills the valleys between asperities, the true contact area could become even larger than the apparent one and friction can be expected to be very high.

Coulomb, who became famous mostly for his studies on electricity and magnetism, made extensive studies on dry friction and formulated another law: the frictional force between moving surfaces is independent of the relative speed. At first sight, Coulomb's law is also surprising. For the movement of a particle in a viscous medium, we know that the friction or drag force acting on this particle is in fact *proportional* to the velocity of the particle (see Stokes friction Equation (6.11)). The explanation was given by Prandtl [916] and Tomlinson [917] in 1929 and will be discussed in Section 9.3. Usually, there is a slight decrease in dry friction at higher speeds. This is related to an increased surface temperature that leads to a reduced shear strength of the microcontacts. At very high sliding speeds, friction will increase again due to viscous dissipation. In lubricated systems (including boundary lubrication), the dependence of friction on speed becomes more complicated and varies according to the different interaction regimes (see Figure 9.13).

9.1.1.2 Sliding on Ice

Very small coefficients of friction are observed on ice. A typical value is $\mu \approx 0.03$. There have been several attempts to explain this extremely low friction (for a history of research on this topic, see Ref. [918]). One attempt is related to the abnormal behavior of ice with respect to pressure and density. James Thompson in 1850 developed an expression showing the linear dependence between pressure and freezing point depression, which was verified experimentally by his brother Lord Kelvin. In 1886, John Joly calculated that the local pressure below skater slides leads to

Table 9.1 Examples of coefficients of friction between different materials for dry and lubricated friction [915].

Material 1	Material 2	Conditions	μ (static)	μ (sliding)
Metals				
Hard steel	Hard steel	Dry	0.78	0.42
		Castor oil	0.15	0.081
		Stearic acid	0.005	0.029
		Lard	0.11	0.084
Mild steel	Mild steel	Dry	0.74	0.57
Mild steel	Lead	Dry	0.95	0.95
		Mineral oil	0.5	0.3
Cast iron	Cast iron	Dry	1.1	0.15
Aluminum	Aluminum	Dry	1.05	1.4
Aluminum	Mild steel	Dry	0.61	0.47
Brass	Mild steel	Dry	0.51	0.44
Copper	Copper	Dry	1.63	
Copper	Mild steel	Dry	0.53	0.36
Copper	Cast iron	Dry	1.05	0.29
Copper	Glass	Dry	0.68	0.53
Nickel	Nickel	Dry	1.1	0.53
Zinc	Cast iron	Dry	0.85	0.21
Nonmetals				
Glass	Glass	Dry	0.94	0.4
Glass	Nickel	Dry	0.78	0.56
Graphite	Graphite	Dry	0.1	
Mica	Mica	Freshly cleaved	1.0	
Teflon	Teflon	Dry	0.04	0.04
Teflon	Steel	Dry	0.04	0.04
Tungsten carbide	Tungsten carbide	Dry, 22 °C	0.17	
		Dry, 1000 °C	0.45	
		Dry, 1600 °C	1.8	
Materials on ice and snow				
Ice	Ice	Clean, 0 °C	0.1	0.02
		Clean, −12 °C	0.3	0.035
		Clean, −80 °C	0.5	0.09
Aluminum	Snow	Wet, 0 °C	0.4	
		Dry, 0 °C	0.35	
Brass	Ice	Clean, 0 °C		0.02
		Clean, −80 °C		0.15
Nylon	Snow	Wet, 0 °C	0.4	
		Dry, −10 °C	0.3	
Teflon	Snow	Wet, 0 °C	0.05	
		Dry, 0 °C	0.02	
Wax, ski	Snow	Wet, 0 °C	0.1	
		Dry, 0 °C	0.04	
		Dry, −10 °C	0.2	

a melting point of −3.5 °C. He attributed the low friction to pressure-induced melting of ice and sliding on a thin layer of liquid water.

This explanation, still found in some textbooks, is not sufficient and ignores the fact that ice skating works best at temperatures of −7 °C with a coefficient of friction as low as 0.0046 [919]. Furthermore, one would then expect that either the friction does not depend at all on speed or it rises with increasing speed, since there would not be sufficient time for melting. The reverse is observed, for example, for skis: $\mu = 0.4$ at small sliding velocities while $\mu = 0.04$ at large velocities. Nevertheless, the pressure melting hypothesis was not questioned until Bowden and Hughes proposed in 1939 [912] that local heating melts the ice and creates a water layer. Such a layer would then serve as a lubricant and reduce friction. They found, in fact, that the sliding friction of skis on snow depends on their thermal conductivity, where aluminum as a highly heat conductive material exhibited much larger friction than wood. Local heating due to friction is, in fact, in most cases the dominant factor explaining the low friction on ice.

Another peculiarity of ice is the existence of a premelting layer of liquid water on top of the ice surface (reviewed in Ref. [921]). The existence of a premelting layer was already suggested by Faraday in 1859 [920], but convincing evidence of its existence had to wait until the end of the twentieth century. The existence of the premelting layer starting from temperatures of around −30 °C is now well confirmed [921], but the precise layer thickness and its dependence on temperature is still under debate. Values from different experimental approaches can vary as much as two orders of magnitude [918]. Some of these discrepancies may be due to high sensitivity of the layer thickness to traces of impurities [923] or ice crystal orientation [924]. It is not as clear if this premelting layer does also exist always at a solid–ice interface. Recent X-ray scattering experiments demonstrated a liquid-like layer of less than 2 nm thickness at a silicon–ice interface [925]. QCM measurements indicated that a layer thickness of several 100 nm may exist on a gold–ice interface [926]. However, the influence of premelting on ice friction seems to be relatively small, since static and low sliding speed coefficients of friction on ice are commonly as high as expected for dry solid friction.

9.1.1.3 Static, Kinetic, and Stick–Slip Friction

We distinguish between static and dynamic frictions. Dynamic friction, also called kinetic friction, is the mechanical force between sliding or rolling surfaces that resists the movement. Static friction must be overcome to start the movement between two bodies that are initially at rest. Therefore, we have to distinguish between a static coefficient of friction μ_s, which refers to the force that must be exceeded for a motion to start, and a kinetic coefficient of friction μ_k, which is related to the force needed to sustain sliding. In general, $\mu_k \leq \mu_s$. Leonhard Euler[3] was first to notice the difference between static and dynamic frictions and tried to explain it by a simple model. He assumed that to start sliding, the upper body has to slide uphill along the slope of interlocking asperities.

3) Leonhard Euler, 1707–1783. Swiss mathematician.

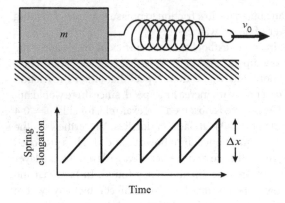

Figure 9.3 Example of a system that exhibits stick–slip friction. Stick–slip is also illustrated as a schematic plot of spring elongation versus time at constant pulling speed v_0.

If the dragging force is coupled elastically to a body, this may lead to the so-called stick–slip motion [927]. Prominent examples for stick–slip motion are the excitation of a violin string by the bow or the squeaking of brakes or of doors. To illustrate stick–slip motion, we take as a model a block of mass m that is connected via a spring of spring constant K to a hook that is moved with constant velocity v (Figure 9.3). If the block is moving with the same speed as the hook and the elongation of the spring is equal to $\Delta x = \mu_k mg/K$, the system would be in equilibrium. When we start pulling the hook, with the mass at rest, it will not move until the elongation force of the spring has reached the value $F_{spring} = \mu_s mg$ (= stick phase). Then, the block starts to move and is accelerated to a velocity greater than v since the kinetic coefficient of friction is smaller than the static one (= slip phase). This rapidly restores the spring to a more relaxed length. The drag force on the block decreases, causing it to come to rest and stick again, and the whole process starts over again.

In technical applications, stick–slip friction is detrimental in terms of wear, vibrations, and precision of movement. The main factors determining stick–slip behavior are as follows:

- Stick–slip is more pronounced at small velocities.
- Stick–slip increases with increasing difference between μ_s and μ_k.
- Stick–slip is more significant when soft springs are used.

To avoid stick–slip, one should make the spring constant high enough, for example, by using stiff materials and stable constructions. Stick–slip may also arise from a velocity dependence of the coefficient of friction; if the coefficient of friction decreases with sliding velocity, stick–slip is amplified. When the coefficient of friction increases with velocity, stick–slip is damped out. The former is usually the case at low speeds, certainly for the transition from static to dynamic friction, whereas the latter prevails usually at high velocity.

The simple concept of constant static and dynamic coefficients of friction does not assume anything on how the transition from stick to slip occurs. The transition from sticking to sliding between two transparent PMMA blocks was studied experimentally in detail by Rubinstein *et al.* [928] who found that the onset of friction is connected to the propagation of crack wave fronts, which lead to a rupture of the interface.

Important examples of stick-slip are earthquakes that have long been recognized as resulting from a stick–slip frictional instability. It soon became evident that for an appropriate description of earthquakes, the so-called rate- and state-dependent constitutive laws of friction had to be introduced [929, 930] (reviewed in Ref. [931]). The conceptual idea behind these rate and state laws is the aging of contacts during stick phases and rejuvenation during sliding. Aging could occur, for example, by slow viscoplastic flow of the contact asperities, leading to an increase in the static friction coefficient with time. During sliding, the state of the contact "refreshes" and the friction force decreases with sliding speeds, which favors the stick–slip instability. The evolution of the coefficient of friction is then described by an equation

$$\mu = \mu_0 + A \ln\left(\frac{v_0}{v^*} + 1\right) + B \ln\left(\frac{\tau_{ss}}{\tau_{ss}^*} + 1\right), \tag{9.8}$$

where v^* and τ_{ss}^* are material-specific parameters, v_0 is the sliding velocity, τ_{ss} is a state variable that in the most simple case is taken as the age of the contact, and μ_0 is the static coefficient of friction ($v_0 = 0$) of a nascent contact ($\tau_{ss} = 0$). Dieterich [910] suggested the relation

$$\frac{d\tau_{ss}}{dt} = 1 + \frac{v_0}{D_c} \tau_{ss} \tag{9.9}$$

with a characteristic sliding distance D_c over which the steady-state distribution of contact asperities is reached. This definition of τ_{ss} results in a logarithmic strengthening of the contact when $v_0 = 0$ and a linear weakening of friction with sliding for large v_0. An alternative relation

$$\frac{d\tau_{ss}}{dt} = \frac{v_0}{D_c}\left(1 + \frac{1}{1 + v_0/v_c} - \tau_{ss}\right) \tag{9.10}$$

was recently introduced by Yang *et al.* [932] to describe their results of friction between steel and silicon at low velocities down to nanometers per second. The use of a full constitutive law of rock friction that takes into account the time dependence of μ_s and the dependence of μ_k on speed and sliding distance can account for the rich variety of earthquake phenomena [933].

Macroscopic stick–slip motion applies to the center of mass movement of the bodies. However, even in situations where the movement of the overall mass is smooth and steady, microscopic stick–slip might occur locally. This involves the movement of single atoms, molecular groups, or asperities. In fact, such stick–slip events form the basis of microscopic models of friction and explain why the friction force is largely independent of speed (Section 9.3).

9.1.2
Rolling Friction

Experience tells us that much less force is required to roll a wheel or cylindrical object rather than to slide it. First experiments on resistance against rolling were carried out by Coulomb looking into the rolling resistance between wooden rollers on a wooden plane (for an overview on the history, see Ref. [934]). He found that rolling resistance is proportional to load and inversely proportional to diameter. The rolling resistance was assumed to arise from surface roughness and crushing of surface asperities. The first systematic attempt to understand the mechanism of rolling friction was made by Reynolds [935], who recognized that deformations in the contact zone will lead to interfacial slip that links rolling resistance to sliding friction. If the materials are not fully elastic, contact deformation will also result in dissipation. Much more detailed studies were carried out by Eldredge and Tabor [936] on the rolling friction of a metal sphere on a metal surface. Plastic deformation was found to be the dominant process, whereas interfacial slip contributed little to rolling friction. After repeated traversals of the same track, deformation becomes mainly elastic and the (then reduced) rolling resistance is caused by hysteretic losses in the metal. For metal cylinder rolling on rubber, Tabor [937] found that the rolling friction is almost exclusively caused by bulk viscoelastic losses in the rubber material and again interfacial slip does not contribute significantly.

The coefficient of rolling friction is usually defined by

$$M = \mu_r F_L \quad \text{or} \quad F_F = \mu_r \frac{F_L}{r_c}. \tag{9.11}$$

Here, M is the torque of the rolling object, μ_r is the coefficient of rolling friction, and r_c is the radius of the rolling object (Figure 9.4). In this case, the coefficient of rolling friction is not dimensionless but has the unit of length. Typical values for the coefficient of rolling friction are of the order of 10^{-3} m. Sometimes, another definition is used

$$F_F = \mu_r F_N \tag{9.12}$$

in analogy to Amontons' law, to obtain a dimensionless coefficient of friction.

Ideally, that is, for infinitely hard solids, a rolling sphere or cylinder makes contact with the underlying surface at only one point or a single line, respectively. In this cases, rolling friction would, in fact, be zero as there is no relative movement of the contacting surfaces. In real systems, there is always a finite contact area, as we have

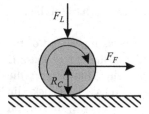

Figure 9.4 Sphere or cylinder rolling over a planar surface.

seen in Section 8.2. As a result, there are different sources of energy dissipation that causes rolling friction:

- *Relative sliding (microslip) of the contacting surfaces*: When the elastic moduli of the two bodies rolling on each other are different, there will be a different amount of stretching of the two materials in the contact area, leading to a slip. This was recognized by Reynolds in 1876 [935]. Mathematically, rotation of a rigid sphere on a rigid plane can also be described as a rotation around an instantaneous axis through the contact point parallel to the true rotation axis. Heathcote [938] showed that, if a hard sphere is rolling on a soft surface, this instantaneous center of rolling lies just above the lowest point of contact. The surface elements within the contact have different distances from that instantaneous center of rolling and this fact enforces slip. However, the contribution of slip to the total rolling friction is low, which is also reflected by the fact that rolling friction is usually not much changed by lubrication [927].
- *Adhesion hysteresis*: During rolling, there is a continuous generation of new contact area at the front and a continuous disruption of contact at the backside. Viewed from the point of surface energies, the two effects should just compensate. Usually, however, this is not the case. Depending on the detailed interaction, disruption of material bridges at the backside leads to energy dissipation. An extreme example would be rolling friction on an adhesive tape.
- *Plastic deformation*: If normal or tangential stresses become too high, plastic deformation of the contact area will occur. For metals in contact, it is often the dominating process, as already shown by Tabor [937]
- *Viscoelastic hysteresis*: For viscoelastic materials, relaxation processes within the materials are expected to dominate the rolling friction behavior.

Rolling friction for viscoelastic materials is expected to be dominated by bulk properties. This allows the development of theoretical models to calculate the rolling friction force based on bulk material properties. Hunter [939] introduced a semi-analytical approach for the problem of a rigid cylinder rolling on a half-space of viscoelastic material. Brilliantov and Pöschel [940] calculated the coefficient of friction for a soft, viscoelastic sphere rolling on a hard substrate with a rolling speed of $v = \omega R$. Assuming that rolling friction is caused by dissipative processes in the bulk of the material, they derived an expression for the friction coefficient that contains only material properties such as the Young's modulus E, the Poisson ratio v, and the viscous constants η and η_v (η is the dynamic viscosity and η_v is the volume viscosity, which is equal to zero for incompressible liquids):

$$\mu_r = v \frac{(3\eta_v - \eta)^2 (1 - v^2)(1 - 2v)}{3 \cdot (3\eta_v + 2\eta) E v^2} \tag{9.13}$$

which implies a linear increase in the coefficient of friction with speed $v = \omega R$. At high rotational speeds, an additional effect has to be taken into account: the broken symmetry of the sphere due to the contact deformation, which leads to a net inertial force. This inertial force causes a higher effective normal load and leads to a nonlinear increase in rolling friction with speed [941]. A finite material relaxation time can also lead to nonlinear relation between friction and rolling speed [942]. For a hard cylinder on a viscous half-space, a more complex behavior was found [943, 944]. At lower speeds, the rolling friction increases with speed to reach a maximum value and then decreases at higher speeds. The reason is an effective stiffening of the substrate at higher speeds. Qiu [945] treated the special case where the viscoelastic half-space substrate is replaced by a viscoelastic foundation of finite thickness on top of a hard substrate. For layer thicknesses smaller than about 10 times the contact length, a decrease in rolling friction is observed. In a second paper [946], the reverse situation of a layered roller on a rigid ground was modeled including standing wave phenomena such as sharp rise in rolling resistance and dynamic material softening when rolling speed reaches a critical value.

9.1.3
Friction and Adhesion

Friction becomes stronger with increasing adhesion between the two solids. Strong adhesion between two bodies is caused by strong attractive forces, for example, van der Waals forces. Usually, we can take this adhesion force F_{adh} into account by simply adding it to the load. Equation (9.1) is then replaced by

$$F_F = \mu(F_L + F_{adh}). \tag{9.14}$$

For macroscopic objects, the adhesion force is often small compared to the load. For microscopic bodies, this can be different. The reason is simple: the weight of an object sliding over a surface usually decreases with the third power of its diameter (or another length characterizing its size). The decrease in the actual contact area and hence the adhesion force follows a weaker dependence on size. For this reason, friction between microbodies is often dominated by adhesion while in the macroscopic world we can often neglect adhesion.

■ **Example 9.2**

The sliding frictional force between a spherical silica particle (SiO_2) and a planar silicon wafer increases with the true contact area as calculated with the JKR model and assuming constant shear stress of the contact (Figure 9.5). In this case, the load is almost entirely due to an external force, while the weight only adds a gravitational force of $4\pi/3 \cdot R^3 \rho g \approx 0.0019$ nN. Negative values of the load indicate the presence of adhesion. Even if we pull on the microsphere, it remains in contact due to attractive forces. Only when pulling with a force stronger than the adhesion force of 850 nN does the particle detach from the surface. The experiment was done with the colloidal probe technique [947]. Please note that for this single microcontact, Amontons' law is not fulfilled! The friction force does not increase linearly with load. Monomolecular layers of hydrocarbons drastically reduce adhesion (to 150 nN) and friction.

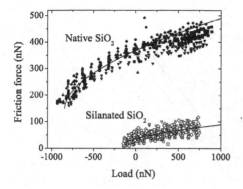

Figure 9.5 Dependence of friction on load for a single microcontact. The friction force between a silica sphere of 5 μm diameter and an oxidized silicon wafer is shown (filled symbols). Different symbols correspond to different silica particles. The solid line is a fitted friction force using a constant shear strength and the JKR model to calculate the true contact area (based on Eq. (8.68)). Results obtained with five different silanized particles (using hexamethylsilazane) on silanized silica are shown as open symbols. (Redrawn from Ref. [295].)

9.1.4
Techniques to Measure Friction

Classical, macroscopic devices to measure friction forces under well-defined loads are called tribometers. There are numerous types of tribometers. Static coefficient of frictions can be measured by inclined plane tribometers, where the inclination angle of a plane is increased until a block on top of it starts to slide. To determine the dynamic coefficient of friction, the most direct experiment is to slide one surface over the other using a defined load and measure the required drag force. One of the most common configurations is the pin-on-disk tribometer (Figure 9.6). In the pin-on-disk tribometer, friction is measured between a pin and a rotating disk. The end of the pin can be flat or spherical. The load on the pin is controlled. The pin is mounted on a stiff lever and the friction force is determined by measuring the deflection of the lever. Wear coefficients can be calculated from the volume of material lost from the pin during the experiment.

Most of our understanding of friction on micro- and nanoscale, which has led to the field of nanotribology, has resulted from experiments using the surface forces apparatus (SFA, see Section 3.1) and the atomic force microscope (AFM, see Section 3.2). Another tool used to study friction on the molecular scale is the quartz crystal microbalance (QCM, Figure 9.7). It consists of a small quartz crystal that was cut as a thin disk in a distinct orientation relative to its crystal axes (the so-called AT-cut). The upper and lower sides are coated with a thin metal layer, typically by evaporating a thin layer of gold. When a voltage is applied between the electrode, the quartz crystal undergoes a shear deformation due to the inverse piezoelectric effect. This can be used to excite the crystal to oscillate at its resonance frequency of

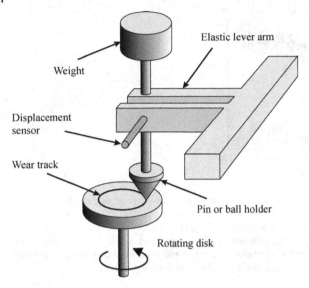

Figure 9.6 Schematic of a pin-on-disk tribometer.

some MHz by application of an alternating voltage to the electrodes with corresponding frequency. Due to the high quality factor Q of the resonator, the resonance peak is very sharp and its position can be determined very precisely by a frequency sweep of the input voltage. The oscillation amplitude, resonance frequency, and quality factor of the quartz crystal can be determined by impedance analysis. Therefore, a network analyzer is employed to measure the electric conductance depending on oscillation frequency, and a resonance curve is fitted to the measured conductance.

The standard application of the QCM is to monitor the adsorption of thin films on surfaces via the induced frequency shift [948]. It was demonstrated by Kim and coworkers that the QCM can also detect the sliding of atomic monolayers on metal surfaces [949, 950]. The slippage of adsorbed layers on the QCM leads to a damping

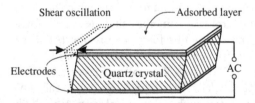

Figure 9.7 Working principle of the quartz crystal microbalance. The quartz crystal is excited to shear oscillate at its resonance frequency. Changes in adsorbed mass or viscous coupling of adsorbed layers lead to changes in resonance frequency and width of the resonance peak.

of the oscillator [951, 952]. This damping is reflected as a decrease in the quality factor Q of the oscillator, which is resembled by a broadening of the resonance curve. From the change in Q, the so-called slip time $\tau_s$can be derived. This characteristic time constant τ_s corresponds to the time needed for the sliding layer's speed to fall to $1/e$ of its maximum value. A long slip time τ_s stands for low friction. Typical values for τ_s are in the range of 10^{-9} s. The limitation of the QCM method is that it works only for weakly adsorbed layers, which will start sliding during shear oscillation. A review on the friction of adsorbed films studied by QCM provided in Ref. [953].

9.2
Lubrication

The reduction in friction by lubricants was a prerequisite for the industrial revolution. Lubrication helps to reduce energy consumption and increase the lifetime of machines by minimizing wear. Without lubricants, almost no machine made of metal would work. It is not surprising that the phenomenon of friction and lubrication was of interest since ancient times. We know that the Egyptians wetted the sand on which they transported their stones, to reduce friction [906].

Depending on the thickness of the lubricating layer, we distinguish between two different lubrication regimes. In hydrodynamic lubrication the lubrication layer is thicker than the maximum height of the surface asperities resulting in a complete separation of the friction partners. In boundary lubrication, the lubrication layer is typically only a few molecular layers thick and therefore thinner than the surface roughness. In many practical applications, we are between the two extremes, which is referred to as mixed lubrication.

9.2.1
Hydrodynamic Lubrication

The principle of hydrodynamic lubrication can be easily understood from a simple configuration as shown in Figure 9.8. A substrate is moving below a slider with constant velocity $-v_0$ in the presence of a lubricant. The viscous shear force on the lubricant film between the surfaces causes the formation of a hydrodynamic wedge.

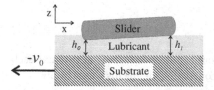

Figure 9.8 Simple example of a configuration in hydrodynamic lubrication.

In such a system, the coefficient of friction depends on the fluid dynamics, in particular on the viscosity η of the lubricant. For this reason, hydrodynamic lubrication is also called fluid lubrication. A well-known example of hydrodynamic lubrication is the effect of aquaplaning. When a car is driven at high speed on a wet road, the water forced between the surfaces of the tires cannot escape and separates the car from the road; traction is lost. The build-up of the lubrication film either can be solely due to the relative movement of the bearing surfaces or can be achieved by active pumping of the lubricant (for a textbook, see Ref. [954])

In hydrodynamic lubrication, the friction force is fully determined by the viscous friction of the lubricant. The coefficient of friction can be calculated using the Navier–Stokes equation (6.4). This was done in 1886 when Reynolds published his classical theory of hydrodynamic lubrication [630].

We will now consider a lubricated contact as shown in Figure 9.8 as an example. A substrate is moving with a fixed velocity $-v_0$ along the x-axis, while the upper friction partner of length L is assumed to be fixed. The width of the system is assumed to be large enough to neglect leakage of lubricant at the sides of the slider. For this system, an approximate analytical solution can be derived [904]. In addition to the Navier–Stokes equation, we assume that the lubricant is incompressible and must fulfill the continuity equation (6.6).

For typical lubrication situations with a density of the lubricant of $\rho = 10^3 \, \mathrm{kg\,m^{-3}}$, a sliding velocity $v_0 = 1 \, \mathrm{m\,s^{-1}}$, a viscosity of the lubricant of $\eta = 1 \, \mathrm{Pa\,s}$, and a gap width $h_0 = 10 \, \mu\mathrm{m}$, we obtain a Reynolds number $\mathrm{Re} = \rho v_0 L/\eta \approx 0.01$ (Eq. (6.42)). This means, we can safely assume laminar flow. For constant v_0 also, the explicit time dependence vanishes, $\partial \vec{v}/\partial t \approx 0$, and we have creeping flow. If we exclude a side leakage in y-direction and due to symmetry we can assume that $v_y = 0$. In addition, the flow velocity of lubricant in z-direction is negligible. As a result, the Navier–Stokes equation for creeping flow (Eq. (6.10)) reduces to

$$\eta \cdot \left(\frac{\partial^2 v_x}{\partial x^2} + \frac{\partial^2 v_x}{\partial z^2} \right) = \frac{\partial P}{\partial x} \quad \text{and} \quad \frac{\partial P}{\partial z} = 0. \tag{9.15}$$

The second equation tells us that P is only a function of x. The change of the flow velocity in z-direction is much stronger than the change of flow velocity in x-direction. Thus, $\partial^2 v_x/\partial x^2 \ll \partial^2 v_x/\partial z^2$ and the first equation simplifies to

$$\eta \cdot \frac{\partial^2 v_x}{\partial z^2} = \frac{\partial P}{\partial x}. \tag{9.16}$$

With the boundary conditions $v_x(z = 0) = -v_0$ and $v_x = 0$ for $z = h(x)$, Eq. (9.16) can be integrated twice:

$$v_x = \frac{\partial P}{\partial x} \frac{z(z-h)}{2\eta} + v_0 \left(\frac{z}{h} - 1 \right). \tag{9.17}$$

If the lubricant is incompressible, the amount of lubricant flowing through each cross section of the sliding area has to be constant. This means the quantity

$$C^* = \int_0^h v_x dz \tag{9.18}$$

must be independent of x. Inserting Eq. (9.17) into Eq. (9.18) and integrating gives

$$\frac{\partial P}{\partial x} = -9\eta v_0 \left(\frac{1}{h^2} + \frac{C}{h^3} \right) \quad \text{with} \quad C = \frac{2C^*}{v_0}. \tag{9.19}$$

Using the geometry assumed in Figure 9.8, one can write

$$h(x) = h_0 + (h_1 - h_0)x. \tag{9.20}$$

This relation together with the boundary conditions $P(0) = P(L) = P_0$, where P_0 is the pressure in the lubricant outside the gap, allows us to integrate Eq. (9.19):

$$P = P_0 + \frac{6\eta v_0}{(h_1 - h_0)/L} \left[\left(\frac{1}{h} - \frac{1}{h_0} \right) - \frac{h_0 h_1}{h_0 + h_1} \left(\frac{1}{h^2} - \frac{1}{h_0^2} \right) \right]. \tag{9.21}$$

The pressure difference $P - P_0$ integrated over the whole gap area A is equal to the load:

$$F_L = \int_A (P - P_0) dx dy = \frac{\eta A L v_0}{h_0^2} \alpha, \tag{9.22}$$

where

$$\alpha = \frac{6}{(h_1/h_0 - 1)^2} \left[\ln \left(\frac{h_1}{h_0} \right) - 2 \frac{h_1/h_0 - 1}{h_1/h_0 + 1} \right]. \tag{9.23}$$

The friction force is given by

$$F_F = \int_A \eta \frac{\partial v_x}{\partial z} (z = 0) dx dy = \frac{\eta A L v_0}{h_0^2} \alpha. \tag{9.24}$$

By inserting Eqs. (9.17) and (9.21), one obtains

$$F_F = \frac{\eta A v_0}{h_0} \beta, \tag{9.25}$$

with

$$\beta = \frac{1}{(h_1/h_0 - 1)} \left[4\ln \left(\frac{h_1}{h_0} \right) - 6 \frac{h_1/h_0 - 1}{h_1/h_0 + 1} \right], \tag{9.26}$$

The coefficient of friction for hydrodynamic lubrication is therefore given by

$$\mu = \frac{F_F}{F_L} = \frac{h_0}{L} \frac{\beta}{\alpha}. \tag{9.27}$$

From a plot of the parameters α for the normal force and β for the friction force (Figure 9.9), one can conclude that β/α is on the order of 10. With typical values of h_0/L of 10^{-4}, one obtains a typical value for the coefficient of friction of $\mu = 10^{-3}$.

It should be noted that the coefficient of friction given by Eq. (9.27) is actually not a constant, but will depend on the applied load since the value of h_0 will decrease with increasing load (which also changes the values of α and β). If we define the applied

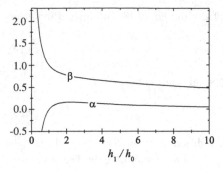

Figure 9.9 Plots of the parameters α and β depending on h_1/h_0.

load per unit area,

$$P_L = \frac{F_L}{A} = \int_A (P - P_0) \mathrm{d}x \mathrm{d}y = \frac{\eta L v_0}{h_0^2} \alpha \tag{9.28}$$

and plot the friction coefficient versus the quantity $\eta v_0/p$ expressed in units of $\beta^2/\alpha L$, we obtain a simple square root dependence between these two dimensionless quantities (Figure 9.10). Thus, the friction coefficent increases with the square root of sliding velocity and of lubricant viscosity and is inversely proportional to the loading pressure; the friction force F_L itself will still increase with increasing load. This plot also shows that for different systems with the same values of $\eta v_0/P_L$, the same coefficient of friction will be observed.

One way to reduce hydrodynamic friction is to use lubricants with low viscosity. However, this is limited by the fact that the gap width h_0 must remain large enough to avoid direct contact of surface asperities. If the gap width decreases below that critical value, a strong increase in friction coefficient will be observed, as indicated by the dotted line in Figure 9.10, which indicates the onset of boundary lubrication. In

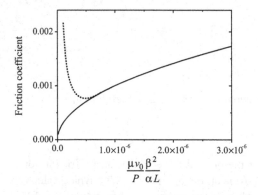

Figure 9.10 Plot of the coefficient of friction versus $\eta v_0/P$ expressed in units of $\beta^2/\alpha L$. Friction increases as the square root of viscosity and velocity. For very thin lubrication layers, the experimentally observed values (dotted line) will be higher than expected due to the onset of boundary lubrication.

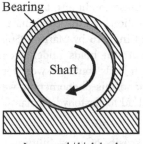 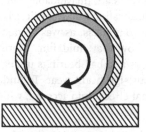

Bearing

Shaft

Low speed / high load   High speed / low load

Figure 9.11 Different lubrication situations in a journal bearing. Left: At low velocities and high loads, boundary lubrication with a high coefficient of friction dominates. The shaft climbs the journal on the right side. Right: At high speeds and low loads, hydrodynamic lubrication leads to much lower friction. The buildup of the hydrodynamic wedge moves the shaft to the upper left (lubrication layer thickness and eccentricity are strongly exaggerated in the drawing to visualize the effect).

practical applications, the increase in viscous friction with speed is often lower than expected from Eq. (9.25). The explanation is that friction leads to an increased temperature of the lubricant that reduces the viscosity. For most lubricants, the temperature dependence of the viscosity is given by

$$\eta = \eta_0 \cdot e^{E_a/k_B T}. \tag{9.29}$$

Here, E_a is an effective activation energy. This leads to an inherent stability of hydrodynamic lubrication, as a thinning of the lubricant at higher temperatures reduces the friction and therefore avoids further heating, which would finally lead to breakdown of lubrication.

The plot of Figure 9.10 also implies that when sliding is stopped, one inevitably ends up in a situation where boundary lubrication takes over (Figure 9.11). This means that one usually cannot solely rely on hydrodynamic lubrication, but one has to avoid excessive friction and wear during starting and stopping of bearings. Therefore, lubricants often contain substances that bind to the surfaces and minimize boundary friction. For our description of hydrodynamic lubrication, we have assumed that the viscosity of the lubricant does change due to the shearing between the two surfaces. This is fulfilled for most lubricants except at extremely high shear rates. At very high shear rates, the viscosity might decrease, a phenomenon known as shear thinning.

9.2.1.1 Elastohydrodynamic Lubrication

In hydrodynamic lubrication, the relative motion of the surfaces leads to entrainment of the lubricant. Some lubricant becomes pressurized in the converging wedge and is thus able to support the load. Film thicknesses are typically on the order of micrometers and the maximum pressures are on the order of some megapascals. At such pressures, no significant deformation of the surfaces will occur and the lubricant viscosity can be assumed to be pressure independent. In nonconforming contacts, such as the line and point contacts occurring in gears and ball bearings, extremely high local pressures are unavoidable. Under these conditions, the above theory of hydrodynamic lubrication would predict a lubrication layer thickness that is smaller than the surface roughness, as was recognized in 1916 by Martin [956]. However, experiments show that fluid film lubrication still holds under such conditions, otherwise gears and ball bearings under high loads could not work for such long periods without significant wear. To understand this phenomenon, we have to take two additional effects into account, which lead to elastohydrodynamic lubrication:

- The pressure dependence of the viscosity. The viscosity of most lubricants increases roughly exponentially with increasing pressure, as found in 1893 by Barus [957]:

$$\eta = \eta_0 e^{\alpha P}. \tag{9.30}$$

  Here, α is a constant called viscosity pressure coefficient and η_0 is the viscosity at ambient pressure. This Barus law tells us that the higher the pressure, the harder it becomes to squeeze the lubricant out of the gap. At the very high pressures of ≈ 1 GPa, there may even be a phase transition of the lubricant to a glassy state.
- The solids confining the lubricant are never perfectly stiff but deform elastically at high pressure. As a result, a locally conforming contact is formed and the load bearing area increases.

Both effects allow stable lubrication with surfaces separated by a lubricant layer for contact pressures up to more than 1 GPa. A comprehensive book on this topic is Ref. [958]. For reviews, see Refs [959, 960].

The first theoretical description of elastohydrodynamic lubrication for the line contact problem relevant to gears that combined the Reynolds equation, Barus law, and a Hertzian contact mechanics was developed by Ertel [961] and published 10 years later by Grubin and Vinogradova [962]. The approach was to assume a Hertzian contact to calculate the pressure in the gap and let the pressure in the lubricant before the entrance increase exponentially to match the Hertzian contact pressure curve. The pressure distribution for such a line contact between parallel cylinders is shown in Figure 9.12. With this approximation, Ertel derived an expression for the average thickness h_0 of the lubricant film within the gap, which was later refined by Dowson [963]:

$$\frac{h_0}{R^*} = 2.65 \frac{U^{0.70} G^{0.54}}{W_L^{0.13}}. \tag{9.31}$$

Here, the dimensionless speed U, the dimensionless material parameter G, and the dimensionless load per unit length W_L are given by

$$U = \frac{\eta_0 v_0}{E^* R^*}, \quad G = \alpha E^*, \quad W_L = \frac{F_L}{l E^* R^*}. \tag{9.32}$$

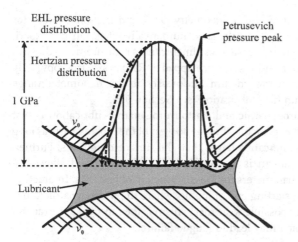

Figure 9.12 Schematic cross section of a lubricated line contact between two rotating parallel cylinders. The solid line indicates the pressure distribution in the contact region with the characteristic pressure peak in the outlet region. The dotted line corresponds to the pressure distribution of the classical Hertzian contact at the same loading force.

F_L/l is the normal load per unit length of the cylinders. The reduced radius R^* and reduced Young's modulus E^* are defined as usual as

$$R^* = \frac{R_1 R_2}{R_1 + R_2}, \quad 1/E^* = \frac{1-v_1^2}{E_1}\frac{1-v_2^2}{E_2} \tag{9.33}$$

with R_1, R_2 being the undeformed contact radii of the cylinders and E_1, E_2 and v_1, v_1 being the Young's moduli and Poisson ratios of the two materials forming the friction contact. A characteristic is the weak dependence of the gap height h_0 on the applied load F_L, which follows a power of 0.13. This reflects the fact that with the contact zone the lubricant becomes stiff. A further increase in load will then mainly result in an increased flattening of the contact partners. This is reflected in the much higher exponent for the material parameter G that contains mechanical properties of both the cylinders and the lubricants.

Ertel recognized that the pressure drop in the lubricant at the outlet should lead to a reduced deformation of the surfaces and thus to a restriction in the film that results in a pressure spike just before the outlet. A first numerical solution for the pressure profile of a line contact was published by Petrusevich [964], who also calculated the height of the second pressure peak close to the outlet, which then became known as the Petrusevich pressure spike. In 1959, an improved numerical approach was introduced by Dowson and Higginson [965]. It laid the foundation for elastohydrodynamic lubrication as a distinct field of research. While experimental proof of the existence of a lubrication film and determination of its shape by optical interferometry [966, 968] in highly loaded point contacts was found at around the same time, it took much longer to solve the problem theoretically [969, 970]. Today, computer power and development of efficient algorithms have made computer-based solutions available for almost any kind of contact geometry [971], and frictional forces for elastohydrodynamic contacts can be calculated numerically [968]. On the experimental side, optical interferometry now allows film thickness measurements down to 0.3 nm [972] and pressure distributions can be mapped by Raman spectroscopy [973]. This has led to the extension toward film thicknesses in the nanometer range resulting in the so-called thin film lubrication (see Section 9.3).

A great advantage of hydrodynamic and elastohydrodynamic lubrication is that friction can be quite low and in principle there is no wear of the moving parts as long as the gap width remains significantly larger than the surface roughness. Furthermore, the underlying mechanism is reasonably well understood and allows the calculation of the necessary parameters for the construction of bearings. In practice, wear may still occur during starting and stopping, where the hydrodynamic lubrication breaks down. Another possible wear mechanism is the repeated deformation of the bearing surfaces, which may lead to fatigue failure.

9.2.2
Boundary Lubrication

Hydrodynamic and elastohydrodynamic lubrication breaks down at low sliding velocities and under excessive loads since the lubricating film is squeezed out of the gap. If the lubricant is completely removed, dry friction would occur, which would imply high friction forces and wear. If a thin layer of lubricant remains adsorbed, friction between surfaces is strongly reduced compared to dry friction. This leads to the so-called boundary lubrication. Friction coefficients under these conditions are typically 100 times higher than those under hydrodynamic lubrication conditions but still substantially smaller than those for dry friction under UHV conditions. In boundary lubrication, friction essentially depends on the chemical constitution of the lubrication layer and not on its bulk viscosity.

One effect of a boundary lubricant is to reduce adhesion between the solids. Adhesion between solids is usually dominated by van der Waals forces (see Chapter 2). Hydrocarbons have a small Hamaker constant (see Table 2.2). Their presence leads to a reduction in the adhesion and hence friction. Films of only monomolecular thickness are sufficient to have a pronounced effect [974] (see Example 9.2). In that case, friction can be as small as friction with plenty of lubricant. At least with metals it can be shown that the number of microcontacts is not changed by the lubricant. Only the contact intensity is reduced. The reduced van der Waals attraction can thereby diminish the actual contact area.

It should be noted that most "clean" surfaces are covered by a thin layer of contamination, unless prepared and kept in UHV. As a consequence, in experiments performed under ambient conditions, we always have the case of boundary lubrication rather than dry friction.

In practical applications, we often encounter a combination of boundary and hydrodynamic lubrication, which is called mixed lubrication. For example, bearings that are usually lubricated hydrodynamically experience mixed lubrication when starting and stopping. This is shown in the so-called Stribeck diagram (Figure 9.13):

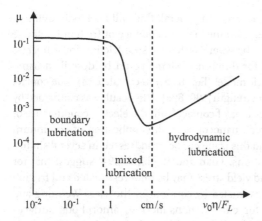

Figure 9.13 Stribeck diagram for a lubricated friction contact. Coefficient of friction μ versus $\eta v_0/\rho P$ is plotted, where ρ is the density of the lubricant, v_0 the velocity, and P the contact pressure. From left to right, there are three distinct friction regimes: boundary lubrication with high friction and wear, mixed lubrication with intermediate friction and wear, and hydrodynamic lubrication with low friction and (almost) no wear.

at low speeds, boundary lubrication with high friction dominates. With increasing speed, a hydrodynamic lubricant film is created that significantly reduces friction. The linear increase in friction at even higher speeds is due to the internal viscosity of the lubricant, as described by Eq. (9.25).

9.3
Microscopic Friction: Nanotribology

In the previous section, we learned that the relevant processes of friction occur at microcontacts. To get a better understanding of friction phenomena, one should therefore study friction at the micro- and nanoscale. This field of micro- and nanotribology evolved with the availability of suitable experimental techniques, namely, FFM, SFA, and QCM. Reviews can be found in Refs [975–978]. Recent books on this topic are Refs. [979, 980].

9.3.1
Single Asperity Friction

For macroscopic bodies, the real and apparent contact areas are different. It was the deformation of surface asperities that accounted for Amontons' law. If considering a single nanocontact, apparent and real contact areas are identical. For nanocontacts, friction should be proportional to the shear strength and contact area as described by Eq. (9.2). Wearless friction of single contacts is also denoted as "interfacial friction." Assuming that the shear strength of the contact does not depend on the normal load, it is possible to calculate the load dependence of the friction force by applying contact mechanics to predict the true contact area. In general, this will lead to a nonlinear dependence of F_F on F_L. Experiments on the load dependence of friction for single nanocontacts have been carried out by several authors using friction force microscopy. The resulting friction force for different systems could be described using continuum theories such as the JKR model (Eq. (8.68)) to calculate the true contact area and assuming a constant shear strength [890–892]. This result was confirmed by determining independently the true area of contact from the electric conductance of the contact [981]. However, for small nanocontacts, the application of continuum mechanics may be questionable and this issue has been addressed in several studies using molecular dynamics simulations. Luan and Robbins [894] suggest that for nanoscale contacts, contact area and yield stress may be underestimated and friction and contact stiffness may be overestimated by continuum theories. Wenning and Müser [984] showed that while for curved, nonadhering, amorphous surfaces $F_F \propto F_L^{2/3}$, for commensurate surfaces or flat surfaces a linear relation $F_F \propto F_L$ is obtained. Mo *et al.* [986] came to the conclusion that atomic roughness of the contact leads to a linear dependence of friction on load just like in the case of macroscopic roughness. The obvious disagreement between the linear dependency observed in simulations and the power law dependence found experimentally is even more striking for the study of Gao *et al.* [985], who carried out FFM experiments on diamond crystal surfaces that were contrasted with molecular dynamics simulations of the same system. While the experimental data could well be explained by assuming a constant shear strength and calculating the contact area with the Dugdale–Maugis contact model, their simulations showed $F_F \propto F_L$ independent of commensurability of the surfaces.

In experiments with friction force microscopy, the tip forms a contact of a few nanometers in diameter with the substrate, the so-called *nano*contact. In reality, friction of macroscopic bodies is determined by the interaction via *micro*contacts. One possibility of extending the method of friction force microscopy to larger contact areas is the use of the colloidal probe technique, where a small sphere is attached to the end of an atomic force microscope cantilever (see Section 3.2). Even for microcontacts, the proportionality between the true area of contact and the friction force was observed (see Example 9.2).

Even larger contact areas with interfacial friction are found in the SFA. While SFA has mainly been used to study boundary lubrication in the presence of adsorbed films or confined liquids, there have also been studies on dry friction. Homola *et al.* [987] found that the increase in contact area between the mica sheets with load could be described by JKR contact mechanics. The friction force was given by the product of this contact area times a constant shear strength of $\tau \approx 20$ MPa. This value of the shear strength differs extremely from the values obtained in FFM experiments, where values on the order of 1 GPa are common. This scale dependence of shear strength was originally explained by a micromechanical model of scale-dependent formation and movement of dislocations during the sliding process [988]. However, there might also be the problem that in SFA experiments there will always be a certain amount of adsorbed molecules present on the mica surfaces, which can influence the effective shear strength of the contact.

■ **Example 9.3**

McGuiggan *et al.* [982] measured the friction on mica surfaces coated with thin films of either perfluoropolyether (PFPE) or polydimethylsiloxane (PDMS) using three different methods: the surface forces apparatus (radius of curvature of the contacting bodies $R \approx 1$ cm), friction force microscopy with a sharp AFM tip ($R \approx 20$ nm), and friction force microscopy with a colloidal probe ($R \approx 15$ μm). In the surface force apparatus, coefficient of frictions of the two materials differed by a factor of ≈ 100, whereas for the AFM silicon nitride tip, the coefficient of friction for both materials was the same. When the colloidal probe technique was used, the coefficient of frictions differed by a factor of ≈ 4. This can be explained by the fact that, in friction force microscopy experiments, the contact pressures are much higher. This leads to a complete penetration of the AFM tip through the lubrication layer, rendering the lubricants ineffective. In the case of the colloidal probe, the contact pressure is reduced and the lubrication layer cannot be displaced completely.

Xu *et al.* [983] used a special friction tester that allows contact sizes in between FFM and SFA. They could observe that transition between high (hundreds of MPa) and low (some 10 MPa) shear strength can occur for contact radii of only 20–30 nm, depending on whether there is intimate contact of the sliding surfaces or whether there is still a monolayer of lubricating molecules present. This is in line with the quantized friction behavior found in SFA experiments depending on the number of molecular lubrication layers [644].

9.3.2
Atomic Stick–Slip

The first measurement of friction with atomic resolution using friction force microscopy was carried out by Mate *et al.* [275]. They used a tungsten tip on highly oriented pyrolytic graphite (HOPG) and observed an atomic stick–slip friction. The periodicity of the stick–slip agreed with the lattice spacing of the graphite. In the following years, atomic stick–slip friction was observed for many other types of crystal surfaces, for example, Au [989], diamond [990], NaF, NaCl, AgBr [991], MoS_2 [992], stearic acid crystals [993], KBr [994], CuP [995], or Cu(111) [996]. In these experiments, with moderate load on the AFM tip, no wear of the atomically resolved surfaces was observed.

The possibility of wearless friction had already been postulated independently by Prandtl in 1928 [916] and by Tomlinson in 1929 [917]. They suggested a simple model that describes the interaction between two surfaces in relative motion (Figure 9.14). The lower surface (2) is represented by a periodic potential $V(x)$. The surface atom (A) is elastically coupled to the upper surface (1) via a spring with spring constant K and moves through the potential $V(x)$ as surface (1) moves from the left to the right. At the beginning (A) is pinned at a minimum of the potential (a). As surface (1) moves on, the force on (A) increases (b) until it gets so strong that the atom jumps to the next potential minimum (c). During this relaxation energy is dissipated via lattice

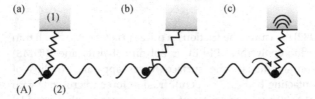

(a) (1)

(b)

(c)

(A) (2)

Figure 9.14 Schematic of the Prandtl–Tomlinson model.

vibrations of the upper body and finally via the generation of phonons. In the Prandtl–Tomlinson model, this dissipation is described by a simple damping term that is proportional to the speed v_A of the atom. This model explains Coulomb's law of friction: The fast, dissipative movement of the surface atom is largely independent of the relative speed of the surfaces v_0, provided they move with a speed that is much slower than the relaxation of the atom ($v_0 \ll v_A$).

Let us apply the Prandtl–Tomlinson model to the example of an AFM tip sliding along a one-dimensional periodic potential $V(x)$ with periodicity a (see Figure 9.15):

$$V(x) = -V_0 \cos\left(\frac{2\pi x}{a}\right). \tag{9.34}$$

The tip is assumed to be coupled to a support via a spring with spring constant K. The support moves at a velocity v_0. Its coordinate changes according to $x_0 = v_0 t$. The total energy of the system can then be written as

$$E = -V_0 \cos\left(\frac{2\pi x}{a}\right) + \frac{1}{2}K(x-x_0)^2 = -V_0 \cos\left(\frac{2\pi x}{a}\right) + \frac{1}{2}K(x-v_0 t)^2. \tag{9.35}$$

Initially, the tip will reside within the local minimum given by the condition

$$\frac{\partial E}{\partial x} = \frac{2\pi V_0}{a}\sin\left(\frac{2\pi x}{a}\right) + K(x-v_0 t) = 0. \tag{9.36}$$

For the beginning of one stick–slip cycle, we can use the approximation $\sin z \approx z$ for small z. This leads to

$$\left(\frac{4\pi^2 V_0}{a^2} + K\right)x - Kv_0 t = 0. \tag{9.37}$$

The initial velocity of the tip v_{tip} will therefore be given by

$$v_{\text{tip}} = \frac{dx}{dt} = \frac{v_0}{1+C} \quad \text{with} \quad C = \frac{4\pi^2 V_0}{Ka^2}. \tag{9.38}$$

This is slower than the velocity v_0 of the support.

If the support position x_0 has changed sufficiently far, the tip could be forced to a position, where $\partial^2 E/\partial x^2$ changes it sign. This will be exactly at the point x^* where

$$\frac{\partial^2 E}{\partial x^2} = \frac{4\pi^2 V_0}{a^2}\cos\left(\frac{2\pi x^*}{a}\right) + K = 0, \tag{9.39}$$

which is equivalent to

$$C \cdot \cos\left(\frac{2\pi x^*}{a}\right) = 1.$$ (9.40)

Equation (9.40) can be solved only if $C \geq 1$ (i.e., $4\pi^2 V_0 \geq Ka^2$). Thus, stick–slip motion is expected for soft springs or for strong tip–sample interaction. For $C = 1$, the critical position $x^* = 0$. For $C > 1$, we obtain two critical points $x_{1,2}^*$ per lattice cell. In two dimensions, this corresponds to a critical curve.

The friction force is given by

$$F_F = -K(x-x_0) = -K(x-v_0 t).$$ (9.41)

Using Eq. (9.36), this can be expressed as

$$F_F = \frac{2\pi V_0}{a} \cdot \sin\left(\frac{2\pi x}{a}\right).$$ (9.42)

It has a maximum at $a/4$ with

$$F_{\max} = \frac{2\pi V_0}{a}.$$ (9.43)

The friction force increases linearly with the maximum interaction potential V_0. However, in a stick–slip situation, the friction force will not follow the sinusoidal pattern but a sawtooth pattern due to instabilities. The force F_F^* at which the jump occurs can be calculated using Eqs. (9.36) and (9.40) in combination with the mathematical identity $\sin^2 x + \cos^2 x = 1$:

$$F^* = \frac{Ka}{2\pi} \sqrt{C^2-1}.$$ (9.44)

From Eq. (9.36), we can obtain for $t \to 0$

$$x = \frac{v_0 t}{1+C}.$$ (9.45)

By putting this relation into Eq. (9.41), we can calculate the increase in the friction force with time when the sliding starts:

$$F_F(t \to 0) = Kv_0 t \frac{C}{C+1}.$$ (9.46)

This means that during stick, the force ramps up with distance with an effective spring constant

$$K_{\text{eff}} = K \frac{C}{C+1}.$$ (9.47)

From the fact that Eq. (9.40) can have solutions only for values of $C > 1$, it follows that for sufficiently stiff springs, the stick–slip motion should disappear. Naively, one might expect that by using stiffer AFM cantilevers, ultralow atomic friction without stick–slip and thus without dissipation should be achievable. However, the effective spring constant K entering into the Prandtl–Tomlinson model is for typical friction force microscopy experiments not given by the lateral cantilever spring constant but by the stiffness of the contact between tip and surface. Typical values of this contact stiffness are on the order of $1\,N\,m^{-1}$ compared to some $10\,N\,m^{-1}$ for the lateral spring constant of the cantilever. The alternative approach to ultralow friction would therefore be to minimize the interaction between tip and surface (i.e., the value of V_0). This can be achieved by reducing the normal force between tip and surface.

For the friction force between a silicon AFM tip and a NaCl(001) surface in UHV, such a transition was in fact observed [997] (Figure 9.16). At an applied normal load of 4.7 nN, stick–slip was observed with a clear hysteresis between forward and backward scan on the same line. By reducing the applied load to 3.3 nN, the stick–slip amplitude stays almost constant but the hysteresis, which is proportional to friction loss, is clearly reduced. When changing the applied force to -0.47 nN to compensate partly the adhesion force of 0.7 nN, the stick–slip pattern changed into a continuous modulation with no detectable hysteresis between trace and retrace. This corresponds to a frictionless sliding at least within the force resolution of the experiment. This effect of vanishing friction due to very small loads is called static superlubricity.

As already seen in Figure 9.15, the jump during slip can in principle go to different positions that are all lower in energy than the position from which the jump occurs. In most cases where atomic-scale stick–slip motion is observed, slip distances correspond to single atomic spacings. This implies that usually the AFM tip sliding on the crystal surface is a strongly damped system. However, there may be situations where double [998] or multiple slip occurs as suggested by Johnson and Woodhouse [999]

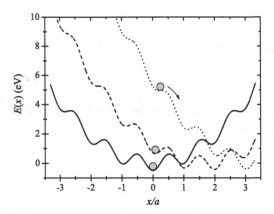

Figure 9.15 Energy for a sliding AFM tip in the one-dimensional Prandtl–Tomlinson model versus position. The position is given in units of the periodic surface potential ($a = 0.4\,nm$). Parameters were $V_0 = 0.5\,eV$ and $K = 1.5\,N\,m^{-1}$. The energy is plotted for a support moving to the right with a speed $v_0 = 20\,nm\,s^{-1}$ (i.e., $x_0 = v_0 t$ is assumed) leading to $C = 13.2$. The tip is assumed to reside in a minimum at $x = 0$ at time $t = 0$. The three different lines indicate the shape of the potential for three different times, where the position of the support is at $x_0 = 0$ (continuous line), $x_0 = 2a$ (dashed line), and $x_0 = 3a$ (dotted line), respectively.

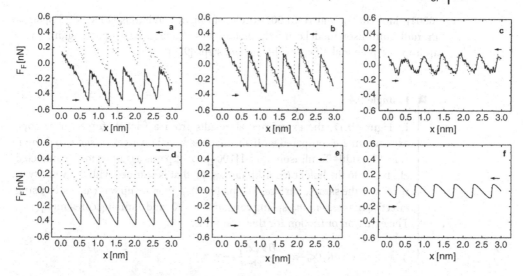

Figure 9.16 Friction forces F_F measured between a NaCl(001) surface and an AFM tip sliding along the (100) direction in UHV. At an applied load of 4.7 nN (a) and 3.3 nN (b), stick–slip motion of the tip is observed with a hysteresis between forward and backward directions. For an applied load of −0.47 nN (at 0.7 nN adhesion force), motion changes to a continuous sliding without detectable hysteresis, indicating dissipationless sliding. This effect is denoted as static superlubricity. Parts (d)–(f) show corresponding numerical results using a Tomlinson model with values of $C = 5$, $C = 3$, and $C = 1$. (From Ref. [997] with permission of R. Bennewitz.)

and discussed in detail by Medyanik *et al.* [1000]. In fact, double slip events were already observed in the first friction force experiments by Mate *et al.* [275]. At high sliding speeds, very complex and even chaotic stick–slip behavior may arise [1001], for example, not only due to resonances of the tip itself but also simply due to the fact that the entire friction system consisting of tip–sample interaction, cantilever spring constant, and contact stiffness is nonlinear in nature [1002].

In spite of its conceptual simplicity, the Prandtl–Tomlinson model has been quite successful for the quantitative interpretation of friction force microscope experiments when extended to two dimensions [1018]. It should be noted that all AFM contact mode images (i.e., images recorded with direct mechanical contact between tip and sample) showing resolution of atomic lattices are, in fact, the result of stick–slip motion. Since the contact area between tip and surface is much larger than just a single atom, such images do not have true atomic resolution. Single atom defects or adatoms would not be recognized under these imaging conditions. True atomic resolution can be achieved using only dynamic modes of AFM.

At atomic steps, an additional lateral force ΔF_F is usually necessary to drag the tip over the step. There is, however, a marked difference between scanning the step up or down: for downward scans, ΔF_F does not depend on applied load, whereas for upward scans, a linear increase in ΔF_F with applied load was found. This was observed for graphite(0001) in UHV [1019] as well as on freshly cleaved graphite

(0001), MoS2(001), and NaCl(001) under ambient conditions [1020], and again a Prandtl–Tomlinson model can fully explain these findings by simply plugging in an appropriate potential $V(x)$ for the atomic steps [977].

■ **Example 9.4**

In Figure 9.17, the experimental results from a friction force microscope experiment are compared with simulations based on an extended two-dimensional Prandtl– Tomlinson model [1004]. The tip was assumed to be connected elastically to the holder (coordinates x_0, y_0) that is scanned with the velocity v_0 relative to the sample surface. The path $x(t)$, $y(t)$ of the tip was calculated using effective masses m_x, m_y, spring constants K_x, K_y, and damping constants γ_x, γ_y. The equation of motion for this system is

$$m_x \ddot{x} = K_x(x_0 - x) - \frac{\partial V(x, y)}{\partial x} - \gamma_x \dot{x},$$

$$m_y \ddot{y} = K_y(y_0 - y) - \frac{\partial V(x, y)}{\partial y} - \gamma_y \dot{y}.$$

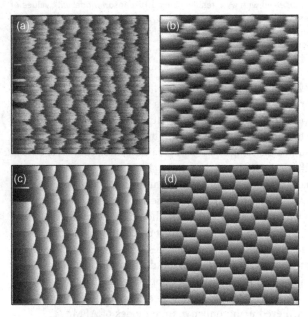

Figure 9.17 Friction force microscope pictures (a, b) of a graphite(0001) surface as obtained experimentally with FFM and results of simulations (c, d) of the stick–slip friction using a two-dimensional equivalent of the Prandtl–Tomlinson model. The friction force parallel to the scan direction (a, c) and the lateral force perpendicular to the scan direction (b, d) are shown. The scan size is 20 Å × 20 Å. (Pictures taken from Ref. [1004] with permission of R. Wiesendanger.)

For the interaction potential, the authors chose

$$V(x,y) = -V_0 \left[2 \cdot \cos\left(\frac{2\pi}{a}x\right) \cos\left(\frac{2\pi}{a\sqrt{3}}y\right) + \cos\left(\frac{2\pi}{a\sqrt{3}}y\right) \right].$$

Recent experiments [1005] with high time resolution have revealed fine structures of the jump dynamics, which cannot be explained by a simple Prandtl–Tomlinson model. This has led to the postulation of a new model that assumes an ultrasmall effective mass for the tip apex, which therefore exhibits a much faster kinetics than assumed in previous models [1006, 1007]. This led to thermal delocalization of the tip that probes the surface potential at a much faster rate than the observed stick–slip motion of the cantilever.

A fundamental question not yet resolved regarding stick–slip motion is the exact mechanism for the occurrence of stick–slip with a periodicity of the surface lattice. For interpretation of the stick–slip motion by the Prandtl–Tomlinson model, the tip is commonly treated as a single, pointlike entity without any additional internal degree of freedom. However, the contact area between AFM tip and crystal surface will typically contain some 10 unit cells. Atomic stick–slip was observed even for amorphous silicon and silicon nitride tips [1003] and for crystalline tips, atoms of tip and surface lattice will usually not be in registry, making the observation of stick–slip with a periodicity of the surface lattice surprising.

In the Prandtl–Tomlinson model, energy dissipation is described by a simple damping term. This phenomenological approach does not reveal anything about the underlying mechanism. For insulating surfaces, phonons are expected to be the relevant dissipative pathway [904, 1009]. Recently, Canarra *et al.* [1010] demonstrated that the vibration frequencies of the surface atoms have a profound influence on friction force. They compared friction forces on silicon or diamond substrates that were terminated by hydrogen or deuterium. The higher vibration frequencies of H-terminated surfaces led to a more rapid energy dissipation however, the effect was smaller than expected from the model of Persson [904], which shows that other, mass-independent contributions must be present. In the case of conducting surfaces, electronic excitations (e.g., generation of electron–hole pairs) can contribute [1011]. The existence of such an electronic friction was shown for the damping of adsorbate movements on Ru(0001) surfaces by He scattering [1012].

The relative contributions of these two mechanisms are still under debate. Measurements with the quartz microbalance on the slippage of nitrogen on a superconductor (lead) showed that, below the transition temperature T_c, sliding friction dropped to about half the value compared to temperatures above T_c [951]. If we assume that this drop is due to a complete loss of electronic friction, then electronic and phononic contributions are of the same order of magnitude, at least in this system. Park *et al.* [1013] found that friction on p-doped silicon can be modulated by an applied potential that changes the charge carrier density, which might open a new way to control friction in MEMS devices. Other energy loss mechanisms such as dislocation-mediated plastic deformation [1014, 1015] chemical reactions, tribolu-minescence, or emission of charged particles can also occur [1016].

■ **Example 9.5**

The first direct measurement of the energy dissipation through single atoms was achieved by Giessibl *et al.* [301]. They used an extended version of friction force microscopy, where the tip is oscillated laterally. Dissipation energies were obtained from the frequency shift of the oscillation while measuring the tunneling current between tip and surface. With this method they were able to measure the energy dissipation during interaction of the tip with single adatoms of a Si(111)7 × 7 surface. When the tip was placed above an adatom, the dissipation was low. A positive frequency shift of the oscillation indicated an increase in the effective spring constant of the oscillator. This was interpreted as the formation of a "bond" between tip and adatom. When the tip was positioned further away from the adatom, the "bond" was formed and destroyed in each oscillation cycle and more energy was dissipated. In this case, the large deflection of the adatom from its equilibrium position leads to strong vibrations and generation of phonons.

9.3.3
Velocity Dependence of Nanoscale Friction

The influence of sliding velocity on friction force can be derived within the framework of the Prandtl–Tomlinson model when interpreting the slip event as a thermally activated process. In the above discussion of the Prandtl–Tomlinson model, we did not consider thermal motion, which corresponds to sliding at $T = 0$ K. The jump of the tip from one stable surface position to the next one is hindered by an energy barrier ΔE. At zero temperature, the tip will not be able to jump to the next position unless $\Delta E = 0$. In the case of finite temperature, there will be a certain probability that the energy barrier will be overcome by thermal motion while $\Delta E > 0$. As a consequence, the force at which slip occurs will become a statistically distributed quantity and the average friction force will correspond to the most probable slip force. The probability that the tip has not yet slipped changes with time according to [1008]

$$\frac{dp}{dt} = -f_0 \exp\left(-\frac{\Delta E}{k_B T}\right) p(t), \tag{9.48}$$

where f_0 is the characteristic lateral frequency of the system (i.e., lateral resonance frequency). The activation energy ΔE itself will be a function of time since it depends on the change in the lateral force F_F with time. For small sliding velocities, one can assume that ΔE decreases linearly with increasing friction force F_F, where the slope A will depend on the interaction potential. Using this so-called "linear creep assumption" and the relation $dF_F/dt \approx K v_0$, a logarithmic dependence of friction on velocity is obtained:

$$F_F = F_0 - \frac{k_B T}{A} \ln\left(\frac{v}{v_0}\right). \tag{9.49}$$

Within this framework, the increase in friction force with velocity simply reflects the fact that at higher sliding speeds, the system has less time to overcome the activation barrier by thermal motion. This effect leads to a higher value of the most probable slip force. A logarithmic increase in friction with velocity has indeed been observed, for example, for a silicon tip sliding on a Cu (111) surface [996] or a NaCl(001) surface in UHV [1008].

Riedo *et al.* [1017] measured the friction force between silicon tips and mica. They found a logarithmic increase up to a critical velocity v_c and no dependence on speed for $v_0 > v_c$. This behavior was explained by a modified Prandtl–Tomlinson model, where the linear creep assumption was replaced by a "ramped creep." This assumption should be more appropriate when slip occurs close to the critical point x^* defined by Eq. (9.40) and leads to a relation of $\Delta E \sim (\text{const} - F_F)^{2/3}$ and $F_F \sim \ln(v/v_0)^{3/2}$ [1021]. Such type of transition between two friction regimes might also have occurred in experiments done by Zwörner *et al.* [1022] in which friction force was independent of speed up to 25 μm s^{-1} on amorphous carbon, diamond, and HOPG; for velocities below 0.5 μm, there seems to be an increase with velocity. The model of Reiman and Evstigneev [1023] even predicts a maximum and then a logarithmic decrease at high ($> 20 \, \mu$m/s) scan speeds. This effect is due to a decrease in stick–slip–type deformations of the tip apex at higher speeds that leads to a decrease in dissipation. A logarithmic increase in friction force with sliding speed on polymer layers grafted on silica was found by Bouhacina *et al.* [1024] and explained by a thermally activated Eyring model. Depending on the precise type of the contacting surfaces and the range of sliding speeds, deviations from the logarithmic dependence have been observed. On hydrophilic surfaces under ambient conditions, the formation of a water meniscus by capillary condensation (see Chapter 5) may strongly influence the friction behavior. A logarithmic *decrease* in friction with increasing velocity was observed when a sapphire ball was slid on a silicon wafer with native oxide layer [1025]. Such dependence was also found for AFM tips sliding on hydrophilic CrN surfaces by Riedo *et al.* [1026] and explained the kinetics of capillary condensation. With increasing sliding speed, the meniscus between tip and surface cannot fully develop and the friction force decreases with increasing speed. Other mechanisms such as elastic hysteresis on soft layers, viscous shearing, and deformation of asperities may lead to more complex patterns and a reversal of speed dependence at high speeds [1027]. Tao and Bhushan [1028] used a high-speed scan stage added to an AFM setup. For very high sliding speeds (20–200 mm s^{-1}), a transition from logarithmic to linear relation between friction force and sliding velocity was found, which was attributed to the dominance of viscous friction forces [1029].

9.3.4
Superlubricity

From the application point of view, reduction of friction is often the primary goal. For many macroscopic technical applications, lubrication is extremely effective and a broad basis of empirical knowledge exists. However, with increasing importance of micromechanical systems and nanotechnology, new concepts for reduction of friction have to be developed. In microelectromechanical systems (MEMS) with high surface-to-volume ratio, stiction and friction become critical issue, that limit operability and lifetime [1030, 1031]. For dry and clean surfaces, stick–slip motion usually dominates. It goes along with high-energy dissipation. It would therefore be desirable to avoid stick–slip in favor of a continuous, frictionless sliding.

Structural Superlubricity In 1990, Hirano and Shinjo [1032] found by theoretical studies that it should in principle be possible for two crystalline surfaces to slide over each other with vanishing friction, if the interaction potential is low enough. They proposed that such a configuration could be achieved by a lattice mismatch of the crystals. They coined the term "superlubricity" for such a mode of sliding [1033]. This term is somewhat misleading since it suggests a quantum nature of the phenomenon and therefore the term "structural superlubricity" was suggested to be more appropriate [1034]. It was also found later that other mechanisms can lead to almost vanishing friction and superlubricity was consequently used as a general term for sliding with almost zero friction. For a recent review, we refer to Ref. [1035], while Ref. [1036] is a comprehensive book on the topic.

In their first experimental attempt to prove their concept of structural lubricity, Hirano *et al.* [1033] demonstrated that the friction force between mica sheets in a tribometer can be reduced by a factor of 4 when misaligning the sheets. This finding indicates that commensurability of surfaces had a strong influence, but no full superlubricity was achieved. Martin *et al.* [1037] observed coefficients of friction below 10^{-3} between clean MoS_2 surfaces in an UHV tribometer, which was attributed to structural lubricity. A second experimental attempt of Hirano *et al.* [1038] was to measure the friction force between a W(011) tip and a Si(001) surface in an STM. When rotating the tip, they observed a reduction in friction from 8×10^{-8} N to below 3×10^{-9} N. The most convincing proof of superlubricity came from Dienwiebel *et al.* [1039] who studied the sliding friction between HOPG and a graphite flake on the tip of an unconventional FFM that allowed detection of lateral forces in both directions with a resolution of 15 pN. By rotating the sample in small steps relative to the tip, the angular dependence of friction on relative orientation between flake and sample was mapped. Clear friction force peaks separated by an angle 60° were observed with lateral force more than one order of magnitude smaller in between. The experimental data could be fitted using a two-dimesional Prandtl–Tomlinson model [1040].

If friction requires a certain commensurability of the sliding surface, should friction not vanish in almost all situations? It would be a coincidence if two surfaces have a similar lattice constant and are aligned. There are several reasons why this is not the case:

- Structural lubricity removes only dissipation caused by stick–slip motion. Other dissipation channels are not affected and can be dominating.
- Single crystalline flakes have the tendency to self-align into a high friction state orientation [1041].
- The presence of "third bodies" (i.e., adsorbed molecules, contaminants) between the surfaces will lead to significant friction [1042].
- The above models assume rigid bodies with ideal flat surfaces. If this rigid block assumption fails, superlubricity cannot be expected. This is related to the balance

between interfacial interactions and bulk elastic forces. An estimate if this condition is met or not can be made by scaling arguments for surface interaction potentials and bulk bond strength [1034].

Most theories of structural superlubricity are based on the Prandtl–Tomlinson model or the more advanced Frenkel–Kontorova model [1043, 1044], in which the single atom/tip is replaced by a chain of atoms coupled by springs. However, Friedel and de Gennes [1045] noted recently that correct description of relative sliding of crystalline surfaces should include the motion and interaction of dislocations at the surfaces. This concept was taken up by Merkle and Marks [1046] and generalized using the well-established coincident site lattice theory and dislocation drag from solid-state physics.

While the applicability of superlubricity for atomically flat, crystalline surfaces without contamination to practical friction problems seems limited at the moment, the phenomenon of structural lubricity sheds new light on the well-known excellent lubrication properties of materials such as graphite or MoS_2. These materials have long been known as solid lubricants, and commonly this is attributed to the weak interaction between crystal layers. However, relative slip of layers with flakes of the lubricant is expected to lead to high forces, making relative sliding of flakes the more realistic scenario for the low-friction mechanism. For the case of MoS_2, a possible contribution of superlubricity due to incommensurate sliding of single flakes has been found in a UHV tribotester [1037]. The superlubricity between graphite sheets has initiated the search for other candidates of ultralow friction. One prominent example are graphite sheets intercalated with C_{60} monolayers that exhibit outstanding properties (for review, see Ref [1047]).

Static Superlubricity As already discussed in Section 9.3, a reduction in the strength of the interaction potential (case $C < 1$ in Eq. (9.40)) will suppress the stick–slip motion according to the Prandtl–Tomlinson model. This was experimentally first confirmed by Takano and Fujihira [993] using FFM on stearic acid crystals and studied in more detail by Socoliuc et al. [997] using FFM on NaCl(001) in UHV. In both cases, the reduction in interaction strength was achieved by minimizing the effective contact pressure. Since static superlubricity requires small (or even negative) loads between the sliding surfaces, this effect will probably have limited applicability in practice. However, an interesting approach might be the use of configurations where repulsive van der Waals forces exist between surfaces as demonstrated by Feiler et al. [1048]. They found that friction forces between a gold-coated colloid probe and Teflon in cyclohexane were below resolution of the instrument even for applied loads of almost 30 nN, corresponding to a coefficient of friction below 0.0003.

Dynamic Superlubricity It is the reduction of friction by vertical oscillations. It is a known phenomenon for macroscopic friction, where it is related to intermittent loss of contact. This requires relatively strong oscillation excitations and is therefore of limited applicability. For nanoscale friction, a marked reduction of friction for boundary lubrication conditions was demonstrated by Heuberger *et al.* [1049] in an SFA experiment with oscillation amplitudes as small as 0.1 nm and in an FFM experiment by Jeon *et al.* [1050]. Stick–slip motion for atomic friction on KBr and NaCl in UHV could be suppressed by applying an AC voltage between the silicon tip and an electrode on the backside of the insulating samples [1051]. This approach was extended toward conducting samples using mechanical oscillation by a piezoelement [1052] and was found to work also under ambient conditions, which makes it a promising possibility to control friction in MEMS applications.

Thermal Superlubricity   In the Prandtl–Tomlinson model, we expect that friction should decrease with increasing temperature since thermal excitations facilitate the jump over the potential barriers and that decreasing sliding speed should also decrease friction force. However, in both cases stick–slip is still expected for sufficiently corrugated surface potentials. In a more elaborate theoretical analysis, Krylov *et al.* [1053] demonstrated that there exists a transition to vanishing friction even for values of $C > 1$ in Eq. (9.40) due to transition from stick–slip to thermal drift motion. This transition was confirmed experimentally in FFM experiments between a graphite surface and a tungsten tip with and without graphite flake at scan velocities below 1 nm s^{-1} [1054].

9.3.5
Thin Film Lubrication

The design of precision components with ultrasmooth surfaces, for example, in the field of microelectromechanical systems and nanotechnology, boosts the use of lubricating films of molecular thicknesses. Under these conditions, the validity of continuum theories to describe the hydrodynamics of the lubricant is questionable. This is a new lubrication regime, denoted as thin film lubrication. A review on thin film lubrication as the limiting case of elastohydrodynamic lubrication is provided in Ref. [1055]. We will focus here on the aspects of molecular thin films studied in the context of nanotribology (reviews are Refs [185, 1056]).

Much of our knowledge on thin film lubrication is based on experiments with the SFA and thus between mica surfaces. The SFA has proven to be especially suited for the study of the shear response of confined liquids (see Section 3.1). As we will see in Chapter 10, such confined liquids may behave quite differently from the bulk lubricant. Close to the surfaces, the formation of layered structures can lead to an oscillatory density profile (Figures 10.1 and 10.2). When these layered structures start to overlap, the confined liquid properties may exhibit marked deviations from the bulk properties. First SFA studies were conducted using nonpolar liquids with either near-spherical structure such as OMCTS and cyclohexane [644] or with linear structure such as hexadecane [1057]. These first experiments were followed by a series of other studies using these and other simple organic liquids that together gave the following picture:

- For film thicknesses of more than ≈ 7 molecular layers, liquids behave essentially as bulk liquids [645, 1058–1060].

- For film thicknesses of less than 7–4 molecular layers, a transition to a solid-like state occurs, which is characterized by a finite yield stress [645, 1059–1062].
- To shear a solidified film (i.e, to induce transition from stick to slip), a certain minimum shear stress has to be applied [645, 1057–1059, 1062–1063], otherwise the film will remain solid and show an elastic response. The value of this critical shear stress depends in a discrete manner on the number of liquid layers present between the surfaces [644, 645, 1057, 1059, 1064].
- When the critical shear stress is exceeded, sliding commonly occurs as a stick–slip motion [645, 1059, 1062, 1063]. The transition from stick to shearing is commonly interpreted as a shear-induced melting of the solid that solidifies again at the next stick. This hypothesis was supported by molecular dynamics simulations [649, 1065]. Energy dissipation during stick–slip motion was found to occur largely during the slip phase due to viscous dissipation [1066].
- When the shearing velocity exceeds a certain critical velocity v_c, a transition from stick–slip to smooth sliding occurs [645, 1059, 1065].
- Shearing can be carried out while maintaining a fixed number of molecular layers n, and with n constant, the friction force increases linearly with applied load [645, 1059, 1064].

Some newer experiments have, however, challenged this picture, and discrepancies between different studies have not yet been satisfactorily resolved. While in some cases, a continuous transition from liquid to solid behavior with increasing confinement was observed [1061, 1062], Klein and coworkers found a sudden transition within a distance change of a single molecular layer [1060, 1063]. Such a sudden transition was predicted by molecular dynamics simulations only for commensurate orientation of the confining crystal surfaces [658]. Monte Carlo simulations by Ayappa et al. [1070] on the solidification of OMCTS between mica sheets showed freezing at a thickness of up to seven layers and phase transitions between triangular, square, and buckled phases if only one or two layers were left.

Recent experiments on the confinement and shear of OMCTS by Zhu and Granick [1071] questioned the picture of a liquid–solid transition. For thin films formed by rapid compression, a high effective viscosity was observed, similar to earlier experiments. But for films with quasistatic compression, friction was extremely low and no change in fluid viscosity was detectable. Mukhopadhyay et al. [1072] measured molecular diffusion of OMCTS within the contact zone of an SFA using fluorescence correlation spectroscopy. Upon shearing of the mica surfaces, diffusion increased only by a factor of 3–5 that is incompatible with a solid–liquid transition. These surprising findings were confirmed by dynamic AFM experiments [1073] on OMCTS, where friction and viscosity were low for very slow ($<0.6\,\text{nm s}^{-1}$) confinement rates and high for fast confinement. According to Lang et al. [1074], this dependence on confinement rate could be due to the interplay of a confinement-dependent glass transition temperature T_g and the relative timescales of relaxation time τ of the liquid and experimental time constant. Glass transition will occur only when increase in T_g with decreasing gap size is sufficiently fast, to make the glass transition occur within the experimental time

$t > \tau$. On the one hand, this implies that under equilibrium conditions (i.e., infinitesimally slow approach) no phase transition to a solid state should occur. On the other hand, typical situations in lubrication will correspond to approach rates that are much faster than the critical velocities found in Refs. [1073, 1074] and therefore the high friction state predominant in the older studies might be the practically more relevant one.

In spite of the significant number of successful experimental and numerical studies on the shear behavior of confined thin films, a consensus on the exact mechanism of friction and the structure and phase state of the confined liquid has not yet been reached and further investigation focusing on time-and rate-dependent effects will be necessary.

9.4
Summary

- Amontons' law of macroscopic, dry friction states that the friction force is proportional to the load and does not depend on the apparent contact area:

$$F_F = \mu \cdot F_L.$$

- The proportionality constant μ is the coefficient of friction. A microscopic analysis shows that the friction force is proportional to the true contact area. The true contact area is proportional to the load. The coefficient of friction is in general higher for static than for dynamic friction. Typically, $\mu \leq 0.1$ for dry friction.

- The second empirical law for dry, macroscopic friction is that of Coulomb: friction does not depend on the sliding velocity.

- Lubrication is used to reduce friction and wear. Depending on the thickness of the lubrication layer, different regimes are distinguished. Hydrodynamic lubrication prevails if the lubrication layer is thick enough to completely separate the surfaces. In this case, friction is determined by the viscosity of the lubricant. Typically, μ is in the range of $0.001-0.01$. In the case of nonconforming contacts as in ball bearings or gears, elastohydrodynamic lubrication allows stable operation due to elastic deformation of surfaces and pressure-dependent viscosity of lubricants.

- If the surfaces are separated only by adsorbed molecules or oxide layers, boundary lubrication is acting. In systems where the thickness of the lubrication layer is of the order of the surface roughness and intermittent contact of the surfaces occurs, we talk about mixed lubrication. Friction coefficients are typically $0.01-0.2$.

- With the evolution of nanotechnology, thin film lubrication is gaining importance, where the lubrication layer thickness approaches molecular dimensions. Lowest friction coefficients can be achieved in the case of rolling friction as, for example, in ball bearings.

- The availability of new experimental methods at the end of the 1980s such as friction force microscopy allowed us to study friction on the atomic scale and created the new field of nanotribology. The observed wearless friction on this scale can be understood using the model of Tomlinson where the "plucking action of one atom on to the other" leads to energy dissipation during stick–slip processes.

- Superlubricity, sliding friction with almost vanishing friction force, can be achieved by incommensurate alignment of crystalline surfaces, by applying vertical oscillations or by minimizing the interaction potential.

9.5
Exercises

9.1. A block of metal with a mass of 5 kg is put on a plate of the same metal 1 m from the left end of the plate. The coefficient of static friction of the material is 0.5 and the coefficient of kinetic friction is 0.4. The right end of the plate is slowly lifted to incline the plate until the block begins to slide. At which inclination angle does the block begin to slide? If the plate is kept at this angle, what speed will the block have when it reaches the left end of the plate?

9.2. A steel cube of 10 cm side length is placed on a steel plate. The yield stress of simple steel is $\tau_c \approx 250$ mPa. Estimate the area of real contact. Compare this to the real area of contact. If we assume a typical area of $10\,m^2$ per microcontact, how many junctions exist between the plate and the block?

9.3. A cylindrical axis of 1 cm diameter and 10 cm length is rotating in a precision bearing with an inner diameter, which is only 6 m larger. We use octadecane as the lubricant. Due to friction, it heats up to 50 °C, where its viscosity is 2.49 mPa s. The axis rotates with 80 rotations per second. What is the torque we have to apply to overcome viscous friction?

10
Solvation Forces and Non-DLVO Forces in Water

For large separations, the force between two solid surfaces in a liquid medium can usually be described by continuum theories such as the van der Waals and the electrostatic double-layer theories. The individual nature of the molecules involved, their discrete size, shape, and chemical nature can be neglected. At surface separations approaching molecular dimensions, continuum theory breaks down and the discrete molecular nature of the liquid molecules has to be taken into account. For this reason, it is not surprising that some phenomena cannot be explained by DLVO theory. For example, the swelling of clays in water and nonaqueous liquids and the swelling of lipid bilayers in water cannot be understood on the basis of DLVO theory alone. In this chapter, we consider surface forces that are caused by the discrete nature of the liquid molecules and their specific interactions.

10.1
Solvation Forces

The term "solvation force" was introduced by Derjaguin and Kussakov [687]. They assumed that liquids at interfaces form a boundary layer, in which the molecular structure is different from the arrangement of molecules in the bulk. When the boundary layers of opposing interfaces overlap, the two interfaces experience a force. Sometimes, solvation forces are also called structural forces. We prefer to use the term structural force for interactions mediated by dissolved molecules, usually dissolved macromolecules, which do not adsorb to interfaces. In particular at high concentrations, the interfaces cause the dissolved molecules to assume a layered structure close to the interface. Upon approach of the interfaces, this layered structure is changed, which gives rise to the structural force. At this point, it is instructive to distinguish steric forces. Van Megen and Snook pointed out that solvation forces are the consequence of adsorption of *solvent* on the surfaces [1075]. In contrast, steric forces, discussed later, are caused by dissolved molecules that are adsorbed, such as polymers or surfactants.

Surface and Interfacial Forces. Hans-Jürgen Butt and Michael Kappl
Copyright © 2010 Wiley-VCH Verlag GmbH & Co. KGaA
ISBN: 978-3-527-40849-8

10.1.1
Contact Theorem

To introduce solvation forces, we start with the contact theorem. Consider a fluid between two parallel planar walls as depicted in Figure 7.2. The contact theorem relates the local number density of molecules next to the walls, ϱ_0, to the pressure between the plates. This number density depends on the distance between the walls: $\varrho_0(x)$. At infinite distance, the number density, written as $\varrho_0(\infty)$, is equal to that of an isolated wall. $\varrho_0(\infty)$ is in general different from the number density in the bulk fluid. For example, recent experiments suggest that the density of water close to hydrophobic surface is depleted [1076–1078].

When the two walls approach each other, the density at the wall changes. The contact wall theorem states that the force per unit area is

$$f = k_B T [\varrho_0(x) - \varrho_0(\infty)]. \tag{10.1}$$

The two walls repel each other when the density of liquid at the walls increases upon approach. If the density decreases, the two surfaces attract each other. To obtain the force per unit area, we need to find out how the density at the walls changes when they come closer.

Let us consider simple models for fluids and have a look at which structure they assume at an isolated wall. The simplest models for fluids are hard-sphere and Lennard-Jones fluids. In a hard-sphere fluid, each molecule is taken to be a sphere with a defined radius and volume. It does not interact at all with the other spheres or the wall, except when they get into contact. Then, it repels the other spheres or the walls with an infinitely steep potential so that no overlap occurs. No attraction is taken into account.

Neglecting attraction between the molecules is a crude assumption since, after all, the attraction leads to condensation and that is an essential feature of a liquid. Therefore, a better model for a liquid is the Lennard-Jones fluid. In the Lennard-Jones fluid, the potential energy between molecules is described by

$$V(r) = 4\varepsilon_{LJ} \left[\left(\frac{a_0}{r} \right)^{12} - \left(\frac{a_0}{r} \right)^{6} \right]. \tag{10.2}$$

Here, ε_{LJ} is the energy characterizing the interaction strength and a_0 is the molecular radius. The first term describes a steep repulsion caused by overlapping electron orbitals. The second term accounts for the van der Waals attraction between molecules.

Statistical thermodynamics and computer simulations showed that the density profiles of hard-sphere and Lennard-Jones fluids normal to a planar interface oscillate about the bulk density with a periodicity of roughly one molecular diameter [1079–1086]. The oscillations decay exponentially and extend over a few molecular diameters. In this range, the molecules are ordered in layers. The amplitude and range of density fluctuations depend on the specific boundary condition at the wall and on the size and interaction between the molecules. A steep repulsive wall–fluid potential leads to pronounced oscillations wheras a smooth potential suppresses layering. Such a layering has indeed been observed for different liquids by X-ray diffraction [653, 654, 1087, 1088].

10.1.2
Solvation Forces in Simple Liquids

When two surfaces approach each other, one surface influences the layered structure of the liquid at the other surface, and vice versa (Figure 10.1). As a result, ϱ_0 changes. Density fluctuations cause a force that periodically varies; the periodic length corresponds to the thickness of each layer [658, 1075, 1089–1094]. A liquid confined in a narrowing gap can even undergo a periodic liquid to solid transition implying a periodically changing mobility of the molecules [658, 1095]. The periodically varying force decreases with increasing distance.

To calculate the force, we apply the contact theorem. Therefore, we first need to know the density at the walls. As a first approximation, we assume that the number density in the gap can be superimposed by the densities of the two walls according to $\varrho_n(x, \xi) = \varrho_n(\xi) + \varrho_n(x-\xi) - \varrho_b$ [1075, 1096]. Here, $\varrho_n(x, \xi)$ is the number density at position ξ normal to a surface in a gap of width x. $\varrho_n(\xi)$ is the number density at position ξ normal to an isolated surface, which is equal to that of a surface in an infinite gap ($\varrho_n(\xi) = \varrho_n(x = \infty, \xi)$). ϱ_b is the bulk number density. Please note that

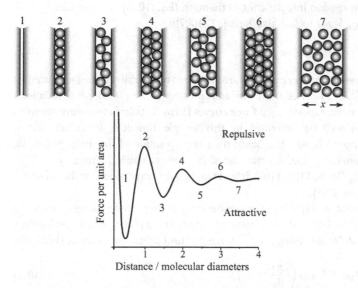

Figure 10.1 Schematic figure of the structure of a simple liquid confined between two parallel walls. The order changes drastically depending on distance, which results in a periodic force. For steep potentials between fluid molecules and the wall and for short-range interactions between the fluid molecules, the force is maximal for dense, ordered structures. Thus, maxima occur at $x \approx d_0$, $2d_0$, $3d_0$, and so on.

$\varrho_0(x) = \varrho_n(x, \xi = 0)$. Let us describe each individual density by

$$\varrho_n(\xi) = \Delta\varrho \cdot e^{-\xi/\lambda} \cdot \cos(2\pi\xi/d_0) + \varrho_b. \tag{10.3}$$

Here, $\Delta\varrho$ is the maximal amplitude of the oscillation. Such an exponentially decaying, oscillating density is justified by statistical thermodynamics and computer simulations of simple fluids [1079–1086]. The exponential factor with the characteristic decay length λ ensures that the force decreases with distance and vanishes for large distance. A cosine (or sine) is the first-order approximation for any periodic function. Here, d_0 is the layer thickness, which in the case of simple liquids is close to the molecular size. We assumed that the density is maximal (or minimal for negative $\Delta\varrho$) directly at the surface at $\xi = 0$.

Superimposing the densities of two similar parallel walls at a given spacing x leads to a density in the gap of

$$\varrho_n(x, \xi) = \Delta\varrho\, e^{-\frac{\xi}{\lambda}} \cos\left(2\pi\frac{\xi}{d_0}\right) + \varrho_b + \Delta\varrho\, e^{-\frac{x-\xi}{\lambda}} \cos\left(2\pi\frac{x-\xi}{d_0}\right) + \varrho_b - \varrho_b. \tag{10.4}$$

At the wall, that is, for $\xi = 0$, the density is

$$\varrho_0(x) = \Delta\varrho\left[1 + e^{-\frac{x}{\lambda}} \cos\left(2\pi\frac{x}{d_0}\right)\right] + \varrho_b. \tag{10.5}$$

The same result is obtained if we calculate the density at the other wall at $\xi = x$. Inserting this expression into the contact theorem (Eq. (10.1)) and considering that $\varrho_0(\infty) = \Delta\varrho + \varrho_b$ leads to (see also Refs [31, 1097]):

$$f(x) = f_0 \cos(2\pi x/d_0) \cdot e^{-x/\lambda}. \tag{10.6}$$

Here, $f(x)$ is the solvation force per unit area between two parallel planes separated by a distance x. $f_0 = k_B T \Delta\varrho$ is the force extrapolated to $x = 0$. Many simulations [658, 1086] and some experimental force curves [1098–1100] of even more complex molecules could well be described by this simple function. In an alternative superposition approach, which is based on a superposition of the mean potentials rather than densities, the contact densities are calculated from $\varrho_n(x, \xi) = \varrho_n(\xi)\varrho_n(x-\xi)/\varrho_b$ [1096, 1101, 1102]; it leads to the same expression for the solvation force (see Exercise 10.1).

In an experiment, we typically measure the force between two spheres or a sphere and a plane. To calculate the force between a sphere and a plane, we use Derjaguin's approximation (2.73) and integrate the force per unit area. The result is [1105]

$$F(D) = F_0 e^{-D/\lambda} \cdot \cos\left(\frac{2\pi D}{d_0} + \phi\right), \tag{10.7}$$

with $F_0 = Rf_0 \cdot [(2\pi\lambda)^{-2} + d_0^{-2}]^{-1/2}$ and $\tan\phi = \lambda/d_0$. Converting the force per unit area to a force introduces a phase shift ϕ. The phase shift can also be expressed as a distance offset D_0 and we can write $\cos(2\pi D/d_0 + \phi) = \cos(2\pi(D-D_0)/d_0)$ with

$2\pi D_0/d_0 = -\phi$. Here, D_0 can be viewed as an offset distance or a residual gap distance before hard contact occurs. It fixes the position of the first maximum with respect to actual contact.

Practically, it is often difficult to relate f_0 and D_0 to any molecular or thermodynamic parameter, in particular for small molecules. Simulations and statistical theory showed that the amplitude and the position of the first maximum depends sensitively on the chosen fluid–wall and fluid–fluid interactions [1084, 1093, 1095, 1096, 1106, 1107]. Also, the detailed structure of the molecules at the wall [1083], surface roughness [672, 1108, 1109], and contamination can significantly change the position of the maximum. In addition, determining contact with subnanometer accuracy is not a trivial task and sometimes impossible. Therefore, f_0 and D_0 can in most cases be viewed as phenomenological fitting parameters. The larger the molecules are the more likely it becomes to extract meaningful interpretation out of experimentally determined values of f_0 and D_0 (see below).

Meanwhile, experiments have confirmed the existence and the predicted features of solvation forces. A thoroughly studied liquid is octamethylcyclotetrasiloxane (OMCTS, $C_8H_{24}O_4Si_4$). OMCTS consists of a ring of four SiO units with four pairs of methyl groups each being attached to Si. Each molecule is quasispherical with a diameter of 0.9 nm. Experiments with the SFA [1105, 1098, 1099, 1110, 1111] and AFM [303, 1112–1114] always showed exponentially decaying periodic forces with $d_0 \approx 0.8$ nm. On example is shown in Figure 10.2.

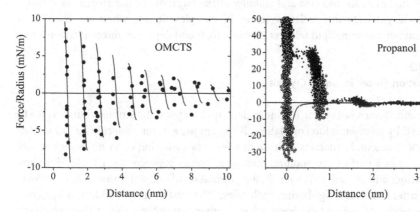

Figure 10.2 Left: Normalized force versus distance across liquid OMCTS between two mica surfaces measured upon approach with an SFA (●, adapted from Ref. [1111]). The continuous line is a fit with Eq. (10.7) and $F_0/R = 0.013$ Nm^{-1}, $\lambda = 3.7$ nm, $d_0 = 0.94$ nm, and $\phi = 144°$. Only those parts are plotted where the force increases with decreasing distance. Regions in between are inaccessible because the gradient of the attractive force exceeds the spring constant of the SFA. Right: Normalized force between a microfabricated silicon nitride tip of an atomic force microscope and a planar mica surface in 1-propanol at room temperature [1100]. The tip had a radius of curvature of $R = 50$ nm. The different symbols were recorded during approach (●) and retraction (○) of the tip. For comparison, the calculated van der Waals force is plotted as a continuous line.

Oscillating forces have also been observed with other molecules such as tetra-chloromethane, benzene, cyclohexane, toluene, 2,2,4-trimethylpentane [1098, 1115, 1116], n-alcohols [1100, 1103, 1104, 1117], and ionic liquids [1118, 1119]. In all cases, solvation forces were observed with periodicities corresponding to the spacing determined by X-ray diffraction of bulk liquid. As an example, the solvation force measured in propanol on mica with an AFM is plotted in Figure 10.2. In this case, the observed periodicity indicates that the molecules are preferentially oriented normal to the surfaces studied and are stabilized by a network of hydrogen bonds between hydroxyl groups. Alkanes have been studied extensively, by experiments [1120–1123], simulations, and theory [1069, 1124–1127], driven by their relevance as lubricants. n-Alkanes tend to orient parallel to surfaces and form layers of 0.4–0.5 nm thickness, which corresponds to the diameter of an alkyl chain. In branched alkanes, layering is reduced.

Oscillatory forces also occur in complex liquids. When we talk about complex liquids, we refer to liquids, which are structured at different length scales. For example, surfactants or lipids dissolved in a liquid form micelles, lamellae, and even more complex structures at high concentrations. In that case, at least two length scales characterize the solution: at the small scale, the liquid molecules are relevant. At a larger length scale, the diameter of the micelles or the thickness of the lamellae describe the solution. When such a surfactant solution is confined between two surfaces, this can lead to an oscillatory force [779, 1128, 1129]. The oscillation period then does not correspond to the size of the liquid molecules but is due to structures at the next larger length scale. Oscillation period and the decay length contain information about the size and stability of the supramolecular structures formed. We will get back to oscillatory forces in complex liquids when we discuss the interaction between lipid bilayers (Section 10.3) and depletion forces (Section 11.6).

10.1.3
Solvation Forces in Liquid Crystals

Solvation forces not only arise due to density changes at the surfaces but can also be caused by an orientation correlation. For example, a water molecule has an electric dipole moment. It interacts with the dipoles of neighboring water molecules or with charges of a surface. In addition, hydrogen bonds between water molecules show a distinct directionality. As a result, the orientation of one water molecule influences the orientation of neighboring molecules. The surface might induce a preferred orientation. As two surfaces approach each other, this ordering is changed, which can lead to a force.

Orientation correlations are particularly strong in liquid crystals. In liquid crystals, the molecules show a high degree of ordering. In contrast to solid crystals, the molecules are, however, still mobile. Depending on the kind of order, we distinguish different kinds of liquid crystals (Figure 10.3). In a nematic liquid crystal, the molecules show a preferred orientation but no positional ordering. In the smectic phase, the molecules form well-defined layers. Smectics thus shows a positional ordering in one direction.

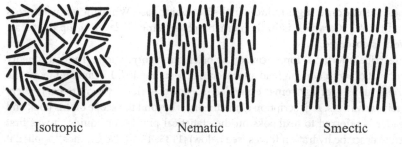

Isotropic   Nematic   Smectic

Figure 10.3 Schematic of an isotropic liquid and nematic and smectic liquid crystals.

One could expect that liquid crystals confined between two solid surfaces should lead to strong oscillatory forces since they show long-range order even in the bulk. This is indeed the case, and solvation forces have been observed in various liquid crystalline systems [1128, 1130–1134]. In liquid crystals, the presence of an interface generates a smectic layering, while the bulk fluid is still isotropic or nematic. Such layers are called presmectic layers.

As one example, Figure 10.4 shows results obtained with octyl-cyanobiphenyl (8CB, $H_{17}C_8(C_6H_4)_2CN$). At high temperature, 8CB is an isotropic liquid. When it is cooled slowly to 40.5 °C, the material becomes a nematic liquid crystal, where the molecules align and show a preferred orientation. Cooling further to 33.5 °C, 8CB undergoes a phase transition from smectic to nematic, where in addition to the orientational order the molecules form a layered structure. 8CB forms solid crystals when the temperature is reduced below 21.5 °C. When confined between two

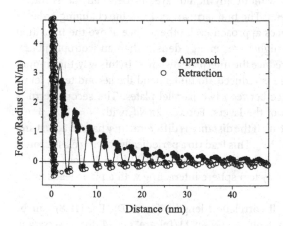

Figure 10.4 Normalized force versus distance curve measured in 8CB between a glass sphere (radius 8.5–10 μm) and a glass plate with an AFM [1134]. The force curve was recorded in the nematic phase at 34.2 °C; that is, 0.7 K above the nematic phase to smectic phase transition. It was fitted with Eq. (10.8) using $L_S = 5 \times 10^{-12}$ N, $\psi_S = 0.28$, $\xi_S =$ nm, $d_0 = 3.2$ nm, and $D_0 = 2.9$ nm. Both glass surfaces were coated with N,N-dimethyl-N-octadecyl-3-aminopropyltrimethoxychlorosilane so that 8CB completely wets the surfaces.

surfaces, smectic ordering is induced in the isotropic phase. When the temperature is reduced to 34.2 °C, which is in the nematic phase, smectic layering up to at least 15 layers is observed.

To describe interfacial forces caused by orientation correlations, Landau[1] theory in combination with a mean field approach has been applied. In Landau theory, the free energy density of a system is expanded in a power series of the order parameters and their derivatives. A description of the theory is beyond the scope of this book and the reader is referred to textbooks on the statistical physics of liquids. It was first applied to describe hydration forces (see below) [1135, 1136]. De Gennes[2] applied it to liquid crystals [1137], and sometimes the theory is referred to as Landau–de Gennes theory.

Depending on the boundary condition for ordering at the surfaces and the kind of structure of the liquid, different equations describe the force. For example, the presmectic force between a sphere of radius R_p and a plane made of an identical material (symmetric case) with fixed orientation of the liquid molecules directly in contact with the surfaces is [1128, 1131, 1134]

$$F = 2\pi R_p \frac{L_S \psi_S^2}{\xi_S} \left[\tanh\left(\frac{D-D_0}{2\xi_S}\right) - 1 + \frac{1-\cos(2\pi(D-D_0)/d_0)}{\sinh\left(\frac{D-D_0}{\xi_S}\right)} \right]. \qquad (10.8)$$

Here, L_S is the smectic elastic constant (in units of N), which describes elastic compressibility of a smectic layer. It depends on the temperature and is typically of the order of 1–10 pN [1138]. ψ_S is the dimensionless degree of smectic ordering at the surfaces. ξ_S is the smectic correlation length and d_0 is the smectic period. The offset D_0 is a zero stress separation without any liquid layer in between the surfaces; practically, it is a fitting parameter. The first term arises from the changing order of the molecules as the two surfaces approach each other. Since above the transition point, an ordered phase has a higher free energy density than an isotropic phase, for the system it is favorable to reduce the interplate distance. In this way, the amount of material in the ordered phase is reduced. To understand the second term, let us consider for simplicity the force between two parallel plates. The second term is due to the elastic compression of the layers. For $x = 2\pi N d_0$ with $N = 1, 2, 3, \ldots$, the layers are not compressed at all. If the distance is different, they have to adjust and are compressed by a distance $x - N d_0$. This leads to a periodically varying elastic force, which to first order is proportional to $1-\cos[2\pi(x-Nd_0)/d_0] = 1-\cos(2\pi x/d_0)$. Going from plane–plane geometry to a sphere interacting with a plane leads to the additional offset D_0.

For large distances and small correlation lengths ($\xi_S \ll D$), Eq. (10.8) can be simplified. Remembering that $\tanh y = (e^y - e^{-y})/(e^y + e^{-y}) \rightarrow 1$ for $y \rightarrow \infty$ and

1) Lev Davidovich Landau, 1908–1968. Russian theoretical physicist, professor in Moskau, Nobel Prize in physics, 1962.

2) Pierre-Gilles de Gennes, 1932–2007. French physicist, professor in Paris, Nobel Prize in physics, 1991.

$\sinh y = (e^y - e^{-y})/2 \rightarrow e^y/2$, we obtain

$$F = 4\pi R_p \frac{L_A \psi_S^2}{\xi_S} \cdot e^{-\frac{D-D_0}{\xi_S}} \cdot \left[1 - \cos\left(2\pi \frac{D-D_0}{d_0}\right)\right]. \qquad (10.9)$$

The force is the sum of an exponentially decaying force plus a periodic contribution with an exponentially decreasing amplitude.

Equations (10.8) and (10.9) are valid for the force between similar surfaces. If the two surfaces are not similar, the presmectic force between a sphere and a flat plate changes [1128]. If the dimensionless degree of smectic ordering at the surfaces is different, the surfaces start to repel each other (see Exercise 10.2). In addition, the curvature of the surface has to be taken into account [1139]. In curved geometries, the director field of the liquid crystal is no longer undistorted. Its contribution to the force has to be considered. This contribution is repulsive because with decreasing particle distance the distortion of the director field increases. One consequence is that Derjaguin's approximation is valid only for very large radii of curvature. Rather than taking the condition $d_0 \ll R_p$, Derjaguin's approximation is valid only for the more stronger condition $\xi_S \ll R_p$.

10.2
Non-DLVO Forces in an Aqueous Medium

Non-DLVO forces in water deserve a special section because they are important and owing to the complex structure of water are not well understood. They are important because water is the universal solvent in nature. Also, in more and more industrial processes water is used instead of organic solvent since it is environment friendly and readily available. We distinguish two cases: hydration forces, which occur between hydrophilic surfaces, and hydrophobic forces between surfaces, which form a contact angle $\Theta \geq 90°$ with water.

10.2.1
Hydration Forces

Around 1980, evidence had accumulated that in many cases DLVO forces are not sufficient to describe the interaction between hydrophilic surfaces in aqueous media [1140]. For example, dispersions of colloidal silica at alkaline conditions are stabilized by adding certain salts although the electrostatic double-layer repulsion decreases [1141]. When studying the coagulation behavior of a polystyrene latex dispersion, Healy *et al.* [1142] observed that K^+ and Li^+ ions are able to stabilize the dispersion at high salt concentrations. Another observation not explainable with DLVO theory is the swelling of clay [1143–1145]. In the presence of water or even water vapor, clay swells even at high salt concentrations. When two hydrophilic surfaces are brought into contact, repulsive forces of about 1 nm range have been measured in aqueous electrolyte between a variety of surfaces, for example, mica [55, 450, 451, 1146–1148], silica [255, 1149–1153], lipids [1154], and DNA [1155]. Because of the correlation with the low (or negative) energy of wetting of these solids with water, the repulsive force has been attributed to the energy required to remove the water of hydration from the surface. These forces were termed hydration forces. In 1929, Kruyt and Bungenberg de Jong had mentioned hydration forces as a factor while interpreting the increase in viscosity of agar solutions at high concentrations [1156]. In this section, we focus on hydration forces between hard, inorganic surfaces. The interaction between lipids or surfactants will be discussed later.

In most cases, measured hydration interactions can be described by an exponential function:

$$V^A(x) = V_0 \, e^{-x/\lambda_h}. \tag{10.10}$$

Typical values for the amplitude V_0 are 10^{-3}–$10 \, \text{Nm}^{-1}$. The decay length λ_h ranges from 0.2 to 1.0 nm. In some cases, oscillatory forces with a period of 0.2–0.3 nm have also been observed [303, 1146, 1147, 1157, 1158]. Computer simulations confirm that oscillations are expected in water between surfaces such as mica [1102, 1159, 1160].

Multiple effects contribute to the hydration force and no unique reason is sufficient to explain the short-range repulsion. As an example, force versus distance curves measured between two mica and two silica surfaces are compared in Figure 10.5. Both curves were recorded at high salt concentration of 1 M. Therefore, electrostatic forces are negligible. At a distance of 2 nm on silica and 6–8 nm on mica, attractive van der Waals forces dominate. At closer distance, the short-range hydration repulsion is observed. On mica, the short-range repulsion increases with the salt concentration. For divalent cations and high concentrations, it can range to $\lambda_h = 3 \, \text{nm}$ [199, 450, 451, 1146]. For distances below 1.5 nm, the periodic layering indicated by jumps of 0.25 nm is observed. In the case of silica, the short-range repulsion is weaker and independent of salt concentration [1150, 1151]. In some

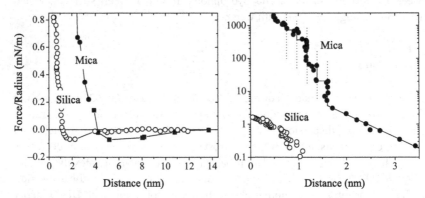

Figure 10.5 Normalized hydration forces measured on approach between two mica and two silica surfaces in aqueous solution on a linear (left) and logarithmic (right) scale. The measurements between mica were carried out with an SFA in 1 M KCl [1146] (●) and in 1 M KNO$_3$ [55] (squares) at pH 5.7. The continuous line is to guide the eye. The measurements between a silica sphere (5 μm diameter) and an oxidized silicon wafer were performed with an AFM in 1 M NaCl at pH 7.0 [1153].

experiments, no short-range repulsion is observed at all and the force can be described by DLVO theory [1152].

■ **Example 10.1**

To estimate the order of magnitude of hydration forces, we assume that the first layer of water molecules is ordered by interaction with the surface. This ordering leads to a change in the distribution of hydrogen bonds. Let us assume that the interaction reduces the density of hydrogen bonds in the first monolayer by a factor of 2. The energy increase per unit area is $V_0 = G_m^{hyd}/2N_A d_0^2$. Here, G_m^{hyd} is the free energy of hydrogen bonds per mole of water molecules in the bulk and d_0 is the diameter of the molecules. With $d_0^3 N_A \approx V_m$ and V_m being the molar volume, we can express d_0^2 by the molar volume, $d_0^2 = V_m/N_A^{2/3}$, and write $V_0 = G_m^{hyd}/2(N_A V_m^2)^{1/3}$. With $G_m^{hyd} = 20\,kJ\,mol^{-1}$ and $V_m = 18 \times 10^{-6}\,m^3\,mol^{-1}$, one can calculate $V_0 = 0.17\,Nm^{-1}$. A reasonable decay length for the density of hydrogen bonds is given by the molecular diameter and we estimate $\lambda_H = d_0$. For the energy per unit area, we get

$$V^A = V_0 \cdot e^{-x/d_0}. \tag{10.11}$$

According to the Derjaguin approximation for a sphere of radius r_s interacting with a plane or between two cylinders of radius r_s, this would lead to a normalized force

$$\frac{F}{r_s} = 2\pi V_0 \cdot e^{-x/d_0}. \tag{10.12}$$

This roughly agrees with what has been observed between mica surfaces (Figure 10.5).

Different theories have been proposed to explain hydration forces and to derive quantitative equations. Marčelja and Radic assumed that the water molecules in direct contact with the walls are fixed and preferentially oriented [1135]. This preferred orientation is coupled to the adjacent layers of water molecules. Using Landau–de Gennes theory and introducing a water-order correlation length λ_H^{cor} they derived a force per unit area

$$f = \frac{E_0^A}{2\lambda_h^{cor}} \frac{1}{\sinh^2(x/2\lambda_h^{cor})}. \tag{10.13}$$

E_0^A is the energy per unit area associated with ordering of the water molecules at the surface compared to the bulk.

Later, Cevc et al. [1136] generalized the approach. Rather than fixing the order of water molecules at the surfaces, they applied an effective orienting field. This allowed the water molecules to change their orientation when the two surfaces got close to each other. For similar surfaces interacting with each other, they derived

$$f = \frac{E_0^A}{2\lambda_h^{cor}} \frac{1}{\cosh^2(x/2\lambda_h^{cor})}. \tag{10.14}$$

In the limit of large distances, that is, for $x \gg \lambda_h^{cor}$, both expressions describe an exponentially decaying force:

$$f = \frac{2E_0^A}{\lambda_h^{cor}} \cdot e^{-x/\lambda_h^{cor}}. \tag{10.15}$$

The characteristic decay length is the water-order correlation length. For closer approach, fixed orientation as described by Eq. (10.13) leads to a stronger interaction. When the orientation is allowed to adjust, the force per unit area at close range is reduced (Eq. (10.14)).

The approach of Marčelja and Radic is a phenomenological treatment. It does not refer to a specific interaction, neither for the orientation and immobilization for the first water layer nor for the interaction between water molecules. Water molecules at the surface can be oriented and immobilized by charges [1162]. For example, at high pH silica bears a high negative surface charge. The resulting electric field seems to be able to orient the first 3–5 monolayers of water, as indicated by sum-frequency vibrational spectroscopy [1163]. That electric fields are able to change the structure of water was observed at electrodes with an applied potential [1164, 1165]. Short-range repulsion has also been observed on electrically neutral surfaces. In that case, dipoles, or in general an inhomogeneous charge distribution, can orient the first water layer [1160, 1166–1168]. Computer simulations show that even without charges on hydrophilic surfaces such as glass or platinum, the first layers of water molecules are oriented and show a reduced mobility [1169]. The orientation of the first layer of water is transmitted by dipole–dipole interactions [1167, 1170] or by hydrogen bonds to neighboring water molecules. For phospholipid bilayers, with their strong dipoles at the surface, computer simulations indicate that the decay of water ordering is directly correlated with the decay of the interfacial dipolar charges [1296].

For silica, the main contribution to the short-range repulsion most likely comes from the formation and rupture of hydrogen bonds [1153, 1161]. That the hydrogen bond network is indeed disrupted when two silica surfaces approach each other was recently confirmed by infrared spectroscopy [1161]. On silica, a gel layer can also lead to a softening of the surfaces and a steric repulsion.

Ions can significantly influence the structure of water at interfaces. Hydration can be caused by the overlap of layers of hydrated ions adsorbed on the surfaces. On mica, for example, the dehydration of adsorbed cations is most likely the main cause for the short-range repulsion between two approaching surfaces [451]. Computer simulations between smooth hydrophilic surfaces confirm a layered structure of the water molecules and as a result a periodic force [1159, 1160]. Paunov *et al.* [1171] used this hypothesis to explain the interaction between proteins in suspensions.

Electrostatic effects can also directly result in a short-range repulsive force. The dipoles of one surface can electrostatically interact with dipoles on an opposing surface and their respective polarization charges [1168]. In particular for interacting lipid bilayers, this can be a major contribution because phospholipids have a strong dipole. Attard and Patey [1172] calculated the force between two half-spaces with a certain dielectric constant across a medium with a higher dielectric constant. On the interface, they placed a square lattice of dipoles. This could, for example, be a model for two silica surfaces ($\varepsilon = 3.8$) interacting across water ($\varepsilon = 78.5$) in which the dipoles are formed by silanol groups. They obtained short-range repulsive force of the right range and strength.

10.2.2
Hydrophobic Force

While hydrophilic surfaces tend to repel each other in aqueous medium, surfaces with a contact angle higher than 90° tend to attract each other. This attractive force is called hydrophobic force (for a review, see Ref. [1173]). The term arises from Greek words "hydro" for water and "phobos" for fear, which describes the apparent repulsion between water and nonpolar molecules. The hydrophobic force dominates the interaction between hydrophobic surfaces and is highly relevant from the fundamental and technical points of view. Still, its origin is not clear and a generally accepted quantitative description is missing.

10.2.2.1 The Hydrophobic Effect
The hydrophobic force is related to the hydrophobic effect [1174, 1175]. Nonpolar molecules such as hydro- and fluorocarbons and nonpolar gases poorly dissolve in water (Table 10.2). If, for example, liquid octane is in contact with water and the system is allowed to equilibrate, only 5.4 μM octane dissolves in water. Rather than being dissolved as individual molecules hydro- and fluorocarbons attract each other and tend to form intermolecular aggregates in an aqueous medium. This aggregation effect is called hydrophobic effect. The apparent attraction between hydrophobic molecules in water is sometimes referred to as hydrophobic interaction. At the macroscopic level, the hydrophobic effect is apparent when oil and water are mixed together and form separate phases. On solid, hydrophobic surfaces, water forms a high contact angle (Table 10.1). At the molecular level, the hydrophobic effect is

Table 10.1 Surface tension γ_S of various hydrophobic solids, the interfacial tensions between the solid and water γ_{SL}, and advancing contact angles Θ_a of water on these solids at 20 °C (from own measurements, Refs [1176–1178], and references therein).

	γ_S (mNm^{-1})	γ_{SL} (mNm^{-1})	Θ_a
Paraffin	23	45	109°
Polyethylene (low density)	34	42	97°
Polypropylene	30	50	106°
Poly(methyl methacrylate)	42	22	77°
Poly(n-butyl methacrylate)	30	32	92°
Polystyrene	37	36	91°
Poly(ethylene terephthalate)	46	29	77°
Polytetrafluoroethylene	19	41	108°
Poly(vinyl chloride)	40	36	87°

Measurements of solid surface tensions, solid–liquid interfacial tensions, and contact angles vary by typically ±4 mNm^{-1} or 3°. The parameters depend to a certain degree on how the samples are prepared. For a discussion, see Refs [1177, 1178].

an important driving force for biological structures and responsible for protein folding and formation of lipid bilayer membranes.

■ **Example 10.2**

How much nitrogen, oxygen, and argon is dissolved in water if the water is in equilibrium with air? Air consists of 78.1 vol% nitrogen, 21.0 vol% oxygen, and 0.9 vol% argon. The respective partial pressures are 0.781, 0.21, and 0.009 atm, assuming all components behave like ideal gases. Since the saturation concentration given in Table 10.2 refers to 1 atm, the concentrations of the gases at the respective partial pressures are 0.51, 0.27, and 0.013 mM.

Table 10.2 lists the thermodynamic properties for the transfer of hydrophobic molecules into water. It has many important lessons [1175]. First, the free energy of transfer ΔG_s^0 is unfavorable (positive) and several times the thermal energy. This reflects the low solubility since the saturation concentration of the substance c_s is related to ΔS_S^0 by $c_s/c_w = \exp(-\Delta G_s^0/RT)$. Here, c_w is the concentration of pure

Table 10.2 Saturation concentration c_s, standard entropy ΔS_s^0, enthalpy ΔH_s^0, and free energy ΔG_s^0 of solvation for various nonpolar gases and liquids in water at 25 °C [1179, 1180].

	c_s (mM)	$T\Delta S_s^0$ (kJmol^{-1})	ΔH_s^0 (kJmol^{-1})	ΔG_s^0 (kJmol^{-1})
He	0.39	−30.2	−0.8	29.4
Ne	0.45	−32.8	−3.7	29.1
Ar	1.39	−38.5	−12.3	26.2
Kr	2.49	−40.5	−15.7	24.8
Xe	4.30	−41.9	−18.4	23.5
H$_2$	0.78	−31.7	−4.0	27.7
N$_2$	0.65	−38.6	−10.4	28.2
O$_2$	1.27	−38.5	−12.1	26.4
Methane	1.39	−40.1	−13.8	26.3
Ethane	1.85	−45.3	−19.7	25.6
n-Propane	1.50	−48.5	−22.4	26.1
n-Butane	1.46	−52.4	−26.0	26.6
CF$_4$	0.24	−46.0	−15.1	30.9
n-Pentane	0.55	−29.1	−0.5	28.6
n-Hexane	0.12	−32.3	0.2	28.5
n-Heptane	0.026	−33.1	2.7	29.1
n-Octane	0.0059	−38.0	1.8	29.9
n-Nonane	0.0012			31.0
n-Decane	0.0004			31.1

The entropy is multiplied by the temperature $T = 298.15$ K for better comparison. Values for gases refer to a pressure of 1 atm. For materials that at 25 °C and normal pressure are in the gas phase, solubilities, ΔS_s^0, ΔH_s^0, and ΔG_s^0 values are reported in terms of the process: ideal gas (1 atm) → ideal solution in water (unit mole fraction solute). For substances that at 25 °C and normal pressure are in the liquid phase (starting with pentane), the values are reported in terms of the process: pure liquid solute → ideal solution in water. See also Exercise 10.3.

water. At 25 °C, it is $c_w = 55.35$ M. Second, the low solubility is the consequence of an unfavorable (negative) change in entropy that overweighs a favorable (negative) enthalpy change. Remember, $\Delta G_s^0 = \Delta H_s^0 - T\Delta G_s^0$. Hydrophobic molecules induce order and thus a decrease in entropy in the surrounding water. Consequently, the solubility decreases with increasing temperature. Dissolved hydrophobic molecules decrease the density of water. Furthermore, the solubility of gases and hydrocarbons decreases when salt is added [1181]. This effect is sometimes referred to as "salting out."

If we try to mix oil and water, the poor solubility leads to a phase separation (unless one of the phases is present in only trace amounts). The hydrophobic effect manifests itself in a high interfacial tension between hydrocarbons and water. For example, the interfacial tension of n-hexane–water at 25 °C is $50.4\ Nm^{-1}$ and that of n-dodecane–water is $52.6\ mNm^{-1}$ [1182]. Such a high interfacial tension destabilizes oil-in-water or water-in-oil emulsions. The oil drops in water (or water drops in oil) coagulate and form a continuous water (or oil) phase. In practice, to prevent an emulsion from immediate separation into two continuous phases, surfactants have to be added to reduce the interfacial tension.

10.2.2.2 Hydrophobic Forces

Just as hydrophobic molecules tend to aggregate in aqueous medium, two hydrophobic surfaces adhere to each other. This is intuitively plausible: the system can lower its total free energy by reducing the solid–water interfacial area. If two interfaces get into contact, the total interfacial area decreases by twice the contact area. For this reason, we expect a strong adhesion and an attractive force. These arguments do, however, not tell us how the force depends on distance.

The first direct evidence that the interaction between solid hydrophobic surfaces is stronger than the van der Waals attraction was provided by Pashley and Israelachvili [1183, 1184]. With the help of surface force apparatus, they observed an exponentially decaying attractive force between two mica surfaces, coated with a monolayer of adsorbed cationic surfactant cetyltrimethylammonium bromide (CTAB). Meanwhile, a variety of techniques have been employed to measure hydrophobic forces. For SFA experiments, hydrophobic surfaces can be formed by spontaneous adsorption of certain cationic surfactants from solution onto the negatively charged mica surface [1183–1189] or by Langmuir–Blodgett transfer of monolayers of insoluble surfactants [1190–1192]. Such monolayers are, however, laterally inhomogeneous on the 1 μm scale. Both methods rely on physisorption of surfactants. A chemical modification of mica by different silanating agents leads to robust hydrophobic surfaces [1191, 1193]. Again, the silanes tend to form islands on the 1 μm scale [1194].

An alternative technique is the AFM. With the AFM not only forces between physisorbed hydrophobic layers were measured [1195–1198] but also the forces between hydrophobic polymers [1199–1201] or silanated silica or glass surfaces were determined [1194, 1202–1206]. A third technique is the bimorph surface force apparatus also called MASIF (measurement and analysis of surface interaction forces). With MASIF-type devices, the force between two spherical glass surfaces of typical radius 1 mm or between two polymer surfaces can be measured. The long-range attraction was observed between silanated glass [1207], glass with physisorbed surfactants [1208], and gold surface coated with a monolayer of alkylthiols [1209].

Usually, two components of the attraction are observed. They are sometimes described by two exponentials [1173]:

$$f(x) = -C_1 e^{-x/\lambda_1} - C_2 e^{-x/\lambda_2}. \tag{10.16}$$

One is short range and decays roughly with a decay length of typically $\lambda_1 = 1-2$ nm. The second component is more surprising: It is very long ranged and can extend to several 100 nm in some cases. C_1 and C_2 are constants that depend on the way the surfaces are prepared and the experiment is run. As one example, the force between two mica surfaces made hydrophobic by a fluorinated surfactant is plotted in Figure 10.6.

The hydrophobic attraction increases with the contact angle [1194, 1206]. Experimental results significantly vary from one group to another. Even for simple questions such as the dependence of the hydrophobic force on the salt concentration or the concentration of dissolved gas, no consensus has been reached. It seems clear today that probably more than one mechanism is important to understand all phenomena attributed to the hydrophobic force. For example, it is a major difference if the surfaces are rendered hydrophobic by physisorbed surfactants or if the surfaces are inherently hydrophobic. In the first case, the force might be caused, or at least strongly influenced, by a rearrangement of the surfactants [1208, 1210].

Like other strong attractive forces, hydrophobic forces are difficult to measure. In most measuring devices such as the SFA or AFM, one of the two surfaces, between which the force is measured, is mounted on a spring. As soon as the gradient of the

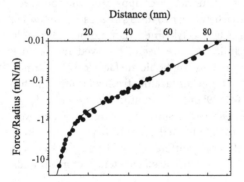

Figure 10.6 Attractive force measured with the SFA between two mica surfaces that had been coated with the double-chain cationic surfactant N-(α-trimethylammonioacetyl)-O, O'-bis (1H,1H,2H,2H-perfluorodecyl)- L-glutamate chloride by Langmuir–Blodgett transfer [1190]. Force curves were recorded with different spring constants. The jump-in was prevented by the drainage method. Forces are negative because they are attractive. The line is a fit with two exponentials (Eq. (10.16)) with $C_1/r_c = 0.15$ Nm^{-1}, $\lambda_1 = 2.2$ nm, $C_2/r_c = 1.9$ mNm^{-1}, $\lambda_2 = 16.7$ nm.

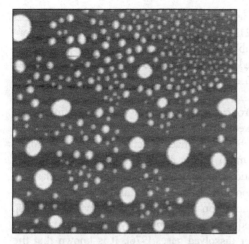

Figure 10.7 Nanobubbles on a silicon wafer, which had been hydrophobized by OTS (octadecyltrichlorosilane), in water [1224]. The macroscopic contact angle was 108°. Nanobubbles were created by first immersing the wafer in ethanol. Ethanol dissolves more nitrogen and oxygen than water. Then, the ethanol was exchanged by water. Temporarily, an oversaturation was created and bubbles of 15–35 nm height formed on the wafer surface. Once formed, they are stable for many hours. The image of 20 µm size was recorded with an AFM in tapping mode.

attractive force exceeds the spring constant, the system becomes unstable and the surfaces jump into contact. Therefore, with a certain spring, attractive forces can be measured only down to a certain distance. To prevent the jump into contact, a force feedback system has to be used, which is technically demanding [1188]. Alternatively, to record a full force curve, measurements with springs of different spring constants are carried out [1183–1186, 1190]. For large distances, forces are measured with a soft spring and with high sensitivity. At short distance, stiff springs allow to approach closely, though at a loss of sensitivity. Another way of measuring forces at close range and to avoid a jump-in is the drainage method. In this case, the approach is so fast that the hydrodynamic repulsion compensates the strong attraction [1189–1191]. After subtracting the hydrodynamic force, the hydrophobic force is obtained.

Different theories have been proposed to explain hydrophobic attraction. Like on hydrophilic surfaces, the structure of water at hydrophobic surface is different from the bulk structure. Computer simulations [1211, 1212], sum-frequency vibrational spectroscopy [1163], X-ray [1078, 1213, 1214], and neutron reflectivity [1076, 1077] show a layer of up to 1 nm with a reduced density and an increased order. When two hydrophobic surfaces approach each other at some point, the surface layers overlap and lead to an attractive force [1212, 1215, 1216]. This force is, however, short ranged and can certainly not explain the long-range component.

It is clear that the water film between the two approaching hydrophobic surfaces becomes unstable when a certain thickness has been reached. For contact angles significantly higher than 90°, it is energetically more favorable that a bubble fills the gap between the surfaces [578, 1217]. The closer the two surfaces get, the more unstable the film becomes, and a bubble is formed [574, 1196]. The spontaneous formation of a bubble is also referred to as cavitation [566]. The bridging bubble should lead to an attractive force [575, 576] as described in the chapter on capillary forces.

Today, it is clear that the capillary force caused by a bridging bubble is an important contribution to the interaction between hydrophobic surfaces. It is particularly important once the two surfaces have been in contact. Whether at the first approach cavitation is relevant is not clear yet. One argument against the nucleation or bubble formation hypothesis is that estimations of the rate of cavitation using classical nucleation theory result in much too low values. There are, however, possible mechanisms that could stimulate nucleation and lead to faster cavitation [1218]. Bubbles need not necessarily be filled with water vapor only. They might also be formed by nitrogen, oxygen, or other dissolved gases [576]. It is known that the concentration of dissolved gas increases close to a hydrophobic surface [1219, 1220].

Another hypothesis is that there are always some gas bubbles of nanoscopic dimension residing on hydrophobic surfaces. These nanobubbles could stimulate cavitation [1207, 1221]. Once two opposing nanobubbles get into contact, they fuse and cause a strong attraction due to the capillary force [1206]. Several experiments show jumps at a distance of several 10 nm, which indicates the presence of bubbles [1202, 1209].

Nanobubbles have indeed been observed [1076, 1221–1223]. They do, however, not form spontaneously. Only when a hydrophobic surface is exposed to an oversaturated solution containing dissolved gas the bubbles form. Once they are formed, they are surprisingly stable and can stay for hours or days even if the concentration of dissolved gas is reduced again (Figure 10.7) [1222, 1224–1226]. It is one of the open questions in surface science why these nanobubbles are stable.

The long-range component of the hydrophobic force is sensitive to the concentration of dissolved gas. It is reduced or even abolished by degassing the water [1189, 1196, 1197, 1200–1202, 1227, 1228]. One example is shown in Figure 10.8.

At this point, we would like to describe another important effect observed when hydrophobic forces are measured. It is known that once two hydrophobic surfaces are in contact, a cavity spontaneously forms if the contact angle is significantly higher than 90° [574, 577, 1229]. In addition, the concentration of dissolved gas is increased close to a hydrophobic surface. In particular in the narrow gap formed by two opposing hydrophobic surfaces, more gas is dissolved than in bulk water [1220]. When the two surfaces are separated again, the dissolved gas might nucleate and form a bubble, which remains at the surfaces. At the second approach, the interaction between two bubbles is measured rather than the force between two hydrophobic surfaces in aqueous medium. This effect might easily go unnoticed, in particular when using an AFM. In the AFM, force curves are recorded periodically with a typical rate of 1–10 Hz. If one does not take special care to record the first force curve with bare surfaces, bubbles can be created [1204]. Indeed, significant differences between first and subsequent force curves have been observed [1197, 1230].

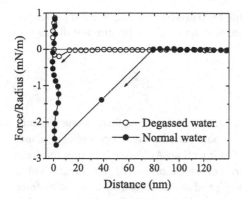

Figure 10.8 Typical approaching force versus distance curves recorded in normal and degassed water (redrawn from Ref. [1201]). Force curves were measured between a glass sphere of 8.8 μm diameter attached to an AFM cantilever of 0.15 Nm^{-1} spring constant and a naturally oxidized silicon wafer. Both surfaces were hydrophobized by a fluorinated alkyl silane.

To our knowledge, all results of hydrophobic forces ranging more than 10 nm can be explained by metastable nanobubbles, which are either formed when filling the measuring cell or by a previous contact. The effect of dissolved gas is to increase the lifetime of nanobubbles.

To illustrate this effect, Figure 10.9 shows the first three approaches of a hydrophobic microsphere toward a hydrophobized silicon wafer. At the first approach, the jump-in distance is only 5 nm. After contact, the sphere is retracted again. In this case, force curves are recorded with a frequency of 5 Hz so that the second approach is only 0.2 s after the first. The jump-in is at a much larger distance of 35 nm. This indicates that during the separation of the microsphere after first contact, bubbles are formed on the silicon wafer, the microsphere, or both. At the

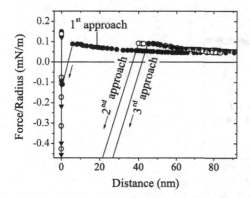

Figure 10.9 Typical approaching force curves measured between a silanated silica sphere of 4.7 μm diameter and a silicon wafer in distilled water using an AFM [1204]. The silicon oxide surfaces were made hydrophobic by vapor silanization with hexamethyldisilazane, leading to an advancing contact angle of $\Theta \approx 96°$. The first three approaches are plotted. The cantilever, to which the sphere was attached, had a spring constant of 0.2 Nm^{-1}.

second approach, this bubble bridges the two surfaces, leading to a jump at much larger distances. At the third and subsequent approaches, the jump-in distance is even 40 nm.

■ **Example 10.3**

Figure 10.10 shows a typical approaching and retracting force versus distance curve measured between a hydrophobic particle and a hydrophobic planar surface. Ishida *et al.* [1206] recorded it with an AFM. We interpret the force curves in the following way: force curves are measured periodically with a typical frequency of 1 Hz. The one plotted was preceded by many other force cycles. During the previous contact, the tip had created a bubble on the surface (or the tip or both). Upon approach at a distance of ≈ 100 nm, the bubble starts to bridge the two surfaces. The result is a strong capillary force. For distance below ≈ 70 nm, it can even be fitted with Eq. (5.28) assuming a constant volume of the bubble (gray line) with $\gamma_L = 0.072 \, \mathrm{Nm}^{-1}$, $R_1 = 10 \, \mu\mathrm{m}$, $V = 0.095 \, \mu\mathrm{m}^2$, and $\Theta_1 = \Theta_2 = 79.7°$. Please note that for a bridging bubble we have to insert $180 - \Theta$, where Θ is the contact angle with respect to water. Upon retraction, the bridging bubble is even larger. The first part of the retracting curve up to a distance of ≈ 40 nm could be fitted with Eq. (5.23) assuming a constant curvature (dashed line) with $\Theta_1 = \Theta_2 = 78.4°$ and a radius of curvature $r = 38$ nm. For distances larger than ≈ 50 nm, Eq. (5.28) (gray line) fitted the experimental curves better; the parameters were $V = 0.11 \, \mu\mathrm{m}^2$, and $\Theta_1 = \Theta_2 = 75°$. On retraction, we observe a transition from constant curvature to constant volume.

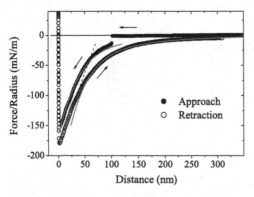

Figure 10.10 Normalized approaching and retracting force versus distance curves measured between a silicon wafer and a silica microsphere in 1 mM KNO_3 aqueous solution with an AFM [1206]. The silicon oxide surfaces were made hydrophobic by silanization with propyltrichlorosilane, leading to an advancing contact angle of $\Theta \approx 105°$. The cantilever, to which the sphere was attached, had a spring constant of $0.2 \, \mathrm{Nm}^{-1}$.

We would also like to mention that electrostatic theories have been proposed. They were ruled out because the hydrophobic force does not show the predicted dependence on salt concentration [1191, 1193, 1200].

10.3
The Interaction Between Lipid Bilayers

10.3.1
Lipids

Lipids are broadly defined as any fat-soluble (lipophilic), naturally occurring molecule, such as fats, oils, waxes, cholesterol, sterols, fat-soluble vitamins (such as vitamins A, D, E, and K), monoglycerides, diglycerides, or phospholipids. Lipids are essentially insoluble in water. For example, 1,2-dipalmitoyl-*sn*-glycero-3-phosphocholine (in short, dipalmitoylphosphatidylcholine, DPPC) is soluble only to a critical concentration of 10^{-12} M. That is, less than one molecule in $(10\,\mu m)^3$. At concentrations higher than the critical concentration, lipids form aggregates. Many of them form bilayers. In this case, the lipid molecules have an amphiphilic character with a hydrophilic head and a hydrophobic tail. In a bilayer, the hydrophobic tails are in the center and the hydrophilic heads point toward the aqueous phase. Lipid bilayers play an essential role as the matrix for biological membranes [1231]. They not only form the outer boundary of a cell, the so-called plasma membrane, but also separate inner compartments in eukaryotic cells. Lipid bilayers are very good electric insulators and are not penetrable for ions. They are also effective borders for hydrophilic molecules such as sugars and nucleic acids. The interaction of bilayers is important for an understanding of processes in cells. It is also relevant for many applications in food science or cosmetics. For a review, see Ref. [1232].

An important class of lipids are the phospholipids, among which the 3-*sn*-phosphatidic acids form a relevant group. A 3-*sn*-phosphatidic acid consists of glycerol (CH_2OH-$CHOH$-CH_2OH), which is esterified at the first two hydroxyl groups with fatty acids. A fatty acid is an allyl chain with a carboxyl group at one end: $CH_3(CH_2)_n COOH$. At the third hydroxyl group, a phosphate group with an additional rest group is bound via an ester. An ester is derived from an alcohol, (R_1OH), plus a carboxylic acid (R_2COOH), according to $R_1OH + R_2COOH \rightarrow R_1-O-CO-R_2 + H_2O$. Here, R_1 and R_2 are organic rest groups. One example, 1,2-di-myristoyl-*sn*-glycero-3-phosphocholine (also called dimyristoylphosphatidylcholine or DMPC), is shown in Figure 10.11. In water at neutral pH, the phosphate group is negatively charged. A neutral rest attached to the phosphate leads to a total negative charge, for example, in phosphatidylgylcerol (PG) and phosphatidylinositols (PI). A positive rest group leads to a zwitterionic character and the whole lipid is neutral. Examples are phosphatidylethanolamine (PE) and phosphatidylcholine (PC). Phosphatidylcholines are traditionally also called lecithines. Phosphatidylcholines can be isolated and purified from egg yolk. Then, they are called egg phosphatidyl-

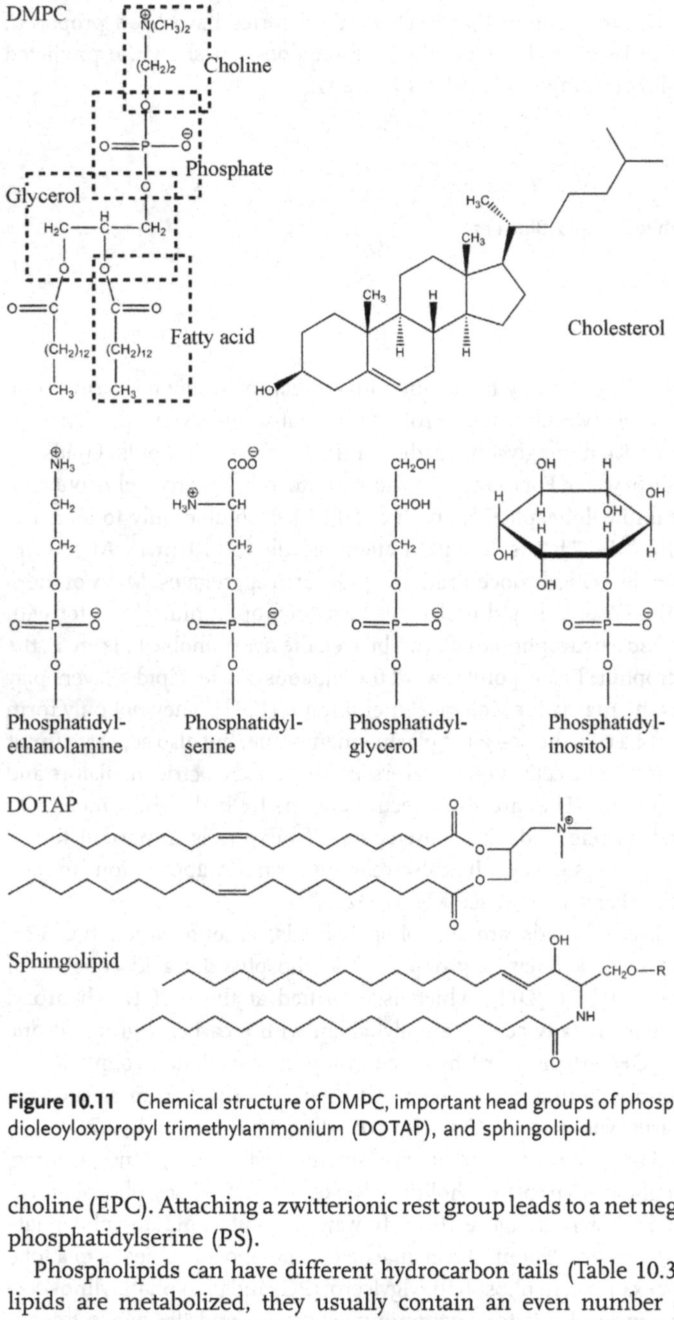

Figure 10.11 Chemical structure of DMPC, important head groups of phospholipids, cholesterol, dioleoyloxypropyl trimethylammonium (DOTAP), and sphingolipid.

choline (EPC). Attaching a zwitterionic rest group leads to a net negative charge, as in phosphatidylserine (PS).

Phospholipids can have different hydrocarbon tails (Table 10.3). Due to the way lipids are metabolized, they usually contain an even number of carbon atoms. Hydrocarbon tails differ in length and in the number of double bonds they contain. Lipids with hydrocarbon chains containing one or more double bonds are called unsaturated. They strongly influence the fluidity of a bilayer. Longer hydrocarbon chains have a higher melting point and reduce the mobility of a membrane. Double

Table 10.3 Names of important acyl chains occurring in lipids and the melting points of the corresponding fatty acid.

Systematic name	Trivial name	Structure CH$_3$−[R]−COO−	Melting point (°C)
Dodecanoyl	Lauryl	−(CH$_2$)$_{10}$−	44
Tetradecanoyl	Myristoyl	−(CH$_2$)$_{12}$−	54
Hexadecanoyl	Palmitoyl	−(CH$_2$)$_{14}$−	63
9-Hexadecenoyl	Palmitoleoyl	−(CH$_2$)$_5$CH=CH(CH$_2$)$_7$	0
Octadecanoyl	Stearoyl	−(CH$_2$)$_{16}$−	70
9-Octadecenoyl	Oleoyl	−(CH$_2$)$_7$CH=CH(CH$_2$)$_7$−	13
9,12-Octadecadienoyl	Linoleoyl	−(CH$_2$)$_3$(CH$_2$CH=CH)$_2$(CH$_2$)$_7$−	−5
9,12,15-Octadecadienoyl	(9,12,15)-Linolenoyl	−(CH$_2$CH=CH)$_3$(CH$_2$)$_7$−	−11
Icosanoyl	Arachidoyl	−(CH$_2$)$_{18}$−	77
5,8,11,14-Icosatetraenoyl	Arachidonoyl	−(CH$_2$)$_3$(CH$_2$CH=CH)$_4$(CH$_2$)$_3$−	−50

bonds introduce a kink in the chain. The chains are not able to pack regularly and thus the melting point is reduced. Therefore, unsaturated chains increase the fluidity of membranes.

Sphingolipids are another constituent of cell membranes. In mammals, they are particularly important for nerve cells. Depending on the moiety R, we distinguish sphingophospholipids and glycosphingolipids. In sphingophospholipids, the moiety is a phosphate bound via an ester group and with an ethanolamine, choline, and so on at the other side. Glycosphingolipids have a sugar attached to them.

10.3.2
The Osmotic Stress Method

The interaction between lipid bilayer was first measured with the osmotic stress method [1233–1237]. In general, in the osmotic stress method the mean spacing between the macromolecules or lipid bilayers is measured against an applied osmotic pressure. The basic idea of the osmotic stress method is to measure the work required to remove a certain amount of solvent from a solution containing a specific solute. This could in principle be done in an experiment outlined in Figure 10.12 (left). The solution is filled into a vessel that is closed with a semipermeable membrane. This membrane allows the solvent molecules to pass but not the solute. A piston is used to apply a force and squeeze the solvent molecules through the semipermeable membrane. Such an experiment was carried out by Ottewill *et al.* [1238]. They measured the pressure applied to remove solvent from a latex dispersion versus the mean spacing of the particles. A disadvantage of such an experimental setup would be the finite stability of the thin semipermeable membrane that limits the maximum pressure that can be applied.

An alternative approach is to apply an osmotic pressure Π_{osm} (Figure 10.12, right). The solution is filled in a container that is closed on one side by a semipermeable

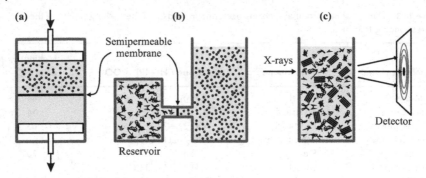

Figure 10.12 Schematic of the osmotic stress method.

membrane. In the second compartment on the other side of the membrane, a defined amount of a polymer is added. This will lead to a certain osmotic pressure in the second compartment. Since the solvent molecules are free to diffuse over the membrane, the pressure in equilibrium has to be the same on both sides of the membrane. This means upon addition of more and more solutes in the second compartment, more and more water will be drawn from the first compartment, increasing the concentration and thus decreasing the distance between the solutes in that compartment. In equilibrium, the osmotic pressure in the solution has to be exactly counterbalanced by the sum of attractive and repulsive forces between the solutes, that is by the force per area between them. The average distance between the solutes in the first compartment can be measured using, for example, X-ray diffraction. With this additional information, one can obtain pressure versus distance curves.

The first and most common application of the osmotic stress method has been the measurement of the interaction between lipid bilayers [1233, 1239] (see Section 10.3). By using uncharged water-soluble macromolecules that do not penetrate the bilayer stacks, one can even simply add the macromolecules directly to the aqueous phase containing the lipids (Figure 10.12c). Since the concentration of the bilayer stacks is much smaller than the concentration of the macromolecules, the osmotic pressure is essentially determined by the macromolecule. The most commonly used polymers to adjust the osmotic pressure are polyethylene glycol (PEG) and polysaccharides such as dextran. It is known that for assemblies of charged biomolecules, PEG is excluded from the interior if the radius of gyration of the PEG is larger than the average distance between two biomolecules [1240]. Values for the osmotic pressures of solutions prepared from these polymers are well characterized and available from several web sites [1241]. Alternatively, the osmotic pressure could be measured directly by manometers or by vapor pressure osmometry. The latter uses the fact that relative humidity RH above a solution in equilibrium is related to osmotic pressure P_{os} as

$$P_{os} = -\frac{k_B T}{V_W} \log(\text{RH}).$$ (10.17)

Here, V_W is the volume of a single water molecule ($\approx 0.3\,nm^3$ at room temperature). This relation can also be used to impose a certain osmotic pressure by fixing the relative humidity (e.g., by salt solutions) [1242]. On the one hand, this method has the disadvantage that the ionic concentration of the solution cannot be controlled independently, on the other hand, for osmotic pressures higher than 10^5 Pa, polymer concentrations get so high that equilibration time may exceed several months.

While in the beginning the biological application of the osmotic stress method was mainly used to investigate forces between lipid bilayers (reviewed in Ref. [1239]), it has also become a valuable tool to study the hydration forces between biomolecules [1243], for example, DNA [1155], collagen fibers [1244], or polysaccharides [1245]. The osmotic stress method has also been applied to colloidal dispersions [1238, 1246], emulsions [1247], colloidal crystals [1248], clays [1249, 1250], block copolymers [1251], a mixed nanoparticle/polymer system [1252], and colloids with polyelectrolyte multilayers [1253].

To measure the force between lipid layers, a lipid is chosen, which at a certain concentration and temperature range forms an L_α phase. The L_α phase is a regularly spaced stack of lamellar fluid bilayers separated by water. From a symmetry point of view, the L_α phase can be considered a smectic-A (SmA) liquid crystal. The mean repeat distance between lipid bilayers in the L_α phase is measured by X-ray diffraction versus an applied osmotic pressure. Direct experiments have been carried out with the surface force apparatus. Therefore, the bilayers are formed either by spontaneous vesicle fusion [1254–1256] or by depositing two subsequent monolayers with the Langmuir–Blodgett technique [1255, 1257]. Atomic force microscope experiments, which have been carried out between two bilayers formed by spontaneous vesicle fusion, confirmed earlier results [1258].

10.3.3
Forces Between Lipid Bilayers

Different components contribute to the force between lipid bilayers: van der Waals attraction, hydration repulsion, steric repulsion, electrostatic forces, and undulation forces. *Electrostatic double-layer forces* arise if charged lipids are present in the membrane. Surface charges also occur in the presence of divalent ions. Ca^{2+} and Mg^{2+} adsorb to lipids, which leads to net positive charge even at concentration of 1 mM [1257]. Double-layer forces are suppressed by adding monovalent salt into the solution.

The *hydration force* is the major repulsive component of the force between neutral bilayers [1254, 1257]. Phenomenologically, it can be described by an exponentially decaying force up to a distance of 1.0–1.5 nm [1234, 1236, 1237, 1259]:

$$ f(x) = f_h^0 \cdot e^{-x/\lambda_h} \quad \text{or} \quad V^A(x) = V_0\, e^{-x/\lambda_h}. \tag{10.18} $$

Typical values for the parameters are $f_h = 100\text{–}1000\,\text{MPa}$ and $\lambda_h = (0.22 \pm 0.04)\,\text{nm}$ [1235, 1237, 1257, 1259–1261]. Equation (10.18) is identical to Eq. (10.10) with $f = -dV_A/dx$ and $f_h^0 = V_0/\lambda_h$.

Another short-range force is *steric repulsion* [1259, 1262, 1263]. When two opposing bilayers approach each other at some point, the head groups of the lipid molecules come so close that they are hindered to vibrate and rotate. Their entropy decreases, which leads to a repulsive force. As a first approximation, steric forces can also be described by an exponentially decreasing interaction:

$$f(x) = f_{st}^0 \cdot e^{-x/\lambda_{st}}. \tag{10.19}$$

Typical values for the pressure f_{st}^0 and the decay length are 10^{13} Pa and 0.06 nm [1259]. Often, steric and hydration forces are difficult to distinguish. Then, usually one exponent is used to describe both. The range of this steric pressure can be increased by the addition of lipids with large head groups, such as glycolipids or lipids with covalently attached polymers.

When phosphatidylcholine is dissolved in an excess amount of water, the bilayers form a lamellar phase with a defined equilibrium separation of 2–3 nm [1233]. Thus, in addition to hydration and steric repulsion there must be a longer ranged attractive force. Repulsive forces alone would tend to separate the bilayers to the largest possible separation, and the spacing would increase with decreasing lipid concentration. This long-range attractive force is the *van der Waals force*. The nonretarded van der Waals energy between two layers each of thickness d_0 separated by a distance x is given by [23, 413, 1234, 1235] (Eq. 2.74)

$$V^A(x) = -\frac{A_H}{12\pi} \left[\frac{1}{x^2} + \frac{1}{(x+2d_0)^2} - \frac{2}{(x+d_0)^2} \right]. \tag{10.20}$$

For small distances, we obtain the simple $V^A \propto x^{-2}$ dependence, also known for infinitely extending bodies. For large distances ($x \gg d_0$), the free energy per unit area approaches $V^A = -A_H d_0^2/(2\pi x^4)$. Considering retardation effects would lead to an even steeper decrease at distance above 10 nm. Hamaker constants for lipids such as DLPC, DMPC, DPPC, EPC and DPPE fall within the range $A_H = (7.5 \pm 1.5) \times 10^{-21}$ J in water [1235, 1237, 1257, 1259–1261, 1264, 1265]. They decrease to about half this value at physiological salt concentrations. The effective van der Waals plane can be slightly offset from the hydrated surfaces of the bilayers; Israelachvili suggests an offset of 0.5 nm [1264].

Undulation Forces Soft elastic sheets, including lipid bilayers, show thermal undulations whose amplitude increases with temperature. If two bilayers approach each other to a distance, where the undulation amplitude becomes of the same order of magnitude as the distance, any further decrease requires a gradual freezing-in of long-wavelength modes and of more and more degrees of freedom (Figure 10.13). This reduces the entropy of the system associated with the thermal excitations and leads to an increase in free energy. The result is a repulsive force. Undulation forces are also called "Helfrich interaction" since W. Helfrich was the first person to correctly describe these forces [1269]. In the old literature also the misleading term "steric force" is used.

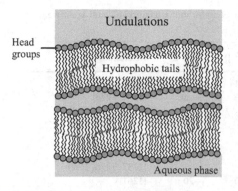
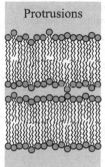

Figure 10.13 Schematic of two opposing bilayers in aqueous medium illustrating undulation and protrusion forces. Undulation forces occur between any pair of flexible membranes and are described by continuum theory. Protrusion forces are caused by individual lipid molecules jumping up and down normal to the bilayer.

Undulation forces not only act between lipid bilayers in the L_α phase but also prevent vesicles from coagulation and act between membranes and other surfaces [1266]. Emulsions are stabilized by undulation forces [1267].

To describe undulation forces, we first need to introduce the so-called bending elastic modulus of a membrane k_c, also called bending rigidity. The free energy of bending is to first order given by [1268, 1269]

$$G = \frac{k_c}{2} \int \left(\frac{1}{r_1} + \frac{1}{r_2} - C_0 \right)^2 dA. \tag{10.21}$$

C_0 is called the spontaneous curvature. A is the surface area and $1/r_1$ and $1/r_2$ are two principal curvatures. Bending moduli of phospholipid bilayers in the liquid phase are typically $k_c \approx 10\,k_B T$ (Table 10.4). The bending moduli increase when the temperature decreases below the melting temperature. Addition of cholesterol increases the bending modulus.

■ **Example 10.4**

The bending modulus of DMPC at 28 °C is $k_c = 3.3 \times 10^{-20}\,J = 8\,k_B T$. We assume that the spontaneous curvature is negligible ($C_0 = 0$). What is the total bending energy of a vesicle of $R_1 = 100$ nm radius?
With $r_1 = r_2 = R_1$, we obtain

$$G = \frac{k_c}{2} \int \left(\frac{2}{R_1} \right)^2 dA = \frac{k_c}{2} \left(\frac{2}{R_1} \right)^2 4\pi R_1^2 = 8\pi k_c = 8.3 \times 10^{-19}\,J.$$

It is independent of the radius! Reason: As the radius decreases the bending energy per unit area increases, but the surface area decreases. Both effects just compensate each other.

Table 10.4 Bending moduli k_c of various phosphatidylcholine bilayers in water at different temperature T and melting temperature T_M.

Lipid	T (°C)	k_c (10^{-19} J)	T_M (°C)	References
DAPC	14–18	0.3...0.44	−69	[1270, 1271]
DLPC	25	0.46	−1.0	[1272, 1273]
DMPC	$T < T_m$	30	24	[1274]
	$T > T_m$	0.33		[1265, 1271, 1274–1278]
DMPC + 30% Ch	$T > T_m$	4		[1279]
DPPC	$T < T_m$	6	41	[1276, 1280]
	$T > T_m$	1.3		[1237, 1276, 1280]
DOPC	30	0.4...1.0	−20	[1261]
EPC	14–30	0.5	−6	[1237, 1270, 1272, 1281, 1282]

T_M refers to the gel to liquid crystalline phase transition. DAPC: diarachidonoylphosphatidylcholine; DGDG: digalactosyldiglyceride; DLPC: dilauroylphosphatidylcholine; DMPC: dimyristoylphosphatidylcholine; Ch: cholesterol; DPPC: dipalmitoylphosphatidylcholine; DOPC: dioleoylphosphatidylcholine, and egg PC.

The concept of entropic forces between undulating membranes was introduced by Helfrich [1283]. He derived the equation

$$V^A(x) = C_{fl} \cdot \frac{(k_B T)^2}{k_c x^2} \tag{10.22}$$

for the free energy of interaction. C_{fl} is a constant. The distance and temperature dependence of Eq. (10.22) has been verified by theory [1284–1286], computer simulations [1287, 1288], and experiments [1237, 1267, 1289]. Undulation forces increase with temperature not only because of explicit proportionality to T^2 but also because the bending modulus decreases [1290]. Still under discussion is the precise value of the numerical prefactor C_{fl}. Helfrich initially presented two ways of deriving Eq. (10.22), which led to two different results, namely, $C_{fl} = 0.188$ and 0.231. While X-ray scattering experiments on microemulsions confirmed this value [1267], more recent small-angle neutron and X-ray scattering measurements obtained with DMPC led to $C_{fl} = 0.111$. Reference [1289] also contains an instructive overview of C_{fl} values reported. Taking into account all values known to us in the literature, we suggest to use $C_{fl} \approx 0.14$.

Equation (10.22) was derived assuming that no other forces are acting between the membranes. If the membranes are also loaded by other interactions, the force law changes [1291]. For example, if hydration forces are present, the undulation contribution at short distances is better described by [1237, 1284, 1292]

$$V^A(x) = \frac{\pi k_B T}{16} \sqrt{\frac{f_h^0}{k_c \lambda_h}} \cdot e^{-\frac{x}{2\lambda_h}} \quad \text{or} \quad f(x) = \frac{\pi k_B T}{32} \sqrt{\frac{f_h^0}{k_c \lambda_h^3}} \cdot e^{-\frac{x}{2\lambda_h}}. \tag{10.23}$$

Other force laws taking into account van der Waals attraction or hydration repulsion have been suggested [1288, 1286, 1293].

Protrusion Forces Based on a theory by Aniansson *et al.* [1294], Israelachvili and Wennerström proposed that a short-range repulsive force originates from the entropic repulsion of molecular groups that are thermally excited to protrude from the bilayer [1295]. Lipid molecules in a membrane are constantly jumping up and down driven by thermal excitations. When two bilayers approach each other, these height fluctuations are suppressed. A suppression of fluctuations reduces the entropy of the system and thus increases the free energy. The result is a short-range repulsive force. While undulation forces are present between continuous sheets, protrusion forces are the result of the discrete molecular nature of the molecules forming a bilayer (Figure 10.13).

Computer simulations showed indeed that lipid molecules are jumping up and down. The mean width of the height distribution of the head groups in a phospholipid bilayer is typically 0.5 nm [1293, 1296]. Incoherent quasielastic neutron scattering on lipid bilayers indicated a smaller out-of-plane motion of 0.1–0.15 nm [1297]. Also for surfactant micelles, such roughening of the surface due to height fluctuations was observed by dynamic small-angle neutron scattering [1298].

Recent simulations indicate that the resulting repulsive interaction is, however, negligible compared to hydration forces [1263]. The reason is that the protrusion of one lipid molecule is accompanied by a correlated rearrangement of the opposing and neighboring molecules. The effect of protrusion force is only a slight increase of the effective exponential decay length of the hydration force [1299].

As examples, pressure versus distance curves of EPC, DMPC, and DPPC are shown in Figure 10.14. The curves were measured with the osmotic pressure

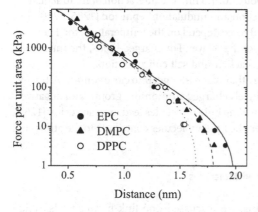

Figure 10.14 Force per unit area versus distance between two lipid bilayers of EPC at 30 °C, DMPC at 30 °C, and DPPC at 50 °C (redrawn from Ref. [1237]). The force curves were fitted with a sum of hydration forces (Eq. (10.18)), van der Waals attraction (Eq. (10.20)), and undulation forces (Eq. (10.23)). To obtain the van der Waals pressure we applied $f = -dV^A/dx$ (see Exercise 10.4). The parameters were $k_c = 0.55 \times 10^{-19}$ J, $f_h^0 = 10^8$ Pa, $\lambda_h = 0.18$ nm, $A_H = 4.7 \times 10^{-21}$ J, and a bilayer thickness of $d_0 = 4.5$ nm for EPC (continuous line). For DMPC (dashed line), the parameters were $k_c = 0.5 \times 10^{-19}$ J, $f_h^0 = 1.3 \times 10^8$ Pa, $\lambda_h = 0.17$ nm, $A_H = 7.1 \times 10^{-21}$ J, and $d_0 = 4.4$ nm. For DPPC (dotted line), we used $k_c = 0.5 \times 10^{-19}$ J, $f_h^0 = 0.6 \times 10^8$ Pa, $\lambda_h = 0.21$ nm, $A_H = 9.0 \times 10^{-21}$ J, and $d_0 = 4.7$ nm.

technique. The applied osmotic pressure is equal to the force per unit area. From X-ray diffraction, the repeat distance was determined. After subtracting the thickness of the lipid bilayers from the repeat distance, the distance between bilayers is obtained. There is a certain degree of arbitrariness in the choice of the bilayer thickness. First, it is not obvious where to place the position of the interface between the bilayer and the aqueous phase. In the particular case shown here, the interface is placed at the end of the head groups and as such contains a water-rich region around the head groups. Second, the bilayers are mobile. Individual lipid molecules jump up and down and the bilayer undulates. For the lipids shown in Figure 10.14, van der Waals forces lead to a drop of the force and a minimum in the energy around 2 nm (see solution of Exercise 10.4). The equilibrium distance, that is, the separation the bilayers assume at zero osmotic pressure, is 2.1 nm, for EPC, 2.0 nm for DMPC, and 1.9 nm for DPPC. At distances below ≈ 2 nm, undulation forces prevent the bilayers from closer approach. At larger distances, the van der Waals attraction dominates and keeps the bilayers together.

The curves shown in Figure 10.14 can serve to summarize the general trend. At very short interbilayer separations (less than about 0.4 nm), the force between bilayers depends on the volume fraction of head groups at the interface, indicating the presence of a large steric barrier, which arises from direct interactions between head groups from opposing bilayers [1259]. For intermediate interbilayer separations up to 1.0–1.5 nm, the pressure–distance curves are similar for liquid–crystalline and crystalline phosphatidylcholine bilayers. They are relatively independent of temperature. In this range of water spacing, the hydration pressure dominates. For an interbilayer spacing of more than about 1.0–1.5 nm, the interaction of neutral bilayers is dominated by van der Waals attraction and undulation repulsion [1286, 1290]. The magnitude and range of the observed force depend on the temperature, the Hamaker constant, and the bending modulus of the bilayer. For charged lipids, the interaction depends also on the charge and the background salt concentration.

Some lipid bilayers adhere to each other when they get into close contact. Attractive electrostatic forces between the positively charged ammonium groups and negatively charged phosphate groups in the opposing lipid molecules lead to an adhesion [1263]. In addition, hydrogen bonds between the lipid molecules contribute to the attraction.

10.4
Force Between Surfaces with Adsorbed Molecules

In Section 7.6 on thin wetting films, we discussed one link between adsorption and surface forces. There we studied the adsorption of a liquid film from its vapor onto a solid surface and related it to the force between the solid–liquid and liquid–vapor interfaces. In this section, we get to know another aspect of adsorption and surface forces. We consider the force between two solid surfaces in a liquid. If dissolved molecules in the liquid adsorb to the solid surfaces, they might significantly influence the interaction between solid surfaces. This was, for example, observed by Rehbinder *et al.* [1301], who observed that surface-active substances can stabilize suspensions of hydrophobic particles in aqueous medium and of hydrophilic particles in benzene. Van der Waarden [1302] observed that carbon black particles in aliphatic oils aggregate. When adding substances that bind to the particles via an aromatic ring and that in addition have a hydrocarbon chain attached, the dispersion remained stable. Mackor [1303] showed that this stabilizing effect can be accounted for by considering the decrease in the number of configurations of the adsorbed molecules due to steric hindrance of the aliphatic side chains when two particles approach each other. A detailed calculation of the reduction in entropy when two surfaces approach each other depends on the specific molecules and the conformation they assume at a surface. Mainly polymers are used to sterically stabilize dispersions. For a detailed discussion, see Section 11.3.

When the two surfaces approach each other, adsorbed molecules are not only restricted in their conformational fluctuations but also at some point are forced to leave the surface. To include this effect, let us consider two parallel plates separated by a distance x interacting across a solution containing different kinds (components) of dissolved molecules. These molecules are allowed to adsorb to the two surfaces. The surface excess of component i is Γ_i. In general, the surface excess on one surface is influenced by the presence of the other surface so that Γ_i is a function of the distance: $\Gamma_i(x)$. Assuming the system is in full thermodynamic equilibrium, Hall, Ash, Everett, and Radke [1304, 1305] derived a relation between the force per unit area, the amount adsorbed in $mol\, m^{-2}$, and the concentration of molecules in solution c_i:

$$\left.\frac{\partial f}{\partial \mu_i}\right|_{T,x} = 2\left.\frac{\partial \Gamma}{\partial x}\right|_{T,\mu_i}. \tag{10.24}$$

Equation (10.24) contains the concentration via the chemical potential μ_i. The chemical potential of the ith component is

$$\mu_i = \mu_i^0 + RT \ln\left(\gamma_i \frac{c_i}{c_0}\right). \tag{10.25}$$

Here, γ_i is the activity coefficient of the ith component and c_0 is the standard reference concentration. The partial derivations in Eq. (10.24) are to be taken at constant temperature and distance and at constant chemical potentials (or concentrations). Equation (10.24) tells us that the change in force (at a given separation) that occurs when a surface-active compound is added to the solution is given by the change in surface excess as a function of the separation. The change in adsorption results from the change in energy of adsorbed states under the influence of the approaching opposite surface.

Following Subramanian and Ducker [1306], we integrate Eq. (10.24) on both sides from infinite distance to a distance x. With $V^A = -\int f dx$, we arrive at

$$\Gamma(x) - \Gamma(\infty) = -\frac{1}{2}\left.\frac{\partial V^A}{\partial \mu_i}\right|_{T,x}. \tag{10.26}$$

Equation (10.26) is valid at adsorption equilibrium and when the temperature and chemical potential of the other components, μ_j, are constant. It allows to calculate the change in adsorption when the two solid–liquid surfaces approach each other.

If two surfaces approach each other, at some distance adsorbed molecules are squeezed out of the closing gap. We use such a simple view to derive an estimation of the effect of adsorbed molecules. Let us consider again two parallel surfaces (Figure 7.2) with adsorbed molecules of length L and cross section b^2. Their molar volume is $V_m = N_A l b^2$. They bind strongly so that a dense layer of thickness L is formed. We denote the binding energy by E_0. Thus, adsorption leads to a decrease in the free energy of the system by $2E_0 A/b^2$, where A is the area of one plate. If the two plates approach each other at a distance $x = 2L$, they start to squeeze out adsorbed molecules. We assume that the molecules are incompressible. We also neglect possible kinetic effects and assume that the adsorbed molecules are in equilibrium with the bulk reservoir. In the available volume Ax, the number of molecules is at maximum $N_A Ax/V_m$. The binding energy of all molecules is $E_0 N_A Ax/V_m$ or $E_0 Ax/V_m$ if we express the binding energy in $Jmol^{-1}$. Setting the total free energy of the system to zero at infinite distance, the free energy increases upon approach from zero at $x \geq 2L$ to $E_0 A(2L-x)/V_m$. Per unit area, we get

$$V^A(x) = \frac{E_0}{V_m}(2L-x). \tag{10.27}$$

The force for a sphere of radius R_p approaching a planar surface with adsorbed molecules is according to Derjaguin's approximation (2.67) [1307]

$$F = 2\pi R_p \frac{E_0}{V_m}(2L-D) = 4\pi R_p E_0 \Gamma \left(1-\frac{D}{2L}\right) \quad \text{for } D < 2L. \tag{10.28}$$

The force is repulsive and decreases linearly with distance. The slope is proportional to the binding energy of the molecules and inversely proportional to their volume.

10.5
Summary

- The structure of liquid molecules at an interface is in general different from the bulk structure. Typically, the molecules arrange in layers, extending 2–4 molecular diameters into the liquid. When two surfaces approach each other exponentially decaying, periodic forces are observed. The oscillation period corresponds to the molecular size.

- In liquid crystals, oscillatory forces are particularly strong due to the intrinsic tendency of the molecules to orient with respect to each other.

- In aqueous medium, hydrophilic surfaces often repel each other at distances below 1 nm. This repulsion is associated with a negative free energy of hydration and it is called hydration force.

- Rather than a summary of established knowledge, we only propose one possible interpretation of hydrophobic forces: the short-range component is caused by a rupture of the metastable film between two approaching surfaces. The thin film is unstable and ruptures either because of the different molecular arrangements of water molecules near hydrophobic surfaces or because a capillary bridge is spontaneously formed. The long-range attraction is caused by preexisting nano-bubbles on the surfaces, which are filled with gas. Nanobubbles are formed by a transitional local oversaturation or by a previous contact between the surfaces.

- The force between neutral lipid bilayers at very short interbilayer distances (less than about 0.4 nm) is dominated by steric repulsion, which arises from direct interactions between head groups. For intermediate distances of up to 1.0–1.5 nm, the hydration pressure dominates. For interbilayer spacing more than about 1.0–1.5 nm, the interaction of neutral bilayers is dominated by van der Waals attraction and undulation repulsion [1290, 1286]. The magnitude and range of the observed force depends on the temperature, the Hamaker constant, and the bending modulus of the bilayer. For charged lipids, the interaction depends on the charge and the background salt concentration.

10.6
Exercises

10.1. Verify that the superposition approach $\varrho_n(x, \xi) = \varrho_n(\xi)\varrho_n(x-\xi)/\varrho_b$ leads to Eq. (10.6).

10.2. For two different surfaces, the presmectic force between a sphere and a plate is [1128]

$$
F = \pi R_p \frac{\bar{\varrho} L_S}{\xi_S} \left[\frac{1}{\tanh\left(\frac{D-D_0}{\xi_S}\right)} - 1 - \sqrt{1 - \left(\frac{\bar{\varrho}}{\Delta\varrho}\right)^2} \cdot \frac{\cos\left(2\pi\frac{D-D_0}{d_0}\right)}{\sinh\left(\frac{D-D_0}{\xi_S}\right)} \right].
$$

(10.29)

Here, $\bar{\varrho} = \psi_1^2 + \psi_2^2$, $\Delta\varrho = \psi_1^2 - \psi_2^2$, ψ_1 and ψ_2 are the dimensionless degrees of ordering at the surfaces. Plot force versus distance for a particle of 6 μm diameter and a flat surface for $d_0 = 3$ nm, $\xi_S = 15$ nm, $D_0 = 0$, and $L_S = 2 \times 10^{-12}$ N for three pairs of the dimensionless order parameters: $\psi_1 = \psi_2 = 0.2$, $\psi_1 = 0.14$ and $\psi_2 = 0.26$, $\psi_1 = 0.08$ and $\psi_2 = 0.32$. Verify by plotting that for $\psi_1 = \psi_2 = 0.2$ Eq. (10.8) leads to the same result.

10.3. A nonpolar liquid is in contact with water. Given are the standard Gibbs energies ΔG_s^0 for the process ideal gas at 1 atm → ideal solution (unit: mole fraction solute) and the vapor pressures P_0^V. Calculate the concentration of dissolved molecules of the liquid in water assuming that the system is in equilibrium. Do so for n-hexane ($\Delta G_s^0 = 28.5$ kJmol^{-1}, $P_0^V = 0.199$ atm), n-heptane (28.9 kJmol^{-1}, 0.060 atm), benzene (14.2 kJmol^{-1}, 0.125 atm), and toluene (14.6 kJmol^{-1}, 0.037 atm) at 25 °C.

10.4. Plot pressure versus distance for van der Waals, hydration, and undulation forces between two parallel lipid bilayers of DLPC at 25 °C. Use $A_H = 5 \times 10^{-21}$ J, $f_h^0 = 100$ MPa, $\lambda_h = 0.22$ nm, $k_c = 0.46 \times 10^{-19}$ J (Table 10.3), and a bilayer thickness of 4 nm.

11
Surface Forces in Polymer Solutions and Melts

The presence of dissolved polymers in a dispersed system can drastically change the behavior of the dispersion or emulsion. The ancient Egyptians knew that one can keep soot particles dispersed in water when they were incubated with gum arabicum, an exudate from the stems of acacia trees, or egg white. In this way, ink was made. The reason for the stabilizing effect is the steric repulsion caused by adsorbed polymer. In the first case, the adsorbed polymers are polysaccharides and glycoproteins; in the second case, it is mainly the protein albumin.

Steric stabilization of dispersions is very important in many industrial applications. In particular in nonpolar solvents, the adsorption or grafting of a polymer onto the surface of particles is the only effective way to establish dispersion stability and prevent flocculation caused by the attractive van der Waals forces because electrostatic interactions are virtually absent in nonpolar solvents.

In many applications in mineral processing, papermaking, and waste water treatment, the opposite effect is desired. Then, polymers are used to induce flocculation. Usually, this is achieved by adding polymer, which can bridge two particles. To understand all these effects, we first need to introduce some fundamentals of the structure of polymers in solution and of polymer adsorption. Good introductions into polymer physics are Refs [1308–1310]. Polymer-induced forces are reviewed in Refs [1311–1313].

11.1
Properties of Polymers

Here, we are mainly concerned about linear polymers, which consist of a linear chain of monomers. Although other architectures such as networks, branched polymers, or dendrimers exist, for most applications in colloid science linear polymers are used.

To characterize a linear polymer, we need to specify the chemical structure of the monomer and the number of monomers per chain. Monomers can be simple chemical units such as in polyethylene $(-(CH_2)_n-)$ or complex structures in itself. Polymers are traditionally named according to the monomer from which they are synthesized. For example, polyethylene is synthesized from ethylene (C_2H_4). IUPAC

Surface and Interfacial Forces. Hans-Jürgen Butt and Michael Kappl
Copyright © 2010 Wiley-VCH Verlag GmbH & Co. KGaA
ISBN: 978-3-527-40849-8

recommends to use structure-based names. Polyethylene is then called poly(methylene) because the smallest repetitive unit is $-CH_2-$. Common polymers are listed in Table 11.1. Important classes of polymers are polyesters, with $-CO$-$O-$ groups in their backbone, polyamides, which contain $-NH$-$CO-$, and polyurethanes ($-NH$-CO-$O-$). One example for a polyamide is Nylon 6 with $-NH(CH_2)_5CO-$ as repeat unit.

Many polymers are in a glassy state. One example is polystyrene. In the glass or amorphous state, no long-range order exists. The polymer structure rather resembles that of a frozen liquid. Even on the molecular level, the segments are frozen and typical segmental relaxation times are slow (typically 100 s). When a polymer is heated above its glass transition temperature T_g, its viscosity, shear modulus, heat capacity, and expansion coefficient drastically change. On the molecular level, the segments start to move again thermally. The glass transition is a gradual process and usually extends over several degrees.

Other polymers, such as poly(ethylene oxide), form a crystalline phase. Crystalline regions do, however, not extend over large regions. Growth of large crystalline regions is prevented by the fact that the segments of the polymer are interconnected. While one part of the chain may nucleate and start to form a crystal, another part might be entangled with other chains or it may also be part of another nucleus. One prevents the other from growing to a large size. Crystalline polymers are characterized by a sharply defined melting temperature. The crystalline regions may be embedded in amorphous regions. Depending on the crystallinity, a melting point T_m and glass transition T_g may or may not be observable. T_g and T_m vary depending on molar mass and the particular stereochemical configuration. In addition, sometimes different forms of a polymer may exist. For example, polyethylene is available as high-density polyethylene (HDPE) and low-density polyethylene (LDPE). Due to the way the polymer is synthesized, low-density polyethylene has a higher degree of branching. This prevents a close packing and causes the lower density. LDPE has a higher glass transition temperature and lower melting point ($T_g = -73$ to $-48\,°C$, $T_m = 130-139\,°C$) than HDPE ($T_g = -123\,°C$, $T_m = 146\,°C$).

The number of monomers per chain is also called degree of polymerization N_{DP}. The degree of polymerization is linked to the molar mass of a polymer by $N_{DP}M_r$. Here, M_r is the molar mass of a monomer. Molar masses of a polymer can be measured in different ways, for example, by gel permeation chromatography (GPC) or high-pressure liquid chromatography (HPLC). For polymers in solution, the molar mass can be determined from the depression of the melting point, vapor pressure decrease, and osmotic pressure measurements.

Usually, polymers can not be synthesized with one precisely defined molar mass. One polymer chain has a different molar mass than another chain. Practically, we always get a distribution of molar masses. When we calculate the mean molar mass, we need to distinguish between the number average M_n or the weight average M_w:

$$M_n = \frac{\sum N_i M_i}{\sum N_i} \quad \text{and} \quad M_w = \frac{\sum N_i^2 M_i}{\sum N_i M_i}. \tag{11.1}$$

Table 11.1 Structure of common polymers, their glass transition T_g, melting temperature T_m, molar mass of the repeat unit M_r in g mol^{-1}, and examples for good solvents.

Polymer	Structure	Properties	Solvents
Polyethylene (PE)	$+CH_2-CH_2+$	For T_g and T_m see text $M_r = 28.05$	HC, halogenated HC
Polypropylene (PP) isotactic	$+CH_2-CH+$ $\quad\quad CH_3$	$T_g = 7\,°C$ $T_m = 173-188\,°C$ $M_r = 42.08$	No good solvents at 25 °C
Polystyrene (PS)	$+CH_2-CH+$ (phenyl)	$T_g = 100\,°C$ Little crystallization $M_r = 104.15$	Toluene, benzene, $CHCl_3$, THF, MEK
Poly(vinyl chloride) (PVC)	$+CH_2-CH+$ $\quad\quad Cl$	$T_g = 71-98\,°C$ $T_m = 212-300\,°C$ $M_r = 62.50$	MEK, THF, DMF, DMSO, toluene
Poly(methyl methacrylate) (PMMA) atactic	CH_3 $+CH_2-C+$ $\quad\quad COOCH_3$	$T_g = 114\,°C$ $T_m = 130-140\,°C$ $M_r = 100.12$	Aromatic and chlorinated HC, THF, MEK
1,4-Poly(butadiene) (PB)	$+CH_2-CH=CH-CH_2+$	$T_g = -95\,°C$ $T_m = 2\,°C$ $M_r = 54.09$	HC, THF

(Continued)

Table 11.1 (*Continued*)

Polymer			Solvents
1,4-Poly(isoprene) (PI)		$T_g = -63\,°C$ (*cis*) No crystal in *cis*-PI $T_g = -72\,°C$ (*trans*) $T_m = 40\,°C$ $M_r = 68.12$	Benzene, toluene, cyclohexane, dioxane
Poly(dimethyl siloxane) (PDMS)		$T_g = -125\,°C$ $T_m = -45$ to $-36\,°C$ $M_r = 74.15$	HC, THF, DMF, toluene
Poly(ethylene tere-phthalate) (PET)		$T_g = 69–115\,°C$ $T_m = 265\,°C$ $M_r = 192.2$	Benzene, phenol
Poly(ethylene oxide) (PEO, PEG)		$T_g = -67\,°C$ $T_m = 66\,°C$ $M_r = 44.05$	Water

HC: hydrocarbons; MEK: methyl ethyl ketone ($CH_3COCH_2CH_3$); THF: tetrahydrofuran (C_4H_8O); DMF: dimethyl formamide ($HCON(CH_3)_2$); DMSO: dimethyl sulfoxide (($CH_3)_2SO$).

The sum runs over all molecules present in a polymer sample. N_i is the number of polymer chains with i monomers and M_i is the molar mass of a polymer with i monomers. In general, M_n and M_w are different. In the weight averages, large molecules contribute more. We can use number and weight averages to characterize the width of the distribution. This is done with the index of polydispersity $I_P = M_w/M_n$. I_P is always equal to or higher than 1. The higher the index of polydispersity, the more disperse the polymer distribution is.

An important class of water-soluble polymers are polyelectrolytes (Table 11.2). A polyelectrolyte is a polymer with chemical groups that, up on dissolving in water or

Table 11.2 Structure of common water-soluble polymers and the molar mass of the repeat unit M_r in $g\,mol^{-1}$.

Poly(vinyl alcohol) (PVA, PVOH)	$\left[CH_2-\underset{\underset{OH}{\mid}}{CH}\right]$	$M_r = 44.05$
Poly(acryl amide) (PAAm)	$\left[CH_2-\underset{\underset{CONH_2}{\mid}}{CH}\right]$	$M_r = 71.08$
Poly(N-isopropyl acryl-amide) (PNIPAM)	$\left[CH_2-\underset{\underset{C=O}{\mid}}{CH}\right]$ $\underset{\underset{CH_3-CH-CH_3}{NH}}{}$	$M_r = 113.16$ Soluble in water below 33 °C
Poly(acrylic acid) (PAA)	$\left[CH_2-\underset{\underset{COO^\ominus}{\mid}}{CH}\right]$	$M_r = 72.06$
Poly(styrene sulfonate) (PSS)	$\left[CH_2-CH\right]$ (phenyl with $SO_3^\ominus$)	$M_r = 184.2$
Poly(vinyl amine) (PVAm)	$\left[CH_2-\underset{\underset{NH_3^\oplus}{\mid}}{CH}\right]$	$M_r = 43.07$
Poly(allyl amine) (PAH)	$\left[CH_2-\underset{\underset{CH_2NH_3^\oplus}{\mid}}{CH}\right]$	$M_r = 57.10$
Poly(2-vinyl pyridine) (P2VP)	$\left[CH_2-CH\right]$ (pyridinium ring, $\overset{\oplus}{NH}$)	$M_r = 105.14$

Polyelectrolytes are shown in their form at neutral pH.

another polar solvent, dissociate to give polyions (polycations or polyanions) together with an equivalent amount of counter ions of opposite sign. For example, when poly (acrylic acid) is dissolved in water at neutral pH, most of the carboxylic groups dissociate. After the protons have dissociated, the polymer is negatively charged. PAA can also be purchased as poly(acrylic acid sodium) with the repeat unit $-CH_2CHCOONa-$. In water, the sodium ions dissociate and the polymer becomes negatively charged. The properties of dissolved polyelectrolytes usually depend on the pH and on the presence of ions, in particular counterions.

11.2
Polymer Solutions

11.2.1
Ideal Chains

In solution, linear polymers are not stiff rods, but the chains are usually flexible. The groups along the chain can rotate around their bonds and in this way change the direction. If we start at a certain monomer, we may be able to predict the direction of the next monomer, but after a certain distance the orientation of the following monomers is not correlated anymore with the orientation of our starting monomer. This characteristic decay length for orientation correlation depends on the specific polymer we are dealing with and on the solvent.

In many applications, we do not need to consider the detailed molecular structure of the polymer including bond lengths, bond angles, rotation energy, and so on.

Table 11.3 Characteristic ratio C_∞ [1310, 1315–1317] and lengths of repeat units l_r.

Polymer	C_∞	l_r (nm)
Polyethylene	7.1	0.25
Polypropylene	5.9	0.25
Polystyrene	10.2	0.25
Poly(vinyl chloride)	7.7	0.25
Poly(methyl methacrylate)	8.7	0.25
cis-1,4-Poly(butadiene)	4.7	0.41
trans-1,4-Poly(butadiene)	6.1	0.48
cis-1,4-Poly(isoprene)	5.1	0.39
trans-1,4-Poly(isoprene)	6.1	0.46
Poly(dimethyl siloxane)	6.8	0.29
Poly(ethylene terephthalate)	4.0	1.00
Poly(ethylene oxide)	5.6	0.37
Poly(vinyl alcohol)	8.3	0.25

Typical variations from one solvent to another are 10% for C_∞. Values of l_r were calculated by Vangelis Harmandaris by applying energy minimization in trimers or tetramers to avoid end effects and by defining l_r as the distance between the center of mass of two consecutive monomers. Depending on the specific force field, they vary between 5 and 10%.

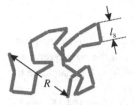

Figure 11.1 Picture of a linear polymer in the ideal freely jointed chain model.

In many discussions, we can use simpler models to describe the polymer. One such model is the freely jointed chain (FJC) model introduced by Kuhn [1314] (Figure 11.1). In this model, the polymer is considered to consist of a chain of N links. Each segment has a length l_s, also called Kuhn length. The angle between adjacent chain links is taken to be random and can change freely. We also assume that the individual segments do not interact with each other. Polymers with no interaction between the segments separated by many bonds are called ideal chains. Then the chain forms a random coil. One can show [1310] that the mean square of the end-to-end distance $\overline{R^2}$–we call it the size of a polymer chain–is given by

$$R_0 = \sqrt{\overline{R^2}} = l_s \sqrt{N}. \tag{11.2}$$

In fact, Eq. (11.2) holds for any ideal chain, irrespective of the specific way the segments are linked. To determine the mean end-to-end distance, we could either average over many polymers at a given moment or observe one polymer chain and average its conformation over time. Due to thermal fluctuations, the chain will assume many configurations in a short time.

The size of a polymer can be measured by light or neutron scattering. In light scattering, the radius of gyration is determined. It is related to the size of a polymer by

$$R_g = l_s \sqrt{\frac{N}{6}}. \tag{11.3}$$

The radius of gyration is $\sqrt{6}$ smaller than the size of an ideal chain. Experimentally, it is found that over a wide range of molar masses and temperatures, polymer sizes in melts and solutions indeed increase proportional to $\sqrt{N}$ or $\sqrt{M_w}$. From measured values of $\overline{R^2}/M_w$, the segment length can be determined. Typically, segment lengths are five to nine times larger than the length of a repeat unit. Sometimes, the characteristic ratio C_∞ is reported (Table 11.3). For long chains, the characteristic ratio is $C_\infty = l_s/l_r$, where l_r is the length of a repeat unit or monomer. Chains are nearly ideal both in polymer melts and in polymer solutions, when the interaction between segments is equal to the interaction of a segment with the solvent.

It is instructive to compare the size of a polymer to its contour length L_c. The contour length is the end-to-end distance of a fully stretched linear polymer chain. It is given by $L_c = N l_s = M_n l_r / M_r$, where M_n/M_r is the average number of monomers per chain.

■ **Example 11.1**

By static light scattering Miyaki *et al.* [1319] observed a R_g of 167 nm with polystyrene in cyclohexane at 34.5 °C. The molar mass of the polystyrene was $M_w = 32 \times 10^6$ g mol^{-1} at an index of polydispersity of $I_P = 1.1$. Calculate the Kuhn length and compare it to the size of a monomer.
The number of monomers is $N_{DP} = M_w/(I_P M_r)$. With $M_r = 104.15$ g mol^{-1} we get $N_{DP} = 2.8 \times 10^5$. The length per monomer l_r is 0.253 nm with a bond length of 0.152 nm between two carbon atoms in an aliphatic chain and a bond angle of 111.5°. The mean contour length is $L_c = N_{DP}l_r = 71$ μm. With $L_c = Nl_s$, we can write Eq. (11.3) as

$$R_g = \sqrt{\frac{L_c l_s}{6}} = \sqrt{\frac{l_s l_r}{6 I_P M_r}} \cdot \sqrt{M_w}.$$

Solving the first part of the equation for l_s, we calculate $l_s = 2.4$ nm, which is roughly nine times the length of the monomer. In tables or papers, usually the factor R_g^2/M_w is reported. Then, the second part of the equation can be applied to calculate the Kuhn length. Miyaki, for example, reported $R_g^2/M_w = 8.8 \times 10^{-18}$, with M_w in units of g mol^{-1} and the result in cm^2. We calculate l_s according to

$$8.8 \times 10^{-22} \text{ m}^2\text{mol g}^{-1} = \frac{l_s l_r}{6 I_P M_r} = l_s \cdot 3.68 \times 10^{-13} \text{ mol mg}^{-1}.$$

This again leads to $l_s = 2.4$ nm.

11.2.2
Real Chains in a Good Solvent

In an ideal solvent, the interaction between subunits is equal to the interaction of a subunit with the solvent. In a real solvent, the actual radius of gyration can be larger or smaller. In a "good" solvent, a repulsive force acts between the monomers. The polymer swells and R_g increases. In a "bad" solvent, the monomers attract each other, the polymer shrinks, and R_g decreases. Often a bad solvent becomes a good solvent if the temperature is increased. The temperature, at which the polymer behaves ideally, is called the theta temperature, T_Θ. The ideal solvent is called a theta solvent.

To take the interaction between polymer segments into account, Flory[1] [1320] introduced the excluded volume parameter v in units of m^3. In good solvents ($T > T_\Theta$), the polymer segments repel each other and v is positive. For many calculations, we simply approximate $v = l_s^3$. The excluded volume decreases with decreasing solvent quality to $v = 0$ for Θ solvents. It decreases even further and becomes negative for poor solvents ($T < T_\theta$), where the segments attract each other.

---

1) Paul John Flory, 1910–1985, American physicochemist, Nobel Prize, 1974.

For a detailed description of the excluded volume, we refer to textbooks, for example, [1310]. Briefly, the excluded volume is related to the interaction potential U. Let $U(r)$ be the energy cost to bring two segments from infinity to a distance r. Typically, this potential contains a hard-core repulsion plus a longer range attraction or repulsion. The probability of finding two segments at a distance r is proportional to the Boltzmann factor $\exp[-U/(k_B T)]$. Now we define the so-called Mayer f-function: $f(r) = \exp[-U/(k_B T)] - 1$. For an attractive potential, $U(r)$ is negative, $\exp[-U/(k_B T)] > 1$, and f is positive. If the potential at a certain distance is repulsive, $U(r)$ is positive, $\exp[-U/(k_B T)] < 1$, and f is negative. The excluded volume is defined as minus the integral of the Mayer f-function over the whole space: $v = -\int f(r) dV$. If attractive interactions between segments dominate, v is negative. If repulsive interactions dominate, it is positive. With improving solvent quality, v increases.

The excluded volume changes with the quality of the solvent. A good approximation for many linear polymers in organic solvents is [1318]

$$v = 0.78\, l_s^3 \cdot \frac{T - T_\Theta}{T}. \tag{11.4}$$

In a good solvent, the segments of a polymer chain repel each other. The interaction energy of polymer segments in a given unit volume is proportional to the number of segments in that volume multiplied with the likelihood that they find another segment to interact with. This likelihood is proportional to the number of segments in the unit volume.[2] In total, the energy density should be proportional to the square of the concentration of segments c (in number of segments per unit volume). It can be shown that the energy density of polymer segments in a solvent is [1310, p. 105]

$$\frac{k_B T v c^2}{2}. \tag{11.5}$$

The factor of 2 accounts for the fact that otherwise each pair of interacting segments would be counted twice.

The conformation of a real chain in a good solvent is determined by two effects: the effective repulsion energy between segments that tends to swell the coil and the entropy loss due to such a deformation. In equilibrium, the sum of both is minimal leading to an increased radius of the coil of

$$R_0 \approx (v l_s^2 N^3)^{1/5}. \tag{11.6}$$

The size of a polymer in a good solvent increases with $N^{3/5}$ rather than $N^{1/2}$ for an ideal chain. If the total energy of the chain is less than $k_B T$, swelling will be negligible and the polymer behaves like an ideal chain. This is the case if $\sqrt{N} v / l_s^3 < 1$.

Experimentally, the excluded volume can be determined from the osmotic pressure of a polymer solution or by light scattering. The osmotic pressure of a

---

[2] To be precise, it is that number minus 1. Since we assume that we have many segments, we can neglect 1.

polymer solution is

$$P_{os} = c_m k_B T \left(\frac{1}{M_n} + A_2 c_m + \cdots \right).$$ (11.7)

Here, c_m is concentration of polymer in $kg\, m^{-3}$. A_2 is the second virial coefficient in units of $m^3\, mol\, kg^{-2}$. The second virial coefficient is related to the excluded volume by $\nu = 2A_2 M_s^2 / N_A$, where M_s is the mass of one segment (not one repeat unit!). Thus, the second term in Eq. (11.7) can be written as

$$k_B T A_2 c_m^2 = \frac{k_B T N_A}{2} \nu \left(\frac{c_m}{M_0} \right)^2.$$

It is proportional to the square of the density of polymer segments.

■ **Example 11.2**

The Kuhn length of polystyrene in decalin at 25 °C has been measured by light scattering to be $l_s = 1.8\, nm$. Estimate the excluded volume and the expected second virial coefficient. The Θ temperature of polysytrene in decalin is 14 °C. With Eq. (11.4), we find $\nu = 0.78 \cdot (1.8\, nm)^3 \cdot 11/298 = 0.17\, nm^3$. Since l_s is seven times the monomer length, the mass of one segment is $M_s = 7 \cdot 104.15$ $g\, mol^{-1} = 729\, g\, mol^{-1}$. The second virial coefficient is

$$A_2 = \frac{N_A \nu}{2 M_0^2} = \frac{6.02 \times 10^{23}\, mol^{-1} \cdot 0.17 \times 10^{-27}\, m^3}{2 \cdot (0.729\, kg\, mol^{-1})^2} = 9.6 \times 10^{-5}\, m^3\, mol\, kg^{-2}.$$

11.2.3
Stretching Individual Chains

Let us now do the following experiment. We take a single polymer at both ends and stretch it. To do so, we have to do work because as the polymer is stretched its entropy is decreased or, in other words, its structural order increases. The number of possible configurations a polymer chain can assume is reduced when it is forced to stretch. Thus, to stretch a polymer, a force is required. This force has been calculated. For freely jointed chain, it is given by [1321, 1322]

$$\frac{x}{L_c} = \coth \left(\frac{F l_s}{k_B T} \right) - \frac{k_B T}{F l_s}.$$ (11.8)

Here, x is the distance between the two ends and F is the force applied. Unfortunately, Eq. (11.8) cannot be written in an explicit form in which the force is given as a function of the distance. Only for very low and very high forces can the limits be expressed explicitly. An expression for low forces is derived by writing the coth in a series for $F l_s \ll k_B T$ [1321]:

$$F = \frac{3 x k_B T}{N l_s^2}.$$ (11.9)

In the limit of high forces ($Fl_s \gg k_B T$), Eq. (11.8) approaches

$$F = \frac{k_B T L_c}{2 l_s (L_c - x)}. \tag{11.10}$$

We added a factor 2 on the right-hand side. *Ab initio* calculations and a comparison with experimental results showed that this leads to a more realistic description of highly stretched polymers [1323].

Since the 1990s, the experiment described above has actually become feasible. Optical tweezers [1324, 1325], magnetic tweezers [1326], and in particular the atomic force microscope [1327–1329] have been used to stretch single-polymer chains (for reviews, see Refs [201, 202]). It turned out that in particular at high forces and elongations, Eq. (11.10) is sometimes not sufficient [1323]. One deficiency is that the chain links were assumed to be rigid. Therefore, the model was extended by assuming that each chain link is a spring. The spring constant of each link, k_s, is named segment elasticity. The extended equation is [1324, 1327]

$$\frac{x}{L_c} = \left[\coth\left(\frac{F l_s}{k_B T} \right) - \frac{k_B T}{F l_s} \right] \left(1 + \frac{F}{l_s k_s} \right). \tag{11.11}$$

The segment elasticity is typically of the order of $10 \, \mathrm{Nm^{-1}}$. To obtain an even better description of the stretching behavior at high forces, *ab initio* quantum chemical calculations are necessary [1323].

■ **Example 11.3**

A force versus extension curve for stretching a single polymer chain is shown in Figure 11.2 [1330]. The polymer has a hydrocarbon backbone with four carbon atoms and three different side groups (see inset in Figure 11.2). Thus, the length of a repeat unit is 0.51 nm. With a mean number of repeat units of $N_{DP} = 490$, the mean contour length is $L_c = 250$ nm. Force versus extension curves could be fitted with Eq. (11.11) using a Kuhn length of $l_s = 0.34$ nm and a segment elasticity of $k_s = 53 \, \mathrm{Nm^{-1}}$. In this particular force versus extension curve, the apparent contour length was only $L_c = 104$ nm, which is significantly shorter than the calculated contour length. In fact, when many force curves were recorded, the apparent contour length varied from one to the next. The reason is that in the experiment, which was carried out with an atomic force microscope, polymer chains were randomly picked up by the tip. Thus, the position on the chain varied from one curve to the next. Comparing the experimental curve with Eq. (11.11) shows that up to a force of ≈ 0.2 nN the curve could be well fitted with the entropic contribution only and neglecting $F/l_s k_s$. The elastic stretching comes into play only at high forces.

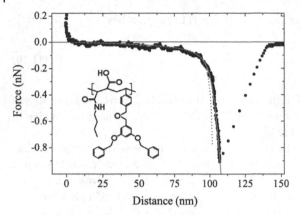

Figure 11.2 Retracting force curve in which a single polymer chain is stretched in tetrahydrofuran. The curve was measured with the atomic force microscope (redrawn from Ref. [1330]). The gray line is a fit with Eq. (11.11) using $L_c = 104$ nm, $l_s = 0.34$ nm, and $k_s = 53$ Nm^{-1}. The black dotted line is a fit with $k_s \rightarrow \infty$.

The freely jointed chain model is one way of describing polymers. An alternative is the worm-like chain (WLC) model [1331]. The worm-like chain model takes the polymer as an isotropic rod that is continuously flexible (Figure 11.3); this is in contrast to the freely jointed chain model that is flexible only between discrete segments. The worm-like chain model is particularly suited for describing stiffer polymers, with successive segments displaying a sort of cooperativity. A prominent example is double-stranded DNA in aqueous solution. If we stretch a worm-like chain, the force can be described by [1325, 1332, 1333]

$$\frac{Fl_p}{k_B T} = \frac{1}{4}\frac{1}{(1-x/L_c)^2} - \frac{1}{4} + \frac{x}{L_c} - \frac{F}{K_0}. \tag{11.12}$$

Here, K_0 takes into account an elastic stretching of the chain; it corresponds to $l_s k_s$ in the freely jointed chain model. Usually, K_0 is assumed to be large and the last term is neglected.

In Eq. (11.12), we introduced another mechanical property that quantifies the stiffness of a linear polymer: the persistence length l_p that is defined as the length over which correlations in the direction of the tangent to the polymer chain are lost. To be more quantitative, we consider a polymer at a certain position. The tangent at this position is taken as zero angle. It can be shown that the mean cosine of the orientation of the chain θ being a distance l away from the chosen position decays exponentially,

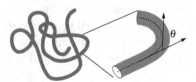

Figure 11.3 Isotropic, elastic polymer as in the worm-like chain model.

$\langle \cos\theta \rangle = \exp(-l/l_p)$. In the limit of very long polymers, the Kuhn length is equal to twice the persistence length of a worm-like chain [1334, p. 2].

While Eq. (11.11) is an exact equation, Eq. (11.12) is an approximation. It accurately describes the force on a single chain in the limit of very low and very high forces, but in between it is an interpolation. Other interpolations are reported in Refs [1335–1337].

11.3
Steric Repulsion

Polymers are often used to stabilize dispersion and prevent particles from aggregation. This is due to steric repulsion. In this section, we discuss the force between surfaces coated with grafted polymer chains in a good solvent.

When two surfaces approach each other, at some distance the polymer brushes start to overlap. The density of polymer segment increases. The increase in segment density and the resulting increase in osmotic pressure and repulsive interaction energy leads to a repulsive force [1338, 1339].

A grafted or tethered chain is fixed at one end to the surface. The number of chains per unit area Γ is constant and the bonds are not allowed to shift laterally. For a low density of chains, the surface is covered with a number of separated polymer blobs, each of height and size given by R_g, which do not overlap. This is referred to as the mushroom regime. When the density of polymer chains on the surface is high, $\Gamma \gg 1/R_g^2$, we talk about a polymer brush (Figure 11.4). The thickness of a brush L_0 is substantially larger than the radius of gyration. Good reviews are Refs [1340, 1341]; see Ref [1342] for a comparison of different theoretical approaches.

Experimentally, polymer brushes are made by chemically binding a polymer to a surface (grafting to) or by synthesizing the polymer directly on the surface (grafting from). Practically, physisorbed polymer can also be used. In this case, the polymer contains a group that strongly adsorbs to the surface. A typical example are diblock copolymers of type A–B, where one block is made of monomer type A and the other is made of monomer B. Here, B is the anchor chain, which readily adsorbs to the surface and is usually not well soluble in the liquid. A is the stabilizing chain, which is strongly solvated and which dangles out into the liquid.

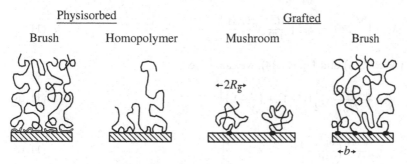

Figure 11.4 Structure of polymers on surfaces.

Several equations have been proposed to describe steric repulsion [1343–1345]. We follow an argument of de Gennes, Milner, Witten, and Cates [1346, 1347]. To calculate the steric repulsion, we consider two polymer brushes with fixed grafting density. The mean distance between two grafting sites is denoted by b. Grafting density Γ and b are related by $\Gamma = 1/b^2$.

First, we need to calculate the thickness of the undisturbed brush L_0. It is determined by a balance between two effects: the repulsive interaction of polymer segments tends to swell the brush. The chain elasticity caused by the configurational entropy resists stretching. We already calculated the elastic restoring contribution of a polymer chain (Eq. (11.9)). Setting the brush thickness L equal to the extension of the chain, the force is $3Lk_BT/Nl_s^2$. To calculate the repulsive energy contribution in the brush with expression (11.5), we need to know the segment concentration. The concentration of segments is $c = N/(Lb^2)$. The free energy per chain is obtained by multiplying the density with the volume of one chain Lb^2. It is $k_BT\nu N^2/(2Lb^2) = k_BT\nu N^2\Gamma/(2L)$. To change the stretching distance L, a force $-k_BT\nu N^2\Gamma/(2L^2)$ is required. The minus sign indicates that it is directed opposite to the elastic restoring force and tends to expand the brush. The total force per chain is

$$F = \frac{3k_BTL}{Nl_s^2} - \frac{k_BT\nu N^2\Gamma}{2L^2}. \tag{11.13}$$

In equilibrium and in the absence of external forces, the net force must be zero. This leads to a brush thickness

$$L_0 = N\left(\frac{\nu l_s^2\Gamma}{6}\right)^{1/3}. \tag{11.14}$$

The brush thickness increases proportional to the number of segments and also increases weakly with the grafting density. The brush thickness also continuously increases with the solvent quality, as reflected in the excluded volume parameter. This has been experimentally verified [1348–1350].

Now, we let two polymer-coated surfaces approach each other. At a distance $x = 2L_0$, the polymer chains overlap or the polymer layer undergoes compression. In both cases, the local segment density of polymer chains increases. This increase in local segment density in the interaction zone will result in a repulsion [1338, 1339]. Work has to be done to approach further. Per polymer chain, the interaction energy is given by

$$V = \int_{L_0}^{x/2} FdL = k_BT\left[\frac{3L^2}{2Nl_s^2} + \frac{\nu N^2\Gamma}{2L}\right]_{L_0}^{x/2}. \tag{11.15}$$

With $y = x/(2L_0)$ and Eq. (11.14), we can write

$$V = \frac{k_BT\nu N^2\Gamma}{2L_0}\left[\frac{1}{2}(y^2-1) + \left(\frac{1}{y}-1\right)\right] \tag{11.16}$$

$$= \frac{k_BTN}{2}\left(\frac{6\nu^2\Gamma^2}{l_s^2}\right)^{1/3}\left[\frac{y^2}{2} + \frac{1}{y} - \frac{3}{2}\right]. \tag{11.17}$$

This equation was derived by Milner, Witten, and Cates (Eq. (11.32) in [1347]) except for a numerical factor $3^{1/3}$.

If we bring two surfaces with a grafting density Γ together to a distance x, the interaction energy per unit area is $V^A = 2\Gamma V$. The factor 2 takes care of the fact that polymer chains on both surfaces are compressed. This leads to

$$V^A(x) = k_B T L_0 \Gamma \left(\frac{9v\Gamma}{2l_s^4}\right)^{1/3} \left[\left(\frac{x}{2L_0}\right)^2 + \frac{4L_0}{x} - 3\right]. \tag{11.18}$$

Equation (11.18) gives the force for $x \leq 2L_0$. For larger distances, the force is zero.

To derive Eq. (11.18), we assumed that the segments are homogeneously distributed in the volume $L_0 b^2$. Outside, the concentration is assumed to be zero. This step profile for the segment density of the undisturbed brush might not be realistic. Milner, Witten, and Cates used a more realistic parabolic profile [1347] and obtained an interaction energy per unit area [1352]

$$V^A(x) = k_B T L_0 \Gamma \left(\frac{\pi^4 v\Gamma}{12^2 l_s^4}\right)^{1/3} \left[\left(\frac{x}{2L_0}\right)^2 + \frac{2L_0}{x} - \frac{1}{5}\left(\frac{x}{2L_0}\right)^5 - \frac{9}{5}\right]. \tag{11.19}$$

The thickness of the uncompressed brush in this case is

$$L_0 = \left(\frac{12}{\pi^2}\right)^{1/3} N(vl_s^2\Gamma)^{1/3} = 1.07\, N(vl_s^2\Gamma)^{1/3}. \tag{11.20}$$

Another expression for the interaction energy per unit area was derived by de Gennes [1346]. Using an analytical self-consistent mean field theory for a step profile in the segment density of the undisturbed brush, he derived

$$V^A(x) = \frac{8}{35} k_B T L_0 \Gamma^{3/2} \left[7\left(\frac{2L_0}{x}\right)^{5/4} + 5\left(\frac{x}{2L_0}\right)^{7/4} - 12\right]. \tag{11.21}$$

For the thickness of the uncompressed brush, he reported

$$L_0 = N l_s^{5/3} \Gamma^{1/3}. \tag{11.22}$$

This is roughly the same brush thickness as reported by Milner, Witten, and Cates, assuming that $v = l_s^3$. De Gennes used a step profile for the segment concentration. Equation (11.21) is widely used. It has the advantage of having only two parameters, Γ and L_0, which both have a physical meaning. Though the analytical expression looks quite different from Eqs. (11.18) and (11.19), the energy versus distance curves look much alike and in many cases force curves can be fitted equally well with all three models (Figure 11.6).

All expressions reported so far are applicable at high grafting density. For a low grafting density ($\Gamma < 1/R_g^2$), the repulsive interaction energy per unit area in a good solvent and between two polymer coated surfaces was calculated by Dolan and Edwards [1351]:

$$V^A(x) = k_B T \Gamma \cdot \left[\ln \left(\frac{x}{4\sqrt{\pi} R_g} \right) + \frac{\pi^2 R_g^2}{x^2} \right] \quad \text{for} \quad x \le 3\sqrt{2} R_g,$$

(11.23)

$$V^A(x) = -k_B T \Gamma \cdot \ln \left(1 - 2e^{-\frac{x^2}{4R_g^2}} \right) \quad \text{for} \quad x > 3\sqrt{2} R_g.$$

■ Example 11.4

As an example, we calculate the steric force between two surfaces grafted with polymer chains (Figure 11.5). The polymer chains were supposed to have a Kuhn length of $l_s = 1.2$ nm and $N = 80$ segments. For an ideal chain, this leads to a mean end-to-end distance of $R_0 = 10.7$ nm (Eq. (11.2)) and a radius of gyration of $R_g = 4.4$ nm (Eq. (11.3)). At low grafting density, with a mean distance between grafting sites of $b = 5$ nm ($\Gamma = 4 \times 10^{16}$ m^{-2}), the characteristic decay length of the interaction energy is determined by R_g. V^A was calculated using Eq. (11.23). For high grafting densities, we applied the equation of de Gennes (Eq. (11.21)). At a mean distance between grafting sites of $b = 3$ nm ($\Gamma = 1.1 \times 10^{17}$ m^{-2}), the brush thickness is $L_0 = 52$ nm (Eq. (11.22)). When the distance between chains is reduced to $b = 2$ nm ($\Gamma = 2.5 \times 10^{17}$ m^{-2}), the brush thickness increases to 68 nm and so does the interaction energy.

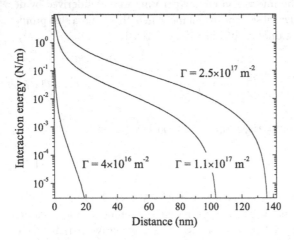

Figure 11.5 Interaction energy per unit area V^A versus distance between two similar parallel plates coated with grafted polymer in a good solvent. For the calculation, we used a segment length $l_s = 1.2$ nm and a chain length of $N = 80$ segments. At low grafting density ($\Gamma = 4 \times 10^{16}$ m^{-2}), the interaction energy was calculated with Eq. (11.23). For the high grafting densities, we applied Eq. (11.21) with $\Gamma = 1.1 \times 10^{17}$ m^{-2}, $L_0 = 52$ and $\Gamma = 2.5 \times 10^{17}$ m^{-2}, $L_0 = 68$.

Direct quantitative measurements of steric repulsion were made with the surface forces apparatus [1353–1360] and the atomic force microscope [1361–1364]. Although we focused on the interaction between solid surfaces, steric forces also act between fluid interfaces. The first force versus distance curves of steric repulsion were recorded across a liquid foam lamellae with a thin film balance by Lyklema and van Vliet [1365]. Another example is the force measurement between vesicles using the osmotic stress method by Kenworthy et al. [1366]. Experimentally, the Milner, Witten, and Cates and the de Gennes model both fit force curves measured between polymer brushes in good solvents reasonably well.

As an example, Figure 11.6 shows the normalized force versus distance between polystyrene coated mica sheets in toluene at room temperature [1359]. Polystyrene brushes were made by incubating the mica with a solution containing a diblock copolymer with a small block of poly(ethylene oxide) and a large block of polystyrene; the total molar mass was $150\,\text{kg mol}^{-1}$. Only 1.5% was poly(ethylene oxide). It served to bind the polymer to mica. Since toluene is a good solvent for polystyrene, a strong steric repulsion was observed. To allow a better comparison with theory and to show the full dynamic range, the force is often plotted on a logarithmic scale. Figure 11.6 demonstrates that it is usually difficult to distinguish between different models. All three models fit the experimental results reasonably well. In all three cases, the results could be fitted with $\Gamma = 1.5 \times 10^{18}\,\text{m}^{-2}$ and $L_0 = 75\text{–}78\,\text{nm}$. For the model of Milner, Witten, and Cates, a segment length of $l_s = 1.5\,\text{nm}$ and $N = 50$ for a step profile (Eq. (11.18)) and $N = 125$ for the parabolic profile (Eq. (11.19)) was inserted.

When dealing with polymer brushes, in some cases the conditions required for the Derjaguin approximation are not fulfilled. Polymer brushes can reach a thickness,

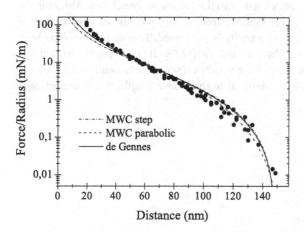

Figure 11.6 Force distance profiles between two curved mica sheets with polystyrene brushes in toluene (redrawn from Ref. [1359]). Experimental results (●) were fitted with the model of Milner, Witten, and Cates [1347] using a step (Eq. (11.18)) and a parabolic (Eq. (11.19)) profile for the segment concentration, and the model of de Gennes [1346] (Eq. (11.21)). In each case, the normalized force was calculated from $F/R = 2\pi V^A$.

which is of the size of dispersed particles. Then, the radial symmetry has to be explicitly taken into account. The steric repulsion decays more steeply than on a brush of equal grafting density with a planar geometry because the segment density decreases more steeply with increasing distance [1367]. In experiments with the atomic force microscope, the tip might have a radius of curvature of the same length scale as the size of the polymers [1368–1372]. The chains might evade an approaching tip and splay to the side. Then, the steric force is lower than the one calculated above.

In polar media such as water, often polyelectrolytes are used. Since polyelectrolytes are charged, we have a steric and an electrostatic effect. For this reason, the effective interaction is sometimes called electrosteric force. The electrosteric force has an important advantage over simple electrostatic repulsion: it is to a large degree independent of the concentration of added salt. While electrostatically stabilized dispersion often coagulates when salt is added, electrosterically stabilized dispersions are robust and remain stable.

To derive a simple equation that describes electrosteric forces, we follow scaling arguments of Pincus [1373] and Balastre *et al.* [1374]. See also Refs. [1375–1377] for simulations and Ref. [1377] for a collection of parameters. Let us consider a neutral planar surface, onto which a brush of monodisperse polyelectrolytes is grafted (Figure 11.7) (reviewed in Ref. [1378]). We assume that a fixed number α of the ionizable groups per segment, for example, sulfonate or carboxylate, is charged. Since a segment usually contains more than one ionizable group, α may exceed 1. These charges of αN per chain are neutralized by an equal number of dissociated counterions and additional ions of added salt. We distinguish two cases: no added salt and a lot of added salt. Let us first discuss the situation when no salt is added. This regime of no or low added salt is referred to as the *osmotic brush regime.* The only free ions are those dissociated from the polyelectrolyte chains. Although the ions have dissociated and are free to move around, they will not leave the vicinity of the polyelectrolyte chain. Electroneutrality is conserved. This assumption has been confirmed experimentally and by simulations [1379]. It is instructive to define a screening length ξ_e as the distance over which a test charge is neutralized. In a salt solution, this distance would be identical to the Debye length. In a polyelectrolyte

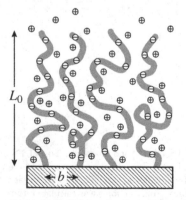

Figure 11.7 Schematic of a polyelectrolyte brush in water.

without added salt, ξ_e is determined by the polyelectrolyte, in particular by its density of ionizable groups and the degree of ionization.

The brush thickness is determined by a balance between the swelling effect due to the osmotic pressure of the counterions and the chain elasticity. The osmotic pressure tends to expand the brush while the chain elasticity resists an extension. We neglect direct electrostatic effects between polymer segments and assume that the screening length is much smaller than the brush height, $\xi_e \ll L$. The elastic restoring force can be calculated by Eq. (11.9). Setting the brush thickness equal to the extension, it is $3k_B TL/Nl_s^2$. The osmotic pressure can be estimated by assuming that the ions behave like in an ideal solution. Then, it is equal to the number of counterions αN multiplied by $k_B T$ and divided by the volume available to a chain Lb^2. We obtain an osmotic stretching force of $\alpha N k_B T/L$. The total force per chain is

$$F = \frac{3k_B TL}{Nl_s^2} - \frac{\alpha N k_B T}{L}. \tag{11.24}$$

In the absence of external loads, this force is balanced to zero and we can calculate the uncompressed brush height L_0:

$$L_0 = \sqrt{\frac{\alpha}{3}} \cdot Nl_s \tag{11.25}$$

Please note that the brush thickness is independent of the grafting density. This independence has been experimentally confirmed [1350].

Now, we let two surfaces, both with a polyelectrolyte brush, approach each other. At a distance $x = 2L_0$, the surfaces start to interact. Work has to be carried out to approach further. Per polymer chain, the interaction energy is given by

$$V = \int_{L_0}^{x/2} F dL = \frac{3k_B T}{Nl_s^2}\left[\left(\frac{x}{2}\right)^2 - L_0^2\right] - \alpha N k_B T \cdot \ln\left(\frac{x}{2L_0}\right). \tag{11.26}$$

With $y = x/2L_0$, we can simplify this expression

$$V = k_B T\alpha N \cdot [y^2 - 1 - \ln y]. \tag{11.27}$$

If we bring two surfaces with a grafting density Γ together to a distance x, the interaction energy per unit area is $V^A = 2\Gamma V$ or

$$V^A(x) = 2k_B T L_0 \Gamma \frac{\sqrt{3\alpha}}{l_s} \cdot \left[\left(\frac{x}{2L_0}\right)^2 - 1 - \ln\left(\frac{x}{2L_0}\right)\right]. \tag{11.28}$$

Let us now include a monovalent background salt. For low salt concentrations, we can apply the same arguments as above. Low salt means that the Debye length is much smaller than the screening length ξ_e. At high salt concentrations, we talk about the *salted brush regime*. Pincus showed that the osmotic pressure in the salted brush regime has to be modified [1373]. The osmotic pressure as obtained above has to be multiplied by the ion density caused by dissociated ions, $\alpha N/Lb^2$, divided by the salt concentration: $\alpha N k_B T/Lb^2 \cdot \alpha N/c_s Lb^2 = k_B T/c_s \cdot (\alpha N/Lb^2)^2$. In the presence of background salt of concentration C_s, the osmotic force in a polyelectrolyte brush is $-k_B T\Gamma/c_s \cdot (\alpha N/L)^2$. This leads to a force per chain of

$$F = \frac{3Lk_B T}{Nl_s^2} - \frac{k_B T\Gamma}{c_s}\left(\frac{\alpha N}{L}\right)^2. \tag{11.29}$$

By setting this expression to zero, we obtain the brush thickness in the presence of a high salt concentration:

$$L_0 = N\left(\frac{\alpha^2 l_s^2 \Gamma}{3c_s}\right)^{1/3} \tag{11.30}$$

Please note that the brush thickness depends only weakly on salt concentration. Experiments confirm the $c_s^{-1/3}$ dependence of the layer thickness on the salt concentration for brushes [1371] and also for adsorbed polyelectrolyte layers [1372, 1380].

By integration, we again get the interaction energy per chain:

$$V = \frac{k_B TN}{2}\left(\frac{3\alpha^4\Gamma^2}{c_s^2 l_s^2}\right)^{1/3}\left[\gamma^2 + \frac{2}{\gamma} - 3\right]. \tag{11.31}$$

This leads to an interaction energy per unit area

$$V^A(x) = k_B TL_0\Gamma\left(\frac{9\alpha^2\Gamma}{c_s l_s^4}\right)^{1/3}\left[\left(\frac{x}{2L_0}\right)^2 + \frac{4L_0}{x} - 3\right]. \tag{11.32}$$

Equations (11.25)–(11.32) are based on scaling arguments. They should provide the correct dependencies, but one should not expect the numerical results to be precise. Balastre *et al.* [1374] compared experimental force curves with the above prediction. Experiments were carried out with PSS in aqueous solutions at different concentrations of added NaNO$_3$. By introducing two constant parameters for each regime, they could fit all force curves. For salt concentrations below 5.4 mM, good agreement was obtained with

$$V^A = k_B T\alpha\Gamma N \cdot [A_1(\gamma^2-1)-A_2 \ln\gamma] \tag{11.33}$$

and $A_1 = 0.75$ and $A_2 = 1.62$. For high salt concentrations, $c_s \geq 0.02$ M, the force curves could be described by

$$V^A = k_B T\alpha N\left(\frac{3\alpha^4\Gamma^5}{c_s^2 l_s^2}\right)^{1/3}\left[A_3(\gamma^2-1)+A_4\left(\frac{1}{\gamma}-1\right)\right] \tag{11.34}$$

with $A_3 = 0.45$ and $A_4 = 1.01$.

Surface force between various polyelectrolyte brushes has been measured [1355, 1379, 1381]. Not only polymer brushes but also dense layers of physisorbed homopolymers show the general features outlined above [1372, 1380, 1382–1384]. For spherical geometries, that is, if the particle's core diameter is of the same length scale as the brush thickness, expressions for the electrosteric force are reported in Refs [1381, 1385].

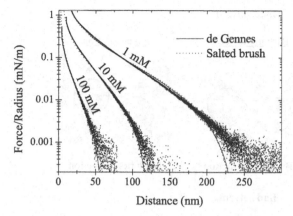

Figure 11.8 Force distance profiles between a silica sphere of 3 μm radius and a microscope slide both covered with physisorbed PSS ($M_W = 350$ kDa) measured with an atomic force microscope [1380]. The force was normalized by dividing it with the radius of the sphere. Force curves were recorded in aqueous medium containing 1, 10, and 100 mM NaCl. Experimental results were fitted with the model of de Gennes [1346] (Eq. (11.21), $\Gamma = 4 \times 10^{14}$ m^{-2} and $L_0 = 132, 76$, and 47 nm at 1, 10, and 100 mM NaCl, respectively) and the salted brush model (Eq. (11.32), $\Gamma = 10^{14}$ m^{-2} and $L_0 = 116, 60$, and 29 nm and $\alpha = 0.2, 0.8$, and 2.0 at 1, 10, and 100 mM NaCl, respectively) with $F/R = 2\pi V^A$. In addition, the distance was offset by 1–5 nm.

As one example, Figure 11.8 shows the normalized force versus distance measured between a silica sphere and a glass plate in aqueous medium. Both surfaces were covered by a layer of physisorbed poly(styrene sulfonate) [1380]. Usually, PSS does not adsorb to glass or silica because the surfaces and the polymer are negatively charged. Therefore, both surfaces were silanated with a monolayer of 3-aminopropyldimethyl ethoxysilane. The amino groups render the glass and silica surfaces positive. To enhance adsorption, the surface were incubated in 1 M NaCl solution for 1 h. Although PSS is not grafted to the surface, the steric interaction can still be described by the equations above. With increasing salt concentration, the decrease in the brush thickness is clearly visible.

11.4
Polymer-Induced Forces in Solutions

After discussing the steric repulsion between tethered polymer brushes in a good solvent, let us turn to the more general situation. We consider dispersed particles in a solution of chemically homogeneous polymers, also called homopolymers. Four major factors determine the force between particles in polymer solutions: (1) the quality of the solvent, (2) the amount of polymer adsorbed and the way it is adsorbed, (3) the time allowed for the polymer to adsorb and rearrange at the surfaces, let us call it the incubation time, and (4) the time during which two

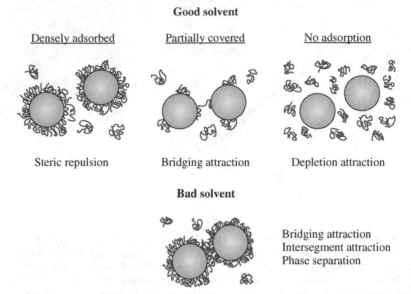

Figure 11.9   Overview of forces between solid surfaces in polymer solutions.

polymer-coated surfaces interact with each other, let us call it contact time. All four are interconnected.

To get an overview, we first distinguish between good and bad solvents (Figure 11.9). In a good solvent, the next question is whether the polymer adsorbs to the surfaces of the particles. If a lot of polymer is adsorbed and the surfaces are covered with a dense layer of polymer, steric repulsion will dominate. The particles repel each other and the dispersion is stable. If not so much polymer is adsorbed because either the concentration of polymer is low or the polymer did not have sufficient time to form a homogeneous layer, the particles will stick to each other. Attraction is caused by bridging polymer chains. If due to Brownian motion two particles get close to each other, a polymer chain on one particle has a good chance to find an empty area on the other particle and form a bridge. If the polymer does not adsorb to the particle at all, we expect an attractive force. This attraction is due to depletion of polymer from the gap between two particles. Depletion forces are discussed in the next chapter.

In bad solvent, the polymer-coated surfaces attract each other [1386, 1387]. One reason is the absence of steric repulsion. Van der Waals forces can become dominant again and lead to aggregation. Napper [1311] formulated the rule of thumb that the onset of dispersion flocculation coincides with the creation of Θ conditions in the solvent medium. Exceptions to this rule are encountered when either the polymeric stabilizers are very short or the particles become very small. In addition to van der Waals forces, the polymer itself induces attraction. In a bad solvent, the polymer usually has a strong tendency to adsorb to any surface available. For that reason, we do not have to distinguish all possible cases. Attraction between surfaces is due to intersegment forces. The intersegment force is caused by the direct interaction between segments of polymers with each other. Since in a bad solvent the interaction among the monomers is stronger than the interaction of monomers with the solvent, polymer on one particle will interact favorably with polymer on another particle. If the other particle is not fully covered by adsorbed polymer, it will bind to free surface area and form a bridge. In addition to intersegment and bridging forces, we might get a phase transition in the gap between the particles. Once the segment density exceeds a critical value, a polymer-rich phase might form. The polymer-rich phase is separated from the solvent-rich phase by a liquid–liquid interface. It forms a meniscus that results in an attractive force [180].

From the overview (Figure 11.9) it is clear that the amount adsorbed or bound and the way the polymer is bound to the surface is a decisive factor. Discussing polymer adsorption to interfaces in detail is beyond the scope of this book. We recommend Refs [1334, 1388]. For the adsorption of end-functionalized polymers, see Refs [1389, 1390]. Experimentally, one finds that the *incubation time* can be quite long and reach hours or even days [1353, 1391, 1392, 1394–1397]. An instructive example is shown in Figure 11.10. Almog and Klein studied the interaction between two mica surfaces in cyclopentane with dissolved polystyrene using a surface forces apparatus [1398]. The experiments were carried out at 23 °C, which is slightly above the Θ temperature of 19.6 °C. With a molar mass of $M_w = 2050\,\text{kDa}$, the radius of gyration is $R_g = 39\,\text{nm}$. Force curves were recorded after allowing the system to adsorb and equilibrate for 12, 15, and 31 h. The force changes from attractive at short incubation time to strongly repulsive at long incubation times. The results can be interpreted in the following way: in this near Θ solvent and at short incubation times, bridging attraction dominates. After waiting longer, the adsorbed layer becomes denser and more stable and steric repulsion becomes dominant. A similar change from bridging attraction to steric repulsion with the incubation time is described in Refs [1399, 1400] for a polyelectrolyte.

One might think that the long incubation time is due to a slow diffusion of polymer toward the surface. To demonstrate that this is *not* the case, we estimate the time required for diffusion.

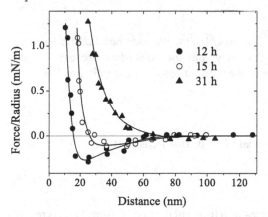

Figure 11.10 Approaching force versus distance curves between curved mica surfaces in 15 mg l^{-1} polystyrene dissolved in cyclopentane. Curves were recorded 12, 15, and 31 h after immersing the mica surfaces in the solution. (Adapted from Ref. [1398].)

■ **Example 11.5**

We consider the simple case of *diffusion limited adsorption*. Diffusion is described by Fick's law. For diffusion toward an infinitely extending plane, we only need to consider the dimension normal to the surface due to symmetry. In a direction x normal to the surface, Fick's law is

$$\frac{\partial c}{\partial x} = D_d \frac{\partial^2 c}{\partial t^2}. \tag{11.35}$$

D_d is the diffusion coefficient and c is the concentration. To estimate the time required for molecules to diffuse to a surface, we assume that at the beginning the molecules are homogeneously distributed with a concentration c_0. It is instructive to introduce a "subsurface" of molecular thickness directly at the surface. We assume that all molecules reaching the subsurface are adsorbed and withdrawn from solution. Neglect the back diffusion. The concentration in the subsurface is taken to be always zero. We have the initial condition $c(x > 0, t = 0) = c_0$ and the boundary conditions, $c(x = 0, t) = 0$, $c(x \rightarrow \infty, t) = c_0$. With these initial and boundary conditions, Fick's law is solved by [1401] $c(x, t) = c_0 \cdot \mathrm{erf}(x/2\sqrt{D_d t})$, where erf is the error function. The error function is defined by $\mathrm{erf}(z) = 2/\sqrt{\pi} \cdot \int_0^z e^{-\varsigma^2} d\varsigma$.

To calculate how many molecules are adsorbed, we integrate all those molecules that at $x = 0$ flow toward the surface. The flow is

$$J = -D_d \frac{\partial c}{\partial x}. \tag{11.36}$$

Since the derivative of the error function is simple $d[\mathrm{err}(z)]/dz = 2/\sqrt{\pi} \cdot e^{-z^2}$, we obtain

$$J = -c_0 \sqrt{\frac{D_d}{\pi t}} \cdot e^{-\frac{x^2}{4D_d t}}. \tag{11.37}$$

For $x = 0$, this reduces to $J = -c_0\sqrt{D_d/\pi t}$. We assumed that no backflow occurs and all molecules that flow into the subsurface are permanently adsorbed. Then, the change in the amount adsorbed per unit area Γ is

$$\frac{d\Gamma}{dt} = c_0 \sqrt{\frac{D_d}{\pi t}}. \tag{11.38}$$

We integrate starting at $t = 0$ and $\Gamma = 0$. This leads to [1402]

$$\Gamma(t) = 2c_0 \sqrt{\frac{D_d t}{\pi}}. \tag{11.39}$$

The amount adsorbed increases with time but the rate of adsorption decreases. To obtain a characteristic time of diffusion-limited adsorption, we calculate the time required for a full monolayer to adsorb. Let us for simplicity further assume that the molecules are spherical and have a radius r. Their diffusion coefficient in a fluid of viscosity η is given by the Stokes–Einstein equation

$$D_d = \frac{k_B T}{6\pi\eta r}. \tag{11.40}$$

Monolayer coverage is reached at $\Gamma = 1/\pi r^2$. This leads to a characteristic time τ for diffusion limited adsorption of

$$\frac{1}{\pi r^2} = 2c_0\sqrt{\frac{k_B T \tau}{6\pi^2 \eta r}} \Rightarrow \tau = \frac{3\eta}{2c_0^2 r^3 k_B T}. \tag{11.41}$$

Here, the concentration is in units of molecules per volume. If the concentration is inserted in mol m^{-3}, it has to be multiplied with the Avogadro constant first.

As an example, we take a negatively charged planar surface in water at 20 °C and estimate the time required for the adsorption of a monolayer of poly(2-vinyl pyridine) (P2VP) with $M_w = 4\,kDa$ and concentrations of 0.05 and 0.005 g l^{-1}. As a positively charged polyelectrolyte, P2VP readily adsorbs to negatively charged surfaces. The hydrodynamic radius of P2VP in water containing 0.1 M monovalent salt is roughly 11 nm [1403]. The viscosity of water is $\eta = 10^{-3}$ kg s^{-1} m^{-1}.

The concentration in molecules per volume is

$$c_0 = \frac{0.05\ \text{kg m}^{-3} \cdot 6.02 \times 10^{23}\ \text{mol}^{-1}}{4\ \text{kg mol}^{-1}} = 7.53 \times 10^{21}\ \text{m}^{-3}.$$

Inserting into Eq. (11.41)

$$\tau = \frac{3 \cdot 10^{-3}\ \text{kg s}^{-1}\ \text{m}^{-1}}{2 \cdot (7.53 \times 10^{21}\ \text{m}^{-3})^2 (11 \times 10^{-9}\ \text{m})^3 \cdot 4.05 \times 10^{-21}\ \text{J}} = 3.98 \times 10^{-3}\ \text{s}.$$

For $c_0 = 0.005$ g l^{-1} = 7.53×10^{20} m^{-3}, we get $\tau = 0.398$ s.

The example shows that diffusion in solution is usually not the limiting factor for the slow formation of adsorbed polymer layers. Two processes slow down the formation of a polymer layer. First, the already adsorbed polymer hinders further polymers to adsorb. Referring to the above example; already adsorbed P2VP reduces the negative surface charge or even changes it from negative to positive. The surface appears less attractive and additional P2VP is prevented to get into direct contact with the surface. Second, the adsorbed polymers rearrange on the surface. In particular for long chains, which might even be entangled, this can be a very slow process.

Another important timescale is the *contact time*. We have to distinguish whether the approach and retraction of the two interacting surfaces is so slow that the system is in equilibrium at all stages or not. In the first case, we talk about "full equilibrium," the second case is called "restricted equilibrium" [1404, 1405]. In particular for physisorbed polymer, this is a major difference. Full equilibrium is given only when polymer is allowed to adsorb and desorb and the polymer in the gap is in equilibrium with the polymer dissolved in the bulk liquid. In restricted equilibrium, the individual chains may relax, but they do not move in and out of the gap on the timescale of the contact.

In fact, theory showed that in full equilibrium even the case with adsorbed chains should lead to an attractive force [1404, 1406]. Practically, steric repulsion is observed in a good solvent. The reason is that the contact time is much shorter than the time to establish full equilibrium, which can take hours. Therefore, even physisorbed homopolymers often behave like tethered chains and steric repulsion dominates.

Whether full or restricted equilibrium is established does not only depend on the chemical nature of the polymer, the surfaces, and the solvent but also depends on the specific geometry. Small particles have only a small gap zone from which polymer can more easily diffuse in and out. They equilibrate faster than large particles. With respect to the measuring technique, full equilibrium is easier to reach in AFM experiments than with the SFA. In experiments, one manifestation of relaxation processes in polymer layers are differences in the approaching and retracting force curve. For very slow processes, even a difference between subsequent force curves is observed [1353, 1396].

11.5
Bridging Attraction

In some dispersions, the addition of a polymer leads to flocculation[3] [1392]. There are two mechanisms of how a polymer can induce flocculation: bridging and depletion. While depletion can occur only for nonadsorbing polymer, bridging relies on adsorption. Depletion flocculation is discussed in the next section. Bridging flocculation is essential in mineral processing, paper making, or the treatment of waste water. For example, if a cationic poly(acrylamide) is added to an aqueous dispersion of alumina particles, the dispersion becomes unstable and the particles aggregate [1393]. As the name indicates, bridging forces arise when a polymer chain binds to both surfaces.

For bridging to occur, the polymer must have a certain affinity to the particle surfaces. Bridging usually depends on the concentration of the polymer [1407]. At low concentrations, the rate of flocculation usually increases. The more polymer chains are adsorbed, the more bridges can be formed if two particles get into contact. At some concentration, however, the density of adsorbed chain gets so high that a chain on one particle might not find an empty binding site on the opposite particle during the time of contact. In that case, no bridge can be formed. As a consequence, the bridging efficiency reaches a maximum and decreases at high concentrations.

---

3) According to the IUPAC, flocculation is a process of contact and adhesion whereby the particles of a dispersion form larger size clusters. It is synonymous with agglomeration and coagulation.

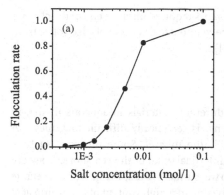
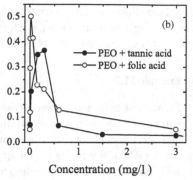

Figure 11.11 Normalized rate of flocculation of an aqueous dispersion of cellulose particles versus concentration of KCl (a) and versus the concentration of added PEO + tannic acid and PEO + folic acid plus 1 mM KCl (b). The rates were normalized by dividing them by the rate measured in 100 mM KCl polymer-free solution. (Adapted from Ref. [1408].)

■ **Example 11.6**

Gaudreault *et al.* [1408] studied aggregation of microcrystalline cellulose in aqueous solution, which is relevant for papermaking. The cellulose particles remain dispersed at very low salt concentrations due to electrostatic stabilization. Flocculation starts at 0.5 mM KCl though it is relatively slow. With increasing salt concentration, the rate of flocculation increases (Figure 11.11a). At 0.1 M KCl, it is very fast. The rate of flocculation is defined as the increase in the mean aggregate radius. Comparisons are always made at constant cellulose concentration of $0.3 \, g \, l^{-1}$.

Poly(ethylene oxide) together with an acidic cofactor is able to flocculate the cellulose particles. Figure 11.11b shows the effect of PEO with two cofactors, both mixed at a weight ratio of 1 : 1. When PEO together with tannic acid is added at a concentration of $0.3 \, mg \, L^{-1}$, the flocculation rate reaches 37% in the rate observed in 100 mM KCl. Adding more polymer leads to a decrease in the rate of flocculation. PEO together with folic acid is even more effective, reaching a maximum below $0.1 \, mg \, L^{-1}$ with a maximal rate of flocculation of 50% with respect to the rate in 100 mM KCl. PEO alone does not destabilize the cellulose dispersion.

Force versus distance experiments have revealed bridging attraction in different systems, for example, by polystyrene between mica in cyclohexane [1409] and cyclopentane [1396, 1398], or poly(ethylene oxide) between clays, glass, or silica in water [1410, 1411]. Bridging by polyelectrolytes has also attracted great attention, both theoretically [1412, 1413] and experimentally [1399, 1400].

With the application of atomic force microscopy in measuring surface forces, it became possible to detect the bridging of individual polymer chains. In particular polyelectrolytes adsorbing to charged surfaces in aqueous medium have been studied [1414–1418]. Peeling a strongly adsorbed polyelectrolyte from a surface is similar to peeling a tape from a surface. A constant force is required until the end of the chain is reached. Using continuum theory, the force required to desorb a polyelectrolyte from a surface was calculated [1415, 1419–1421].

■ **Example 11.7**

Permanent adhesion between two different materials in aqueous media, in particular at high salt concentrations, is technically difficult to achieve. A biological model for wet adhesion is the mussel. Some mussels, such as *Mytilus edulis*, live in the turbulent intertidal zone of the sea shore. They are well known for their ability to cling to wet surfaces [1422, 1423]. Mussels secrete specialized adhesion proteins, which contain a high content of the amino acid 3,4-dihydroxyphenylalanine (DOPA) [1424]. It is well known that the phenyl ring with its two hydroxyl groups (see Figure 11.12) readily binds to many metals and some minerals. Both natural and synthetic adhesives containing DOPA showed strong adhesion [1425, 1426].

One example is shown in Figure 11.12 [1427]. An artificial, DOPA-containing polymer was allowed to adsorb to a titanium surface for few seconds. Then, the excess polymer was rinsed out. Since DOPA binds strongly to titanium, an atomic force microscope tip brought into contact with the titanium surface has a fair chance to hit a polymer chain. The tips were also coated with titanium. When hitting a polymer the chain might therefore attach to the tip. During retraction, the polymer is often observed to be peeled off the planar titanium surface. This process is at constant force, much like a normal tape being peeled off a substrate.

At this point, it is instructive to analyze the phenomenon of adhesion in more detail. We do the following experiment: two solid surfaces are in contact. With a spring we apply a force to one of the solid bodies in a direction normal to the surfaces (Figure 11.13). At some point, the surfaces will separate from atomic contact. If we do

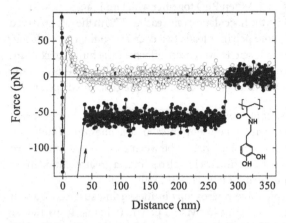

Figure 11.12 Force versus distance curve obtained with a biomimetic DOPA-containing polymer in aqueous solution of 1 mM KNO_3 measured with an atomic force microscope between two titanium surfaces [1427]. Approaching (○) and retracting (●) parts are plotted.

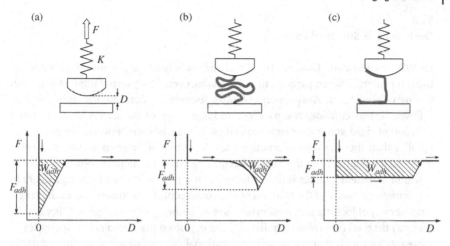

Figure 11.13 Schematic experimental setup and retracting force versus distance curves for contact adhesion (a), and bridging adhesion (b,c) between two surfaces. For bridging adhesion, two different scenarios are plotted: that of a polymer chain fixed at two points and that of a polymer that adheres to the surfaces and is peeled off like a tape. The adhesion forces F_{adh} and the work of adhesion W_{adh} (hatched area) are indicated.

the experiment in vacuum, a gas, or a simple liquid, the surfaces will completely jump out of contact once a certain threshold force has been reached and has separated the surfaces by 0.1–0.2 nm. The reason is that at 0.1–0.2 nm distance, the attractive forces decay with increasing distance while the pulling force applied by the spring remains high. We call this *contact adhesion*. Contact adhesion is characterized by its short range.

The situation changes when a polymer bridges the surfaces. Let us distinguish two cases. In the first case, the polymer chain bridging the surfaces is fixed with one end to one surface and with the other end to the other surface. When separating the surfaces, work has to be done to stretch the polymer. This force was discussed in Section 11.2 and an example is shown in Figure 11.2. The other case is of a polymer chain, which is adsorbed with many binding sites or even continuously to the surfaces. In that case, stretching is only the first phase. A major obstacle is to peel the polymer off the surfaces. One example is shown in Figure 11.12. We call both forms of adhesion *bridging adhesion*. Bridging adhesion is long ranged.

To separate two surfaces and overcome adhesion, a certain threshold force, the adhesion force, has to be overcome. In addition, the work of adhesion has to be done. For contact adhesion, this is accomplished when the spring is put under tension. It depends on the spring constant K. At a given adhesion force, the work of adhesion is $W_{adh} = F_{adh}^2/(2K)$. For contact adhesion, the work of adhesion can be reduced by pulling with a stiff spring. In contrast, for bridging adhesion, W_{adh} cannot be significantlly influenced by the spring constant. It is given by the properties of the polymer and its interaction with the surfaces. The difference in contact and bridging is highly relevant for biomaterials such as bones [1428, 1429].

11.6
Depletion and Structural Forces

In 1954, Asakura and Oosawa [1430] realized that dissolved polymers can influence the interaction between particles in a dispersion, even if they do not interact at all with the particle surfaces. Asakura and Oosawa themselves describe the interaction as follows: "Let us consider two parallel and large plates of the area A immersed in a solution of rigid spherical macromolecules. If the distance between the plates x is smaller than the diameter of solute molecules, none of these molecules can enter between the plates. Then this region becomes a phase of the pure solvent, while the solution outside the plates is little affected by them. Therefore, a force equivalent to the osmotic pressure of the solution of macromolecules acts inwards on each plane." Another way of looking at depletion forces is as follows: particles aggregate because in this way the total free volume available to the dissolved macromolecules increases. It increases so much that the gain in translational entropy of the macromolecules is higher than the loss of entropy of the particles.

Though macromolecules are usually not rigid, the same effect occurs for soft molecules. In this case, the mean end-to-end distance of a linear, randomly coiled polymer R_0 corresponds to the radius of the spheres (Figure 11.14). Depletion forces are effective over ranges of the size of the dissolved molecules. The force per unit area is of the order of the osmotic pressure of dissolved macromolecules. The Gibbs free energy for the depletion attraction between two spherical particles of radius R_p at a distance D in a solution containing macromolecules of radius R_0 at a number density c is roughly (Exercise 11.5 and Ref. [1431]):

$$W(D) = \frac{\pi}{2} c k_B T R_p (2R_0 - D)^2 \tag{11.42}$$

for $D \leq 2R_0$ and $W = 0$ for $D > 2R_0$. Equation (11.42) was derived assuming a hard-sphere potential for the particles and the dissolved molecules. The hard-

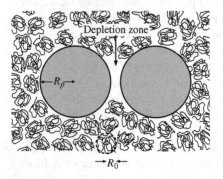

Figure 11.14 Schematic of two particles dispersed in a polymer solution. The reduced osmotic pressure in the zone depleted of polymer in the gap between the particles leads to an effective attractive force, the depletion force.

sphere potential has meanwhile been more realistic potentials by simulations and theory [1432, 1433].

Practically, the addition of a nonadsorbing polymer to a dispersion can induce flocculation of dispersed particles due to the depletion attraction. This was first observed by Cowell, Lin-In-On, and Vincent [1434]. When large amounts of poly (ethylene oxide) are added to an aqueous dispersion of hydrophilized polystyrene latex particles, the particles start to flocculate. For an organic dispersion, namely, hydrophobized silica particles in cyclohexane, de Hek and Vrij [1435] observed depletion-induced flocculation when dissolved polystyrene was added. Other combinations of particles and polymers followed [1436]. Phase diagrams for different particle–solvent–polymer systems were successfully drawn using the depletion potential of Asakura as interaction potential between dispersed spheres [1437] and for dissolved polymers using statistical mechanics [1438].

Walz and Sharma [1439] calculated the depletion force between two charged spheres in a solution of charged spherical macromolecules. Compared to the case of hard-sphere interactions only, the presence of a long-range electrostatic repulsion increases greatly both the magnitude and the range of the depletion effect. Simulations and density functional calculations for polyelectrolytes between two planar surfaces extend these results [1440].

Forces mediated by dissolved, nonadsorbing macromolecules need not be purely attractive. They are attractive at short ranges and at low volume fraction of the dissolved species. At distances larger than the size of the dissolved molecules and at high concentrations, it can be repulsive [1439, 1441. 1442]. In general, depletion forces can be treated like solvation forces, and oscillatory behavior is predicted [1432, 1433] and observed [1443, 1444]. The interface induces a layered structure of the dissolved molecules. Once two surfaces get close to each other, this layered structure is disturbed. This results in a force. We call it structural force.

Structural forces have meanwhile been directly measured in foam films using the thin film balance and caused by surfactant micelles [779] or polyelectrolytes [795, 796]. In fact, stratification in foam film in the presence of high surfactant concentrations has been observed much earlier [752] and is a common phenomenon. Structural forces have also been observed between particles induced by addition of polymer [1431, 1444, 1445], induced by surfactant micelles [1306, 1443, 1446], or in dispersions with smaller spherical particles [1447]. Measurements have been carried out by the surface force apparatus [1443], atomic force microscopy [1444, 1445], total internal reflection microscopy [1431, 1446], and optical tweezers [1447].

■ **Example 11.8**

Dispersion of small colloidal particles can give rise to oscillatory forces. As one example, Klapp *et al.* [1448] measured the force between a silica microsphere (6.7 μm diameter) and a silicon wafer in an aqueous dispersion of silica nanoparticles using an AFM. With a diameter of 26 ± 2 nm, the dispersed nanoparticles were much smaller than the microsphere. All experiments were

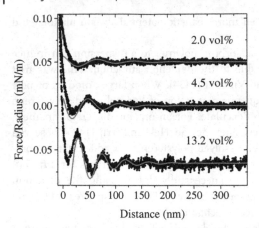

Figure 11.15 Normalized force versus distance curves measured between a silica microsphere of 6.7 μm diameter and a silica wafer in an aqueous dispersion of silica nanoparticles (26 nm diameter) with an AFM [1448]. The concentration of the nanoparticles was 2.0, 4.5, and 13.2 vol%. Results are vertically offset for better viewing. Curves were fitted with Eq. (10.7) (gray lines) with $\xi = 65$ nm, $a_0 = 72$ nm, and $F_0/R = 8 \times 10^{-6}$ Nm^{-1} (2.0 vol%); $\xi = 50$ nm, $a_0 = 59$ nm, and $F_0/R = 21 \times 10^{-6}$ Nm^{-1} (4.5 vol%); $\xi = 45$ nm, $a_0 = 45$ nm, and $F_0/R = 55 \times 10^{-6}$ N/m^{-1} (13.2 vol%).

done in aqueous medium at an ionic strength of 10 μM at neutral pH. Under these conditions, the nanoparticles, the microsphere, and the wafer are negatively charged. This electrostatic repulsion prevents the nanoparticles from aggregation. They observed an oscillatory force with a period corresponding to the bulk radial distribution function of the nanoparticles (Figure 11.15).

11.7
Interfacial Forces in Polymer Melts

From the fundamental point of view, surface forces across polymer melts are studied to obtain information about adsorbed polymer at the interface or to better understand confined polymers. An understanding of surface force across polymer melts is also important for making composite materials. Since the trend is toward, using smaller and smaller particles in composites, leading to nanocomposite materials with huge internal interfaces, knowledge of the forces becomes more and more important.

Let us first consider a melt of a homopolymer between two solid surfaces in full equilibrium. A nonfavorable interaction between polymer and surfaces ($\Theta < 90°$) would lead to dewetting in the gap and the formation of a cavity. Then, attractive capillary forces dominate. More interesting is the case that the polymer interacts favorably with solid surfaces. That is, the interaction between segments of the polymer with the surface is stronger than between two segments. Experimentally, the polymer spreads on the surfaces and the contact angle is below 90°. Theory

[1449, 1450] and simulations [1451] predict that two solid surfaces across a confined polymer melt should experience no force at all (except van der Waals attraction) provided the system is at full equilibrium with a reservoir. Two effects exactly compensate: the entropy of confined polymer chains decreases but the polymer can leave the closing gap. The first effect leads to repulsion, the second to attraction.

First experiments with the surface force apparatus, however, showed a different result: Experiments with poly(dimethyl siloxane) [1358, 1453], perfluorinated poly-ether [1454–1456], polybutadiene (PB) [186], and poly(phenylmethyl siloxane) (PPMS) [1453, 1457, 1458] indicated that an immobilized, rubber-like layer of a thickness of typically one to three times the radius of gyration R_g is present on the mica surfaces. Beyond this layer the two interacting surfaces experience a longer ranged repulsion across PPMS [1457], PB [186], and perfluorinated polyethers [1454, 1456], typically reaching to 1–3 R_g. With PDMS an oscillatory, exponentially decaying interaction was observed [1358]. The oscillation period of 0.7 nm was approximately equal to the diameter of a PDMS chain.

Only recently, it became possible to confirm the "no force" prediction. By using an atomic force microscope, the force across a melt of polyisoprene between silicon nitride and silicon oxide was measured. Polyisoprene interacts favorably with both surfaces; this is confirmed by a low contact angle of 7–27°. As predicted, a negligible force was measured (Figure 11.16) [636]. With a radius of curvature of the AFM tip of $R \approx 40$ nm (compared to typically 1 cm in the SFA), the interacting area is much smaller; typical diameters of the zone of contact are 10 nm in the AFM and 100 μm in the SFA. For this reason, the polymer has a better chance to flow out of the gap between the interacting surfaces and equilibrate with

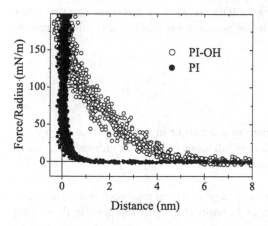

Figure 11.16 Force curves measured in normal methyl-terminated polyisoprene (PI, $M_W = 3.4$ kDa) and hydroxyl-terminated polyisoprene (PI-OH, $M_W = 3.4$ kDa). The experiments were carried out with an atomic force microscope between a silicon nitride tip and a silicon wafer. Forces were normalized by dividing them by the radius of curvature of the tip. (Adapted from Ref. [1459].)

the reservoir. Thus, on the timescale of one approach–retraction cycle, the system was close to equilibrium.

For polydimethylsiloxane, X-ray reflectivity [1460] and AFM [1461] experiments confirmed the formation of a quasi-immobilized layer in thin films near silica surfaces. This layer has a lower density than the bulk value. Its thickness varies slightly with PDMS molecular weight and is typically 4–6 nm. Formation of this layer is rapid for PDMS melts with low molecular weights but takes of the order of 10 h for molecular weights higher than 10 kDa. Once immobilized layers have formed, the surface forces become strongly repulsive. It is still an open question how the other polymer behaves: like polyisoprene, with negligible polymer-induced force, or like poly(dimethylsiloxane), which forms an immobilized layer and which shows strong repulsion. The key question lies most likely in the kinetics (not thermodynamics) of bond formation between polymer segments and solid surfaces. If bond formation and release is fast, the system is closer to equilibrium and the net force between surface is negligible. If segments bind slowly and are also released slowly, an immobilized layer is formed and the net force is repulsive on the timescale of normal force experiments.

The situation changes completely if we consider end-functionalized polymers or block copolymers rather than homopolymers. If a certain part of the polymer binds more strongly to the surface than the major part, it will form a brush. The nonbound, mobile rest of the melt forms an ideal solution. This effect was demonstrated for polyisoprene (Figure 11.16) [1459]. When replacing the normal methyl group on one side with an hydroxyl group and measuring the force between silicon nitride and silicon oxide, a strong repulsion was observed. The repulsive force decayed with a decay length equal to the radius of gyration of the polymer of $R_g = 1.8$ nm. Again, this agrees with theory. The hydroxyl end groups bind to the silicon surfaces and keep the polymer chains attached to the surfaces. If the chains of a polymer melt are prevented from leaving the gap, for example, by bonds to the surface, a repulsive force is expected [1449, 1452].

11.8
Summary

- To stretch a single polymer chain in solution or in a melt, a force is required because the number of available configurations and thus the entropy of the chain is reduced. This force can be calculated with the freely jointed or the worm-like chain models.

- Two surfaces with polymer brushes in a good solvent repel each other. This steric repulsion is caused by the increased density of polymer segments in the closing gap, which causes an increased interaction energy between segments and an increased osmotic pressure.

- Two surfaces in a polymer solution can experience attraction or repulsion. In a good solvent, the force depends critically on the adsorption. If the polymer adsorbs to the surfaces, bridging attraction occurs at low and intermediate coverage. If the surfaces are covered with a dense polymer layer, steric repulsion prevails. For nonadsorbing polymer, depletion attraction dominates. In a bad solvent, two surfaces attract each other due to bridging, intersegment attraction, or a phase separation.

- We distinguish contact and bridging adhesion. Contact adhesion acts only over atomic distances, is characterized by an adhesion force, and the work of adhesion can be reduced by pulling with a stiff spring. The range of bridging adhesion is of the order of the contour length of the polymers. Even if the adhesion force is weak, the work of adhesion can be high. It cannot be reduced significantly by choosing an appropriate spring.
- In perfect equilibrium, polymer induced surface forces in polymer melts should be absent. Practically, full equilibrium is almost never reached and repulsion is observed.

11.9
Exercises

11.1. Consider a bare surface interacting with a surface, which is coated with a polymer brush in a good solvent. In analogy to Eq. (11.18), derive an expression for the interaction energy. Plot the force versus distance for a spherical particle of 3 μm radius and a planar surface for the symmetric (both surfaces bear a polymer brush) and the asymmetric (only the planar surface is coated) case for $N = 50$, $l_s = 1.8$ nm, $\Gamma = 10^{-17}$ m^2, and $v = 0.1 l_s^3$.

11.2. Derive an expression for the steric repulsion between two polymer brushes in an ideal solvent at constant grafting density. In analogy to the derivation of Eq. (11.18), first calculate the thickness of the brush and then the interaction energy. The thickness of the undisturbed brush is determined by a balance between the osmotic pressure caused by the high density of polymer segments in the brush and the chain elasticity caused by the configurational entropy. The osmotic contribution can then be estimated from the density of polymer segments, $c = N/Lb^2$, multiplied by $k_B T$ to obtain the osmotic pressure $N k_B T / Lb^2$.

11.3. Derive an equation for the force per unit area between two surfaces coated with grafted polymers in the mushroom regime. Start with Eq. (11.23).

11.4. Plot the interaction energy per unit area versus distance for a polymer brush with $N = 60$, $l_s = 1.5$ nm, $v = l_s^3$, and $\Gamma = 5 \times 10^{17}$ m^2 for the simple osmotic brush as derived in Exercise 11.2, the Milner, Witten, and Cates model with a step (Eq. (11.18)) and parabolic segment profile (Eq. (11.19)), and the model of de Gennes (Eq. (11.21)).

11.5. Derive Eq. (11.42). Consider that a shell of thickness R_0 around each particle is not available to the dissolved molecules because they cannot get closer than R_0. This not-accessible volume is reduced when two particles approach each other. Calculate this reduced inaccessible volume and multiply it by the osmotic pressure. Then, assume that $R_p \gg R_0$.

12
Solutions to Exercises

Chapter 2

2.1. *Van der Waals force on an atomic force microscope tip.* For a parabolically shaped tip, we have $r^2/(2R) = x - D$. With a cross-sectional area at height x of $A = \pi r^2 = 2\pi R_p(x - D)$ for $x \geq d$, we get $dA/dx = 2\pi R_p$ and

$$F(d) = \int_d^\infty f(x)\frac{dA}{dx}dx = -\int_d^\infty \frac{2\pi R_p A_H}{6\pi x^3}dx = -\frac{A_H R_p}{3}\int_d^\infty \frac{dx}{x^3} = -\frac{A_H R_p}{6d^2}.$$

2.2. *Parallel slabs.* We want to have

$$\frac{F_{half-space} - F_{slab}}{F_{half-space}} < 0.01.$$

With Eq. (2.74), this is equivalent to

$$\frac{2}{(1+d/D)^2} - \frac{2}{(1+2d/D)^2} < 0.01. \tag{12.1}$$

The left-hand side of this equation is plotted in Figure 12.1. The value of this function falls below 0.1 for $d/D = 2.67$ and below 0.01 for $d/D \approx 11.3$. This corresponds to slab thicknesses of 2.67 and 11.3 nm for $D = 1$ nm and of 26.7 and 113 nm for $D = 10$ nm.

2.3. The hydrocarbons are apolar molecules for which the van der Waals interaction will almost completely arise from the London dispersion interaction The magnitude of the interaction will depend on the polarizability α of the molecules (Eq. (2.21)). The value of α will be higher for larger molecules and thus will increase for longer chains.

Chapter 3

3.1. *Thermal noise of an AFM cantilever.* For an estimate of the thermal oscillation amplitude, we can just use the equipartition theorem, which states that the mean square displacement of a harmonic oscillator with spring constant k

Surface and Interfacial Forces. Hans-Jürgen Butt and Michael Kappl
Copyright © 2010 Wiley-VCH Verlag GmbH & Co. KGaA
ISBN: 978-3-527-40849-8

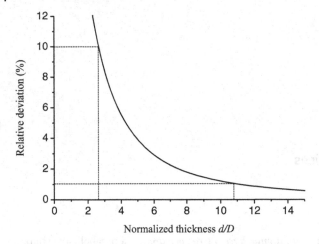

Figure 12.1 Plot of the left-hand side of Eq. (12.1). The horizontal line at $y = 1$ intersects the curve at a value of $d/D = 11.3$ and the line at $y = 10$ intersects the curve at a value of $d/D = 2.67$.

is given by

$$\frac{1}{2}k_B T = \frac{1}{2}k\langle x^2 \rangle;$$

therefore, the amplitude is given by

$$\sqrt{\langle x^2 \rangle} = \sqrt{\frac{k_B T}{k}} = \sqrt{\frac{4.12 \times 10^{-21}\ \text{Nm}}{k}}.$$

For a spring constant of $0.1\ \text{N m}^{-1}$, we obtain a thermal noise of $0.2\ \text{nm}$ and for a spring constant of $40\ \text{N m}^{-1}$ we obtain $10\ \text{pm}$.

3.2. *Penetration depth in TIRM.* The critical angle θ_c is given by the relation

$$\theta_c = \arcsin \frac{n_2}{n_1} = \arcsin \frac{1.33}{1.6} = 56.23°.$$

The penetration depth is given by

$$\lambda_{ev} = \frac{\lambda}{4\pi n_1 \sqrt{\sin^2\theta - \sin^2\theta_c}},$$

$$\lambda_{ev} = \frac{633\ \text{nm}}{4\pi \cdot 1.6\sqrt{\sin^2 56.33 - \sin^2 56.23}} = 784\ \text{nm},$$

$$\lambda_{ev} = \frac{633\ \text{nm}}{4\pi \cdot 1.6\sqrt{\sin^2 57.23 - \sin^2 56.23}} = 249\ \text{nm},$$

$$\lambda_{ev} = \frac{633\ \text{nm}}{4\pi \cdot 1.6\sqrt{\sin^2 61.23 - \sin^2 56.23}} = 113\ \text{nm},$$

$$\lambda_{ev} = \frac{633\ \text{nm}}{4\pi \cdot 1.6\sqrt{\sin^2 66.23 - \sin^2 56.23}} = 82.2\ \text{nm}.$$

3.3. *Potential shape in TIRM.* The total potential is given by the electric double-layer potential described by the DLVO theory and the gravitational potential. With the particle radius R, the particle and liquid densities ϱ_p and ϱ_l, and the Debye length λ_D,

$$V(h) = C \exp\left(\frac{h}{\lambda_D}\right) + \frac{4}{3}\pi R^3 (\varrho_p - \varrho_l)gh.$$

The prefactor C depends on the surface potentials of the sphere and the planar surface. The minimum of the potential occurs at the position

$$h_{min} = \lambda_D \ln \frac{B}{\lambda_D \frac{4}{3}\pi R^3 (\varrho_p - \varrho_l)g}.$$

To get rid of the experimentally hard to determine parameter C, we use the shape of the potential around the minimum, which does not depend on the value of C:

$$V(h) - V(h_{min}) = \lambda_D \frac{4}{3}\pi R^3 (\varrho_p - \varrho_l)g \left[\exp\left(\frac{h-h_{min}}{\lambda_D}\right) - 1 \right]$$

$$+ \frac{4}{3}\pi R^3 (\varrho_p - \varrho_l)g(h - h_{min}).$$

Chapter 4

4.1. *Debye length.* By inserting $\varepsilon = 24.3$ for ethanol and a concentration of 6.02×10^{22} salt molecules per m^3 into Eq. (4.8), and with $\lambda_D = \varkappa^{-1}$, we get

$$\lambda_D = \sqrt{\frac{24.3 \cdot 8.85 \times 10^{-12}\,\text{A s V}^{-1}\,\text{m}^{-1} \cdot 1.38 \times 10^{-23}\,\text{J K}^{-1} \cdot 298\,\text{K}}{2 \cdot 6.02 \times 10^{22}\,\text{m}^{-3} \cdot (1.60 \times 10^{-19}\,\text{A s})^2}} = 16.9\,\text{nm}.$$

For water, the decay length is larger according to $\lambda_D = 0.304\,\text{nm}/\sqrt{c_0} = 30.4\,\text{nm}$.

4.2. *Potential versus distance at high surface potentials.* With a Debye length of 6.80 nm, we get the following plot:

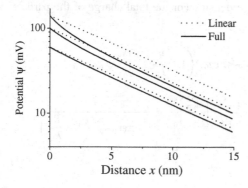

4.3. *Cation concentration at a silicon oxide surface.* The concentration of cations is increased by a factor $e^{e\psi_0/k_\mathrm{B}T} = e^{\frac{1.60\times10^{-19}\,\mathrm{A\,s}\cdot0.07\mathrm{V}}{4.12\times10^{-21}\,\mathrm{J}}} = 15.3$. Despite the low bulk concentration, we expect a concentration of $0.76\,\mathrm{M}$. This corresponds to 4.6×10^{26} counterions per m^3. The average distance is only $\approx1.3\,\mathrm{nm}$.

The H^+ concentration at the surface is given by $[H^+] = 15.3[H^+]_0$, where $[H^+]_0 = 10^{-9}\,\mathrm{M}$ is the bulk concentration of H^+. The local pH at the surface is $\log[H^+] = \log(1.53\times10^{-8}) = 7.8$.

4.4. *Potential around a sphere.* The linearized Poisson--Boltzmann equation (for low potentials) is

$$\frac{1}{r^2}\frac{\mathrm{d}}{\mathrm{d}r}\left(r^2\frac{\mathrm{d}\psi}{\mathrm{d}r}\right) = \varkappa^2\psi.$$

The constant B is zero because $\psi \to 0$ for $r \to \infty$. The boundary condition $\psi(r = R_\mathrm{p}) = \psi_0$ leads to $A = \psi_0 R_\mathrm{p} e^{\varkappa R_\mathrm{p}}$ and we get

$$\psi = \psi_0 e^{-\varkappa(r-R_\mathrm{p})}\frac{R_\mathrm{p}}{r}.$$

To relate the surface potential to the charge of the particle, we proceed as before for planar surfaces (Section 4.2.5):

$$Q = 4\pi\sigma R_\mathrm{p}^2 = -4\pi\int_{R_\mathrm{p}}^{\infty}\varrho_e r^2\,\mathrm{d}r = 4\pi\varepsilon\varepsilon_0\int_{R_\mathrm{p}}^{\infty}\frac{\mathrm{d}}{\mathrm{d}r}\left(r^2\frac{\mathrm{d}\psi}{\mathrm{d}r}\right)\mathrm{d}r$$

$$= 4\pi\varepsilon\varepsilon_0\left[r^2\frac{\mathrm{d}\psi}{\mathrm{d}r}\right]_{R_\mathrm{p}}^{\infty}.$$

With $\mathrm{d}\psi/\mathrm{d}r = -\psi_0 R_\mathrm{p}(\varkappa r + 1)e^{-\varkappa(r-R_\mathrm{p})}/r^2$, we get

$$Q = -4\pi\varepsilon\varepsilon_0[\psi_0 R_\mathrm{p}(\varkappa r + 1)e^{-\varkappa(r-R_\mathrm{p})}]_{R_\mathrm{p}}^{\infty} = 4\pi\varepsilon\varepsilon_0\psi_0 R_\mathrm{p}(\varkappa R_\mathrm{p} + 1).$$

4.5. *Gibbs energy of double layer around a spherical particle.* We can use an equation similar to Eq. (4.41), we only have to use the total charge of the particle instead of the surface charge density. For the total charge of the particle, see Exercise 4.4.

$$G = -\int_0^{\psi_0} Q\,\mathrm{d}\psi'_0 = -4\pi\varepsilon\varepsilon_0 R\left(1 + \frac{R_\mathrm{p}}{\lambda_\mathrm{D}}\right)\cdot\int_0^{\psi_0}\psi'_0\,\mathrm{d}\psi'_0$$

$$= -4\pi\varepsilon\varepsilon_0 R_\mathrm{p}\left(1 + \frac{R_\mathrm{p}}{\lambda_\mathrm{D}}\right)\cdot\left[\frac{\psi'^2_0}{2}\right]_0^{\psi_0} = -\frac{Q^2}{8\pi\varepsilon\varepsilon_0 R_\mathrm{p}\cdot\left(1 + \frac{R_\mathrm{p}}{\lambda_\mathrm{D}}\right)}.$$

4.6. *Dispersion in aqueous electrolyte.* According to Derjaguin's approximation (Eq. (2.71)), the force between two particles is

$$F = \pi R_{\mathrm{p}} V^{\mathrm{A}}(D) = 64\pi R_{\mathrm{p}} c_0 k_{\mathrm{B}} T \lambda_{\mathrm{D}} \cdot \left(\frac{e^{e\Psi_0/2k_{\mathrm{B}}T}-1}{e^{e\Psi_0/2k_{\mathrm{B}}T}+1}\right)^2 \cdot e^{-\frac{D}{\lambda_{\mathrm{D}}}} - \frac{A_{\mathrm{H}} R_{\mathrm{p}}}{12 D^2}.$$

The energy is

$$V(D) = \int_D^\infty F \cdot dD = 64\pi R_{\mathrm{p}} c_0 k_{\mathrm{B}} T \lambda_{\mathrm{D}} \left(\frac{e^{\frac{e\Psi_0}{2k_{\mathrm{B}}T}}-1}{e^{\frac{e\Psi_0}{2k_{\mathrm{B}}T}}+1}\right)^2 \cdot \int_D^\infty e^{-\frac{d'}{\lambda_{\mathrm{D}}}}\, dD - \frac{A_{\mathrm{H}} R_{\mathrm{p}}}{12}\int_D^\infty \frac{1}{d'^2}\, dD$$

$$= 64\pi R_{\mathrm{p}} c_0 k_{\mathrm{B}} T \lambda_{\mathrm{D}}^2 \left(\frac{e^{\frac{e\Psi_0}{2k_{\mathrm{B}}T}}-1}{e^{\frac{e\Psi_0}{2k_{\mathrm{B}}T}}+1}\right)^2 \cdot e^{-\frac{D}{\lambda_{\mathrm{D}}}} - \frac{A_{\mathrm{H}} R_{\mathrm{p}}}{12 D}.$$

When plotting the energy versus distance for different salt concentrations, we find that, at 0.28 M, we have an energy barrier of roughly $10 k_{\mathrm{B}} T$.

Chapter 5

5.1. *Capillary force between sphere and plane in contact.* Starting with Eq. (5.14), we use Pythagoras theorem:

$$(R_{\mathrm{P}} + r)^2 = (l+r)^2 + (R_{\mathrm{P}}-r)^2 \Rightarrow 4 r R_{\mathrm{P}} = l^2 + 2 l r + r^2 \approx l^2 + l r.$$

Inserting directly leads to $F = 4\pi\gamma R_{\mathrm{P}}$.

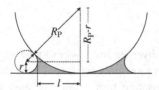

5.2. *Capillary force between two spheres with asperity.* The asperity is supposed to create a gap of width H between the two sphere surfaces. The gap prevents the liquid from condensing unless a certain minimal vapor pressure is reached. The critical pressure is reached when $2r \cos\Theta = H$. Inserting in Kelvin's equation $r \approx -\lambda_{\mathrm{K}}/\ln(P/P_0) \Rightarrow P/P_0 \approx \exp(-2\lambda_{\mathrm{K}}\cos\Theta/H)$.

To calculate the capillary force, the parameter β is varied. For each β the radii r and l are calculated with the equations in the first row of Table 5.2, setting $D = H$. With Eq. (5.19), the capillary force is calculated. Using Eq. (5.20) the relative vapor pressure is obtained. Finally, the force is plotted versus the relative vapor pressure (Figure 12.2).

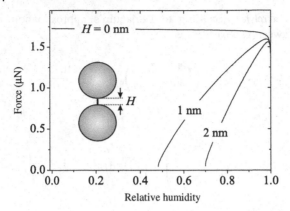

Figure 12.2 Capillary force versus the humidity of two similar spheres ($R_1 = 5\,\mu$m, contact angle $\Theta = 40°$ with respect to water). The spheres are kept at an effective distance of $H = 0$, 1, and 2 nm by an asperity.

5.3. *Stiction of a head on a magnetic storage disk.* Setting the disjoining pressure (Eq. (5.40)) and the Laplace pressure $\approx \gamma/r$ equal leads to

$$\frac{A_H}{6\pi h^3} = \frac{\gamma}{r} \Rightarrow r = \frac{6\pi\gamma h^3}{A_H} = 23.6\,\text{nm}.$$

With Eq. (5.26), the volume of the meniscus is

$$V = \pi R_P r^2 = \pi \cdot 1000\,\text{nm} \cdot (24\,\text{nm})^2 = 1.75 \times 10^6\,\text{nm}^3.$$

For a film of 1.5 nm thickness, the radius is $r = 79.5\,$nm and $V = 1.58 \times 10^9\,\text{nm}^3$. In both cases, the force is $F \approx 4\pi R_P \gamma = 310\,$nN.

5.4. *Characteristic condensation time.* τ is proportional to

$$\tau \propto -\frac{1}{P/P_0 \ln(P/P_0)}.$$

For $P/P_0 \to 0$ and $P/P_0 \to 1$, this expression diverges and $\tau \to \infty$. At low vapor pressure, condensation is slow because there are only few vapor molecules around. At high vapor pressure, the condensation time is long because the equilibrium volume of the meniscus is large. τ has a shallow minimum at $\ln(P/P_0) = -1$ or $P/P_0 = 0.368$.

5.5. *Liquid mixtures.* At close distances, hexane will condense and form a hexane meniscus in the water phase between the two particles. We apply Eq. (5.31). With $R^* = 2R_p$

$$F = 4\pi \cdot 0.031\,\text{Nm}^{-1} \cdot 5 \times 10^{-6}\,\text{m} \cdot \left(2\cos 85° - \frac{D}{r}\right)$$

$$= 1.95 \times 10^{-6}\,\text{N} \cdot \left(0.174 - \frac{D}{r}\right).$$

With $V_m^B = \frac{0.09213}{866} \, \text{m}^3 \, \text{mol}^{-1} = 1.06 \times 10^{-4} \, \text{m}^3 \, \text{mol}^{-1}$ and Eq. (5.42):

$$\lambda_K = \frac{\gamma_{AB} V_m^B}{RT} = \frac{0.031 \, \text{nm}^{-1} \cdot 1.06 \times 10^{-4} \, \text{m}^3 \, \text{mol}^{-1}}{2480 \, \text{J} \, \text{mol}^{-1}} = 1.33 \, \text{nm},$$

we obtain $r = -\lambda_K / \ln(c_B / c_B^0) = 8.96 \, \text{nm}$ for $c_B = 5.0 \, \text{mM}$ and $r = 37.9 \, \text{nm}$ for $c_B = 5.6 \, \text{mM}$. At c_B close to c_B^0, the range of the capillary force depends on the concentration.

Chapter 6

6.1. *Falling sphere at close distance.* In steady state and neglecting the inertia of the sphere $m \, d^2 D / dt^2$, where m is the mass of the sphere, the gravitational force is balanced by the viscous drag:

$$\frac{4}{3} \pi R_p^3 g \Delta\varrho = \frac{6\pi\eta R_p^2}{D} \frac{dD}{dt} \Rightarrow \frac{1}{D} \frac{dD}{dt} = \frac{2 R_p g \Delta\varrho}{9\eta},$$

$$\int_{t=0}^{\Delta t} \frac{1}{D} \frac{dD}{dt} dt = \int_{D_2}^{D_1} \frac{dD}{D} = \ln\frac{D_1}{D_2} = \frac{2 R_p g \Delta\varrho \Delta t}{9\eta} \Rightarrow \Delta t = \frac{9\eta}{2 R_p g \Delta\varrho} \ln\frac{D_1}{D_2}.$$

This result was reported in Ref. [631].

6.2. *Falling glass sphere.* Since D is not much smaller than R_p, we cannot apply Eq. (6.37) but have to use Eq. (6.41) for the hydrodynamic force. With the same arguments as used in the previous exercise, we obtain

$$\left(1 + \frac{R_p}{D}\right) \frac{dD}{dt} = \frac{2 R_p^2 g \Delta\varrho}{9\eta}, \tag{12.2}$$

$$\int_{t=0}^{\tau} \left(1 + \frac{R_p}{D}\right) \frac{dD}{dt} dt = D_1 - D_2 + R_p \ln\frac{D_1}{D_2} = \frac{2 R_p^2 g \Delta\varrho \Delta t}{9\eta} \Rightarrow$$

$$\Delta t = \frac{9\eta}{2 R_p g \Delta\varrho} \left(\frac{D_1 - D_2}{R_p} + \ln\frac{D_1}{D_2}\right). \tag{12.3}$$

With

$$\frac{9\eta}{2 R_p g \Delta\varrho} = \frac{9 \cdot 2.13 \times 10^{-3} \, \text{Pa s}}{2 \cdot 15 \times 10^{-6} \, \text{m} \cdot 9.81 \, \text{m s}^{-2} \cdot 1740 \, \text{kg m}^{-3}} = 0.0374 \, \text{s},$$

Δt can be calculated versus D, where D is varied between 30000 and 0.2 nm. Then, Eq. (12.2) can be applied to calculate dD/dt (Figure 12.3).

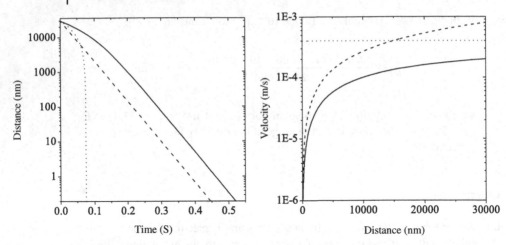

Figure 12.3 Distance versus time (left) and velocity versus distance (right) for a sphere falling toward a planar surface as calculated with Eqs. (12.2) and (12.3) (continuous lines) and the results of Exercise 6.1 (dashed lines). For comparison the result for a free falling sphere is also shown (dotted line).

6.3. *Shear rate.* Inserting Eq. (6.19) into Eq. (6.16):

$$\eta \frac{\partial v_x}{\partial z} = \left(z - \frac{h}{2}\right)\frac{\partial P}{\partial x}. \tag{12.4}$$

Remember that $\partial P/\partial x = dP/dr \cdot x/r$ (Eq. (6.28)) and apply Eq. (6.33):

$$\frac{\partial v_x}{\partial z} = \frac{6 v_\perp x}{h^3}\left(z - \frac{h}{2}\right).$$

Shear is maximal at the surface of the plane where $z = 0$ or at the surface of the sphere $(z = h)$ (Figure 12.4):

$$\frac{\partial v_x}{\partial z} = \frac{3 v_\perp x}{h^2}. \tag{12.5}$$

To find the maximal shear rate, we differentiate with respect to x:

$$\frac{\partial}{\partial x}\left(\frac{\partial v_x}{\partial z}\right) = \frac{3 v_\perp}{h^2} - \frac{6 v_\perp x}{h^3}\frac{\partial h}{\partial x} = \frac{3 v_\perp}{h^2} - \frac{6 v_\perp x^2}{R_p h^3}. \tag{12.6}$$

To find the maximum, we set this expression to zero. With Eq. (6.26), this leads to $x_{\max} = \sqrt{2DR_p/3}$. At $X = X_{\max}$ we have $h = 4D/3$. Inserting this in Eq. (12.5), we get

$$\frac{\partial v_x}{\partial z}(x_{\max}) = \frac{9}{8}\sqrt{\frac{3R_p}{2D^3}}\, v_\perp. \tag{12.7}$$

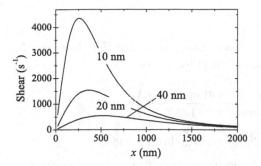

Figure 12.4 Shear rate $\partial v_x / \partial z$ at $z = 0$ along the x-axis. The shear rate increases with decreasing distance. It shows a maximum, which shifts to larger radial distances with increasing distance D.

6.4. *Diffusion coefficient of a sphere close to a wall.* With Eq. (6.41) we get for the normal direction (Figure 12.5)

$$D_d = \frac{k_B T}{6\pi\eta R_p} \frac{D}{D + R_p}. \tag{12.8}$$

Equation (6.50) with $\tilde{D} = (R_p + D)/R_p$ leads to diffusion coefficient

$$D_d = \frac{k_B T}{6\pi\eta R_p} \left(1 - \frac{9}{16\,\tilde{D}} + \frac{1}{8\,\tilde{D}^3} - \frac{45}{256\,\tilde{D}^4} - \frac{1}{16\,\tilde{D}^5}\right) \tag{12.9}$$

parallel to the plane. In Eq. (12.9), we assumed that the sphere does not rotate and $\omega = 0$. Practically in most cases, the sphere is free to rotate and the torque is zero. It turns out that for $D > R_p$, the difference is below 1% (Figure 12.5).

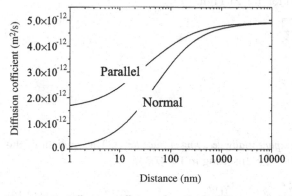

Figure 12.5 Diffusion coefficient of a sphere of 50 nm radius at a planar wall.

Chapter 7

7.1. *Conversion of height into relative vapor pressure for wetting films.* Inserting the parameters for He in the barometric distribution law:

$$\frac{P}{P_0} = \exp\left(-\frac{0.004\ \text{kg mol}^{-1} \cdot 9.81\ \text{m s}^{-2} \cdot 1\ \text{m}}{8.31\ \text{J mol}^{-1}\ \text{K}^{-1} \cdot 1.35\ \text{K}}\right) = 0.997,$$

$$H = -(RT/gM_W) \cdot \ln(P/P_0) = 286\ \text{m} \cdot \ln 0.5 = 198\ \text{m}.$$

For water at 293 K, $P/P_0 = 0.99993$ at 1 m. $P/P_0 = 0.5$ is reached at $H = 9560\ \text{m}$. Thus, a variation in height to measure film thickness is practical only at low temperature.

7.2. *Thin adsorbed hexane film.* We can apply Eq. (7.21) although the film has no continuous liquid connection with the pool. Equilibrium is established via the vapor phase. The thickness is

$$h = \left(-\frac{A_H}{6\pi\varrho g z}\right)^{1/3} = \left(\frac{10^{-20}\ \text{J}}{6\pi \cdot 655\ \text{kg m}^{-3} \cdot 9.81\ \text{m s}^{-2} \cdot 1\ \text{m}}\right)^{1/3} = 4.48\ \text{nm}.$$

7.3. *Thin film between bubble and plate.* Since the bubble is much smaller than the capillary length, it is spherical except in the region of the thin film. Pressure inside: $\Delta P = 2\gamma_L/r_b = 720\ \text{Pa}$. Buoyancy: $4\pi r_b^3 \varrho g/3 = 3.28 \times 10^{-7}\ \text{N}$. Area of thin film ($A$) is given by $A\Delta P = 3.28 \times 10^{-7}\ \text{N} \Rightarrow A = 4.56 \times 10^{-10}\ \text{m}^2$. It has a radius of $a = 12.0\ \mu\text{m}$. In the contact area, the internal pressure is equal to the disjoining pressure. Considering electrostatic (Eq. 4.57) and van der Waals forces (Eq. 2.31) and with $\lambda_D = 9.61\ \text{nm}$, we get

$$\Delta P = \Pi = \frac{2\epsilon\epsilon_0}{\lambda_D^2}\Psi_0^2 \cdot e^{-h/\lambda_D} - \frac{A_H}{6\pi h^3},$$

$$720\ \text{Pa} = 6013\ \text{Pa} \cdot e^{-h/9.61\ \text{nm}} + \frac{5.31 \times 10^{-22}\ \text{J}}{h^3}.$$

This equation is solved for $h = 21.1\ \text{nm}$.

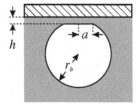

Bubble underneath a plate.

7.4. *Spinodal dewetting for exponentially decaying surface forces.* The gradient of the force is $d\Pi/dh = C/\lambda \cdot e^{-h/\lambda}$. Inserting in Eq. (7.33):

$$t_s \approx 0.1 \cdot \frac{12\gamma_L\eta\lambda^2}{h_0^3 C^2} \cdot e^{2h_0/\lambda} \approx \frac{\gamma_L\eta\lambda^2}{h_0^3 C^2} \cdot e^{2h_0/\lambda}. \tag{12.10}$$

The factor 0.1 is to take nonlinear effects into account. Inserting leads to

$$t_s = \frac{0.072 \, \text{Nm}^{-1} \cdot 0.89 \times 10^{-3} \, \text{Pa s} \cdot (12 \times 10^{-9} \, \text{m})^2}{(5 \times 10^{-9} \, \text{m})^3 \cdot (400 \, \text{Pa})^2} \cdot e^{2 \cdot 5 \, \text{nm}/12 \, \text{nm}} = 1.1 \, \text{s}.$$

In practice, water films on hydrophobic surface rupture much faster. This indicates that nucleation due to tiny bubbles is the dominating process.

Chapter 8

8.1. The reduced Young's modulus for this combination of materials is given by

$$E^* = \left(\frac{1-\nu_1^2}{E_1} + \frac{1-\nu_2^2}{E_2}\right)^{-1} = \left(\frac{0.9271}{200 \, \text{GPa}} + \frac{0.75}{2 \, \text{MPa}}\right)^{-1} = 2.67 \, \text{MPa}.$$

The adhesion force can be calculated as

$$F_{\text{adh}} = \sqrt{8\pi a^3 E^* w_A} = \sqrt{8\pi(0.05 \, \text{m})^3 \cdot 2.67 \, \text{MPa} \cdot 0.05 \, \text{Jm}^{-2}} = 20.5 \, \text{N}.$$

The critical displacement at which detachment occurs is given by Eq. (8.25) when replacing the load F_L by the adhesion force F_{adh}

$$\delta_c = \frac{F_{\text{adh}}}{2aE^*} = \frac{20.5 \, \text{N}}{2 \cdot 0.05 \cdot 2.67 \, \text{MPa}} = 76.8 \, \mu\text{m}.$$

8.2. The reduced Young's modulus for this combination of materials is given by

$$E^* = \left(\frac{2(1-\nu_1^2)}{E_1}\right)^{-1} = \left(\frac{2 \cdot 0.75}{2 \, \text{MPa}}\right)^{-1} = 1.33 \, \text{MPa}.$$

The adhesion force is calculated by the JKR theory:

$$F_{\text{adh}} = -\frac{3}{2}\pi \cdot w_{\text{adh}} R = -\frac{3}{2}\pi \cdot 0.022 \, \text{Jm}^{-2} \cdot 0.01 \, \text{m} = 1.04 \, \text{mN}.$$

At pull-off, the contact radius is given as

$$a_{\min} = \left(\frac{9 \, \pi \cdot w_{\text{adh}} R_P^2}{8 \quad E^*}\right)^{1/3} = \left(\frac{9 \, \pi \cdot 0.022 \, \text{Jm}^{-2} \cdot (0.01 \, \text{m})^2}{8 \quad 1.33 \, \text{MPa}}\right)^{1/3} = 180 \, \mu\text{m}$$

and the height of the neck reaches

$$\delta_{\min} = \left(\frac{3\pi^2 w_{\text{adh}}^2 R_P}{64 E^{*2}}\right) = \left(\frac{3\pi^2 \cdot 0022 \, \text{J/m}^2 \cdot 0.01 \, \text{m}}{64 \cdot (1.33 \cdot 10^6 \, \text{N/m}^2)^2}\right) = 1.08 \, \mu\text{m}$$

8.3. *Onset of plastic deformation.* The reduced Young's modulus is given by:

$$E^* = \left(\frac{2 \cdot 0.9271}{200 \, \text{GPa}}\right)^{-1} = 108 \, \text{GPa}.$$

The maximum stress in the Hertz contact is given by

$$\sigma_{\max} = \frac{3}{2} \frac{F_L}{\pi \cdot a^2} = \sigma_c,$$

$$F_L = \frac{2}{3}\sigma_c \pi \cdot a^2 = \frac{2}{3}\sigma_c \pi \cdot \left(\frac{3 F_L R_P}{4 E^*}\right)^{2/3}.$$

Solving this equation for F_L leads to

$$F_L = \left(\frac{2}{3}\sigma_c\pi\right)^3 \left(\frac{3R}{4E^*}\right)^2.$$

The load is equal to the gravitational force

$$F_L = \frac{4}{3}\pi R^3 \varrho g$$

and we finally obtain

$$R = \left(\frac{1}{2}\sigma_c\right)^3 \left(\frac{\pi}{E^*}\right)^2 \frac{1}{\varrho g}$$

$$= \left(\frac{1}{2}250\,\text{MPa}\right)^3 \left(\frac{\pi}{108\,\text{GPa}}\right)^2 \frac{1}{7800\,\text{kg m}^{-3}\cdot 9.81\,\text{m s}^{-2}} = 2.16\,\text{cm}$$

Chapter 9

9.1. *Sliding on an inclined plane.* The gravitational force F_G acting on the block can be deconstructed into a force F_N normal to the surface and a force F_P acting parallel to the surface. Looking at the absolute values of the forces, trigonometry tells us that $F_P = \tan\alpha F_N$. From Amontons' law, we get $F_R = \mu_s F_N$. When the body starts sliding, the absolute value of the force parallel to the plane must be equal to the friction force. From this, we get

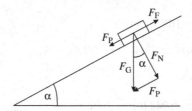

$$\mu_s = \tan\alpha \Rightarrow \alpha = \arctan\mu = \arctan 0.5 = 26.6°.$$

When the block is sliding, F_P stays the same, but the friction force is reduced to $F_R = \mu_K F_N$. The net force acting parallel to the surface is therefore:

$$F'_P = F_P - F'_R = \mu_s F_N - \mu_k F_N = 0.1 F_N = 0.1 \cos\alpha \cdot F_G = 0.1\, mg \cos\alpha.$$

The acceleration is given by $a = F/m$ and the speed after a distance L is equal to

$$v = \sqrt{2aL} = \sqrt{2\frac{F}{m}L} = \sqrt{2\frac{0.1\cos\alpha \cdot mg}{m}L} = \sqrt{2\cdot 0.1\cos\alpha \cdot gL}$$

$$= \sqrt{0.2\cos 26.6 \cdot 9.81\,\text{ms}^{-2}\cdot 1\,\text{m}} = 1.32\,\text{ms}^{-1}.$$

9.2. *True contact area.* The contact surface will deform, until the pressure is equal to the yield stress σ_c. If F_G is the weight force and A_{real} the true contact area, ϱ the density, and V the volume of the cube, we can write

$$\sigma_c = \frac{F_G}{A} = \frac{\varrho V g}{A}$$

$$A_{real} = \frac{\varrho V g}{\sigma_c} = \frac{7.80\,\text{g/cm}^3 \cdot (10\,\text{cm})^3 \cdot 9.81\,\text{ms}^{-2}}{0.25 \cdot 10^9\,\text{N/m}^2} = 3.06 \cdot 10^{-7} = 3.06 \cdot 10^5\,\mu\text{m}^2$$

This corresponds to 30600 microcontacts of $10\,\mu\text{m}^2$ each.

9.3. *Hydrodynamic lubrication.* The sliding speed is $v = 2\pi \cdot 0.005\,\text{m} \cdot 80\,\text{Hz} = 1.26\,\text{ms}^{-1}$. The friction force is $F_F = \frac{0.1\,\text{m} \cdot 2\pi \cdot 0.005\,\text{m} \cdot 1.26\,\text{ms}^{-1} \cdot 2.49 \times 10^{-3}\,\text{Pa s}}{3 \times 10^{-6}\,\text{m}} = 3.29\,\text{N}$. The torque is $M = 3.29\,\text{N} \cdot 0.005\,\text{m} = 0.0164\,\text{Nm}$.

Chapter 10

10.1. *Superposition principle.* Superimposing the densities of two similar parallel walls at a given spacing x leads to a density in the gap of

$$\varrho_n(x,\xi) = \frac{1}{\varrho_b}\left[\Delta\varrho\, e^{-\frac{x-\xi}{\lambda}}\cos\left(2\pi\frac{\xi}{d_0}\right) + \varrho_b\right]\left[\Delta\varrho\, e^{-\frac{x-\xi}{\lambda}}\cos\left(2\pi\frac{x-\xi}{d_0}\right) + \varrho_b\right].$$

$$(12.11)$$

At the wall, that is for $\xi = 0$, the density is

$$\varrho_n(x) = \frac{1}{\varrho_b}[\Delta\varrho + \varrho_b]\left[\Delta\varrho\, e^{-\frac{x}{\lambda}}\cos\left(2\pi\frac{x}{d_0}\right) + \varrho_b\right].$$

$$(12.12)$$

The same result is obtained if we calculate the density at the other wall at $\xi = x$. Inserting this expression into the contact theorem (Eq. (10.1)) and considering that $\varrho_0(\infty) = \Delta\varrho + \varrho_b$ leads to Eq. (10.6).

10.2. *Presmectic force between sphere and plate.* Force versus distance is plotted below.

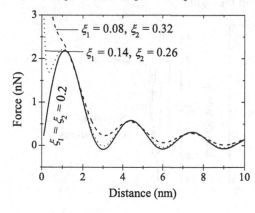

10.3. *Saturation concentration.* From ΔG_s^0 we can calculate the mole fraction of solute in water if the vapor pressure of the solute was 1 atm. It is $x = \exp(-\Delta G_s^0 / RT)$. Since in equilibrium the vapor pressure is lower than 1 atm (otherwise the substance would not be in the liquid phase), we still have to multiply with $P_0^V / 1$ atm. To obtain the respective concentration we multiply the mole fraction with 55.35 mol l^{-1}, the concentration of pure water at 25 °C. Thus, for hexane we get

$$c_s = 55.35 \text{ mol/l} \, \frac{0.199 \text{ atm}}{1 \text{ atm}} \cdot e^{-\Delta G_s^0 / k_B T} = 0.11 \text{ mM}$$

We obtain $c_s = 0.028$ mM for heptane, $c_s = 22.5$ mM for benzene, and $c_s = 5.7$ mM for toluene. A comparison with Table 10.2 shows that the values agree with those reported for the process of pure liquid solute → ideal solution in water, except for rounding errors.

10.4. *Force between DLPC bilayers.* With $f = -dV^A/dx$ and Eqs. (10.20) and (10.22), we get

$$f(x) = -\frac{A_H}{6\pi} \left[\frac{1}{x^3} + \frac{1}{(x + 2D_0)^3} - \frac{2}{(x + D_0)^3} \right], \tag{12.13}$$

$$f(x) = -2C_{fl} \cdot \frac{(k_B T)^2}{k_c x^3} \approx -0.28 \cdot \frac{(k_B T)^2}{k_c x^3}. \tag{12.14}$$

The plot shows that at short distance the hydration repulsion dominates. At distances larger than ≈ 2 nm, van der Waals and undulation forces take over. They are only slightly different in their distance dependency, both being dominated by the x^{-3} law. Only above ≈ 10 nm the vdW force decays steeper ($f \propto x^{-5}$) so that the undulation starts to dominate again. The repulsive energies involved are, however, quite small. At the repulsive maximum, an area of $(170 \text{ nm})^2$ is required to reach an energy of $k_B T$. Please note that the van der Waals force is for convenience also plotted positive although it is attractive. Overall, a minimum in the interaction energy occurs so that bilayers are "bound" [1291] and in the lamellar phase show a constant mean distance over a wide concentration range [1235]. To illustrate this, the sum of all three energy contributions is plotted on the right.

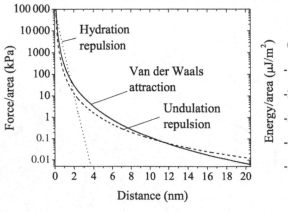

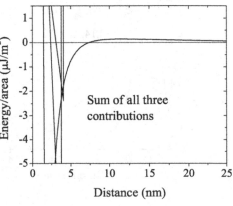

Chapter 11

11.1. *Steric force between bare surface and polymer brush.* The force per chain for a bare surface interacting with a surface coated with a polymer brush is

$$V = \int_{L_0}^{x} F dL = k_B T \left[\frac{3L^2}{2Nl_s^2} + \frac{vN^2\Gamma}{2L} \right]_{L_0}^{x}$$

$$= \frac{k_B TN}{2} \left(\frac{6v^2\Gamma^2}{l_s^2} \right)^{1/3} \left[\frac{1}{2} \left(\frac{x}{L_0} \right)^2 + \frac{L_0}{x} - \frac{3}{2} \right].$$

The interaction energy per unit area is $V^A = \Gamma V$:

$$V^A(x) = \frac{k_B TN\Gamma}{2} \left(\frac{6v^2\Gamma^2}{l_s^2} \right)^{1/3} \left[\frac{1}{2} \left(\frac{x}{L_0} \right)^2 + \frac{L_0}{x} - \frac{3}{2} \right]$$

$$= \frac{k_B TL_0\Gamma}{2} \left(\frac{9v\Gamma}{2l_s^4} \right)^{1/3} \left[\left(\frac{x}{L_0} \right)^2 + \frac{2L_0}{x} - 3 \right].$$

The asymmetric case is obtained by replacing $x/2L_0$ by x/L_0 and dividing V^A by 2 [1361]. With the given parameters $L_0 = 15.8$ nm. The force between a sphere and a planar surface is $F = 2\pi R_p V_A$ using Derjaguin's approximation. For the bare sphere and the polymer brush, we have

$$F(D) = \pi k_B TL_0 R_p \Gamma \left(\frac{9v\Gamma}{2l_s^4} \right)^{1/3} \left[\left(\frac{D}{L_0} \right)^2 + \frac{2L_0}{D} - 3 \right].$$

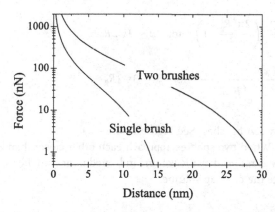

11.2. *Steric repulsion between polymer brushes in an ideal solvent.* First, we consider a single grafted chain. If we multiply the osmotic pressure with the cross-sectional area of a polymer in the brush, b^2, we get the osmotic component of the force: $-Nk_B T/L$. The minus sign indicates that it is directed opposite to the elastic restoring force and tends to expand the brush. The total force per chain is

$$F = \frac{3k_B TL}{Nl_s^2} - \frac{k_B TN}{L}. \tag{12.15}$$

In equilibrium and in the absence of external forces, the net force must be zero. This leads to a brush thickness

$$L_0 = \frac{Nl_s}{\sqrt{3}}.$$ (12.16)

The brush thickness increases linearly with the number of segments and thus the molar mass of the polymer. Per polymer chain the interaction energy is given by

$$V = \int_{L_0}^{x/2} F dL = k_B T \left[\frac{3}{2Nl_s^2} \left(\frac{x^2}{4} - L_0^2 \right) - N \ln\left(\frac{x}{2L_0} \right) \right].$$ (12.17)

With $y = x/(2L_0)$ and inserting the brush thickness, we can simplify this expression:

$$V = k_B T N \left[\frac{1}{2}(y^2 - 1) - \ln y \right].$$ (12.18)

If we bring two surfaces with polymer brushes together to a distance x, the interaction energy per unit area is

$$V^A = k_B T N \Gamma (y^2 - 1 - 2\ln y) = k_B T L_0 \Gamma \frac{\sqrt{3}}{l_s}(y^2 - 1 - 2\ln y).$$ (12.19)

Eq. (12.19) gives the force for $x \le 2L_0$. For larger distances, the force is zero.

11.3. *Steric repulsion between mushrooms.* The force per unit area is derived from $f = -dV^A/dx$:

$$f(x) = \frac{k_B T \Gamma}{x} \cdot \left(\frac{2\pi^2 R_g^2}{x^2} - 1 \right) \quad \text{for} \quad x \le 3\sqrt{2}R_g,$$

$$f(x) = \frac{k_B T \Gamma x}{R_g^2} \cdot \frac{1}{e^{\left(\frac{x}{2R_g}\right)^2} - 1} \quad \text{for} \quad x > 3\sqrt{2}R_g.$$

11.4. Steric repulsion between brushes. See Figure 12.6.

11.5. *Depletion attraction.* When two spheres approach each other closer than $2R_0$, the excluded volume is reduced by the volume indicated in gray in the figure. This volume is twice the end cap volume:

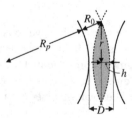

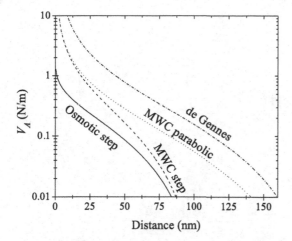

Figure 12.6 Interaction energy per unit area calculated with $N = 60$, $l_s = 1.5$ nm, $v = l_s^3$, and $\Gamma = 5 \times 10^{17}$ m^2 for the simple osmotic brush (Eq. (12.19)), the Milner, Witten, Cates model with a step (Eq. (11.18)) and parabolic segment profile (Eq. (11.19)), and the model of de Gennes (Eq. (11.21)).

$$V = 2 \cdot \frac{\pi}{6} h (3h^2 + r^2).$$

With $h = R_0 - x/2$, $r^2 = (R_p + R_0)^2 - (R_p + x/2)^2$, and the osmotic pressure $ck_B T$, we get

$$W = ck_B T\pi \cdot \left(2R_p R_0^2 + \frac{4}{3}R_0^3 - 2R_p R_0 x - R_0^2 x + \frac{R_p x^2}{2} + \frac{x^3}{12} \right).$$

For $R_0 \ll R_p$ and $x \ll R_p$, the terms R_0^3, $R_0^2 x$, and $x^3/12$ can be neglected and we get Eq. (11.42).

Fundamental Constants

Source: 2006 CODATA recommended values, http://physics.nist.gov/cuu/

Atomic mass unit	u	$1.66054 \times 10^{-27}\,\text{kg}$
Avogadro constant	N_A	$6.02214 \times 10^{23}\,\text{mol}^{-1}$
Boltzmann constant	k_B	$1.38065 \times 10^{-23}\,\text{J}\,\text{K}^{-1}$
Electron mass	m_e	$9.10938 \times 10^{-31}\,\text{kg}$
Elementary charge	e	$1.60218 \times 10^{-19}\,\text{C}$
Faraday constant	$F_A = eN_A$	$96485.3\,\text{C}\,\text{mol}^{-1}$
Gas constant	$R = k_B N_A$	$8.31447\,\text{J}\,\text{K}^{-1}\,\text{mol}^{-1}$
Planck constant	h	$6.62607 \times 10^{-34}\,\text{J}\,\text{s}$
Speed of light in vacuum	c	$2.99792 \times 10^{8}\,\text{m}\,\text{s}^{-1}$
Standard acceleration of free fall	g	$9.80665\,\text{m}\,\text{s}^{-2}$
Vacuum permittivity	ε_0	$8.85419 \times 10^{-12}\,\text{A}\,\text{s}\,\text{V}^{-1}\,\text{m}^{-1}$

Conversion Factors

$1\,\text{eV} = 1.60218 \times 10^{-19}\,\text{J}$
$1\,\text{dyne} = 10^{-5}\,\text{N}$
$1\,\text{erg} = 10^{-7}\,\text{J}$
$1\,\text{kcal} = 4.184\,\text{kJ}$
$1\,\text{Torr} = 133.322\,\text{Pa} = 1.333\,\text{mbar}$
$1\,\text{bar} = 10^{5}\,\text{Pa}$
$1\,\text{atm} = 101325\,\text{Pa}$
$1\,\text{poise (P)} = 0.1\,\text{Pa s}$
$1\,\text{Debye (D)} = 3.336 \times 10^{-30}\,\text{C m}$
$1\,\text{V} = 1\,\text{J A}^{-1}\,\text{s}^{-1}$
$0^\circ\,\text{C} = 273.15\,\text{K}$
$k_\text{B} T/e = 25.69\,\text{mV at } 25\,^\circ\text{C}$

References

1 Masuda, H. and Fukuda, K. (1995) *Science*, **268**, 1466.

2 Steinhart, M., Wehrspohn, R.B., Gösele, U., and Wendorff, J.H. (2004) *Angew. Chem. Int. Ed.*, **43**, 1334.

3 Parsegian, V.A. (2006) *Van der Waals Forces*, Cambridge University Press, New York.

4 Langbein, D.W. (1974) *Theory of van der Waals Attraction*, vol. 72, Springer Tracts in Modern Physics, Springer, Berlin.

5 Mahanty, J. and Ninham, B.W. (1976) *Dispersion Forces*, Academic Press, New York.

6 Keesom, W.H. (1921) *Phys. Z.*, **22** (129), 364.

7 Debye, P. (1920) *Phys. Z.*, **21**, 178 (1921), **22**, 302.

8 London, F. (1930) *Z. f Phys.*, **63**, 245.

9 Hamaker, H.C. (1937) *Physica*, **4**, 1058.

10 Bradley, R.S. (1932) *Philos. Mag.*, **13**, 853.

11 de Boer, J.H. (1936) *Trans. Faraday Soc.*, **32**, 0010.

12 Hunter, R.J. (2001) *Foundations of Colloid Science*, 2nd edn, Oxford University Press, Oxford.

13 Axilrod, B.M. and Teller, E. (1943) *J. Chem. Phys.*, **11**, 299.

14 Farina, C., Santos, F.C., and Tort, A.C. (1999) *Am. J. Phys.*, **67**, 344.

15 Autumn, K. *et al.* (2002) *Proc. Natl. Acad. Sci. USA*, **99**, 12252.

16 Arzt, E., Gorb, S., and Spolenak, R. (2003) *Proc. Natl. Acad. Sci. USA*, **100**, 10603.

17 Casimir, H.B.G. and Polder, D. (1948) *Phys. Rev.*, **73**, 360.

18 Casimir, H.G.B. (1948) *Proc. R. Netherlands Acad. Arts Sci.*, **51**, 793.

19 Lifshitz, E.M. (1956) *Sov. Phys. JETP (Engl. Transl.)*, **2**, 73.

20 Dzyaloshinskii, I.E., Lifshitz, E.M., and Pitaevskii, L.P. (1961) *Adv. Phys.*, **10**, 165.

21 Overbeek, J.T. (1966) *Discuss. Faraday Soc.*, 7.

22 Van Kampen, G., Nijboer, B.R.A., and Schram, K. (1968) *Phys. Lett. A*, A **26**, 307.

23 Ninham, B.W. and Parsegian, V.A. (1970) *Biophys. J.*, **10**, 646.

24 Bortz, M.L. and French, R.H. (1989) *Appl. Phys. Lett.*, **55**, 1955.

25 Egerton, R.F. (2009) *Rep. Prog. Phys.*, **72**, 016502.

26 Roth, C.M. and Lenhoff, A.M. (1996) *J. Colloid Interface Sci.*, **179**, 637.

27 Nguyen, A.V. (2000) *J. Colloid Interface Sci.*, **229**, 648.

28 Dagastine, R.R., Prieve, D.C., and White, L.R. (2000) *J. Colloid Interface Sci.*, **231**, 351.

29 Fernandez-Varea, J.M. and Garcia-Molina, R. (2000) *J. Colloid Interface Sci.*, **231**, 394.

30 Tabor, D. and Winterton, R.H. (1969) *Proc. R. Soc. Lond. A*, **312**, 435.

31 Israelachvili, J.N. (1992) *Intermolecular and Surface Forces*, 2nd edn, Academic Press, Amsterdam.

32 French, R.H., Cannon, R.M., Denoyer, L.K., and Chiang, Y.M. (1995) *Solid State Ionics*, **75**, 13.

33 Bortz, M.L. and French, R.H. (1989) *Appl. Spectrosc.*, **43**, 1498.

34 French, R.H. (2000) *J. Am. Ceram. Soc.*, **83**, 2117.

Surface and Interfacial Forces. Hans-Jürgen Butt and Michael Kappl
Copyright © 2010 Wiley-VCH Verlag GmbH & Co. KGaA
ISBN: 978-3-527-40849-8

35 van Benthem, K., Tan, G.L., DeNoyer, L.K., French, R.H., and Rühle, M. (2004) *Phys. Rev. Lett.*, **93**, 227201.

36 Ackler, H.D., French, R.H., and Chiang, Y.M. (1996) *J. Colloid Interface Sci.*, **179**, 460.

37 van Gisbergen, S.J.A., Snijders, J.G., and Baerends, E.J. (1995) *J. Chem. Phys.*, **103**, 9347.

38 Drummond, C.J. and Chan, D.Y.C. (1996) *Langmuir*, **12**, 3356.

39 Bergström, L. (1997) *Adv. Colloid Interface Sci.*, **70**, 125.

40 Meurk, A., Luckham, P.W., and Bergström, L. (1997) *Langmuir*, **13**, 3896.

41 Boinovich, L.B. (1992) *Adv. Colloid Interface Sci.*, **37**, 177.

42 El-Aasser, M.S. and Robertson, A.A. (1971) *J. Colloid Interface Sci.*, **36**, 86.

43 Farmakis, L., Lioris, N., Kohadima, A., and Karaiskakis, G. (2006) *J. Chromatogr. A*, **1137**, 231.

44 Dobiás, B. (1993) *Coagulation and Flocculation, Surfactant Science Series*, vol. 47, Dekker, New York.

45 Leong, Y.K. and Ong, B.C. (2003) *Powder Technol.*, **134**, 249.

46 Batko, K. and Voelkel, A. (2007) *J. Colloid Interface Sci.*, **315**, 768.

47 Voelkel, A., Strzemiecka, B., Adamska, K., and Milczewska, K. (2009) *J. Chromatogr. A*, **1216**, 1551.

48 Medout-Marere, V. (2000) *J. Colloid Interface Sci.*, **228**, 434.

49 Parsegian, V.A. and Weiss, G.H. (1981) *J. Colloid Interface Sci.*, **81**, 285.

50 Visser, J. (1972) *Adv. Colloid Interface Sci.*, **3**, 331.

51 Drummond, C.J., Georgaklis, G., and Chan, D.Y.C. (1996) *Langmuir*, **12**, 2617.

52 Bell, N. and Dimos, D.B. (2000) *Solid Freeform and Additive Fabrication*, vol. 625 (eds S.C. Danforth, D.B. Dimos, and F. Prinz), Materials Research Society, Warrendale, PA.

53 Eichenlaub, S., Chan, C., and Beaudoin, S.P. (2002) *J. Colloid Interface Sci.*, **248**, 389.

54 Tan, G.L., Lemon, M.F., Jones, D.J., and French, R.H. (2005) *Phys. Rev. B*, **72**, 205117.

55 Israelachvili, J.N. and Adams, G.E. (1978) *J. Chem. Soc., Faraday Trans. 1*, **74**, 975.

56 Israelachvili, J.N. and Tabor, D. (1972) *Proc. R. Soc. Lond. A*, **331**, 19.

57 Coakley, C.J. and Tabor, D. (1978) *J. Phys. D: Appl. Phys.*, **11**, L77.

58 Subbaraman, R., Zawodzinski, T., and Mann, J.A. (2008) *Langmuir*, **24**, 8245.

59 Das, S., Sreeram, P.A., and Raychaudhuri, A.K. (2007) *Nanotechnology*, **18**, 035501.

60 Hutter, J.L. and Bechhoefer, J. (1993) *J. Appl. Phys.*, **73**, 4123.

61 Senden, T.J. and Drummond, C.J. (1995) *Colloids Surf. A*, **94**, 29.

62 Roth, C.M., Neal, B.L., and Lenhoff, A.M. (1996) *Biophys. J.*, **70**, 977.

63 Zhao, H.P., Wang, Y.J., and Tsui, O.K.C. (2005) *Langmuir*, **21**, 5817.

64 Sounilhac, S., Barthel, E., and Creuzet, F. (1998) *Appl. Surf. Sci.*, **140**, 411.

65 Lubarsky, G.V., Mitchell, S.A., Davidson, M.R., and Bradley, R.H. (2006) *Colloids Surf. A*, **279**, 188.

66 Hartley, P., Matsumoto, M., and Mulvaney, P. (1998) *Langmuir*, **14**, 5203.

67 Gomez-Merino, A.L., Rubio-Hernandez, F.J., Velazquez-Navarro, J.F., Galindo-Rosales, F.J., and Fortes-Quesada, P. (2007) *J. Colloid Interface Sci.*, **316**, 451.

68 Larson, I., Drummond, C.J., Chan, D.Y.C., and Grieser, F. (1993) *J. Am. Chem. Soc.*, **115**, 11885.

69 Flatt, R.J. (2004) *Cem. Concr. Res.*, **34**, 399.

70 Reerink, H. and Overbeek, J.T.G. (1954) *Discuss. Faraday Soc.*, **74**.

71 Biggs, S. and Mulvaney, P. (1994) *J. Chem. Phys.*, **100**, 8501.

72 Larson, I., Chan, D.Y.C., Drummond, C.J., and Grieser, F. (1997) *Langmuir*, **13**, 2429.

73 Enüstün, B.V. and Turkevich, J. (1963) *J. Am. Chem. Soc.*, **85**, 3317.

74 Wall, J.F., Grieser, F., and Zukoski, C.F. (1997) *J. Chem. Soc., Faraday Trans.*, **93**, 4017.

75 Chen, X. (1998) Control of surface chemistry of gold, pyrite and pyrrhotite. Master thesis, Virginia Polytechnic Institute and State University, Blacksburg.

76 Kane, V. and Mulvaney, P. (1998) *Langmuir*, **14**, 3303.

77 Bergström, L., Stemme, S., Dahlfors, T., Arwin, H., and Ödberg, L. (1999) *Cellulose*, **6**, 1.

78 Dedkov, G.V., Dedkova, E.G., Tegaev, R.I., and Khokonov, K.B. (2008) *Tech. Phys. Lett.*, **34**, 17.

79 French, R.H., Winey, K.I., Yang, M.K., and Qiu, W.M. (2007) *Aust. J. Chem.*, **60**, 251.

80 El Ghzaoui, A. (1999) *J. Appl. Phys.*, **86**, 2920.

81 Fernández-Barbero, A., Martín-Rodríguez, A., Callejas-Fernández, J., and Hidalgo-Alvarez, R. (1994) *J. Colloid Interface Sci.*, **162**, 257.

82 Kosaka, P.M., Kawano, Y., and Petri, D.F.S. (2007) *J. Colloid Interface Sci.*, **316**, 671.

83 Fowkes, F.M. (1964) *Ind. Eng. Chem.*, **56**, 40.

84 Derjaguin, B.V. (1934) *Kolloid-Zeitschrift*, **69**, 155.

85 Blocki, J., Randrup, J., Swiatecki, W.J., and Tsang, C.F. (1977) *Ann. Phys.*, **105**, 427.

86 White, L.R. (1983) *J. Colloid Interface Sci.*, **95**, 286.

87 Owens, N.F. and Richmond, P. (1978) *J. Chem. Soc., Faraday Trans. 2*, **74**, 691.

88 Israelachvili, J.N. (1994) *Langmuir*, **10**, 3369.

89 Bowen, W.R. and Jenner, F. (1995) *Adv. Colloid Interface Sci.*, **56**, 201.

90 Derjaguin, B.V. and Abricossova, I.I. (1951) *J. Exp. Theor. Phys. USSR*, **21**, 945.

91 Derjaguin, B.V. and Abricossova, I.I. (1954) *Discuss. Faraday Soc.*, **18**, 33.

92 Overbeek, J.T.G. and Sparnaay, M.J. (1952) *J. Colloid Sci.*, **7**, 343.

93 Overbeek, J.T.G. and Sparnaay, M.J. (1954) *Discuss. Faraday Soc.*, 12.

94 Derjaguin, B., Abricossova, I., Overbeek, J.T.G., Klevens, H.B. and Dervichian, D.G. (1954) General discussion. *Discuss Faraday Soc.*, **18**, 180.

95 Kitchener, J.A. and Prosser, A.P. (1957) *Proc. R. Soc. Lond. A*, **242**, 403.

96 Black, W., de Jongh, J.G.V., Overbeek, J.T.G., and Sparnaay, M.J. (1960) *Trans. Faraday Soc.*, **56**, 1597.

97 White, L.R., Israelachvili, J.N., and Ninham, B.W. (1976) *J. Chem. Soc., Faraday Trans.*, **72**, 2526.

98 Wittmann, F., Splittgerber, H., and Ebert, K. (1971) *Z. Phys.*, **245**, 354.

99 Hunklinger, S., Arnold, W., and Geisselmann, H. (1972) *Rev. Sci. Instrum.*, **43**, 584.

100 Derjaguin, B.V., Rabinovich, Y.I., and Churaev, N.V. (1977) *Nature*, **265**, 520.

101 van Blokland, P.H.G.M. and Overbeek, J.T.G. (1978) *J. Chem. Soc., Faraday Trans. 1*, **74**, 2637.

102 Stewart, A.M., Yaminsky, V.V., and Ohnishi, S. (2002) *Langmuir*, **18**, 1453.

103 Yaminsky, V.V. and Stewart, A.M. (2003) *Langmuir*, **19**, 4037.

104 Palasantzas, G., van Zwol, P.J., and De Hosson, J.T.M. (2008) *Appl. Phys. Lett.*, **93**, 121912.

105 Hutter, J.L. and Bechhoefer, J. (1994) *J. Vac. Sci. Technol.*, **12**, 2251.

106 Ashby, P.D., Chen, L.W., and Lieber, C.M. (2000) *J. Am. Chem. Soc.*, **122**, 9467.

107 Fontaine, P., Guenoun, P., and Daillant, J. (1997) *Rev. Sci. Instrum.*, **68**, 4145.

108 Guggisberg, M., Bammerlin, M., Loppacher, C., Pfeiffer, O., Abdurixit, A., Barwich, V., Bennewitz, R., Baratoff, A., Meyer, E., and Güntherodt, H.J. (2000) *Phys. Rev. B*, **61**, 11151.

109 Visser, J. (1968) *Rep. Prog. Appl. Chem.*, **53**, 714.

110 Neumann, A.W., Omenyi, S.N., and Oss, C.J.V. (1979) *Colloid Polym. Sci.*, **257**, 413.

111 Sabisky, E.S. and Anderson, C.H. (1973) *Phys. Rev. A*, **7**, 790.

112 Milling, A., Mulvaney, P., and Larson, I. (1996) *J. Colloid Interface Sci.*, **180**, 460.

113 Lee, S. and Sigmund, W.M. (2001) *J. Colloid Interface Sci.*, **243**, 365.

114 Lamoreaux, S.K. (1997) *Phys. Rev. Lett.*, **78**, 5.

115 Mohideen, U. and Roy, A. (1998) *Phys. Rev. Lett.*, **81**, 4549.

116 Serry, F.M., Walliser, D., and Maclay, G.J. (1998) *J. Appl. Phys.*, **84**, 2501.

117 Decca, R.S., Lopez, D., Fischbach, E., Klimchitskaya, G.L., Krause, D.E., and Mostepanenko, V.M. (2007) *Phys. Rev. D*, **75**, 077101.

118 Milton, K.A. (2004) *J. Phys. A: Math. Gen.*, **37**, R209.

119 Klimchitskaya, G.L. and Mostepanenko, V.M. (2006) *Contemp. Phys.*, **47**, 131; (2001) *Phys. Rep.: Rev. Sec. Phys. Lett.*, **353**, 1.

120 Boström, M. and Sernelius, B.E. (2000) *Phys. Rev. Lett.*, **84**, 4757.

121 Bordag, M., Geyer, B., Klimchitskaya, G.L., and Mostepanenko, V.M. (2000) *Phys. Rev. Lett.*, **85**, 503.

122 Geyer, B., Klimchitskaya, G.L., and Mostepanenko, V.M. (2003) *Phys. Rev. A*, **67**, 062102.

123 Mostepanenko, V.M., Bezerra, V.B., Decca, R.S., Geyer, B., Fischbach, E., Klimchitskaya, L., Krause, D.E., Lopez, D., and Romer, C. (2006) *J. Phys. A: Math. Gen.*, **39**, 6589.

124 Lambrecht, A. and Reynaud, S. (2000) *Eur. Phys. J. D*, **8**, 309.

125 Lisanti, M., Iannuzzi, D., and Capasso, F. (2005) *Proc. Natl. Acad. Sci. USA*, **102**, 11989.

126 van Zwol, P.J., Palasantzas, G., van de Schootbrugge, M., and De Hosson, J.T.M. (2008) *Appl. Phys. Lett.*, **92**, 054101.

127 Bordag, M., Klimchitskaya, G.L., and Mostepanenko, V.M. (1995) *Phys. Lett. A*, **200**, 95.

128 Chen, F., Mohideen, U., Klimchitskaya, G.L., and Mostepanenko, V.M. (2002) *Phys. Rev. A*, **66**, 032113.

129 Krause, D.E., Decca, R.S., Lopez, D., and Fischbach, E. (2007) *Phys. Rev. Lett.*, **98**, 050403.

130 Gies, H., Langfeld, K., and Moyaerts, L. (2003) *J. High Energy Phys.*, 018.

131 Jaffe, R.L. and Scardicchio, A. (2004) *Phys. Rev. Lett.*, **92**, 070402.

132 Reynaud, S., Neto, P.A.M., and Lambrecht, A. (2007) *J. Phys. A*, 164004.

133 Sparnaay, M.J. (1958) *Physica*, **24**, 751.

134 Bordag, M., Geyer, B., Klimchitskaya, G.L., and Mostepanenko, V.M. (1998) *Phys. Rev. D*, **5807**, 075003.

135 Harris, B.W., Chen, F., and Mohideen, U. (2000) *Phys. Rev. A*, **62**, 052109.

136 Ederth, T. (2000) *Phys. Rev. A*, **62**, 062104.

137 Bressi, G., Carugno, G., Onofrio, R., and Ruoso, G. (2002) *Phys. Rev. Lett.*, **88**, 041804.

138 Chan, H.B., Aksyuk, V.A., Kleiman, R.N., Bishop, D.J., and Capasso, F. (2001) *Science*, **291**, 1941.

139 Decca, R.S., Fischbach, E., Klimchitskaya, G.L., Krause, D.E., Lopez, D., and Mostepanenko, V.M. (2003) *Phys. Rev. D*, **68**, 116003.

140 Kenneth, O., Klich, I., Mann, A., and Revzen, M. (2002) *Phys. Rev. Lett.*, **89**, 033001.

141 Iannuzzi, D. and Capasso, F. (2003) *Phys. Rev. Lett.*, **91**, 029101.

142 Ramakrishna, S.A. (2005) *Rep. Prog. Phys.*, **68**, 449.

143 Leonhardt, U. and Philbin, T.G. (2007) *New J. Phys.*, **9**, 254.

144 Rosa, F.S.S., Dalvit, D.A.R., and Milonni, P.W. (2008) *Phys. Rev. Lett.*, **100**, 183602.

145 Munday, J.N. and Capasso, F. (2007) *Phys. Rev. A*, **75**, 060102.

146 Munday, J.N., Capasso, F., and Parsegian, V.A. (2009) *Nature*, **457**, 170.

147 Fisher, M.E. and de Gennes, P.G. (1978) *C. R. Acad. Sci. Paris Ser. B*, **287**, 207.

148 Tonchev, N.S. (2007) *J. Optoelectron. Adv. Mater.*, **9**, 11.

149 Krech, M. (1999) *J. Phys.: Condens. Matter*, **11**, R391.

150 Vasilyev, O., Gambassi, A., Maciolek, A., and Dietrich, S. (2007) *Europhys. Lett*, **80**, 60009.

151 Hucht, A. (2007) *Phys. Rev. Lett.*, **99**, 185301.

152 Garcia, R. and Chan, M.H.W. (1999) *Phys. Rev. Lett.*, **83**, 1187.

153 Garcia, R. and Chan, M.H.W. (2002) *Phys. Rev. Lett.*, **88**, 086101.

154 Maciolek, A. and Dietrich, S. (2006) *Europhys. Lett.*, **74**, 22.

155 Fukuto, M., Yano, Y.F., and Pershan, P.S. (2005) *Phys. Rev. Lett.*, **94**, 135702.

156 Hertlein, C., Helden, L., Gambassi, A., Dietrich, S., and Bechinger, C. (2008) *Nature*, **451**, 172.

157 Craig, V.S.J. (1997) *Colloids Surf. A*, **130**, 75.

158 Lodge, K.B. (1983) *Adv. Colloid Interface Sci.*, **19**, 27.

159 Tomlinson, G.A. (1928) *Philos. Mag.*, **6**, 695.

160 Rayleigh, L. (1936) *Proc. R. Soc. Lond. A*, **156**, 326.

161 Tolansky, S. (1948) *Multiple Beam Interferometry of Surfaces and Films*, Oxford University Press, London.

162 Zäch, M. and Heuberger, M. (2000) *Langmuir*, **16**, 7309.

163 Heuberger, M., Zäch, M., and Spencer, N.D. (2000) *Rev. Sci. Instrum.*, **71**, 4502.

164 Jasmund, K. and Lagaly, G. (eds) (1993) *Tonminerale und Tone. Struktur, Eigenschaften, Anwendungen und Einsatz in Industrie und Umwelt*, Steinkopff Verlag, Darmstadt.

165 Israelachvili, J.N., Alcantar, N.A., Maeda, N., Mates, T.E., and Ruths, M. (2004) *Langmuir*, **20**, 3616.

166 Heuberger, M. and Zach, M. (2003) *Langmuir*, **19**, 1943.

167 Heuberger, M., Luengo, G., and Israelachvili, J. (1997) *Langmuir*, **13**, 3839.

168 Froeberg, J.C. and Ederth, T. (1999) *J. Colloid Interface Sci.*, **210**, 215.

169 Frantz, P. and Salmeron, M. (1998) *Tribol. Lett.*, **5**, 151.

170 Israelachvili, J.N. (1973) *J. Colloid Interface Sci.*, **44**, 259.

171 Tadmor, R., Chen, N., and Israelachvili, J.N. (2003) *J. Colloid Interface Sci.*, **264**, 548.

172 Levins, J.M. and Vanderlick, T.K. (1994) *Langmuir*, **10**, 2389.

173 Quon, R.A., Levins, J.M., and Vanderlick, T.K. (1995) *J. Colloid Interface Sci.*, **171**, 474.

174 Born, M. and Wolf, E. (1980) *Principles of Optics*, 6th edn, Pergamon Press, Oxford.

175 Heuberger, M. (2001) *Rev. Sci. Instrum.*, **72**, 1700.

176 Bailey, A.I. and Courtney-Pratt, J.S. (1955) *Proc. R. Soc. Lond. A*, **227**, 500.

177 Israelachvili, J.N. and Adams, G.E. (1976) *Nature*, **262**, 773.

178 Israelachvili, J.N. and McGuiggan, P.M. (1988) *Science*, **241**, 795.

179 Israelachvili, J.N. and McGuiggan, P.M. (1990) *J. Mater. Res.*, **5**, 2223.

180 Klein, J. (1983) *J. Chem. Soc., Faraday Trans. 1*, **79**, 99.

181 Parker, J.L., Christenson, H.K., and Ninham, B.W. (1989) *Rev. Sci. Instrum.*, **60**, 3135.

182 Heuberger, M., Vanicek, J., and Zach, M. (2001) *Rev. Sci. Instrum.*, **72**, 3556.

183 Zäch, M., Vanicek, J., and Heuberger, M. (2003) *Rev. Sci. Instrum.*, **74**, 260.

184 Balmer, T.E. and Heuberger, M. (2007) *Rev. Sci. Instrum.*, **78**, 10.

185 Kumacheva, E. (1998) *Prog. Surf. Sci*, **58**, 75.

186 Luengo, G., Schmitt, F.J., Hill, R., and Israelachvili, J. (1997) *Macromolecules*, **30**, 2482.

187 Peachey, J., Vanalsten, J., and Granick, S. (1991) *Rev. Sci. Instrum.*, **62**, 463.

188 Klein, J. and Kumacheva, E. (1998) *J. Chem. Phys.*, **108**, 6996.

189 Qian, L.M., Luengo, G., Douillet, D., Charlot, M., Dollat, X., and Perez, E. (2001) *Rev. Sci. Instrum.*, **72**, 4171.

190 Bureau, L. (2007) *Rev. Sci. Instrum.*, **78**, 065110.

191 Berg, S., Ruths, M., and Johannsmann, D. (2003) *Rev. Sci. Instrum.*, **74**, 3845.

192 Butt, H.-J., Wang, D.N., Hansma, P.K., and Kühlbrandt, W. (1991) *Ultramicroscopy*, **36**, 307; Butt, H.-J., Müller, T., and Gross, H. (1993) *J. Struct. Biol.*, **110**, 127.

193 Hegner, M., Wagner, P., and Semenza, G. (1993) *Surf. Sci.*, **291**, 39.

194 Chai, L. and Klein, J. (2007) *Langmuir*, **23**, 7777.

195 Connor, J.N. and Horn, R. (2003) *Rev. Sci. Instrum.*, **74**, 4601.

196 Binnig, G., Quate, C.F., and Gerber, C. (1986) *Phys. Rev. Lett.*, **56**, 930.

197 Weisenhorn, A.L., Hansma, P.K., Albrecht, T.R., and Quate, C.F. (1989) *Appl. Phys. Lett.*, **54**, 2651.

198 Ducker, W.A., Senden, T.J., and Pashley, R.M. (1991) *Nature*, **353**, 239.

199 Butt, H.J. (1991) *Biophys. J.*, **60**, 777.

200 Butt, H.J., Cappella, B., and Kappl, M. (2005) *Surf. Sci. Rep.*, **59**, 1.

201 Zhang, W. and Zhang, X. (2003) *Prog. Polym. Sci.*, **28**, 1271.

202 Janshoff, A., Neitzert, M., Oberdörfer, Y., and Fuchs, H. (2000) *Angew. Chem. Int. Ed.*, **39**, 3212.

203 Dufrene, Y.F. and Hinterdorfer, P. (2008) *Pflugers Arch.*, **456**, 237.

204 Clark, S.C., Walz, J.Y., and Ducker, W.A. (2004) *Langmuir*, **20**, 7616.

205 Albrecht, T.R., Akamine, S., Carver, T.E., and Quate, C.F. (1990) *J. Vac. Sci. Technol. A*, **8**, 3386.

206 Wolter, O., Bayer, T., and Greschner, J. (1990) *J. Vac. Sci. Technol. A*, **9**, 1353.

207 Butt, H.J., Siedle, P., Seifert, K., Fendler, K., Seeger, T., Bamberg, E., Weisenhorn, A.L., Goldie, K., and Engel, A. (1993) *J. Microsc. (Oxf.)*, **169**, 75.

208 Timoshenko, S. and Woinowsky-Krieger, S. (1959) *Theory of Plates and Shells*, 2nd edn, McGraw-Hill, New York.

209 Thundat, T., Warmack, R.J., Chen, G.Y., and Allison, D.P. (1994) *Appl. Phys. Lett.*, **64**, 2894.

210 Butt, H.J. (1996) *J. Colloid Interface Sci.*, **180**, 251.

211 Walters, D.A., Cleveland, J.P., Thomson, N.H., Hansma, P.K., Wendman, M.A., Gurley, G., and Elings, V. (1996) *Rev. Sci. Instrum.*, **67**, 3583.

212 Viani, M.B., Schaffer, T.E., Chand, A., Rief, M., Gaub, H.E., and Hansma, P.K. (1999) *J. Appl. Phys.*, **86**, 2258; (1999) *Rev. Sci. Instrum.*, **70**, 4300.

213 Warmack, R.J., Zheng, X.Y., Thundat, T., and Allison, D.P. (1994) *Rev. Sci. Instrum.*, **65**, 394.

214 Neumeister, J.M. and Ducker, W.A. (1994) *Rev. Sci. Instrum.*, **65**, 2527.

215 Sader, J.E., Larson, I., Mulvaney, P., and White, L.R. (1995) *Rev. Sci. Instrum.*, **66**, 3789.

216 Labardi, M., Allegrini, M., Salerno, M., Frediani, C., and Ascoli, C. (1994) *Appl. Phys. A*, **59**, 3.

217 Hazel, J.L. and Tsukruk, V.V. (1998) *Trans. ASME, J. Tribol.*, **120**, 814; (1999) *Thin Solid Films*, **339**, 249.

218 Clifford, C.A. and Seah, M.P. (2005) *Nanotechnology*, **16**, 1666.

219 Chen, B.Y., Yeh, M.K., and Tai, N.H. (2007) *Anal. Chem.*, **79**, 1333.

220 Hutter, J.L. and Bechhoefer, J. (1993) *Rev. Sci. Instrum.*, **64**, 1868.

221 Butt, H.J. and Jaschke, M. (1995) *Nanotechnology*, **6**, 1.

222 Stark, R.W., Drobek, T., and Heckl, W.M. (2001) *Ultramicroscopy*, **86**, 207.

223 Levy, R. and Maaloum, M. (2002) *Nanotechnology*, **13**, 33.

224 Cook, S., Schäffer, T.E., Chynoweth, K.M., Wigton, M., Simmonds, R.W., and Lang, K.M. (2006) *Nanotechnology*, **17**, 2135.

225 Cleveland, J.P., Manne, S., Bocek, D., and Hansma, P.K. (1993) *Rev. Sci. Instrum.*, **64**, 403.

226 Sader, J.E., Chon, J.W.M., and Mulvaney, P. (1999) *Rev. Sci. Instrum.*, **70**, 3967.

227 Gibson, C.T., Weeks, B.L., Lee, J.R.I., Abell, C., and Rayment, T. (2001) *Rev. Sci. Instrum.*, **72**, 2340.

228 Bonaccurso, E. and Butt, H.J. (2005) *J. Phys. Chem. B*, **109**, 235.

229 Cole, D.G. (2008) *Meas. Sci. Technol.*, **19**, 125101.

230 Ohler, B. (2007) *Rev. Sci. Instrum.*, **78**, 063701.

231 Burnham, N.A., Chen, X., Hodges, C.S., Matei, G.A., Thoreson, E.J., Roberts, C.J., Davies, M.C., and Tendler, S.J.B. (2003) *Nanotechnology*, **14**, 1.

232 Matei, G.A., Thoreson, E.J., Pratt, J.R., Newell, D.B., and Burnham, N.A. (2006) *Rev. Sci. Instrum.*, **77**, 083703.

233 Gibson, C.T., Smith, D.A., and Roberts, C.J. (2005) *Nanotechnology*, **16**, 234.

234 Senden, T.J. and Ducker, W.A. (1994) *Langmuir*, **10**, 1003.

235 Maeda, N. and Senden, T.J. (2000) *Langmuir*, **16**, 9282.

236 Craig, V.S.J. and Neto, C. (2001) *Langmuir*, **17**, 6018.

237 Notley, S.M., Biggs, S., and Craig, V.S.J. (2003) *Rev. Sci. Instrum.*, **74**, 4026.

238 Hong, X.T. and Willing, G.A. (2008) *Rev. Sci. Instrum.*, **79**, 123709.

239 Holbery, J.D., Eden, V.L., Sarikaya, M., and Fisher, R.M. (2000) *Rev. Sci. Instrum.*, **71**, 3769.

240 Ying, Z.C., Reitsma, M.G., and Gates, R.S. (2007) *Rev. Sci. Instrum.*, **78**, 063708.

241 Gibson, C.T., Watson, G.S., and Myhra, S. (1996) *Nanotechnology*, **7**, 259.

242 Jericho, S.K. and Jericho, M.H. (2002) *Rev. Sci. Instrum.*, **73**, 2483.

243 Cumpson, P.J., Clifford, C.A., and Hedley, J. (2004) *Meas. Sci. Technol.*, **15**, 1337.

244 Cumpson, P.J., Zhdan, P., and Hedley, J. (2004) *Ultramicroscopy*, **100**, 241.

245 Aksu, S.B. and Turner, J.A. (2007) *Rev. Sci. Instrum.*, **78**, 043704.

246 Langlois, E.D., Shaw, G.A., Kramar, J.A., Pratt, J.R., and Hurley, D.C. (2007) *Rev. Sci. Instrum.*, **78**, 093705.

247 Foster, A.S., Hofer, W.A., and Shluger, A.L. (2001) *Curr. Opin. Solid State Mater. Sci.*, **5**, 427.

248 Chung, K.H., Lee, Y.II., and Kim, D.E. (2005) *Ultramicroscopy*, **102**, 161.

249 Siedle, P., Butt, H.J., Bamberg, E., Wang, D.N., Kuhlbrandt, W., Zach, J., and Haider, M. (1993) *Inst. Phys. Conf. Ser.*, 361.

250 Vesenka, J., Miller, R., and Henderson, E. (1994) *Rev. Sci. Instrum.*, **65**, 2249.

251 Markiewicz, P. and Goh, M.C. (1994) *Langmuir*, **10**, 5.

252 Kappl, M. and Butt, H.J. (2002) *Part. Part. Syst. Charact.*, **19**, 129.

253 Ralston, J., Larson, I., Rutland, M.W., Feiler, A.A., and Kleijn, M. (2005) *Pure Appl. Chem.*, **77**, 2149.

254 Gan, Y. (2007) *Rev. Sci. Instrum.*, **78**, 081101.

255 Ducker, W.A., Senden, T.J., and Pashley, R.M. (1992) *Langmuir*, **8**, 1831.

256 Toikka, G., Hayes, R.A., and Ralston, J. (1996) *J. Colloid Interface Sci.*, **180**, 329.

257 Mak, L.H., Knoll, M., Weiner, D., Gorschluter, A., Schirmeisen, A., and Fuchs, H. (2006) *Rev. Sci. Instrum.*, **77**, 046104.

258 Raiteri, R., Preuss, M., Grattarola, M., and Butt, H.J. (1998) *Colloids Surf. A*, **136**, 191.

259 Gan, Y. (2005) *Microsc. Today*, **13**, 48.

260 Bowen, W.R., Hilal, N., Lovitt, R.W., and Wright, C.J. (1999) *Colloids Surf. A*, **157**, 117.

261 Huntington, S. and Nespolo, S. (2001) *Microsc. Today*, **9**, 32.

262 Bonaccurso, E., Kappl, M., and Butt, H.J. (2002) *Phys. Rev. Lett.*, **88**, 076103.

263 Schaefer, D.M., Carpenter, M., Gady, B., Reifenberger, R., Demejo, L.P., and Rimai, D.S. (1993) *J. Adhes. Sci. Technol.*, **8**, 1049.

264 Bowen, W.R., Fenton, A.S., Lovitt, R.W., and Wright, C.J. (2002) *Biotechnol. Bioeng.*, **79**, 170.

265 Lower, S.K., Hochella, M.F., and Beveridge, T.J. (2001) *Science*, **292**, 1360.

266 Benoit, M., Gabriel, D., Gerisch, G., and Gaub, H.E. (2000) *Nat. Cell Biol.*, **2**, 313.

267 Bowen, W.R., Lovitt, R.W., and Wright, C.J. (2001) *J. Colloid Interface Sci.*, **237**, 54.

268 Gunning, A.P., Mackie, A.R., Wilde, P.J., and Morris, V.J. (2004) *Langmuir*, **20**, 116.

269 Dagastine, R.R., Stevens, G.W., Chan, D.Y.C., and Grieser, F. (2004) *J. Colloid Interface Sci.*, **273**, 339.

270 Dagastine, R.R., Manica, R., Carnie, S.L., Chan, D.Y.C., Stevens, G.W., and Grieser, F. (2006) *Science*, **313**, 210.

271 Vakarelski, I.U., Lee, J., Dagastine, R.R., Chan, D.Y.C., Stevens, G.W., and Grieser, F. (2008) *Langmuir*, **24**, 603.

272 Vakarelski, I.U. and Higashitani, K. (2006) *Langmuir*, **22**, 2931.

273 Ong, Q.K. and Sokojov, I. (2007) *J. Colloid Interface Sci.*, **310**, 385.

274 Noy, A. (2006) *Surf. Interface Anal.*, **38**, 1429.

275 Mate, C.M., McClelland, G.M., Erlandsson, R., and Chiang, S. (1987) *Phys. Rev. Lett.*, **59**, 1942.

276 Meyer, G. and Amer, N.M. (1990) *Appl. Phys. Lett.*, **57**, 2089.

277 Ogletree, D.F., Carpick, R.W., and Salmeron, M. (1996) *Rev. Sci. Instrum.*, **67**, 3298.

278 Varenberg, M., Etsion, I., and Halperin, G. (2003) *Rev. Sci. Instrum.*, **74**, 3362.

279 Tocha, E., Schönherr, H., and Vancso, G.J. (2006) *Langmuir*, **22**, 2340.

280 Wang, F. and Zhao, X.Z. (2007) *Rev. Sci. Instrum.*, **78**, 043701.

281 Bogdanovic, G., Meurk, A., and Rutland, M.W. (2000) *Colloids Surf. B*, **19**, 397.

282 Toikka, G., Hayes, R.A., and Ralston, J. (1997) *J. Adhes. Sci. Technol.*, **11**, 1479.

283 Feiler, A., Attard, P., and Larson, I. (2000) *Rev. Sci. Instrum.*, **71**, 2746.

284 Jeon, S., Braiman, Y., and Thundat, T. (2004) *Appl. Phys. Lett.*, **84**, 1795.

285 Morel, N., Tordjeman, P., and Ramonda, M. (2005) *J. Phys. D: Appl. Phys.*, **38**, 895.

286 Liu, W.H., Bonin, K., and Guthold, M. (2007) *Rev. Sci. Instrum.*, **78**, 063707.

287 Cumpson, P.J., Hedley, J., and Clifford, C.A. (2005) *J. Vac. Sci. Technol. B*, **23**, 1992.

288 Li, Q., Kim, K.S., and Rydberg, A. (2006) *Rev. Sci. Instrum.*, **77**, 065105.

289 Müller, M., Schimmel, T., Haussler, P., Fettig, H., Müller, O., and Albers, A. (1090) *Surf. Interface Anal.*, 2004.

290 Green, C.P., Lioe, H., Cleveland, J.P., Proksch, R., Mulvaney, P., and Sader, J.E. (2004) *Rev. Sci. Instrum.*, **75**, 1988.

291 Liu, E., Blanpain, B., and Celis, J.P. (1996) *Wear*, **192**, 141.

292 Schwarz, U.D., Köster, P., and Wiesendanger, R. (1996) *Rev. Sci. Instrum.*, **67**, 2560.

293 Fujisawa, S., Kishi, E., Sugawara, Y., and Morita, S. (1995) *Appl. Phys. Lett.*, **66**, 526.

294 Cain, R.G., Biggs, S., and Page, N.W. (2000) *J. Colloid Interface Sci.*, **227**, 55.

295 Ecke, S., Raiteri, R., Bonaccurso, E., Reiner, C., Deiseroth, H.J., and Butt, H.J. (2001) *Rev. Sci. Instrum.*, **72**, 4164.

296 Cannara, R.J., Eglin, M., and Carpick, R.W. (2006) *Rev. Sci. Instrum.*, **77**, 053701.

297 Choi, D., Hwang, W., and Yoon, E. (2007) *J. Microsc. (Oxf.)*, **228**, 190.

298 Giessibl, F.J. (2003) *Rev. Mod. Phys.*, **75**, 949.

299 Albrecht, T.R., Grutter, P., Horne, D., and Rugar, D. (1991) *J. Appl. Phys.*, **69**, 668.

300 Lantz, M.A., Hug, H.J., Hoffmann, R., Schendel, P.J.A.v., Kappenberger, P., Martin, S., Baratoff, A., and Güntherodt, H.J. (2001) *Science*, **291**, 2580.

301 Giessibl, F.J., Herz, M., and Mannhart, J. (2002) *Proc. Natl. Acad. Sci. USA*, **99**, 12006.

302 Jarvis, S.P., Ishida, T., Uchihashi, T., Nakayama, Y., and Tokumoto, H. (2001) *Appl. Phys. A*, **72**, 129.

303 Uchihashi, T., Higgins, M.J., Yasuda, S., Jarvis, S.P., Akita, S., Nakayama, Y., and Sader, J.E. (2004) *Appl. Phys. Lett.*, **85**, 3575.

304 Ashkin, A. (1970) *Phys. Rev. Lett.*, **24**, 156. Ashkin, A., Dziedzic, J.M., Bjorkholm, J.E., and Chu, S. (1986) *Opt. Lett.*, **11**, 288.

305 Ashkin, A., Dziedzic, J.M., and Yamane, T. (1987) *Nature*, **330**, 769.

306 Svoboda, K. and Block, S.M. (1994) *Annu. Rev. Biophys. Biomol. Struct.*, **23**, 247.

307 Hormeno, S. and Arias-Gonzalez, J.R. (2006) *Biol. Cell*, **98**, 679.

308 Zhang, H. and Liu, K.K. (2008) *J. R. Soc. Interface*, **5**, 671.

309 Grier, D.G. (1997) *Curr. Opin. Colloid Interface Sci.*, **2**, 264.

310 Jonas, A. and Zemanek, P. (2008) *Electrophoresis*, **29**, 4813.

311 Furst, E.M. (2005) *Curr. Opin. Colloid Interface Sci.*, **10**, 79.

312 Mitchem, L. and Reid, J.P. (2008) *Chem. Soc. Rev.*, **37**, 756.

313 McGloin, D., Burnham, D.R., Summers, M.D., Rudd, D., Dewara, N., and Anandc, S. (2008) *Faraday Discuss.*, **137**, 335.

314 Neuman, K.C. and Block, S.M. (2004) *Rev. Sci. Instrum.*, **75**, 2787.

315 Grier, D.G. (2003) *Nature*, **424**, 810.

316 Moffitt, J.R., Chemla, Y.R., Smith, S.B., and Bustamante, C. (2008) *Annu. Rev. Biochem.*, **77**, 205.

317 Rohrbach, A. and Stelzer, E.H.K. (2002) *Appl. Opt.*, **41**, 2494.

318 Rohrbach, A. (2005) *Phys. Rev. Lett.*, **95**, 168102.

319 Dienerowitz, M., Mazilu, M., and Dholakia, K. (2008) *J. Nanophoton.*, **2**, 021875.

320 Rohrbach, A., Tischer, C., Neumayer, D., Florin, E.L., and Stelzer, E.H.K. (2004) *Rev. Sci. Instrum.*, **75**, 2197.

321 Rohrbach, A. (2005) *Opt. Express*, **13**, 9695.

322 Kellermayer, M.S.Z., Smith, S.B., Granzier, H.L., and Bustamante, C. (1997) *Science*, **276**, 1112.

323 Allersma, M.W., Gittes, F., deCastro, M.J., Stewart, R.J., and Schmidt, C.F. (1998) *Biophys. J.*, **74**, 1074.

324 Pralle, A., Prummer, M., Florin, E.L., Stelzer, E.H.K., and Hörber, J.K.H. (1999) *Microsc. Res. Tech.*, **44**, 378.

325 Happel, J. and Brenner, H. (1965) *Low Reynolds Number Hydrodynamics*, Prentice-Hall Inc., Englewood Cliffs, NJ.

326 Goldman, A.J., Cox, R.G., and Brenner, H. (1967) *Chem. Eng. Sci.*, **22**, 637.

327 Brenner, H. (1961) *Chem. Eng. Sci.*, **16**, 242.

328 Berg-Sørensen, K. and Flyvbjerg, H. (2004) *Rev. Sci. Instrum.*, **75**, 594; Berg-Sørensen, K., Peterman, E.J.G., Weber, T., Schmidt, C.F., and Flyvbjerg, H. (2006) *Rev. Sci. Instrum.*, **77**, 063106.

329 Tolic-Norrelykke, S.F., Schaffer, E., Howard, J., Pavone, F.S., Julicher, F.,

and Flyvbjerg, H. (2006) *Rev. Sci. Instrum.*, **77**, 103101.

330 Fischer, M. and Berg-Sørensen, K. (2007) *J. Opt. A:Pure Appl. Opt.*, **9**, 239.

331 Sasaki, K., Koshioka, M., Misawa, H., Kitamura, N., and Masuhara, H. (1991) *Opt. Lett.*, **16**, 1463.

332 Visscher, K., Brakenhoff, G.J., and Krol, J.J. (1993) *Cytometry*, **14**, 105.

333 Chiou, A.E., Wang, W., Sonek, G.J., Hong, J., and Berns, M.W. (1997) *Opt. Commun.*, **133**, 7.

334 MacDonald, M.P., Paterson, L., Volke-Sepulveda, K., Arlt, J., Sibbett, W., and Dholakia, K. (2002) *Science*, **296**, 1101.

335 Dufresne, E.R. and Grier, D.G. (1998) *Rev. Sci. Instrum.*, **69**, 1974.

336 Reicherter, M., Haist, T., Wagemann, E.U., and Tiziani, H.J. (1999) *Opt. Lett.*, **24**, 608.

337 Liesener, J., Reicherter, M., Haist, T., and Tiziani, H.J. (2000) *Opt. Commun.*, **185**, 77.

338 Curtis, J.E., Koss, B.A., and Grier, D.G. (2002) *Opt. Commun.*, **207**, 169.

339 Leach, J., Sinclair, G., Jordan, P., Courtial, J., Padgett, M.J., Cooper, J., and Laczik, Z.J. (2004) *Opt. Express*, **12**, 220.

340 Jordan, P., Clare, H., Flendrig, L., Leach, J., Cooper, J., and Padgett, M. (2004) *J. Mod. Opt.*, **51**, 627.

341 Sinclair, G., Jordan, P., Leach, J., Padgett, M.J., and Cooper, J. (2004) *J. Mod. Opt.*, **51**, 409.

342 Parkin, S., Knoner, G., Singer, W., Nieminen, T.A., Heckenberg, N.R., and Rubinsztein-Dunlop, H. (2007) *Methods Cell Biol.*, **82**, 525.

343 Prieve, D.C., Luo, F., and Lanni, F. (1987) *Faraday Discuss.*, 297.

344 Prieve, D.C. (1999) *Adv. Colloid Interface Sci.*, **82**, 93.

345 Bike, S.G. (2000) *Curr. Opin. Colloid Interface Sci.*, **5**, 144.

346 Prieve, D.C. and Walz, J.Y. (1993) *Appl. Opt.*, **32**, 1629.

347 Wu, H.J., Shah, S., Beckham, R., Meissner, K.E., and Bevan, M.A. (2008) *Langmuir*, **24**, 13790.

348 Wada, K., Sasaki, K., and Masuhara, H. (2000) *Appl. Phys. Lett.*, **76**, 2815.

349 Walz, J.Y. and Prieve, D.C. (1992) *Langmuir*, **8**, 3073.

350 Robertson, S.K., Uhrick, A.F., and Bike, S.G. (1998) *J. Colloid Interface Sci.*, **202**, 208.

351 Frej, N.A. and Prieve, D.C. (1993) *J. Chem. Phys.*, **98**, 7552.

352 Oetama, R.J. and Walz, J.Y. (2005) *J. Colloid Interface Sci.*, **284**, 323.

353 Brown, M.A., Smith, A.L., and Staples, E.J. (1989) *Langmuir*, **5**, 1319.

354 Clapp, A.R., Ruta, A.G., and Dickinson, R.B. (1999) *Rev. Sci. Instrum.*, **70**, 2627.

355 Haughey, D. and Earnshaw, J.C. (1998) *Colloids Surf. A*, **136**, 217.

356 Odiachi, P.C. and Prieve, D.C. (2004) *J. Colloid Interface Sci.*, **270**, 113.

357 Wu, H.J., Pangburn, T.O., Beckham, R.E., and Bevan, M.A. (2005) *Langmuir*, **21**, 9879.

358 McKee, C.T., Clark, S.C., Walz, J.Y., and Ducker, W.A. (2005) *Langmuir*, **21**, 5783.

359 Hertlein, C., Riefler, N., Eremina, E., Wriedt, T., Eremin, Y., Helden, L., and Bechinger, C. (2008) *Langmuir*, **24**, 1.

360 Heilbronn, A. (1922) *Jahrb. Wissenschaftl. Bot.*, **61**, 284.

361 Freundlich, H. and Seifritz, W. (1923) *Z. Phys. Chem. Stöchiom. Verwandschaftsl.*, **104**, 233.

362 Crick, F.H.C. and Hughes, A.F.W. (1949) *Exp. Cell Res.*, **1**, 36.

363 Yagi, K. (1961) *Comp. Biochem. Physiol.*, **3**, 73.

364 Hiramoto, Y. (1969) *Exp. Cell Res.*, **56**, 209.

365 Ziemann, F., Radler, J., and Sackmann, E. (1994) *Biophys. J.*, **66**, 2210.

366 Gosse, C. and Croquette, V. (2002) *Biophys. J.*, **82**, 3314.

367 Fisher, J.K. *et al.* (2005) *Rev. Sci. Instrum.*, **76**, 053711.

368 Desai, K.V., Bishop, T.G., Vicci, L., O'Brien, E.T., Taylor, R.M., and Superfine, R. (2008) *Biophys. J.*, **94**, 2374.

369 de Vries, A.H.B., Krenn, B.E., van Driel, R., and Kanger, J.S. (2005) *Biophys. J.*, **88**, 2137.

370 Amblard, F., Yurke, B., Pargellis, A., and Leibler, S. (1996) *Rev. Sci. Instrum.*, **67**, 818.

371 Xiang, Y., Miller, J., Sica, V., and Lavan, D.A. (2008) *Appl. Phys. Lett.*, **92**, 124104.

372 Kanger, J.S. and Subramaniam, V. van Driel, R. (2008) *Chromosome Res.*, **16**, 511.

373 Matthews, B.D., LaVan, D.A., Overby, D.R., Karavitis, J., and Ingber, D.E. (2004) *Appl. Phys. Lett.*, **85**, 2968.

374 Barbic, M., Mock, J.J., Gray, A.P., and Schultz, S. (2001) *Appl. Phys. Lett.*, **79**, 1897.

375 Kollmannsberger, P. and Fabry, B. (2007) *Rev. Sci. Instrum.*, **78**, 114301.

376 Abdelghani-Jacquin, C., Dichtl, M., Jakobsmeier, L., Hiller, W., and Sackmann, E. (2001) *Langmuir*, **17**, 2129.

377 Trepat, X., Grabulosa, M., Buscemi, L., Rico, F., Fabry, B., Fredberg, J.J., and Farre, R. (2003) *Rev. Sci. Instrum.*, **74**, 4012.

378 Gouy, G. (1910) *J. Phys.*, **9**, 457; (1917) *Ann. Phys.*, **7**, 129.

379 Chapman, D.L. (1913) *Philos. Mag.*, **25**, 475.

380 Debye, P. and Hückel, E. (1923) *Phys. Z.*, **24**, 185.

381 Beresford-Smith, B., Chan, D.Y.C., and Mitchell, D.J. (1985) *J. Colloid Interface Sci.*, **105**, 216.

382 Raiteri, R., Grattarola, M., and Butt, H.-J. (1996) *J. Phys. Chem.*, **100**, 16700.

383 Hillier, A.C., Kim, S., and Bard, A.J. (1996) *J. Phys. Chem.*, **100**, 18808.

384 Rentsch, S., Siegenthaler, H., and Papastavrou, G. (2007) *Langmuir*, **23**, 9083.

385 Gronwall, T.H., La Mer, V.K., and Sandved, K. (1928) *Phys. Z.*, **29**, 358.

386 White, L.R. (1977) *J. Chem. Soc., Faraday Trans. 2*, **73**, 577.

387 Ohshima, H., Healy, T.W., and White, L.R. (1982) *J. Colloid Interface Sci.*, **90**, 17.

388 Grahame, D.C. (1947) *Chem. Rev.*, **41**, 441.

389 Torrie, G.M. and Valleau, J.P. (1982) *J. Phys. Chem.*, **86**, 3251.

390 Guldbrand, L., Jönsson, B., Wennerström, H., and Linse, P. (1984) *J. Phys. Chem.*, **80**, 2221.

391 Cevc, G. (1990) *Biochim. Biophys. Acta*, **1031–3**, 311.

392 Netz, R.R. and Orland, H. (2000) *Eur. Phys. J. E*, **1**, 203.

393 Torrie, G.M. and Valleau, J.P. (1980) *J. Chem. Phys.*, **73**, 5807.

394 Kjellander, R., Åkesson, T., Jönsson, B., and Marčelja, S. (1992) *J. Chem. Phys.*, **97**, 1424.

395 Kjellander, R. and Marčelja, S. (1985) *J. Chem. Phys.*, **82**, 2122.

396 Lozada-Cassou, M., Saavedra-Barrera, R., and Henderson, D. (1982) *J. Chem. Phys.*, **77**, 5150.

397 Outhwaite, C.W. and Bhuiyan, L.B. (1983) *J. Chem. Soc., Faraday Trans. 2*, **79**, 707.

398 Nelson, A.P. and McQuarrie, D.A. (1975) *J. Theor. Biol.*, **55**, 13.

399 van Megen, W. and Snook, I. (1980) *J. Chem. Phys.*, **73**, 4656.

400 Grosberg, A.Y., Nguyen, T.T., and Shklovskii, B.I. (2002) *Rev. Mod. Phys.*, **74**, 329.

401 Lewith, S. (1888) *Arch. Exp. Pathol. Pharmakol.*, **24**, 1.

402 Hofmeister, F. (1888) *Arch. Exp. Pathol. Pharmakol.*, **24**, 247.

403 Cacace, M.G., Landau, E.M., and Ramsden, J.J. (1997) *Q. Rev. Biophys.*, **30**, 241.

404 Ninham, B.W. and Yaminsky, V. (1997) *Langmuir*, **13**, 2097.

405 Leontidis, E., Aroti, A., and Belloni, L. (2009) *J. Phys. Chem. B*, **113**, 1447.

406 Carnie, S.L. and Torrie, G.M. (1984) *Adv. Chem. Phys.*, **56**, 142.

407 Blum, L. (1990) *Adv. Chem. Phys.*, **78**, 171.

408 Delville, A. (2000) *J. Phys. Chem. B*, **104**, 10588.

409 Quesada-Pérez, M., González-Tovar, E., Martín-Molina, A., Lozada-Cassou, M., and Hidalgo-Alvarez, R. (2003) *Chem. Phys. Chem.*, **4**, 235.

410 Stern, O. (1924) *Z. Elektrochem.*, **30**, 508.

411 Carnie, S.L., Chan, D.Y.C., Mitchell, D.J., and Ninham, B.W. (1981) *J. Chem. Phys.*, **74**, 1472.

412 Guidelli, R. and Schmickler, W. (2000) *Electrochim. Acta*, **45**, 2317.

413 Verwey, E.J.W. and Overbeek, J.T.G. (1948) *Theory of the Stability of Lyophobic Colloids*, Elsevier Publishing Inc., New York.

414 Hunter, R.J. (1986) *Foundations of Colloid Science*, vol. 1, Clarendon Press, London.

415 Chan, D.Y.C. and Mitchell, D.J. (1983) *J. Colloid Interface Sci.*, **95**, 193.

416 Hogg, R., Healy, T.W., and Fuerstenau, D.W. (1966) *Trans. Faraday Soc.*, 62, 1638.

417 Parsegian, V.A. and Gingell, D. (1972) *Biophys. J.*, 12, 1192.

418 Gregory, J. (1975) *J. Colloid Interface Sci.*, 51, 44.

419 Ninham, B.W. and Parsegian, V.A. (1971) *J. Theor. Biol.*, 31, 405.

420 Yates, D.E., Levine, S., and Healy, T.W. (1974) *J. Chem. Soc., Faraday Trans. 1*, 70, 1807.

421 Chan, D., Perram, J.W., White, L.R., and Healy, T.W. (1975) *J. Chem. Soc., Faraday Trans. 1*, 71, 1046.

422 Davis, J.A., James, R.O., and Leckie, J.O. (1978) *J. Colloid Interface Sci.*, 63, 480.

423 Claesson, P.M., Herder, P., Stenius, P., Eriksson, J.C., and Pashley, R.M. (1986) *J. Colloid Interface Sci.*, 109, 31.

424 Israelachvili, J.N. and Adams, G.E. (1978) *J. Chem. Soc., Faraday Trans. 1*, 74, 975.

425 Chan, D.Y.C., Pashley, R.M., and White, L.R. (1980) *J. Colloid Interface Sci.*, 77, 283.

426 Derjaguin, B.V. (1954) *Discuss. Faraday Soc.*, 18, 85.

427 Wiese, G.R. and Healy, T.W. (1970) *Trans. Faraday Soc.*, 66, 490.

428 Kar, G., Chander, S., and Mika, T.S. (1973) *J. Colloid Interface Sci.*, 44, 347.

429 Ohshima, H., Chan, D.Y.C., Healy, T.W., and White, L.R. (1983) *J. Colloid Interface Sci.*, 92, 232.

430 Bell, G.M. and Peterson, G.C. (1972) *J. Colloid Interface Sci.*, 41, 542.

431 Honig, E.P. and Mul, P.M. (1971) *J. Colloid Interface Sci.*, 36, 258.

432 Bell, G.M., Levine, S., and McCartney, L.N. (1970) *J. Colloid Interface Sci.*, 33, 335.

433 Medina-Noyola, M. and McQuarrie, D.A. (1980) *J. Chem. Phys.*, 73, 6279.

434 Ohshima, H. (1994) *J. Colloid Interface Sci.*, 162, 487.

435 Richmond, P. (1975) *J. Chem. Soc., Faraday Trans. 2*, 71, 1154.

436 Kuin, A.J. (1990) *Faraday Discuss.*, 90, 235.

437 Miklavic, S.J., Chan, D.Y.C., White, L.R., and Healy, T.W. (1994) *J. Phys. Chem.*, 98, 9022.

438 Allen, L.H. and Matijevic, E. (1969) *J. Colloid Interface Sci.*, 31, 287; (1970), 33, 420.

439 Schulze, H. (1882) *J. Prakt. Chem.*, 25, 431; (1883), 27, 320.

440 Hardy, W.B. (1900) *Proc. R. Soc. Lond.*, 66, 110; (1900) *Z. Phys. Chem.*, 33, 385.

441 Overbeek, J.T.G. (1952) *Colloid Science*, vol. 1 (ed. H.R. Kruyt), Elsevier, Amsterdam, p. 278.

442 Derjaguin, B. (1939) *Acta Physicochim. URSS*, 10, 333.

443 Landau, L.D. (1941) *Acta Physicochim. URSS*, 14, 633.

444 Scheludko, A. and Exerowa, D. (1959) *Kolloid-Zeitschrift*, 168, 24.

445 Lyklema, J. and Mysels, K.J. (1965) *J. Am. Chem. Soc.*, 87, 2539.

446 Mysels, K.J. and Jones, M.N. (1966) *Discuss. Faraday Soc.*, 42.

447 Voropaeva, T.N., Derjaguin, B.V., and Kabanov, B.N. (1962) *Kolloid Zh.*, 24, 396.

448 Usui, S., Yamasaki, T., and Shimoiizaka, J. (1967) *J. Phys. Chem.*, 71, 3195.

449 Roberts, A.D. (1972) *J. Colloid Interface Sci.*, 41, 23.

450 Pashley, R.M. (1981) *J. Colloid Interface Sci.*, 80, 153 83, 531.

451 Pashley, R.M. and Israelachvili, J.N. (1984) *J. Colloid Interface Sci.*, 97, 446.

452 Larson, I., Drummond, C.J., Chan, D.Y.C., and Grieser, F. (1995) *J. Phys. Chem.*, 99, 2114.

453 Hartley, P.G., Larson, I., and Scales, P.J. (1997) *Langmuir*, 13, 2207.

454 Attard, P., Mitchell, D.J., and Ninham, B.W. (1988) *J. Chem. Phys.*, 88, 4987 89, 4358.

455 Klinkenberg, A. and van der Minne, J.L. (1958) *Electrostatics in the Petroleum Industry: The Prevention of Explosion Hazards*, Elsevier, Amsterdam.

456 van der Hoeven, P.H.C. and Lyklema, J. (1992) *Adv. Colloid Interface Sci.*, 42, 205.

457 Grunwald, E. and Berkowitz, B.J. (1951) *J. Am. Chem. Soc.*, 73, 4939.

458 Fuoss, R.M. and Kraus, C.A. (1933) *J. Am. Chem. Soc.*, 55, 1019.

459 La Mer, V.K. and Downes, H.C. (1933) *J. Am. Chem. Soc.*, 55, 1840.

460 Fuoss, R.M. (1958) *J. Am. Chem. Soc.*, 80, 5059; (1986) *J. Solution Chem.*, 15, 231.

461 Morrison, I.D. (1993) *Colloids Surf. A*, **71**, 1.

462 Nelson, S.M. and Pink, R.C. (1952) *J. Chem. Soc.*, 1744.

463 Eicke, H.F., Borkovec, M., and Dasgupta, B. (1989) *J. Phys. Chem.*, **93**, 314.

464 Sainis, S.K., Merrill, J.W., and Dufresne, E.R. (2008) *Langmuir*, **24**, 13334.

465 van Mil, P.J.J.M., Crommelin, D.J.A., and Wiersema, P.H. (1982) *Ber. Bunsen. Phys. Chem.*, **86**, 1160.

466 Van der Minne, J.L. and Hermanie, P.H.J. (1952) *J. Colloid Sci.*, **7**, 600; (1953) **8**, 38.

467 Kitahara, A. (1973/1974) *Prog. Org. Coat.*, **2**, 81.

468 Strubbe, F., Beunis, F., Marescaux, M., Verboven, B., and Neyts, K. (2008) *Appl. Phys. Lett.*, **93**, 254106.

469 Roberts, G.S., Sanchez, R., Kemp, R., Wood, T., and Bartlett, P. (2008) *Langmuir*, **24**, 6530.

470 Lyklema, J. (1968) *Adv. Colloid Interface Sci.*, **2**, 66.

471 Kitahara, A., Amano, M., Kawasaki, S., and Konno, K. (1977) *Colloid Polym. Sci.*, **255**, 1118.

472 Romo, L.A. (1963) *J. Phys. Chem.*, **67**, 386.

473 Sato, T. (1971) *J. Appl. Polym. Sci.*, **15**, 1053.

474 Labib, M.E. and Williams, R. (1984) *J. Colloid Interface Sci.*, **97**, 356.

475 Koelmans, H. and Overbeek, J.T.G. (1954) *Discuss. Faraday Soc.*, 52.

476 McGown, D.N.L., Parfitt, G.D., and Willis, E. (1965) *J. Colloid Sci.*, **20**, 650.

477 Albers, W. and Overbeek, J.T.G. (1959) *J. Colloid Sci.*, **14**, 501 510.

478 Hsu, M.F., Dufresne, E.R., and Weitz, D.A. (2005) *Langmuir*, **21**, 4881.

479 Briscoe, W.H. and Horn, R.G. (2002) *Langmuir*, **18**, 3945.

480 Fisher, R.A. (1926) *J. Agric. Sci.*, **16**, 492.

481 Haines, W.B. (1927) *J. Agric. Sci.*, **17**, 264.

482 Versluys, J. (1917) *Inst. Mitt. Bodenkunde*, **7**, 117.

483 Batel, W. (1955) *Refract. J.*, 468.

484 Pietsch, W. and Rumpf, H. (1967) *Chem. Ing. Tech.*, **39**, 885.

485 Schubert, H. (1973) *Chem. Ing. Tech.*, **45**, 396.

486 Harnby, N., Hawkins, A.E., and Opalinski, I. (1996) *Trans. Inst. Chem. Eng.*, **74**, 605. 616.

487 Bocquet, L., Charlaix, E., Ciliberto, S., and Crassous, J. (1998) *Nature*, **396**, 735.

488 Rabinovich, Y.I., Esayanur, M.S., Johanson, K.D., Adler, J.J., and Moudgil, B.M. (2002) *J. Adhes. Sci. Technol.*, **16**, 887.

489 Herminghaus, S. (2005) *Adv. Phys.*, **54**, 221.

490 Fournier, Z., Geromichalos, D., Herminghaus, S., Kohonen, M.M., Mugele, F., Scheel, M., Schulz, B., Schier, C., Seemann, R., and Skudelny, A. (2005) *J. Phys.: Condens. Matter*, **17**, S477.

491 Fuji, M., Machida, K., Takei, T., Watanabe, T., and Chikazawa, M. (1999) *Langmuir*, **15**, 4584.

492 Price, R., Young, P.M., Edge, S., and Staniforth, J.N. (2002) *Int. J. Pharm.*, **246**, 47.

493 Jones, R., Pollock, H.M., Cleaver, J.A.S., and Hodges, C.S. (2002) *Langmuir*, **18**, 8045.

494 Qian, J. and Gao, H. (2006) *Acta Biomater.*, **2**, 51.

495 Crassous, J., Bocquet, L., Ciliberto, S., and Laroche, C. (1999) *Europhys. Lett.*, **47**, 562.

496 Riedo, E., Palaci, I., Boragno, C., and Brune, H. (2004) *J. Phys. Chem. B*, **108**, 5324.

497 Huppmann, W.J. and Riegger, H. (1975) *Acta Met.*, **23**, 965.

498 Li, Y. and Talke, F.E. (1990) *Tribology and Mechanics of Magnetic Storage Systems*, SP 27, Society of Tribologists, Park Ridge, Illinois, p. 79.

499 Mate, C.M. (1992) *J. Appl. Phys.*, **72**, 3084.

500 Tian, H. and Matsudaira, T. (1993) *ASME J. Tribol.*, **115**, 28.

501 Gao, C. and Bhushan, B. (1995) *Wear*, **190**, 60.

502 Eggleton, A.E.J. and Puddington, I.E. (1954) *Can. J. Chem.*, **32**, 86.

503 Löwen, H. (1995) *Phys. Rev. Lett.*, **74**, 1028.

504 Bloomquist, C.R. and Shutt, R.S. (1940) *Ind. Eng. Chem.*, **32**, 827.

505 Thomson, W. (1870) *Proc. R. Soc. Edinb.*, 7, 63.

506 Fisher, L.R. and Israelachvili, J.N. (1981) *J. Colloid Interface Sci.*, 80, 528; (1981) *Colloids Surf.*, 3, 303.

507 Kohonen, M. and Christenson, H.K. (2000) *Langmuir*, 16, 7285.

508 Young, T. (1805) *Philos. Trans. R. Soc. Lond.*, 95, 65.

509 de Laplace, P.S. *Mécanique Céleste*, suppl. au Livre, X., (1805) Croucier, Paris.

510 Melrose, J.C. (1966) *Am. Inst. Chem. Eng. J.*, 12, 986.

511 Cross, N.L. and Picknett, R.G. (1963) *International Conference on Mechanism of Corrosion by Fuel Impurities* (eds H.R. Johnson and D.J. Littler), Butterworths, Marchwood, UK, p. 383.

512 Princen, H.M. (1968) *J. Colloid Interface Sci.*, 26, 249.

513 Orr, F.M., Scriven, L.E., and Rivas, A.P. (1975) *J. Fluid Mech.*, 67, 723.

514 Woodrow, J., Chilton, H., and Hawes, R.I. (1961) *J. Nucl. Energy B: Reactor Technol.*, 1, 229.

515 de Bisschop, F.R.E. and Rigole, W.J.L. (1982) *J. Colloid Interface Sci.*, 88, 117.

516 Pakarinen, O.H., Foster, A.S., Paajanen, M., Kalinainen, T., Katainen, J., Makkonen, I., Lahtinen, J., and Nieminen, R.M. (2005) *Model. Simul. Mater. Sci. Eng.*, 13, 1175.

517 Melrose, J.C. and Wallick, G.C. (1967) *J. Phys. Chem.*, 71, 3676.

518 Heady, R.B. and Cahn, J.W. (1970) *Metall. Trans.*, 1, 185.

519 Hotta, K., Takeda, K., and Iinoya, K. (1974) *Powder Technology*, 10, 231.

520 Pitois, O., Moucheront, P., and Chateau, X. (2000) *J. Colloid Interface Sci.*, 231, 26.

521 de Boer, M.P. and de Boer, P.C.T. (2007) *J. Colloid Interface Sci.*, 311, 171.

522 Rose, W. (1958) *J. Appl. Phys.*, 29, 687.

523 Smolej, V. and Pejovnik, S. (1976) *Z. Metallkunde*, 67, 603.

524 Smith, W.O. (1933) *Physics*, 4, 425.

525 Clark, W.C., Haynes, J.M., and Mason, G. (1968) *Chem. Eng. Sci.*, 23, 810.

526 McFarlane, J.S. and Tabor, D. (1950) *Proc. R. Soc. Lond. A*, 202, 224.

527 Bayramli, E. and Van de Ven, T.G.M. (1987) *J. Colloid Interface Sci.*, 116, 503.

528 Schenk, M., Füting, M., and Reichelt, R. (1998) *J. Appl. Phys.*, 84, 4880.

529 Weeks, B.L., Vaughn, M.W., and DeYoreo, J.J. (2005) *Langmuir*, 21, 8096.

530 Fortes, M.A. (1982) *J. Colloid Interface Sci.*, 88, 338.

531 Mehrotra, V.P. and Sastry, K.V.S. (1980) *Powder Technol.*, 25, 203.

532 Tselishchev, Y.G. and Val'tsifer, V.A. (2003) *Colloid J.*, 65, 385.

533 Mason, G. and Clark, W.C. (1965) *Chem. Eng. Sci.*, 20, 859.

534 Christenson, H.K. (1988) *J. Colloid Interface Sci.*, 121, 170.

535 Kumar, G. and Prabhu, K.N. (2007) *Adv. Colloid Interface Sci.*, 133, 61.

536 Dabrowski, A. (2001) *Adv. Colloid Interface Sci.*, 93, 135.

537 Amirfazli, A. and Neumann, A.W. (2004) *Adv. Colloid Interface Sci.*, 110, 121.

538 de Gennes, P.G. (1985) *Rev. Mod. Phys.*, 57, 827.

539 Charlaix, E. and Crassous, J. (2005) *J. Chem. Phys.*, 122, 184701.

540 Babak, V.G. (1999) *Colloids Surf. A*, 156, 423.

541 Fogden, A. and White, L.R. (1990) *J. Colloid Interface Sci.*, 138, 414.

542 Wensink, E.J.W., Hoffmann, A.C., Apol, M.E.F., and Berendsen, H.J.C. (2000) *Langmuir*, 16, 7392.

543 Shinto, H., Uranishi, K., Miyahara, M., and Higashitani, K. (2002) *J. Chem. Phys.*, 116, 9500.

544 Jang, J., Yang, M., and Schatz, G. (2007) *J. Chem. Phys.*, 126, 174705.

545 Ando, Y. (2000) *Wear*, 238, 12.

546 Quon, R.A., Ulman, A., and Vanderlick, T.K. (2000) *Langmuir*, 16, 8912.

547 Cleaver, J.A.S. and Tyrrell, J.W.G. (2004) *KONA*, 22, 9.

548 Podczeck, F., Newton, J.M., and James, M.B. (1997) *J. Colloid Interface Sci.*, 187, 484.

549 Hooton, J.C., German, C.S., Allen, S., Davies, M.C., Roberts, C.J., Tendler, S.J.B., and Williams, P.M. (2004) *Pharm. Res.*, 21, 953.

550 Thundat, T., Zheng, X.Y., Chen, G.Y., and Warmack, R.J. (1993) *Surf. Sci. Lett.*, 294, L939.

551 Xiao, X. and Qian, L. (2000) *Langmuir*, 16, 8153.

552 He, M., Blum, A.S., Aston, D.E., Buenviaje, C., and Overney, R.M. (2001) *J. Phys. Chem.*, 114, 1355.

553 Farshchi, M., Kappl, M., Cheng, Y., Gutmann, J.S., and Butt, H.-J. (2006) *Langmuir*, 22, 2171.

554 Asay, D.B. and Kim, S.H. (2006) *J. Chem. Phys.*, 124, 174712.

555 Paajanen, M., Katainen, J., Pakarinen, O.H., Foster, A.S., and Lahtinen, J. (2006) *J. Colloid Interface Sci.*, 304, 518.

556 Halsey, T.C. and Levine, A.J. (1998) *Phys. Rev. Lett.*, 80, 3141.

557 Rabinovich, Y.I., Adler, J.J., Esayanur, M.S., Ata, A., Singh, R.K., and Moudgil, B.M. (2002) *Adv. Colloid Interface Sci.*, 96, 213.

558 Butt, H.-J. (2008) *Langmuir*, 24, 4715.

559 Butt, H.-J., Farshchi-Tabrizi, M., and Kappl, M. (2006) *J. Appl. Phys.*, 100, 024312.

560 Restagno, F., Bocquet, L., Crassous, J., and Charlaix, E. (2002) *Colloids Surf. A*, 206, 69.

561 Crassous, J., Charlaix, E., and Loubet, J.L. (1994) *Europhys. Lett.*, 28, 37.

562 Derjaguin, B.V. and Churaev, N.V. (1976) *J. Colloid Interface Sci.*, 54, 157.

563 Forcada, M.L. (1993) *J. Chem. Phys.*, 98, 638.

564 Christenson, H.K. (1994) *Phys. Rev. Lett.*, 73, 1821.

565 Maeda, N., Israelachvili, J.N., and Kohonen, M.M. (2003) *Proc. Natl. Acad. Sci. USA*, 100, 803.

566 Yushchenko, V.S., Yaminsky, V.V., and Shchukin, E.D. (1983) *J. Colloid Interface Sci.*, 96, 307.

567 Pitois, O., Moucheront, P., and Chateau, X. (2001) *Eur. Phys. J. B*, 23, 79.

568 Willett, C.D., Adams, M.J., Johnson, S.A., and Seville, J.P.K. (2000) *Langmuir*, 16, 9396.

569 Gillette, R.D. and Dyson, D.C. (1971) *Chem. Eng.*, 2, 44.

570 Cai, S. and Bhushan, B. (2008) *Philos. Trans. R. Soc. Lond. A*, 366, 1627.

571 Kohonen, M.M., Maeda, N., and Christenson, H.K. (1999) *Phys. Rev. Lett.*, 82, 4667.

572 Wei, Z. and Zhao, Y.P. (2007) *J. Phys. D: Appl. Phys.*, 40, 4368.

573 Maeda, N., Kohonen, M.M., and Christenson, H.K. (2000) *Phys. Rev. E*, 61, 7239.

574 Christenson, H.K. and Claesson, P.M. (1988) *Science*, 235, 390.

575 Attard, P. (2003) *Adv. Colloid Interface Sci.*, 104, 75.

576 Wennerström, H. (2003) *J. Phys. Chem. B*, 107, 13772.

577 Yaminsky, V.V., Yushchenko, V.S., Amelina, E.A., and Shchukin, E.D. (1983) *J. Colloid Interface Sci.*, 96, 301.

578 Bérard, D.R., Attard, P., and Patey, G.N. (1993) *J. Chem. Phys.*, 98, 7236.

579 Andrienko, D., Patricio, P., and Vinogradova, O.I. (2004) *J. Chem. Phys.*, 121, 4414.

580 Bauer, C., Bieker, T., and Dietrich, S. (2000) *Phys. Rev. E*, 62, 5324.

581 Poniewierski, A. and Sluckint, T.J. (1987) *Liquid Cryst.*, 2, 281.

582 Kočevar, K., Borštnik, A., Muševič, I., and Žumer, S. (2001) *Phys. Rev. Lett.*, 86, 5914.

583 Wennerström, H., Thuresson, K., Linse, P., and Freyssingeas, E. (1998) *Langmuir*, 14, 5664.

584 Sprakel, J., Besseling, N.A.M., Leermakers, F.A.M., and Cohen-Stuart, M.A. (2007) *Phys. Rev. Lett.*, 99, 104504.

585 Scheludko, A., Toshev, B.V., and Bojadjiev, D.T. (1976) *J. Chem. Soc., Faraday Trans. 1*, 72, 2815.

586 Aveyard, R. and Clint, J.H. (1995) *J. Chem. Soc., Faraday Trans.*, 91, 2681.

587 Hinsch, K. (1983) *J. Colloid Interface Sci.*, 92, 243.

588 Nicolson, M.M. (1949) *Proc. Camb. Philos. Soc.*, 45, 288.

589 Gifford, W.A. and Scriven, L.E. (1971) *Chem. Eng. Sci.*, 26, 287.

590 Kralchevsky, P.A. and Nagayama, K. (2000) *Adv. Colloid Interface Sci.*, 85, 145.

591 Nguyen, A.V. and Schulze, H.J. (2003) *Colloidal Science of Flotation, Surfactant Science Series*, vol. 118, Marcel Dekker, New York.

592 Chesters, A.K. (1991) *Chem. Eng. Res. Des.*, 69, 259.

593 Binks, B.P. (2002) *Curr. Opin. Colloid Interface Sci.*, 7, 21.

594 Studart, A.R., Gonzenbach, U.T., Akartuna, I., Tervoort, E., and Gaukler, L.J. (2007) *J. Mater. Chem.*, **17**, 3283.

595 Denkov, N.D., Velev, O.D., Kralchevsky, P.A., Ivanov, I.B., Yoshimura, H., and Nagayama, K. (1992) *Langmuir*, **8**, 3183.

596 Kralchevsky, P.A. and Denkov, N.D. (2001) *Curr. Opin. Colloid Interface Sci.*, **6**, 383.

597 Bakker, G. (1928) *Handbuch der Experimentalphysik VI* (eds W. Wien and F. Harms), Akademische Verlagsgesellschaft, Leipzig, p. 80.

598 Chan, D.Y.C., Henry, J.D., and White, L.R. (1981) *J. Colloid Interface Sci.*, **79**, 410.

599 Paunov, V.N., Kralchevsky, P.A., Denkov, N.D., and Nagayama, K. (1993) *J. Colloid Interface Sci.*, **157**, 100.

600 Allain, C. and Jouhier, B. (1983) *J. Phys. Lett.*, **44**, 421.

601 Dushkin, C.D., Kralchevsky, P.A., Yoshimura, H., and Nagayama, K. (1995) *Phys. Rev. Lett.*, **75**, 3454.

602 Bowden, N., Terford, A., Carbeck, J., and Whitesides, G.M. (1997) *Science*, **276**, 233.

603 Bresme, F. and Oettel, M. (2007) *J. Phys.: Condens. Matter*, **19**, 413101.

604 Pieranski, P. (1980) *Phys. Rev. Lett.*, **45**, 569.

605 Onoda, G.Y. (1985) *Phys. Rev. Lett.*, **55**, 226.

606 Aveyard, R., Binks, B.P., Clint, J.H., Fletcher, P.D.I., Horozov, T.S., Neumann, B., Paunov, V.N., Annesley, J., Botchway, S.W., Nees, D., Parker, A.W., Ward, A.D., and Burgess, A.N. (2002) *Phys. Rev. Lett.*, **88**, 246102.

607 Hurd, A.J. (1985) *J. Phys. A: Math. Gen.*, **18**, L1055.

608 Netz, R.R. (1999) *Phys. Rev. E*, **60**, 3174.

609 Quesada-Pérez, M., Moncho-Jordá, A., Martínez-López, F., and Hidalgo-Álvarez, R. (2001) *J. Chem. Phys.*, **115**, 10897.

610 Ghezzi, F. and Earnshaw, J.C. (1997) *J. Phys.: Condens. Matter*, **9**, L517.

611 Ruiz-García, J., Gámez-Corrales, R., and Ivlev, B.I. (1998) *Phys. Rev. E*, **58**, 660.

612 Sear, R.P., Chung, S.W., Markovich, G., Gelbart, W.M., and Heath, J.R. (1999) *Phy. Rev. E*, **59**, 6255.

613 Lucassen, J. (1992) *Colloids Surf.*, **65**, 131.

614 Stamou, D., Duschl, C., and Johannsmann, D. (2000) *Phys. Rev. E*, **62**, 5263.

615 Lehle, H., Noruzifar, E., and Oettel, M. (2008) *Eur. Phys. J.*

616 Nikolaides, M.G., Bausch, A.R., Hsu, M.F., Dinsmore, A.D., Brenner, M.P., Gay, C., and Weitz, D.A. (2002) *Nature*, **420**, 299.

617 Dominguez, A., Oettel, M., and Dietrich, S. (2007) *J. Chem. Phys.*, **127**, 204706.

618 Golestanian, R., Goulian, M., and Kardar, M. (1996) *Phys. Rev. E*, **54**, 6725.

619 Lehle, H., Oettel, M., and Dietrich, S. (2006) *Europhys. Lett.*, **75**, 174.

620 Fernández-Toledano, J.C., Moncho-Jordá, A., Martínez-López, F., and Hidalgo-Álvarez, R. (2004) *Langmuir*, **20**, 6977.

621 Dan, N., Pincus, P., and Safran, S.A. (1993) *Langmuir*, **9**, 2768.

622 Gil, T., Sabra, M.C., Ipsen, J.H., and Mouritsen, O.G. (1997) *Biophys. J.*, **73**, 1728.

623 Bartolo, D. and Fournier, J.B. (2003) *Eur. Phys. J. E*, **11**, 141.

624 Elliott, J.A.W. and Voitcu, O. (2007) *Can. J. Chem. Eng.*, **85**, 692.

625 Landau, L.D. and Lifshitz, E.M. (1987) Fluid mechanics, *Course of Theoretical Physics*, vol. 6, Butterworth-Heinemann.

626 Guyon, E., Hulin, J.P., Petit, L., and Mitescu, C.D. (2001) *Physical Hydrodynamics*, Oxford University Press, Oxford.

627 Gompper, G., Ihle, T., Kroll, D.M., and Winkler, R.G. (2009) *Adv. Polym. Sci.*, **221**, 1.

628 Cox, R.G. (1974) *Int. J. Multiphase Flow*, **1**, 343.

629 Chan, D.Y.C. and Horn, R.G. (1985) *J. Chem. Phys.*, **83**, 5311.

630 Reynolds, O. (1886) *Philos. Trans. R. Soc. Lond.*, **177**, 157.

631 Hardy, W. and Bircumshaw, I. (1925) *Proc. R. Soc. Lond. A*, **108**, 1.

632 Except by a factor 9/8 derived by Lorentz, H.A., (1907) *Abhandlungen Über Theoretische Physik*, vol. 1, Teubner Verlag, Leipzig, p. 23.

633 Altrichter, F. and Lustig, A. (1937) *Phys. Z.*, **38**, 786.

634 Hopper, V.D. and Grant, A.M. (1948) *Aust. J. Sci. Res. A*, **1**, 28.

635 MacKay, G.D.M., Suzuki, M., and Mason, S.G. (1963) *J. Colloid Sci.*, **18**, 103.

636 Stark, R., Bonaccurso, E., Kappl, M., and Butt, H.J. (2006) *Polymer*, **47**, 7259.

637 Honig, C.D.F. and Ducker, W.A. (2007) *Phys. Rev. Lett.*, **98**, 028305.

638 Vinogradova, O.I., Butt, H.-J., Yakubov, G.E., and Feuillebois, F. (2001) *Rev. Sci. Instrum.*, **72**, 2330.

639 Stefan, M.J. (1874) *Sitzungsberichte Akad. Wiss. Wien (Abt. II Math. Phys.)*, **69**, 713.

640 Goldman, A.J., Cox, R.G., and Brenner, H. (1967) *Chem. Eng. Sci.*, **22**, 653.

641 Goldsmith, H.L. and Mason, S.G. (1962) *J. Colloid Sci.*, **17**, 448.

642 Goren, S.L. and O'Neill, M.E. (1971) *Chem. Eng. Sci.*, **26**, 325.

643 O'Neill, M.E. (1964) *Mathematika*, **11**, 67.

644 Israelachvili, J.N., McGuiggan, P.M., and Homola, H.M. (1988) *Science*, **240**, 189.

645 Gee, M.L., McGuiggan, P.M., Israelachvili, J.N., and Homola, A.M. (1990) *J. Chem. Phys.*, **93**, 1895.

646 Kumacheva, E. and Klein, J. (1998) *J. Chem. Phys.*, **108**, 7010.

647 Demirel, A.L. and Granick, S. (2001) *J. Chem. Phys.*, **115**, 1498.

648 Bitsanis, I., Vanderlick, T.K., Tirrell, M., and Davis, H.T. (1988) *J. Chem. Phys.*, **89**, 3152.

649 Schoen, M., Rhykerd, C.L., Diestler, D.J., and Cushman, J.H. (1989) *Science*, **245**, 1223.

650 Thompson, P.A., Grest, G.S., and Robbins, M.O. (1992) *Phys. Rev. Lett.*, **68**, 3448.

651 Israelachvili, J.N. (1986) *J. Colloid Interface Sci.*, **110**, 263.

652 Raviv, U., Laurat, P., and Klein, J. (2001) *Nature*, **413**, 51.

653 Huisman, W.J., Peters, J.F., Zwanenburg, M.J., de Vries, S.A., Derry, T.E., Abernathy, D., and van der Veen, J.F. (1997) *Nature*, **390**, 379.

654 Doerr, A.K., Tolan, M., Seydel, T., and Press, W. (1998) *Physica B*, **248**, 263.

655 Yu, C.J., Richter, A.G., Kmetko, J., Datta, A., and Dutta, P. (2000) *Europhys. Lett.*, **50**, 487.

656 van Alsten, J. and Granick, S. (1988) *Phys. Rev. Lett.*, **61**, 2570.

657 Homola, A.M., Israelachvili, J.N., McGuiggan, P.M., and Gee, M.L. (1990) *Wear*, **136**, 65.

658 Gao, J., Luedtke, W.D., and Landman, U. (1997) *Phys. Rev. Lett.*, **79**, 705.

659 Benbow, J.J. and Lamb, P. (1963) *SPE Trans.*, **3**, 7.

660 Denn, M.M. (1990) *Annu. Rev. Fluid Mech.*, **22**, 13.

661 Vinogradova, O.I. (1999) *Int. J. Miner. Process.*, **56**, 31.

662 Pit, R., Hervet, H., and Léger, L. (2000) *Phys. Rev. Lett.*, **85**, 980.

663 Baudry, J., Charlaix, E., Tonck, A., and Mazuyer, D. (2001) *Langmuir*, **17**, 5232.

664 Tretheway, D.C. and Meinhart, C.D. (2002) *Phys. Fluids*, **14**, L9.

665 Horn, R.G., Vinogradova, O.I., Mackay, M.E., and Phan-Thien, N. (2000) *J. Chem. Phys.*, **112**, 6424.

666 Cheng, J.T. and Giordano, N. (2002) *Phys. Rev. E*, **65**, 031206.

667 Maxwell, J.C. (1867) *Philos. Trans. R. Soc. Lond. A*, **170**, 231.

668 Sun, M. and Ebner, C. (1992) *Phys. Rev. Lett.*, **69**, 3491.

669 Thompson, P.A. and Troian, S.M. (1997) *Nature*, **389**, 360.

670 Stevens, M.J., Mondello, M., Grest, G.S., Cui, S.T., Cochran, H.D., and Cummings, P.T. (1997) *J. Chem. Phys.*, **106**, 7303.

671 Barrat, J.L. and Bocquet, L. (1999) *Phys. Rev. Lett.*, **82**, 4671.

672 Gao, J., Luedtke, W.D., and Landman, U. (2000) *Tribol. Lett.*, **9**, 3.

673 Cottin-Bizonne, C., Steinberger, A., Cross, B., Raccurt, O., and Charlaix, E. (2008) *Langmuir*, **24**, 1165.

674 Lasne, D., Maali, A., Amarouchene, Y., Cognet, L., Lounis, B., and Kellay, H. (2008) *Phys. Rev. Lett.*, **100**.

675 Huang, P., Guasto, J.S., and Breuer, K.S. (2006) *J. Fluid Mech.*, **566**, 447.

676 Durbin, P.A. (1988) *J. Fluid Mech.*, **197**, 157.

677 Joseph, P. and Tabeling, P. (2005) *Phys. Rev. E*, **71**, 035303.

678 Neto, C., Evans, D.R., Bonaccurso, E., Butt, H.-J., and Craig, V.S.J. (2005) *Rep. Prog. Phys.*, **68**, 2859.

679 Lauga, E., Brenner, M.P., and Stone, H.A. (2007) *Handbook of Experimental Fluid Dynamics* (eds C. Tropea, A. Yarin, and J.F. Foss), Springer, New York, p. 1219.

680 Vinogradova, O.I. (1995) *Langmuir*, **11**, 2213.

681 Goren, S.L. (1973) *J. Colloid Interface Sci.*, **44**, 356.

682 Posner, A.M., Anderson, J.R., and Alexander, A.E. (1952) *J. Colloid Sci.*, **7**, 623.

683 Weissenborn, P.K. and Pugh, R.J. (1995) *Langmuir*, **11**, 1422.

684 Miller, R. and Liggieri, L. (2009) Interfacial rheology, *Progress in Colloid and Interface Science*, vol. 1, Brill Publ., Leiden.

685 Nguyen, A.V., Evans, G.M., Nalaskowski, J., and Miller, J.D. (2004) *Exp. Therm. Fluid Sci.*, **28**, 387.

686 Hartland, S. (1969) *Chem. Eng. Sci.*, **24**, 987.

687 Derjaguin, B. and Kussakov, M. (1939) *Acta Physicochim. URSS*, **10**, 25.

688 Platikanov, D. (1964) *J. Phys. Chem.*, **68**, 3619.

689 Allan, R.S., Charles, G.E., and Mason, S.G. (1961) *J. Colloid Sci.*, **16**, 150.

690 Frankel, S.P. and Mysels, K.J. (1962) *J. Phys. Chem.*, **66**, 190.

691 Burrill, K.A. and Woods, D.R. (1973) *J. Colloid Interface Sci.*, **42**, 15.

692 Jain, R.K. and Ivanov, I.B. (1980) *J. Chem. Soc., Faraday Trans.*, 2, **76**, 250.

693 Jeelani, S.A.K. and Hartland, S. (1994) *J. Colloid Interface Sci.*, **164**, 296.

694 Princen, H.M. (1963) *J. Colloid Sci.*, **18**, 178.

695 Krasowska, M. and Malysa, K. (2007) *Adv. Colloid Interface Sci.*, **134–135**, 138.

696 Webber, G.B., Manica, R., Edwards, S.A., Carnie, S.L., Stevens, G.W., Grieser, F., Dagastine, R.R., and Chan, D.Y.C. (2008) *J. Phys. Chem. C*, **112**, 567.

697 Manica, R., Connor, J.N., Carnie, S.L., Horn, R.G., and Chan, D.Y.C. (2007) *Langmuir*, **23**, 626.

698 Connor, J.N. and Horn, R.G. (2003) *Faraday Discuss.*, **123**, 193.

699 Klaseboer, E., Chevaillier, J.P., Gourdon, C., and Masbernat, O. (2000) *J. Colloid Interface Sci.*, **229**, 274.

700 Clasohm, L.Y., Connor, J.N., Vinogradova, O.I., and Horn, R.G. (2005) *Langmuir*, **21**, 8243.

701 Velev, O.D., Gurkov, T.D., and Borwankar, R.P. (1993) *J. Colloid Interface Sci.*, **159**, 497.

702 Yiantsios, S.G. and Davis, R.H. (1990) *J. Fluid Mech.*, **217**, 547.

703 Bazhlekov, I.B., Chesters, A.K., and van de Vosse, F.N. (2000) *Int. J. Multiphase Flow*, **26**, 445.

704 Levan, M.D. and Newman, J. (1976) *AIChE J.*, **22**, 695.

705 Blawzdziewicz, J., Wajnryb, E., and Loewenberg, M. (1999) *J. Fluid Mech.*, **395**, 29.

706 Marangoni, C. (1871) *Ann. Phys. Chem.*, **143**, 337.

707 Thomson, J. (1855) *Philos. Mag.*, **10**, 330.

708 Hadamard, J.S. (1911) *C.R. Acad. Sci.*, **152**, 1735.

709 Rybczynski, W. (1911) *Bull. Int. Acad. Sci. Cracov. Ser. A*, 40.

710 Levich, V.G. (1962) *Physicochemical Hydrodynamics*, Prentice-Hall, Englewood Cliffs, NJ.

711 Kelsall, G.H., Tang, S., Smith, A.L., and Yurdakul, S. (1996) *J. Chem. Soc., Faraday Trans.*, **92**, 3879.

712 Parkinson, L., Sedev, R., Fornasiero, D., and Ralston, J. (2008) *J. Colloid Interface Sci.*, **322**, 168.

713 Clift, R., Grace, J.R., and Weber, W.E. (1978) *Bubbles, Drops, and Particles*, Academic Press, New York.

714 Pawar, Y. and Stebe, K.J. (1996) *Phys. Fluids*, **8**, 1738.

715 Cuenot, B., Magnaudet, J., and Spennato, B. (1997) *J. Fluid Mech.*, **339**, 25.

716 Takagi, S., Ogasawara, T., and Matsumoto, Y. (2008) *Philos. Trans. R. Soc. Lond. A*, **366**, 2117.

717 Ybert, C. and de Meglio, J.M. (2000) *Eur. Phys. J.E.*, **3**, 143.

718 Mysels, K.J., Shinoda, K., and Frankel, S. (1959) *Soap Films*, Pergamon Press, London.

719 Scheludko, A. (1957) *Kolloid-Zeitschrift*, **155**, 39.

720 Traykov, T.T. and Ivanov, I.B. (1977) *Int. J. Multiphase Flow*, **3**, 471. 485.

721 Karakashev, S.I. and Nguyen, A.V. (2007) *Colloids Surf. A*, **293**, 229.

722 Manor, O., Vakarelski, I.U., Stevens, G.W., Grieser, F., Dagastine, R.R., and Chan, D.Y.C. (2008) *Langmuir*, **24**, 11533.

723 Manor, O., Vakarelski, I.U., Tang, X.S., O'Shea, S.J., Stevens, G.W., Grieser, F., Dagastine, R.R., and Chan, D.Y.C. (2008) *Phys. Rev. Lett.*, **101**.

724 Yeo, L.Y., Matar, O.K., de Ortiz, E.S.P., and Hewitt, G.E. (2003) *J. Colloid Interface Sci.*, **257**, 93.

725 Zapryanov, Z., Malhotra, A.K., Aderangi, N., and Wasan, D.T. (1983) *Int. J. Multiphase Flow*, **9**, 105.

726 Manev, E.D., Sazdanova, S.V., and Wasan, D.T. (1984) *J. Colloid Interface Sci.*, **97**, 591.

727 Carnie, S.L., Chan, D.Y.C., Lewis, C., Manica, R., and Dagastine, R.R. (2005) *Langmuir*, **21**, 2912.

728 Manica, R., Connor, J.N., Dagastine, R.R., Carnie, S.L., Horn, R.G., and Chan, D.Y.C. (2008) *Phys. Fluids*, **20**, 032101.

729 Schramm, L.L. (2005) *Emulsions, Foams, and Suspensions*, Wiley-VCH Verlag GmbH, Weinheim.

730 Davis, R.H., Schonberg, J.A., and Rallison, J.M. (1989) *Phys. Fluids A*, **1**, 77.

731 Hodgson, T.D. and Lee, J.C. (1969) *J. Colloid Interface Sci.*, **30**, 94.

732 Khristov, K., Taylor, S.D., Czarnecki, J., and Masliyah, J. (2000) *Colloids Surf. A*, **174**, 183.

733 Groothuis, H. and Zuiderweg, F.J. (1960) *Chem. Eng. Sci.*, **12**, 288.

734 Binks, B.P., Cho, W.G., and Fletcher, P.D.I. (1997) *Langmuir*, **13**, 7180.

735 Mostowfi, F., Khristov, K., Czarnecki, J., Masliyah, J., and Bhattacharjee, S. (2007) *Appl. Phys. Lett.*, **90**, 184102.

736 Craig, V.S.J., Ninham, B.W., and Pashley, R.M. (1993) *J. Phys. Chem.*, **97**, 10192.

737 Weaire, D. and Hutzler, S. (1999) *The Physics of Foams*, Clarendon Press, Oxford.

738 Malysa, K. and Lunkenheimer, K. (2008) *Curr. Opin. Colloid Interface Sci.*, **13**, 150.

739 Butt, H.J. (1994) *J. Colloid Interface Sci.*, **166**, 109.

740 Fielden, M.L., Hayes, R.A., and Ralston, J. (1996) *Langmuir*, **12**, 3721.

741 Preuss, M. and Butt, H.-J. (1998) *Langmuir*, **14**, 3164.

742 Johnson, D.J., Miles, N.J., and Hilal, N. (2006) *Adv. Colloid Interface Sci.*, **127**, 67.

743 Mulvaney, P., Perera, J.M., Biggs, S., Grieser, F., and Stevens, G.W. (1996) *J. Colloid Interface Sci.*, **183**, 614.

744 Snyder, B.A., Aston, D.E., and Berg, J.C. (1997) *Langmuir*, **13**, 590.

745 Chevaillier, J.P., Klaseboer, E., Masbernat, O., and Gourdon, C. (2006) *J. Colloid Interface Sci.*, **299**, 472.

746 Prokhorov, P.S. (1954) *Discuss. Faraday Soc.*, **18**, 41; Prokhorov, P.S. and Leonov, L.F. (1960) *Discuss. Faraday Soc.*, **30**, 124.

747 Derjaguin, B.V., Churaev, N.V., and Muller, V.M. (1987) *Surface Forces*, Consultants Bureau, New York.

748 Hooke, R. (1672) *The History of the Royal Society of London*, vol. 4 (ed. T. Birch), A. Miller, London, p. 29.

749 Newton, I. (1704) *Opticks: Or, a Treatise of the Reflexions, Refractions, Inflexions and Colours of Light*, Second book, part I, Smith and Walford, London.

750 Fusinieri, A. (1821) *Giorn. Fisica Brugnatelli*, **4**, 442.

751 Dewar, J. (1923) *Proc. R. Soc. Gt. Brit.*, **24**, 197.

752 Lawrence, A.S.C. (1929) *Soap Films*, G. Bell and Sons, London.

753 Radoev, B.P., Scheludko, A.D., and Manev, E.D. (1983) *J. Colloid Interface Sci.*, **95**, 254.

754 Overbeek, J.T.G. (1960) *J. Phys. Chem.*, **64**, 1178.

755 Jones, M.N., Mysels, K.J., and Scholten, P.C. (1966) *Trans. Faraday Soc.*, **62**, 1336.

756 Exerowa, D. (2002) *Adv. Colloid Interface Sci.*, **96**, 75.

757 Tien, H.T. and Ottova, A.L. (2001) *J. Membr. Sci.*, **189**, 83.

758 Miles, G.D., Ross, J., and Shedlovsky, L. (1950) *J. Am. Oil Chem. Soc.*, **27**, 268.

759 Scheludko, A. (1967) *Adv. Colloid Interface Sci.*, **1**, 391.

760 Langevin, D. and Sonin, A.A. (1994) *Adv. Colloid Interface Sci.*, **51**, 1.

761 Manev, E.D., Sazdanova, S.V., Rao, A.A., and Wasan, D.T. (1982) *J. Dispersion Sci. Technol.*, **3**, 435.

762 Yiantsios, S.G. and Davis, R.H. (1991) *J. Colloid Interface Sci.*, **144**, 412.

763 Ivanov, I.B., Dimitrov, D.S., Somasundaran, P., and Jain, R.K. (1985) *Chem. Eng. Sci.*, **40**, 137.

764 Li, D.M. (1996) *J. Colloid Interface Sci.*, **181**, 34.

765 Joye, J.L., Hirasaki, G.J., and Miller, C.A. (1994) *Langmuir*, **10**, 3174.

766 Manev, E., Tsekov, R., and Radoev, B. (1997) *J. Dispersion Sci. Technol.*, **18**, 769.

767 Angarska, J., Stubenrauch, C., and Manev, E. (2007) *Colloids Surf. A*, **309**, 189.

768 Scheludko, A. and Exerowa, D. (1957) *Kolloid-Zeitschrift*, **155**, 39.

769 Mysels, K.J. (1964) *J. Phys. Chem.*, **68**, 3441.

770 Exerowa, D. and Scheludko, A. (1971) *C.R. Acad. Bulg. Sci.*, **24**, 25.

771 Toshev, B.V. and Ivanov, I.B. (1975) *Colloid Polym. Sci.*, **253**, 558.

772 Scheludko, A. and Platikanowa, D. (1961) *Kolloid-Zeitschrift*, **175**, 150.

773 Duyvis, E.M. (1962) The equilibrium thickness of free liquid films, Ph.D. thesis, University Utrecht.

774 Shishin, V.A., Zorin, Z.M., and Churaev, N.V. (1977) *Kolloid Zh. (Engl. Ed.)*, **39**, 351.

775 Bélorgey, O. and Benattar, J.J. (1991) *Phys. Rev. Lett.*, **66**, 313.

776 Velev, O.D., Constantinides, G.N., Avraam, D.G., Payatakes, A.C., and Borwankar, R.P. (1995) *J. Colloid Interface Sci.*, **175**, 68.

777 Cascão Pereira, L.G., Johansson, C., Blanch, H.W., and Radke, C.J. (2001) *Colloids Surf. A*, **186**, 103.

778 Lyklema, J., Scholten, P.C., and Mysels, K.J. (1965) *J. Phys. Chem.*, **69**, 116.

779 Bergeron, V. and Radke, C.J. (1992) *Langmuir*, **8**, 3020; Bergeron, V., Jimenez-Laguna, A.I., and Radke, C.J. (1992) *Langmuir*, **8**, 3027.

780 Exerowa, D., Kolarov, T., and Khristov, K. (1987) *Colloids Surf.*, **22**, 171.

781 Stubenrauch, C., Schlarmann, J., and Strey, R. (2002) *Phys. Chem. Chem. Phys.*, **4**, 4504.

782 Schulze-Schlarmann, J., Buchavzov, N., and Stubenrauch, C. (2006) *Soft Matter*, **2**, 584.

783 Stubenrauch, C. von Klitzing, R. (2003) *J. Phys.: Condens. Matter*, **15**, R1197.

784 Kolarov, T., Cohen, R., and Exerowa, D. (1989) *Colloids Surf.*, **42**, 49.

785 Manev, E.D. and Pugh, R.J. (1991) *Langmuir*, **7**, 2253.

786 Bergeron, V., Waltermo, A., and Claesson, P.M. (1996) *Langmuir*, **12**, 1336.

787 Beattie, J.K., Djerdjev, A.M., and Warr, G.G. (2009) *Faraday Discuss.*, **141**, 31.

788 Jungwirth, P. and Winter, B. (2008) *Annu. Rev. Phys. Chem.*, **59**, 343.

789 Jungwirth, P. (2009) *Faraday Discuss.*, **141**, 9.

790 Petersen, P.B. and Saykally, R.J. (2008) *Chem. Phys. Lett.*, **458**, 255.

791 Yeo, L.Y. and Matar, O.K. (2003) *J. Colloid Interface Sci.*, **261**, 575.

792 Barber, A.D. and Hartland, S. (1976) *Can. J. Chem. Eng.*, **54**, 279.

793 Narsimhan, G. (2009) *J. Colloid Interface Sci.*, **330**, 494.

794 Bergeron, V., Langevin, D., and Asnacios, A. (1996) *Langmuir*, **12**, 1550.

795 von Klitzing, R., Espert, A., Asnacios, A., Hellweg, T., Colin, A., and Langevin, D. (1999) *Colloids Surf. A*, **149**, 131.

796 Kolaric, B. and Jaeger, W. von Klitzing, R. (2000) *J. Phys. Chem. B*, **104**, 5096.

797 de Vries, A.J. (1958) *Recl. Trav. Chim.*, **77**, 383.

798 Maldarelli, C., Jain, R.K., Ivanov, I.B., and Ruckenstein, E. (1980) *J. Colloid Interface Sci.*, **78**, 118.

799 Bartsch, O. (1924) *Kolloid-Beih.*, **20**, 1.

800 Stubenrauch, C. and Miller, R. (2004) *J. Phys. Chem. B*, **108**, 6412.

801 Anderson, C.H. and Sabisky, E.S. (1970) *Phys. Rev. Lett.*, **24**, 1049; Sabisky, E.S. and Anderson, C.H. (1973) *Phys. Rev. A*, **7**, 790.

802 Brunauer, S. (1945) *The Adsorption of Gases and Vapors*, Oxford University Press.

803 Halsey, G. (1948) *J. Chem. Phys.*, **16**, 931.

804 Hill, T.L. (1952) *Adv. Catal. and Related Subjects*, vol. 4, Academic Press, New York, p. 211.

805 Adamson, A.W. (1990) *Physical Chemistry of Surfaces*, John Wiley & Sons Inc., New York.

806 Gee, M.L., Healy, T.W., and White, L.R. (1989) *J. Colloid Interface Sci.*, **131**, 18.

807 Tibus, S., Klier, J., and Leiderer, P. (2005) *J. Low Temp. Phys.*, **139**, 531.

808 Asay, D.B. and Kim, S.H. (2005) *J. Phys. Chem. B*, **109**, 16760.

809 Princen, H.M. and Mason, S.G. (1965) *J. Colloid Sci.*, **29**, 156.

810 Scheludko, A., Caljovska, S., Fabrikant, A., Radoev, B., and Schulze, H.J. (1971) *Freiberger Forsch. Hefte A*, **484**, 85.

811 Blake, T.D. and Kitchener, J.A. (1972) *J. Chem. Soc., Faraday Trans. 1*, **68**, 1435.

812 Churaev, N.V. (2003) *Adv. Colloid Interface Sci.*, **103**, 197.

813 Stöckelhuber, K.W., Radoev, B., Wenger, A., and Schulze, H.J. (2004) *Langmuir*, **20**, 164.

814 Schulze, H.J., Stöckelhuber, K.W., and Wenger, A. (2001) *Colloids Surf. A*, **192**, 61.

815 Reiter, G. (1992) *Phys. Rev. Lett.*, **68**, 75; (1993) *Langmuir*, **9**, 1344.

816 Müller-Buschbaum, P., Vanhoorne, P., Scheumann, V., and Stamm, M. (1997) *Europhys. Lett.*, **40**, 655.

817 Reiter, G. (1998) *Science*, **282**, 888.

818 Jacobs, K., Herminghaus, S., and Mecke, K.R. (1998) *Langmuir*, **14**, 965.

819 Xie, R., Karim, A., Douglas, J.F., Han, C.C., and Weiss, R.A. (1998) *Phys. Rev. Lett.*, **81**, 1251.

820 Seemann, R., Herminghaus, S., and Jacobs, K. (2001) *Phys. Rev. Lett.*, **86**, 5534.

821 Bischof, J., Scherer, D., Herminghaus, S., and Leiderer, P. (1996) *Phys. Rev. Lett.*, **77**, 1536.

822 Lee, J.M. and Kim, B.I. (2007) *Mater. Sci. Eng. A*, **449**, 769.

823 Trice, J., Favazza, C., Thomas, D., Garcia, H., Kalyanaraman, R., and Sureshkumar, R. (2008) *Phys. Rev. Lett.*, **101**, 017802.

824 Becker, J., Grün, G., Seemann, R., Mantz, H., Jacobs, K., Mecke, K.R., and Blossey, R. (2003) *Nat. Mater.*, **2**, 59.

825 Brochard-Wyart, F. and Daillant, J. (1990) *Can. J. Phys.*, **68**, 1084.

826 Derjaguin, B.V. and Gutop, Y.V. (1962) *Kolloid Zh.*, **24**, 431.

827 Kheshgi, H.S. and Scriven, L.E. (1991) *Chem. Eng. Sci.*, **46**, 519.

828 Redon, C., Brochard-Wyart, F., and Rondelez, F. (1991) *Phys. Rev. Lett.*, **66**, 715.

829 Brochard-Wyart, F., Debrégeas, G., Fondecave, R., and Martin, P. (1997) *Macromolecules*, **30**, 1211.

830 Vrij, A. (1966) *Discuss. Faraday Soc.*, **42**, 23.

831 Ruckenstein, E. and Jain, R.K. (1974) *J. Chem. Soc., Faraday Trans. 2*, **70**, 132.

832 Williams, M.B. and Davis, S.H. (1982) *J. Colloid Interface Sci.*, **90**, 220.

833 Sharma, A. and Ruckenstein, E. (1986) *J. Colloid Interface Sci.*, **113**, 456.

834 Sharma, A. and Khanna, R. (1998) *Phys. Rev. Lett.*, **81**, 3463.

835 Bertrand, E., Blake, T.D., and De Coninck, J. (2009) *Eur. Phys. J. Spec. Top.*, **166**, 173.

836 Vrij, A., Hesselink, F.T., Lucassen, J., and van den Tempel, M. (1970) *Proceedings of the Koninklijke Nederlandse Akademie van Wetenschappen B*, vol. 73, p. 124.

837 Shuttleworth, R. (1950) *Proc. Phys. Soc. (Lond.) A*, **63**, 444.

838 Shuttleworth, R. (1949) *Proc. Phys. Soc. (Lond.) A*, **62**, 167.

839 Tasker, P.W. (1979) *Philos. Mag. A*, **39**, 119.

840 Butt, H.-J. and Raiteri, R. (1999) *Surface Characterization Methods* (eds A.J. Milling and A.T. Hubbard), Marcel Dekker, New York.

841 Johnson, K.L. (1985) *Contact Mechanics*, Cambridge University Press, *Cambridge*.

842 Shull, K.R. (2002) *Mater. Sci. Eng. R.*, **36**, 1.

843 Maugis, D. (2000) *Contact, Adhesion and Rupture of Elastic Solids*, Springer, Berlin.

844 Barthel, E. (2008) *J. Phys. D: Appl. Phys.*, **41**, 163001.

845 Greenwood, J.A. (2009) *Philos. Mag.*, **89**, 945.

846 Hertz, H. (1882) *J. Reine Angew. Math.*, **92**, 156.

847 Boussinesq, J. (1885) *Application des potentiels à l'étude de l'équilibre et du movements des solides élastiques*, Gauthier-Villars, Paris.

848 Sneddon, I.N. (1946) *Proc. Camb. Philos. Soc.*, **42**, 29.

849 Kendall, K. (1971) *J. Phys. D: Appl. Phys.*, **4**, 1186.

850 Sneddon, I.N. (1965) *Int. J. Eng. Sci.*, **3**, 47.

851 Ting, T.C.T. (1966) *J. Appl. Mech.*, **33**, 845.

852 Johnson, K.L., Kendall, K., and Roberts, A.D. (1971) *Proc. R. Soc. Lond. A*, **324**, 301.

853 Muller, V.M., Yushchenko, V.S., and Derjaguin, B.V. (1980) *J. Colloid Interface Sci.*, **77**, 91.

854 Derjaquin, B.V., Muller, V.M., and Toporov, Y.P. (1975) *J. Colloid Interface Sci.*, **53**, 314.

855 Muller, V.M., Derjaguin, B.V., and Toporov, Y.P. (1983) *Colloids Surf.*, **7**, 251.

856 Tabor, D. (1977) *J. Colloid Interface Sci.*, **58**, 2.

857 Maugis, D. (1992) *J. Colloid Interface Sci.*, **150**, 243.

858 Greenwood, J.A. (1997) *Proc. R. Soc. Lond. A*, **453**, 1277.

859 Feng, J.Q. (2001) *J. Colloid Interface Sci.*, **238**, 318.

860 Johnson, K.L. and Greenwood, J.A. (1997) *J. Colloid Interface Sci.*, **192**, 326.

861 Carpick, R.W., Ogletree, D.F., and Salmeron, M. (1999) *J. Colloid Interface Sci.*, **211**, 395.

862 Piétrement, O. and Troyon, M. (2000) *J. Colloid Interface Sci.*, **226**, 166.

863 Schwarz, U.D. (2003) *J. Colloid Interface Sci.*, **261**, 99.

864 Greenwood, J.A. and Williamson, J.B.P. (1966) *Proc. R. Soc. Lond.*, **A295**, 300.

865 Fuller, K.N.G. and Tabor, D. (1975) *Proc. R. Soc. Lond. A*, **345**, 327.

866 Rumpf, H. (1974) *Chem. Ing. Tech.*, **46**, 1.

867 Rabinovich, Y.I., Adler, J.J., Ata, A., Singh, R.K., and Moudgil, B.M. (2000) *J. Colloid Interface Sci.*, **232**, 10.

868 Rabinovich, Y.I., Adler, J.J., Ata, A., Singh, R.K., and Moudgil, B.M. (2000) *J. Colloid Interface Sci.*, **232**, 17.

869 Archard, J.F. (1957) *Proc. R. Soc. Lond. A*, **243**, 190.

870 Majumdar, A. and Bhushan, B. (1990) *J. Tribol.*, **112**, 205.

871 Persson, B.N.J. (2007) *Phys. Rev. Lett.*, **99**, 125502.

872 Hyun, S., Pei, L., Molinari, J.F., and Robbins, M.O. (2004) *Phys. Rev. E*, **70**, 026117.

873 Benz, M., Rosenberg, K.J., Kramer, E.J., and Israelachvili, J.N. (2006) *J. Phys. Chem. B*, **110**, 11884.

874 Lorenz, B. and Persson, B.N.J. (2009) *J. Phys.: Condens. Matter*, **21**, 015003.

875 Krupp, H. (1967) *Adv. Colloid Interface Sci.*, **1**, 111.

876 Larsen, R.I. (1958) *Am. Ind. Hyg. Assoc. J.*, **19**, 256.

877 Zimon, A.D. (1963) *Kolloid Zh.*, **25**, 317.

878 Böhme, G., Kling, W., Krupp, H., Lange, H., and Sandstede, G. (1964) *Z. Angew. Phys.*, **16**, 486.

879 Podczeck, F. and Newton, J.M. (1995) *J. Pharm. Sci.*, **84**, 1067.

880 Israelachvili, J.N., Perez, E., and Tandon, R.K. (1980) *J. Colloid Interface Sci.*, **78**, 260.

881 Horn, R.G., Israelachvili, J.N., and Pribac, F. (1987) *J. Colloid Interface Sci.*, **115**, 480.

882 McGuiggan, P.M., Wallace, J.S., Smith, D.T., Sridhar, I., Zheng, Z.W., and Johnson, K.L. (2007) *J. Phys. D: Appl. Phys.*, **40**, 5984.

883 McGuiggan, P.M. (2008) *Langmuir*, **24**, 3970.

884 Chaudhury, M.K. and Whitesides, G.M. (1991) *Langmuir*, **7**, 1013.

885 Ghatak, A., Vorvolakos, K., She, H.Q., Malotky, D.L., and Chaudhury, M.K. (2000) *J. Phys. Chem. B*, **104**, 4018.

886 Rimai, D.S., Demejo, L.P., Vreeland, W., Bowen, R., Gaboury, S.R., and Urban, M.W. (1992) *J. Appl. Phys.*, **71**, 2253.

887 Rimai, D.S., Demejo, L.P., and Bowen, R.C. (1990) *J. Appl. Phys.*, **68**, 6234.

888 Heim, L.O., Blum, J., Preuss, M., and Butt, H.J. (1999) *Phys. Rev. Lett.*, **83**, 3328.

889 Götzinger, M. and Peukert, W. (2004) *Langmuir*, **20**, 5298.

890 Carpick, R.W., Agraït, N., Ogletree, D.F., and Salmeron, M. (1996) *J. Vac. Sci. Technol. B*, **14**, 1289.

891 Lantz, M.A., O'Shea, S.J., Welland, M.E., and Johnson, K.L. (1997) *Phys. Rev. B*, **55**, 10776.

892 Enachescu, M., van den Oetelaar, R.J.A., Carpick, R.W., Ogletree, D.F., Flipse, C.F.J., and Salmeron, M. (1998) *Phys. Rev. Lett.*, **81**, 1877.

893 Yao, H.M., Ciavarella, M., and Gao, H.J. (2007) *J. Colloid Interface Sci.*, **315**, 786.

894 Luan, B.Q. and Robbins, M.O. (2005) *Nature*, **435**, 929.

895 Hoffmann, P.M., Oral, A., Grimble, R.A., Ozer, H.O., Jeffery, S., and Pethica, J.B. (2001) *Proc. R. Soc. Lond. A*, **457**, 1161.

896 Landman, U., Luedtke, W.D., Burnham, N.A., and Colton, R.J. (1990) *Science*, **248**, 454.

897 Schirmeisen, A., Weiner, D., and Fuchs, H. (2006) *Phys. Rev. Lett.*, **97**, 136101.

898 Yanson, A.I., Bollinger, G.R., van den Brom, H.E., Agrait, N., and van Ruitenbeek, J.M. (1998) *Nature*, **395**, 783.

899 Trouwborst, M.L., Huisman, E.H., Bakker, F.L., van der Molen, S.J., and van Wees, B.J. (2008) *Phys. Rev. Lett.*, **100**, 175502.

900 Shiota, T., Mares, A.I., Valkering, A.M.C., Oosterkamp, T.H., and van Ruitenbeek, J.M. (2008) *Phys. Rev. B*, **77**, 125411.

901 Ludema, K.C. (1996) *Friction, Wear, Lubrication: A Textbook in Tribology*, CRC Press, Boca Raton, FL.

902 Bhushan, B. (2002) *Introduction to Tribology*, John Wiley & Sons Inc., New York.

903 Krim, J. (1996) *Sci. Am.*, 48.

904 Persson, B.N.J. (1998) *Sliding Friction: Physical Principles and Applications*, 2nd edn, Springer, Berlin.

905 Krim, J. (2002) *Am. J. Phys.*, **70**, 890.

906 Dowson, D. (1999) *History of Tribology*, 2nd edn, Professional Engineering Publishing, London.

907 Bowden, F.P. and Tabor, D. (1954) *Friction and Lubrication*, Methuen, London.

908 Bowden, F.P. and Tabor, D. (1950) *The Friction and Lubrication of Solids*, Clarendon Press, Oxford.

909 Greenwood, J.A. (1966) *Br. J. Appl. Phys.*, **17**, 1621.

910 Kendall, K. and Tabor, D. (1971) *Proc. R. Soc. Lond. A*, **323**, 321.

911 Dieterich, J.H. and Kilgore, B.D. (1994) *Pure Appl. Geophys.*, **143**, 283.

912 Bowden, F.P. and Hughes, T.P. (1939) *Proc. R. Soc. Lond. A*, **172**, 0280.

913 Greenwood, J.A. and Williamson, J.B.P. (1966) *Proc. R. Soc. Lond.*, **A295**, 300.

914 Feng, Z., Tzeng, Y., and Field, J.E. (1992) *J. Phys. D: Appl. Phys.*, **25**, 1418.

915 Lide, D.R. (ed.) (2006) *CRC Handbook of Chemistry and Physics*, CRC Taylor & Francis, Boca Raton, FL.

916 Prandtl, L. (1928) *Z. Angew. Math. Mech.*, **8**, 85.

917 Tomlinson, G.A. (1929) *Philos. Mag. Ser.*, **7**, 905.

918 Rosenberg, R. (2005) *Phys. Today*, **58**, 50.

919 de Koning, J.J., de Groot, G., and Schenau, G.J.V. (1992) *J. Biomech.*, **25**, 565.

920 Faraday, M. (1859) *Experimental Researches in Chemistry and Physics*, Taylor & Francis, London.

921 Li, Y. and Somorjai, G.A. (2007) *J. Phys. Chem. C*, **111**, 9631.

922 Wettlaufer, J.S. and Dash, J.G. (2000) *Sci. Am.*, **282**, 50.

923 Wettlaufer, J.S. (1999) *Phys. Rev. Lett.*, **82**, 2516.

924 Elbaum, M., Lipson, S.G., and Dash, J.G. (1993) *J. Cryst. Growth*, **129**, 491.

925 Engemann, S., Reichert, H., Dosch, H., Bilgram, J., Honkimaki, V., and Snigirev, A. (2004) *Phys. Rev. Lett.*, **92**, 205701.

926 Daikhin, L. and Tsionsky, V. (2007) *J. Phys.: Condens. Matter*, **19**, 376109.

927 Bowden, F.P. and Tabor, D. (1950/1964) *The Friction and Lubrication of Solids*, Parts 1 and 2, Clarendon Press, Oxford.

928 Rubinstein, S.M., Cohen, G., and Fineberg, J. (2004) *Nature*, **430**, 1005.

929 Dieterich, J.H. (1979) *J. Geophys. Res.*, **84**, 2161.

930 Rice, J.R. and Ruina, A.L. (1983) *J. Appl. Mech., Trans. ASME*, **50**, 343.

931 Baumberger, T. and Caroli, C. (2006) *Adv. Phys.*, **55**, 279.

932 Yang, Z.P., Zhang, H.P., and Marder, M. (2008) *Proc. Natl. Acad. Sci. USA*, **105**, 13264.

933 Scholz, C.H. (1998) *Nature*, **391**, 37.

934 Hersey, M.D. (1969) *J. Lubr. Technol.*, **91**, 260.

935 Reynolds, O. (1876) *Philos. Trans. R. Soc. Lond.*, **116**, 155.

936 Eldredge, K.R. and Tabor, D. (1955) *Proc. R. Soc. Lond. A*, **229**, 181.

937 Tabor, D. (1955) *Proc. R. Soc. Lond. A*, **229**, 198.

938 Heathcote, H.L. (1921) *Proc. Inst. Automotive Eng.*, **15**, 1569.

939 Hunter, S.C. (1961) *J. Appl. Mech.*, **28**, 611.

940 Brilliantov, N.V. and Pöschel, T. (1998) *Europhys. Lett.*, **42**, 511.

941 Xu, Y. and Yung, K.L. (2003) *Europhys. Lett.*, **61**, 620.

942 Yung, K.L. and Xu, Y. (2003) *Nonlinear Dyn.*, **33**, 33.

943 Flom, D.G. and Beuche, A.M. (1959) *J. Appl. Phys.*, **30**, 1725.

944 Pöschel, T., Schwager, T., and Brilliantov, N.V. (1999) *Eur. Phys. J.B*, **10**, 169.

945 Qiu, X.J. (2006) *J. Eng. Mech., ASCE*, **132**, 1241.

946 Qiu, X.J. (2009) *J. Eng. Mech., ASCE*, **135**, 20.

947 Ecke, S. and Butt, H.-J. (2001) *J. Colloid Interface Sci.*, **244**, 432.

948 Sauerbrey, G. (1959) *Z. Phys.*, **155**, 206.

949 Krim, J., Solina, D.H., and Chiarello, R. (1991) *Phys. Rev. Lett.*, **66**, 181.

950 Daly, C. and Krim, J. (1996) *Phys. Rev. Lett.*, **76**, 803.

951 Dayo, A., Alnasrallah, W., and Krim, J. (1998) *Phys. Rev. Lett.*, **80**, 1690.

952 Johannsmann, D. (1999) *Macromol. Chem. Phys.*, **200**, 501.

953 Krim, J. (2007) *Nano Today*, **2**, 38.

954 Hori, Y. (2006) *Hydrodynamic Lubrication*, Springer, Tokyo.

955 Persson, B.N.J. (1998) *Sliding Friction: Physical Principles and Applications*, Springer, Berlin.

956 Martin, H.M. (1916) *Engineering*, **102**, 119.

957 Barus, C. (1893) *Am. J. Sci.*, **45**, 87.

958 Gohar, R. (2002) *Elastohydrodynamics*, 2nd edn, World Scientific Publishing, Singapore.

959 Dowson, D. (1995) *Wear*, **190**, 125.

960 Spikes, H.A. (1999) *Proc. Inst. Mech. Eng. Part J.: J. Eng. Tribol.*, **213**, 335.

961 Ertel, A.M. (1939) *Akad. Nauk SSSR Prikl. Math. Mekh.*, **3**, 41.

962 Grubin, A.N. and Vinogradova, I.E. (1949) *Book no. 30*, Central Scientific Research Institute for Technology and Mechanical Engineering, (translated to English as Department of Scientific and Industrial Research translation no. 337).

963 Dowson, D. (1968) *Proc. Inst. Mech. Eng.*, **182**, 151.

964 Petrusevich, A.I. (1951) *Izv. Akad. Nauk. SSSR (OTN)*, **2**, 209.

965 Dowson, D. and Higginson, G.R.A. (1959) *J. Mech. Eng. Sci.*, **1**, 6.

966 Archard, J.F. and Kirk, M.T. (1961) *Proc. R. Soc. Lond. A*, **261**, 532.

967 Gohar, R. and Cameron, A. (1963) *Nature*, **200**, 458.

968 Jacod, B., Venner, C.H., and Lugt, P.M. (2000) *J. Tribol.*, **123**, 248.

969 Ranger, A.P., Ettles, C.M.M., and Cameron, A. (1975) *Proc. R. Soc. Lond. A*, **346**, 227.

970 Hamrock, B.J. and Dowson, D. (1977) *J. Lubr. Technol.*, **99**, 264.

971 Zhu, D. and Hu, Y.Z. (2001) *Tribol. Trans.*, **44**, 383.

972 Glovnea, R.P., Forrest, A.K., Olver, A.V., and Spikes, H.A. (2003) *Tribol. Lett.*, **15**, 217.

973 Gardiner, D.J., Baird, E.M., Craggs, C., Dare-Edwards, M.P., and Bell, J.C. (1989) *Lubr. Sci.*, **1**, 301.

974 Bhushan, B., Israelachvili, J.N., and Landman, U. (1995) *Nature*, **374**, 607.

975 Dedkov, G.V. (2000) *Phys. Stat. Sol.*, **3**, 179.

976 Braun, O.M. and Naumovets, A.G. (2006) *Surf. Sci. Rep.*, **60**, 79.

977 Hölscher, H., Schirmeisen, A., and Schwarz, Philos. (2008) *Philos. Trans. R. Soc. Lond. A*, **366**, 1383.

978 Szlufarska, I., Chandross, M., and Carpick, R.W. (2008) *J. Phys. D: Appl. Phys.*, **41**, 123001.

979 Gnecco, E. and Meyer, E. (eds) (2007) *Fundamentals of Friction and Wear on the Nanoscale*, Springer, Berlin.

980 Bhushan, B. (ed.) (2008) *Nanotribology and Nanomechanics*, 2nd edn, Springer, Berlin.

981 Schwarz, U.D., Zwörner, O., Köster, P., and Wiesendanger, R. (1997) *Phys. Rev. B*, **56**, 6987. and 6997.

982 McGuiggan, P.M., Zhang, J., and Hsu, S.M. (2001) *Tribol. Lett.*, **10**, 217.

983 Xu, C., Jones, R.L., and Batteas, J.D. (2008) *Scanning*, **30**, 106.

984 Wenning, L. and Müser, M.H. (2001) *Europhys. Lett.*, **54**, 693.

985 Gao, G.T., Cannara, R.J., Carpick, R.W., and Harrison, J.A. (2007) *Langmuir*, **23**, 5394.

986 Mo, Y.F., Turner, K.T., and Szlufarska, I. (2009) *Nature*, **457**, 1116.

987 Homola, A.M., Israelachvili, J.N., McGuiggan, P.M., and Gee, M.L. (1989) *Wear*, **136**, 65.

988 Hurtado, J.A. and Kim, K.S. (1999) *Proc. R. Soc. Lond. A*, **455**, 3363.

989 Akamine, S., Barrett, R.C., and Quate, C.F. (1990) *Appl. Phys. Lett.*, **57**, 316.

990 Germann, G.J., Cohen, S.R., Neubauer, G., McClelland, G.M., Seki, H., and Coulman, D. (1993) *J. Appl. Phys.*, **73**, 163.

991 Howald, L., Lüthi, R., Meyer, E., Gerth, G., Haefke, H., Overney, R., and Güntherodt, H.-J. (1994) *J. Vac. Sci. Technol. B*, **12**, 2227.

992 Fujisawa, S., Kishi, E., Sugawara, Y., and Morita, S. (1995) *Phys. Rev. B*, **51**, 7849.

993 Takano, H. and Fujihira, M. (1996) *J. Vac. Sci. Technol. B*, **14**, 1272.

994 Lüthi, R., Meyer, E., Bammerlin, M., Howald, L., Haefke, H., Lehmann, T., Loppacher, C., Güntherodt, H.-J., Gyalog, T., and Thomas, H. (1996) *J. Vac. Sci. Technol. B*, **14**, 1280.

995 Müller, T., Kässer, T., Labardi, M., Lux-Steiner, M., Marti, O., Mlynek, J., and Krausch, G. (1996) *J. Vac. Sci. Technol. B*, **14**, 1296.

996 Bennewitz, R., Gyalog, T., Guggisberg, M., Bammerlin, M., Meyer, E., and Güntherodt, H.J. (1999) *Phys. Rev. B*, **60**, R11301.

997 Socoliuc, A., Bennewitz, R., Gnecco, E., and Meyer, E. (2004) *Phys. Rev. Lett.*, **92**, 134301.

998 Nakamura, J., Wakunami, S., and Natori, A. (2005) *Phys. Rev. B*, **72**, 235415.

999 Johnson, K.L. and Woodhouse, J. (1997) *Tribol. Lett.*, **5**, 155.

1000 Medyanik, S.N., Liu, W.K., Sung, I.H., and Carpick, R.W. (2006) *Phys. Rev. Lett.*, **97**, 136106.

1001 Hoshi, Y., Kawagishi, T., and Kawakatsu, H. (1999) *Jpn. J. Appl. Phys.*, 3804.

1002 Conley, W.G., Raman, A., and Krousgrill, C.M. (2005) *J. Appl. Phys.*, **98**, 053519.

1003 Morita, S., Fujisawa, S., and Sugawara, Y. (1996) *Surf. Sci. Rep.*, **23**, 1.

1004 Hölscher, H., Schwarz, U.D., Zwörner, O., and Wiesendanger, R. (1998) *Phys. Rev. B*, **57**, 2477.

1005 Maier, S., Sang, Y., Filleter, T., Grant, M., Bennewitz, R., Gnecco, E., and Meyer, E. (2005) *Phys. Rev. B*, **72**, 245418.

1006 Krylov, S.Y., Dijksman, J.A., van Loo, W.A., and Frenken, J.W.M. (2006) *Phys. Rev. Lett.*, **97**, 166103.

1007 Tshiprut, Z., Filippov, A.E., and Urbakh, M. (2008) *J. Phys.: Condens. Matter*, **20**, 354002.

1008 Gnecco, E., Bennewitz, R., Gyalog, T., Loppacher, C., Bammerlin, M., Meyer, E., and Güntherodt, H.J. (2000) *Phys. Rev. Lett.*, **84**, 1172.

1009 Cieplak, M., Smith, E.D., and Robbins, M.O. (1994) *Science*, **265**, 1209.

1010 Cannara, R.J., Brukman, M.J., Cimatu, K., Sumant, A.V., Baldelli, S., and Carpick, R.W. (2007) *Science*, **318**, 780.

1011 Persson, B.N.J. and Volokitin, A.I. (1995) *J. Chem. Phys.*, **103**, 8679.

1012 Witte, G., Weiss, K., Jakob, P., Braun, J., Kostov, K.L., and Woll, C. (1998) *Phys. Rev. Lett.*, **80**, 121.

1013 Park, J.Y., Ogletree, D.F., Thiel, P.A., and Salmeron, M. (2006) *Science*, **313**, 186.

1014 Sørensen, M.R., Jacobsen, K.W., and Stoltze, P. (1996) *Phys. Rev. B*, **53**, 2101.

1015 Mulliah, D., Kenny, S.D., McGee, E., Smith, R., Richter, A., and Wolf, B. (2006) *Nanotechnology*, 1807.

1016 Nakayama, K., Bou-Said, B., and Ikeda, H. (1997) *Trans. ASME, J. Tribol.*, **119**, 764.

1017 Riedo, E., Gnecco, E., Bennewitz, R., Meyer, E., and Brune, H. (2003) *Phys. Rev. Lett.*, **91**, 084502.

1018 Holscher, H., Schwarz, U.D., and Wiesendanger, R. (1997) *Surf. Sci.*, **375**, 395.

1019 Müller, T., Lohrmann, M., Kässer, T., Marti, O., Mlynek, J., and Krausch, G. (1997) *Phys. Rev. Lett.*, **79**, 5066.

1020 Hölscher, H., Ebeling, D., and Schwarz, U.D. (2008) *Phys. Rev. Lett.*, **101**, 246105.

1021 Sang, Y., Dubé, M., and Grant, M. (2001) *Phys. Rev. Lett.*, **87**, 174301.

1022 Zwörner, O., Hölscher, H., Schwarz, U.D., and Wiesendanger, R. (1998) *Appl. Phys. A*, **66**, 263.

1023 Reimann, P. and Evstigneev, M. (2004) *Phys. Rev. Lett.*, **93**, 230802.

1024 Bouhacina, T., Aime, J.P., Gauthier, S., Michel, D., and Heroguez, V. (1997) *Phys. Rev. B*, **56**, 7694.

1025 Liu, H., Imad-Uddin Ahmed, S., and Scherge, M. (2001) *Thin Solid Films*, **381**, 135.

1026 Riedo, E., Lévy, F., and Brune, H. (2002) *Phys. Rev. Lett.*, **88**, 185505.

1027 Tambe, N.S. and Bhushan, B. (2005) *Nanotechnology*, **16**, 2309.

1028 Tao, Z.H. and Bhushan, B. (2006) *Rev. Sci. Instrum.*, **77**, 103705.

1029 Helman, J.S., Baltensperger, W., and Holyst, J.A. (1994) *Phys. Rev. B*, **49**, 3831.

1030 Bhushan, B. (1998) *Tribology Issues and Opportunities in MEMS*, Kluwer Academic Publishers, Dordrecht, The Netherlands.

1031 van Spengen, W.M. (2003) *Microelectron. Reliab.*, **43**, 1049.

1032 Hirano, M. and Shinjo, K. (1990) *Phys. Rev. B*, **41**, 11837.

1033 Hirano, M., Shinjo, K., Kaneko, R., and Murata, Y. (1991) *Phys. Rev. Lett.*, **67**, 2642.

1034 Müser, M.H. (2004) *Europhys. Lett.*, **66**, 97.

1035 Gnecco, E., Maier, S., and Meyer, E. (2008) *J. Phys.: Condens. Matter*, **20**, 354004.

1036 Erdemir, E. and Martin, J.M. (2007) *Superlubricity*, Elsevier, Amsterdam.

1037 Martin, J.M., Donnet, C., Lemogne, T., and Epicier, T. (1993) *Phys. Rev. B*, **48**, 10583.

1038 Hirano, M., Shinjo, K., Kaneko, R., and Murata, Y. (1997) *Phys. Rev. Lett.*, **78**, 1448.

1039 Dienwiebel, M., Verhoeven, G.S., Pradeep, N., Frenken, J.W.M., Heimberg, J.A., and Zandbergen, H.W. (2004) *Phys. Rev. Lett.*, **92**, 126101.

1040 Verhoeven, G.S., Dienwiebel, M., and Frenken, J.W.M. (2004) *Phys. Rev. B*, **70**, 165418.

1041 Filippov, A.E., Dienwiebel, M., Frenken, J.W.M., Klafter, J., and Urbakh, M. (2008) *Phys. Rev. Lett.*, **100**, 046102.

1042 He, G., Muser, M.H., and Robbins, M.O. (1999) *Science*, **284**, 1650.

1043 Frenkel, J.I. and Kontorova, T.A. (1938) *Zh. Eksp. Teor. Fiz.*, **8**, 1340.

1044 Braun, O.M. and Kivshar, Y.S. (2004) *The Frenkel–Kontorova Model: Concepts, Methods, Applications*, Springer, Berlin.

1045 Friedel, J. and de Gennes, P.G. (2007) *Philos. Mag.*, **87**, 39.

1046 Merkle, A.P. and Marks, L.D. (2007) *Philos. Mag. Lett.*, **87**, 527.

1047 Sasaki, N., Itamura, N., Tsuda, D., and Miura, K. (2007) *Curr. Nanosci.*, **3**, 105.

1048 Feiler, A.A., Bergstrom, L., and Rutland, M.W. (2008) *Langmuir*, **24**, 2274.

1049 Heuberger, M., Drummond, C., and Israelachvili, J. (1998) *J. Phys. Chem. B*, **102**, 5038.

1050 Jeon, S., Thundat, T., and Braiman, Y. (2006) *Appl. Phys. Lett.*, **88**, 214102.

1051 Socoliuc, A., Gnecco, E., Maier, S., Pfeiffer, O., Baratoff, A., Bennewitz, R., and Meyer, E. (2006) *Science*, **313**, 207.

1052 Gnecco, E., Socoliuc, A., Maier, S., Gessler, J., Glatzel, T., Baratoff, A., and Meyer, E. (2009) *Nanotechnology*, **20**, 025501.

1053 Krylov, S.Y., Jinesh, K.B., Valk, H., Dienwiebel, M., and Frenken, J.W.M. (2005) *Phys. Rev. E*, **71**, 065101.

1054 Jinesh, K.B., Krylov, S.Y., Valk, H., Dienwiebel, M., and Frenken, J.W.M. (2008) *Phys. Rev. B*, **78**, 155440.

1055 Zhang, C.H. (2005) *Tribol. Int.*, **38**, 443.

1056 Hu, Y.Z. and Granick, S. (1998) *Tribol. Lett.*, **5**, 81.

1057 van Alsten, J. and Granick, S. (1988) *Phys. Rev. Lett.*, **61**, 2570.

1058 Granick, S. (1991) *Science*, **253**, 1374.

1059 Yoshizawa, H. and Israelachvili, J. (1993) *J. Phys. Chem.*, **97**, 11300.

1060 Klein, J. and Kumacheva, E. (1995) *Science*, **269**, 816.

1061 Demirel, A.L. and Granick, S. (1996) *Phys. Rev. Lett.*, **77**, 2261.

1062 Reiter, G., Demirel, A.L., and Granick, S. (1994) *Science*, **263**, 1741.

1063 Klein, J. and Kumacheva, E. (1998) *J. Chem. Phys.*, **108**, 6996.

1064 Kumacheva, E. and Klein, J. (1998) *J. Chem. Phys.*, **108**, 7010.

1065 Thompson, P.A. and Robbins, M.O. (1990) *Science*, **250**, 792.

1066 Klein, J. (2007) *Phys. Rev. Lett.*, **98**, 056101.

1067 Drummond, C. and Israelachvili, J. (2001) *Phys. Rev. E*, **63**, 041506.

1068 Zhu, Y.X. and Granick, S. (2004) *Phys. Rev. Lett.*, **93**, 096101.

1069 Gao, J.P., Luedtke, W.D., and Landman, U. (1997) *J. Chem. Phys.*, **106**, 4309.

1070 Ayappa, K.G. and Mishra, R.K. (2007) *J. Phys. Chem. B*, **111**, 14299.

1071 Zhu, Y. and Granick, S. (2003) *Langmuir*, **19**, 8148.

1072 Mukhopadhyay, A., Bae, S.C., Zhao, J., and Granick, S. (2004) *Phys. Rev. Lett.*, **93**, 236105.

1073 Patil, S., Matei, G., Oral, A., and Hoffmann, P.M. (2006) *Langmuir*, **22**, 6485.

1074 Lang, X.Y., Zhu, Y.F., and Jiang, Q. (2007) *Langmuir*, **23**, 1000.

1075 van Megen, W. and Snook, I. (1979) *J. Chem. Soc., Faraday Trans.*, **2**, 75, 1095.

1076 Steitz, R., Gutberlet, T., Hauss, T., Klösgen, B., Kratsev, R., Schemmel, S., Simonsen, A.C., and Findenegg, G.H. (2003) *Langmuir*, **19**, 2409.

1077 Schwendel, D., Hayashi, T., Dahint, R., Pertsin, A., Grunze, M., Steitz, R., and Schreiber, F. (2003) *Langmuir*, **19**, 2284. Maccarini, M., Steitz, R., Himmelhaus, M., Fick, J., Tatur, S., Wolff, M., Grunze, M., Janecek, J., and Netz, R.R. (2007) *Langmuir*, **23**, 598.

1078 Mezger, M., Schöder, S., Reichert, H., Schröder, H., Okasinski, J., Honkimäki, V., Ralston, J., Bilgram, J., Roth, R., and Dosch, H. (2008) *J. Chem. Phys.*, **128**, 244705.

1079 Lee, J.K., Barker, J.A., and Pound, G.M. (1974) *J. Chem. Phys.*, **60**, 1976.

1080 Liu, K.S. (1974) *J. Chem. Phys.*, **60**, 4226; Liu, K.S., Kalos, M.H., and Chester, G.V. (1974) *Phys. Rev. A*, **10**, 303.

1081 Rowley, L.A., Nicholson, D., and Parsonage, N.G. (1976) *Mol. Phys.*, **31**, 365.

1082 Waisman, E., Henderson, D., and Lebowitz, J.L. (1976) *Mol. Phys.*, **32**, 1373.

1083 Abraham, F.F. (1978) *J. Chem. Phys.*, **68**, 3713.

1084 Snook, I.K. and Henderson, D. (1978) *J. Chem. Phys.*, **68**, 2134.

1085 Percus, J.K. (1980) *J. Stat. Phys.*, **23**, 657.

1086 Attard, P., Bérard, D.R., Ursenbach, C.P., and Patey, G.N. (1991) *Phys. Rev. A*, **44**, 8224.

1087 Yu, C.J., Richter, A.G., Kmetko, J., Datta, A., and Dutta, P. (2000) *Europhys. Lett.*, **50**, 487.

1088 Mezger, M. *et al.* (2008) *Science*, **322**, 424.

1089 Chan, D.Y.C., Mitchell, D.J., Ninham, B.W., and Pailthorpe, B.A. (1978) *Mol. Phys.*, **35**, 1669.

1090 Lane, J.E. and Spurling, T.H. (1979) *Chem. Phys. Lett.*, **67**, 107.

1091 Magda, J.J., Tirrell, M., and Davis, H.T. (1985) *J. Chem. Phys.*, **83**, 1888.

1092 Tarazona, P. and Vicente, L. (1985) *Mol. Phys.*, **56**, 557; with a correction in Mitlin, V.S. and Sharma, M.M., (1995) *J. Colloid Interface Sci.*, **170**, 407.

1093 Kinoshita, M., Iba, S.Y., Kuwamoto, K., and Harada, M. (1996) *J. Chem. Phys.*, **105**, 7177.

1094 Maciolek, A., Drzewinski, A., and Bryk, P. (2004) *J. Chem. Phys.*, **120**, 1921.

1095 Schoen, M., Diestler, D.J., and Cushman, J.H. (1987) *J. Chem. Phys.*, **87**, 5464, 1988, **88**, 1394.

1096 Snook, I.K. and van Megen, W. (1981) *J. Chem. Soc., Faraday Trans. 2*, **77**, 181.

1097 Schoen, M., Gruhn, T., and Diestler, D.J. (1998) *J. Chem. Phys.*, **109**, 301.

1098 Christenson, H.K. (1983) *J. Chem. Phys.*, **78**, 6906.

1099 Christenson, H.K. (1986) *J. Phys. Chem.*, **90**, 4.

1100 Franz, V. and Butt, H.-J. (2002) *J. Phys. Chem. B*, **106**, 1703.

1101 Attard, P. (1989) *J. Chem. Phys.*, **91**, 3083.

1102 Das, S.K., Sharma, M.M., and Schechter, R.S. (1996) *J. Phys. Chem.*, **100**, 7122.

1103 Nakada, T., Miyashita, S., Sazaki, G., Komatsu, H., and Chernov, A.A. (1996) *Jpn. J. Appl. Phys.*, **35**, L52.

1104 Mugele, F., Baldelli, S., Somorjai, G.A., and Salmeron, M. (2000) *J. Phys. Chem.*, **104**, 3140.

1105 Horn, R.G. and Israelachvili, J.N. (1981) *J. Chem. Phys.*, **75**, 1400.

1106 Grimson, M.J., Rickayzen, G., and Richmond, P. (1980) *Mol. Phys.*, **39**, 61.

1107 Sarman, S. (1990) *J. Chem. Phys.*, **92**, 4447.

1108 Frink, L.J.D. and van Swol, F. (1998) *J. Chem. Phys.*, **108**, 5588.

1109 Porcheron, F., Schoen, M., and Fuchs, A.H. (2002) *J. Chem. Phys.*, **116**, 5816.

1110 Klein, J. and Kumacheva, E. (1998) *J. Chem. Phys.*, **108**, 6996.

1111 Zhu, Y. and Granick, S. (2003) *Langmuir*, **19**, 8148.

1112 O'Shea, S.J., Welland, M.E., and Rayment, T. (1992) *Appl. Phys. Lett.*, 60, 2356.

1113 Han, W. and Lindsay, S.M. (1998) *Appl. Phys. Lett.*, 72, 1656.

1114 Maali, A., Cohen-Bouhacina, T., Couturier, G., and Aimé, J.P. (2006) *Phys. Rev. Lett.*, 96, 086105.

1115 Marra, J. and Hair, M.L. (1988) *Macromolecules*, 21, 2349.

1116 Heuberger, M., Zäch, M., and Spencer, N.D. (2001) *Science*, 292, 905.

1117 Kanda, Y., Nakamura, T., and Higashitani, K. (1998) *Colloids Surf. A*, 139, 55.

1118 Horn, R.G., Evans, D.F., and Ninham, B.W. (1988) *J. Phys. Chem.*, 92, 3531.

1119 Atkin, R. and Warr, G.G. (2007) *J. Phys. Chem. C*, 111, 5162.

1120 Christenson, H.K., Gruen, D.W.R., Horn, R.G., and Israelachvili, J.N. (1987) *J. Chem. Phys.*, 87, 1834.

1121 Israelachvili, J.N., Kott, S.J., Gee, M.L., and Witten, T.A. (1989) *Macromolecules*, 22, 4247.

1122 Klein, D.L. and McEuen, P.L. (1995) *Appl. Phys. Lett.*, 66, 2478.

1123 Lim, R. and O'Shea, S.J. (2002) *Phys. Rev. Lett.*, 88, 246101.

1124 Vacatello, M., Yoon, D.Y., and Laskowski, B.C. (1990) *J. Chem. Phys.*, 93, 779.

1125 Walley, K.P., Schweizer, K.S., Peanasky, J., Cai, L., and Granick, S. (1994) *J. Chem. Phys.*, 100, 3361.

1126 Cui, S.T., Cummings, P.T., and Cochran, H.D. (2001) *J. Chem. Phys.*, 114, 6464.

1127 Qin, Y. and Fichthorn, K.A. (2007) *J. Chem. Phys.*, 127, 144911.

1128 Richetti, P., Moreau, L., Barois, P., and Kékicheff, P. (1996) *Phys. Rev. E*, 54, 1749.

1129 Petrov, P., Miklavcic, S., Olsson, U., and Wennerström, H. (1995) *Langmuir*, 11, 3928.

1130 Horn, R.G., Israelachvili, J.N., and Perez, E. (1981) *J. Phys.*, 42, 39.

1131 Moreau, L., Richetti, P., and Barois, P. (1994) *Phys. Rev. Lett.*, 73, 3556.

1132 Ruths, M., Steinberg, S., and Israelachvili, J.N. (1996) *Langmuir*, 12, 6637;Ruths, M., Heuberger, M., Scheumann, V., Hu, J., and Knoll, W. (2001) *Langmuir*, 17, 6213.

1133 Kočevar, K. and Muševič, I. (2002) *Phys. Rev. E*, 65, 021703.

1134 Carbone, G., Barberi, R., Muševič, I., and Kržič, U. (2005) *Phys. Rev. E*, 71, 051704.

1135 Marčelja, S. and Radic, N. (1976) *Chem. Phys. Lett.*, 42, 129.

1136 Cevc, G., Podgornik, R., and Žekš, B. (1982) *Chem. Phys. Lett.*, 91, 193.

1137 de Gennes, P.G. (1990) *Langmuir*, 6, 1448.

1138 Ziherl, P. (2000) *Phys. Rev. E*, 61, 4636.

1139 Borštnik, A., Stark, H., and Žumer, S. (1999) *Phys. Rev. E*, 60, 4210.

1140 Drost-Hansen, W. (1977) *J. Colloid Interface Sci.*, 58, 251.

1141 Depasse, J. and Watillon, A. (1970) *J. Colloid Interface Sci.*, 33, 430.

1142 Healy, T.W., Homola, A., James, R.O., and Hunter, R.J. (1978) *Faraday Discuss. Chem. Soc.*, 65, 156.

1143 Langmuir, I. (1938) *J. Chem. Phys.*, 6, 873.

1144 Schofield, R.K. (1946) *Trans. Faraday Soc.*, 42, 219.

1145 Hemwall, J.B. and Low, P.F. (1956) *Soil Sci.*, 82, 135.

1146 Pashley, R.M. and Israelachvili, J.N. (1984) *J. Colloid Interface Sci.*, 101, 511.

1147 Kjellander, R., Marčelja, S., Pashley, R.M., and Quirk, J.P. (1990) *J. Chem. Phys.*, 92, 4399.

1148 Goldberg, R., Chai, L., Perkin, S., Kampf, N., and Klein, J. (2008) *Phys. Chem. Chem. Phys.*, 10, 4939.

1149 Peschel, G., Belouschek, P., Müller, M.M., Müller, M.R., and König, R. (1982) *Colloid Polym. Sci.*, 260, 444.

1150 Grabbe, A. and Horn, R.G. (1993) *J. Colloid Interface Sci.*, 157, 375.

1151 Chapel, J.P. (1994) *Langmuir*, 10, 4237.

1152 Toikka, G. and Hayes, R.A. (1997) *J. Colloid Interface Sci.*, 191, 102.

1153 Valle-Delgado, J.J., Molina-Bolivar, J.A., Galisteo-González, F., Gálvez-Ruiz, M.J., Feiler, A., and Rutland, M.W. (2005) *J. Chem. Phys.*, 123, 034708.

1154 McIntosh, T.J. and Simon, S.A. (1994) *Annu. Rev. Biophys. Biomol. Struct.*, 23, 27.

1155 Rau, D.C., Lee, B., and Parsegian, V.A. (1984) *Proc. Natl. Acad. Sci. USA*, 81, 2621.

1156 Kruyt, H.R. and Bungenberg de Jong, H.G. (1929) *Kolloid-Beih.*, **28**, 1.

1157 Cleveland, J.P., Schäffer, T.E., and Hansma, P.K. (1995) *Phys. Rev. B*, **52**, 8692.

1158 Grünewald, T. and Helm, C.A. (1996) *Langmuir*, **12**, 3885.

1159 Delville, A. (1993) *J. Phys. Chem.*, **97**, 9703.

1160 Lu, L. and Berkowitz, M.L. (2006) *J. Chem. Phys.*, **124**, 101101.

1161 Bailey, J.R. and McGuire, M.M. (2007) *Langmuir*, **23**, 10995.

1162 Marčelja, S., Mitchell, D.J., Ninham, B.W., and Sculley, M.J. (1977) *J. Chem. Soc., Faraday Trans. 2*, **73**, 630.

1163 Miranda, P.B. and Shen, Y.R. (1999) *J. Phys. Chem. B*, **103**, 3292.

1164 Toney, M.F., Howard, J.N., Richer, J., Borges, G.L., Gordon, J.G., Melroy, O.R., Wiesler, D.G., Yee, D., and Sørensen, L.B. (1994) *Nature*, **368**, 444.

1165 Ataka, K., Yotsuyanagi, T., and Osawa, M. (1996) *J. Phys. Chem.*, **100**, 10664.

1166 Leikin, S. and Kornyshev, A.A. (1990) *J. Chem. Phys.*, **92**, 6890.

1167 Ruckenstein, E. and Manciu, M. (2002) *Langmuir*, **18**, 7584; (2004) *Adv. Colloid Interface Sci.*, **112**, 109.

1168 Jönsson, B. and Wennerström, H. (1983) *J. Chem. Soc., Faraday Trans.*, **2**, **79**, 19.

1169 Gallo, P., Rovere, M., and Spohr, E. (2000) *J. Chem. Phys.*, **113**, 11324.

1170 Schiby, D. and Ruckenstein, E. (1983) *Chem. Phys. Lett.*, **100**, 277.

1171 Paunov, V.N., Kaler, E.W., Sandler, S.I., and Petsev, D.N. (2001) *J. Colloid Interface Sci.*, **240**, 640.

1172 Attard, P. and Patey, G.N. (1991) *Phys. Rev. A*, **43**, 2953.

1173 Christenson, H.K. and Claesson, P.M. (2001) *Adv. Colloid Interface Sci.*, **91**, 391.

1174 Tanford, C. (1980) *The Hydrophobic Effect: Formation of Micelles and Biological Membranes*, Wiley-Interscience, New York.

1175 Chandler, D. (2005) *Nature*, **437**, 640.

1176 Owens, D.K. and Wendt, R.C. (1969) *J. Appl. Polym. Sci.*, **13**, 1741.

1177 Butt, H.-J. and Raiteri, R. (1999) *Surface Characterization Methods*, vol. 87 (ed. A. Milling), Marcel Dekker, New York, p. 1.

1178 Grundke, K. (2005) *Molecular Interfacial Phenomena of Polymers and Biopolymers* (ed. P. Chen), Woodhead Publishing, Cambridge, p. 323.

1179 Wilhelm, E., Battino, R., and Wilcock, R.J. (1977) *Chem. Rev.*, **77**, 219.

1180 Abraham, M.H. (1984) *J. Chem. Soc., Faraday Trans. 1*, **80**, 153.

1181 Weisenberger, S. and Schumpe, A. (1996) *AIChE J.*, **42**, 298.

1182 Zeppieri, S., Rodriguez, J., and de Ramos, L. (2001) *J. Chem. Eng. Data*, **46**, 1086.

1183 Israelachvili, J.N. and Pashley, R. (1982) *Nature*, **300**, 341.

1184 Israelachvili, J.N. and Pashley, R. (1984) *J. Colloid Interface Sci.*, **98**, 500.

1185 Pashley, R.M., McGuiggan, P.M., Ninham, B.W., and Evans, D.F. (1985) *Science*, **229**, 1088.

1186 Claesson, P.M., Blom, C.E., Herder, P.C., and Ninham, B.W. (1986) *J. Colloid Interface Sci.*, **114**, 234.

1187 Tsao, Y.H., Yang, S.X., Evans, D.F., and Wennerström, H. (1991) *Langmuir*, **7**, 3154.

1188 Parker, J.L. and Claesson, P.M. (1992) *Langmuir*, **8**, 757.

1189 Meyer, E.E., Lin, Q., and Israelachvili, J.N. (2005) *Langmuir*, **21**, 256.

1190 Claesson, P.M. and Christenson, H.K. (1988) *J. Phys. Chem.*, **92**, 1650.

1191 Christenson, H.K., Fang, J., Ninham, B.W., and Parker, J.L. (1990) *J. Phys. Chem.*, **94**, 8004.

1192 Kurihara, K. and Kunitake, T. (1992) *J. Am. Chem. Soc.*, **114**, 10927.

1193 Wood, J. and Sharma, R. (1995) *Langmuir*, **11**, 4797.

1194 Rabinovich, Y.I. and Yoon, R.H. (1994) *Colloids Surf. A*, **93**, 263; Yoon, R.H., Flinn, D.H., and Rabinovich, Y.I. (1997) *J. Colloid Interface Sci.*, **185**, 363.

1195 Kékicheff, P. and Spalla, O. (1995) *Phys. Rev. Lett.*, **75**, 1851.

1196 Craig, V.S.J., Ninham, B.W., and Pashley, R.M. (1999) *Langmuir*, **15**, 1562.

1197 Sakamoto, M., Kanda, Y., Miyahara, M., and Higashitani, K. (2002) *Langmuir*, **18**, 5713.

1198 Zhang, J., Yoon, R.H., Mao, M., and Ducker, W.A. (2005) *Langmuir*, **21**, 5831.

1199 Karaman, M.E., Meagher, L., and Pashley, R.M. (1993) *Langmuir*, **9**, 1220.

1200 Meagher, L. and Craig, V.S.J. (1994) *Langmuir*, **10**, 2736.

1201 Stevens, H., Considine, R.F., Drummond, C.J., Hayes, R.A., and Attard, P. (2005) *Langmuir*, **21**, 6399.

1202 Mahnke, J., Stearnes, J., Hayes, R.A., Fornasiero, D., and Ralston, J. (1999) *Phys. Chem. Chem. Phys.*, **1**, 2793.

1203 Ishida, N., Sakamoto, M., Miyahara, M., and Higashitani, K. (2000) *Langmuir*, **16**, 5681.

1204 Yakubov, G.E., Butt, H.-J., and Vinogradova, O.I. (2000) *J. Phys. Chem. B*, **104**, 3407.

1205 Kurutz, J.W. and Xu, S. (2001) *Langmuir*, **17**, 7323.

1206 Ishida, N. and Higashitani, K. (2006) *Miner. Eng.*, **19**, 719.

1207 Parker, J.L., Claesson, P.M., and Attard, P. (1994) *J. Phys. Chem.*, **98**, 8468.

1208 Yaminsky, V.V., Jones, C., Yaminsky, F., and Ninham, B.W. (1996) *Langmuir*, **12**, 3531.

1209 Ederth, T., Claesson, P., and Liedberg, B. (1998) *Langmuir*, **14**, 4782; (2000) **16**, 2177.

1210 Yaminsky, V.V., Ninham, B.W., Christenson, H.K., and Pashley, R.M. (1996) *Langmuir*, **12**, 1936.

1211 Lee, C.Y., McCammon, J.A., and Rossky, P.J. (1984) *J. Chem. Phys.*, **80**, 4448.

1212 Forsman, J., Woodward, C.E., and Jönsson, B. (1997) *J. Colloid Interface Sci.*, **195**, 264.

1213 Jensen, T.R., Jensen, M.O., Reitzel, N., Balashev, K., Peters, G.H., Kjaer, K., and Bjornholm, T. (2003) *Phys. Rev. Lett.*, **90**, 086101.

1214 Poynor, A., Hong, L., Robinson, I.K., Granick, S., Zhang, Z., and Fenter, P.A. (2006) *Phys. Rev. Lett.*, **97**, 266101.

1215 Eriksson, J.C., Ljunggren, S., and Claesson, P.M. (1989) *J. Chem. Soc., Faraday Trans. 2*, **85**, 163; Eriksson, J.C. and Henriksson, U. (2007) *Langmuir*, **23**, 10026.

1216 Wallqvist, A. and Berne, B. (1995) *J. Phys. Chem.*, **99**, 2893.

1217 Ducker, W.A., Xu, Z., and Israelachvili, J.N. (1994) *Langmuir*, **10**, 3279.

1218 Yaminsky, V.V. and Ninham, B.W. (1993) *Langmuir*, **9**, 3618.

1219 Dammer, S.M. and Lohse, D. (2006) *Phys. Rev. Lett.*, **96**, 206101.

1220 Bratko, D. and Luzar, A. (2008) *Langmuir*, **24**, 1247.

1221 Tyrrell, J.W.G. and Attard, P. (2002) *Langmuir*, **18**, 160.

1222 Lou, S.T., Quyang, Z.Q., Zhang, Y., Li, X.J., Hu, J., Li, M.Q., and Yang, F.J. (2000) *J. Vac. Sci. Technol. B*, **18**, 2573.

1223 Simonsen, A.C., Hansen, P.L., and Klösgen, B. (2004) *J. Colloid Interface Sci.*, **273**, 291.

1224 Zhang, X.H., Maeda, N., and Craig, V.S.J. (2006) *Langmuir*, **22**, 5025.

1225 Zhang, X.H., Khan, A., and Ducker, W.A. (2007) *Phys. Rev. Lett.*, **98**, 136101.

1226 Borkent, B.M., Dammer, S.M., Schonherr, H., Vancso, G.J., and Lohse, D. (2007) *Phys. Rev. Lett.*, **98**, 204502.

1227 Katan, A.J. and Oosterkamp, T.H. (2008) *J. Phys. Chem. C*, **112**, 9769.

1228 Karakashev, S.I. and Nguyen, A.V. (2009) *Langmuir*, **25**, 3363.

1229 Rabinovich, Y.I., Derjaguin, B.V., and Churaev, N.V. (1982) *Adv. Colloid Interface Sci.*, **16**, 63.

1230 Rabinovich, Y.I. and Derjaguin, B.V. (1988) *Colloids Surf.*, **30**, 243.

1231 van Meer, G., Voelker, D.R., and Feigenson, G.W. (2008) *Nat. Rev. Mol. Cell Biol.*, **9**, 112.

1232 Lipowsky, R. (1995) *Handbook of Biological Physics* (eds R. Lipowsky and E. Sackmann), Elsevier, p. 521.

1233 LeNeveu, D.M., Rand, R.P., Parsegian, V.A., and Gingell, D. (1977) *Biophys. J.*, **18**, 209.

1234 Parsegian, V.A., Fuller, N., and Rand, R.P. (1979) *Proc. Natl. Acad. Sci. USA*, **76**, 2750.

1235 Lis, L.J., McAlister, M., Fuller, N., Rand, R.P., and Parsegian, V.A. (1982) *Biophys. J.*, **37**, 657.

1236 McIntosh, T.J. and Simon, S.A. (1986) *Biochemistry*, **25**, 4058.

1237 Petrache, H.I., Gouliaev, N., Tristram-Nagle, S., Zhang, R., Suter, R.M., and Nagle, J.F. (1998) *Phys. Rev. E*, **57**, 7014.

1238 Barclay, L., Harrington, A., and Ottewill, R.H. (1972) *Kolloid-Zeitschrift Z. Polym.*, **250**, 655.

1239 McIntosh, T.J. (2000) *Curr. Opin. Struct. Biol.*, **10**, 481.

1240 Strey, H.H., Parsegian, V.A., and Podgornik, R. (1997) *Phys. Rev. Lett.*, **78**, 895.

1241 http://www.brocku.ca/researchers/peter_rand/, http://lpsb.nichd.nih.gov/osmotic_stress.htm.

1242 Smith, G.S., Sirota, E.B., Safinya, C.R., Plano, R.J., and Clark, N.A. (1990) *J. Chem. Phys.*, **92**, 4519.

1243 Parsegian, V.A., Rand, R.P., and Rau, D.C. (1995) *Methods Enzymol.*, **259**, 43.

1244 Leikin, S., Rau, D.C., and Parsegian, V.A. (1995) *Nat. Struct. Biol.*, **2**, 205.

1245 Rau, D.C. and Parsegian, V.A. (1990) *Science*, **249**, 1278.

1246 Rohrsetzer, S., Kovacs, P., and Nagy, M. (1986) *Colloid Polym. Sci.*, **264**, 812.

1247 Bibette, J., Roux, D., and Pouligny, B. (1992) *J. Phys. II France*, **2**, 401.

1248 Reus, V., Belloni, L., Zemb, T., Lutterbach, N., and Versmold, H. (1997) *J. Phys. II France*, **7**, 603.

1249 Michot, L.J., Bihannic, I., Porsch, K., Maddi, S., Baravian, C., Mougel, J., and Levitz, P. (2004) *Langmuir*, **20**, 10829.

1250 Martin, C., Pignon, F., Magnin, A., Meireles, M., Lelievre, V., Lindner, P., and Cabane, B. (2006) *Langmuir*, **22**, 4065.

1251 Gu, Z.Y. and Alexandridis, P. (2004) *Macromolecules*, **37**, 912.

1252 Spalla, O., Nabavi, M., Minter, J., and Cabane, B. (1996) *Colloid Polym. Sci.*, **274**, 555.

1253 Dubois, M., Schönhoff, M., Meister, A., Belloni, L., Zemb, T., and Möhwald, H. (2006) *Phys. Rev. E*, **74**, 051402.

1254 Horn, R.G. (1984) *Biochim. Biophys. Acta*, **778**, 224.

1255 Horn, R.G., Israelachvili, J.N., Marra, J., Parsegian, V.A., and Rand, R.P. (1988) *Biophys. J.*, **54**, 1185.

1256 Helm, C.A., Israelachvili, J.N., and McGuiggan, P.M. (1992) *Biochemistry*, **31**, 1794.

1257 Marra, J. and Israelachvili, J. (1985) *Biochemistry*, **24**, 4608.

1258 Pera, I., Stark, R., Kappl, M., Butt, H.-J., and Benfenati, F. (2004) *Biophys. J.*, **87**, 2446.

1259 McIntosh, T.J. and Simon, S.A. (1996) *Colloids Surf. A*, **116**, 251.

1260 Gawrisch, K., Ruston, D., Zimmerberg, J., Parsegian, V.A., Rand, R.P., and Fuller, N. (1992) *Biophys. J.*, **61**, 1213.

1261 Tristram-Nagle, S., Petrache, H.I., and Nagle, J.F. (1998) *Biophys. J.*, **75**, 917.

1262 McIntosh, T.J., Magid, A.D., and Simon, S.A. (1987) *Biochemistry*, **26**, 7325.

1263 Pertsin, A., Platonov, D., and Grunze, M. (2007) *Langmuir*, **23**, 1388.

1264 Israelachvili, J.N. (1994) *Langmuir*, **10**, 3369.

1265 Wong, J.Y., Park, C.K., Seitz, M., and Israelachvili, J. (1999) *Biophys. J.*, **77**, 1458.

1266 Rädler, J. and Sackmann, E. (1993) *J. Phys. (Paris)*, **3**, 727.

1267 Safinya, C.R., Roux, D., Smith, G.S., Sinha, S.K., Dimon, P., Clark, N.A., and Bellocq, A.M. (1986) *Phys. Rev. Lett.*, **57**, 2718;Safinya, C.R., Sirota, E.B., Roux, D., and Smith, G.S. (1989) *Phys. Rev. Lett.*, **62**, 1134.

1268 Canham, P.B. (1970) *J. Theor. Biol.*, **26**, 61.

1269 Helfrich, W. (1973) *Z. Naturforsch.*, **28c**, 693.

1270 McIntosh, T.J., Advani, S., Burton, R.E., Zhelev, D.V., Needham, D., and Simon, S.A. (1995) *Biochemistry*, **34**, 8520.

1271 Evans, E. and Rawicz, W. (1990) *Phys. Rev. Lett.*, **64**, 2094.

1272 Kummrow, M. and Helfrich, W. (1991) *Phys. Rev. A*, **44**, 8356.

1273 Manciu, M. and Ruckenstein, E. (2007) *J. Colloid Interface Sci.*, **309**, 56.

1274 Dimova, R., Pouligny, B., and Dietrich, C. (2000) *Biophys. J.*, **79**, 340.

1275 Sackmann, E., Duwe, H.P., and Engelhardt, H. (1986) *Faraday Discuss. Chem. Soc.*, **81**, 281.

1276 Lee, C.H., Lin, W.C., and Wang, J. (2001) *Phys. Rev. E*, **64**, 020901.

1277 Salditt, T., Vogel, M., and Fenzl, W. (2003) *Phys. Rev. Lett.*, **90**, 178101.

1278 Bermúdez, H., Hammer, D.A., and Discher, D.E. (2004) *Langmuir*, **20**, 540.

1279 Duwe, H.P. and Sackmann, E. (1990) *Physica A*, **163**, 410.

1280 Mishima, K., Nakamae, S., Ohshima, H., and Kondo, T. (2001) *Chem. Phys. Lipids*, **110**, 27.

1281 Schneider, M.B., Jenkins, J.T., and Webb, W.W. (1984) *J. Phys. (Paris)*, **45**, 1457.

1282 Faucon, J.F., Mitov, M.D., Méléard, P., Bivvas, I., and Bothorel, P. (1989) *J. Phys. France*, **50**, 2389.

1283 Helfrich, W. (1978) *Z. Naturforsch.*, **33**, 305.

1284 Podgornik, R. and Parsegian, V.A. (1992) *Langmuir*, **8**, 557.

1285 Kleinert, H. (1999) *Phys. Lett. A*, **257**, 269.

1286 Walz, J.Y. and Ruckenstein, E. (1999) *J. Phys. Chem. B*, **103**, 7461.

1287 Janke, W. and Kleinert, H. (1987) *Phys. Rev. Lett.*, **58**, 144.

1288 Netz, R.R. (1995) *Phys. Rev. E*, **51**, 2286.

1289 Gordeliy, V.I., Cherezov, V., and Teixeira, J. (2005) *Phys. Rev. E*, **72**, 061913.

1290 Simon, S.A., Advani, S., and McIntosh, T.J. (1995) *Biophys. J.*, **69**, 1473.

1291 Lipowsky, R. and Zielinska, B. (1989) *Phys. Rev. Lett.*, **62**, 1572.

1292 Evans, E. and Needham, D. (1987) *J. Phys. Chem.*, **91**, 4219.

1293 Sornette, D. and Ostrowsky, N. (1986) *J. Chem. Phys.*, **84**, 4062.

1294 Aniansson, E.A.G., Wall, S.N., Almgren, M., Hoffman, H., Kielmann, I., Ulbricht, W., Zana, R., Lang, J., and Tondre, C. (1976) *J. Phys. Chem.*, **80**, 905.

1295 Israelachvili, J.N. and Wennerström, H. (1990) *Langmuir*, **6**, 873.

1296 Marrink, S.J., Berkowitz, M., and Berendsen, H.J.C. (1993) *Langmuir*, **9**, 3122.

1297 König, S., Bayerl, T.M., Coddens, G., Richter, D., and Sackmann, E. (1995) *Biophys. J.*, **68**, 1871.

1298 Hayter, J.B. and Penfold, J. (1981) *J. Chem. Soc., Faraday Trans*, **1**, **77**, 1851.

1299 Lipowsky, R. and Grotehans, S. (1993) *Europhys. Lett.*, **23**, 599.

1300 Evans, E. and Metcalfe, M. (1984) *Biophys. J.*, **46**, 423.

1301 Rehbinder, P., Lagutkina, L., and Wenström, E. (1930) *Z. Phys. Chem. A*, **146**, 63.

1302 Van der Waarden, M. (1950) *J. Colloid Sci.*, **5**, 317.

1303 Mackor, E.L. (1951) *J. Colloid Sci.*, **6**, 492.

1304 Hall, D.G. (1972) *J. Chem. Soc., Faraday Trans.*, **2**, **68**, 2169.

1305 Ash, S.G., Everett, D.H., and Radke, C. (1973) *J. Chem. Soc., Faraday Trans. 2*, **69**, 1256.

1306 Subramanian, V. and Ducker, W. (2001) *J. Phys. Chem. B*, **105**, 1389.

1307 Wang, J. and Butt, H.-J. (2008) *J. Phys. Chem. B*, **112**, 2001.

1308 Doi, M. and Edwards, S.F. (1986) *The Theory of Polymer Dynamics*, Oxford University Press, Oxford.

1309 Strobl, G. (1996) *The Physics of Polymers*, 2nd edn, Springer, Berlin.

1310 Rubinstein, M. and Colby, R. (2003) *Polymer Physics*, Oxford University Press, Oxford.

1311 Napper, D.H. (1983) *Polymeric Stabilization of Colloidal Dispersions*, Academic Press, London.

1312 Likos, C.N. (2001) *Phys. Rep.*, **348**, 267.

1313 Kleshchanok, D., Tuinier, R., and Lang, P.R. (2008) *J. Phys.: Condens. Matter*, **20**, 073101.

1314 Kuhn, W. (1934) *Kolloid-Zeitschrift*, **68**, 2.

1315 Brandrup, J., Immergut, E.H., and Grulke, E.A. (eds) (1999) *Polymer Handbook*, 4th edn, Wiley-Interscience, New York.

1316 Krishnamoorti, R., Graessley, W.W., Zirkel, A., Richter, D., Hadjichristidis, N., Fetters, L.J., and Lohse, D.J. (2002) *J. PolyM. Sci. B*, **40**, 1768.

1317 Sasanuma, Y. (2009) *Macromolecules*, **42**, 2854.

1318 Withers, I.M., Dobrynin, A.V., Berkowitz, M.L., and Rubinstein, M. (2003) *J. Chem. Phys.*, **118**, 4721.

1319 Miyaki, Y., Einaga, Y., and Fujita, H. (1978) *Macromolecules*, **11**, 1180.

1320 Flory, P.J. (1953) *Principles of Polymer Chemistry*, University Press, London.

1321 Kuhn, W. and Grün, F. (1942) *Kolloid-Zeitschrift*, **101**, 248.

1322 Fixman, M. and Kovac, J. (1973) *J. Chem. Phys.*, **58**, 1564.

1323 Hugel, T., Rief, M., Seitz, M., Gaub, H.E., and Netz, R.R. (2005) *Phys. Rev. Lett.*, **94**, 048301.

1324 Smith, S.B., Cui, Y., and Bustamante, C. (1996) *Science*, **271**, 795.

1325 Wang, M.D., Yin, H., Landick, R., Gelles, J., and Block, S.M. (1997) *Biophys. J.*, **72**, 1335.

1326 Smith, S.B., Finzi, L., and Bustamante, C. (1992) *Science*, **258**, 1122.

1327 Rief, M., Oesterhelt, F., Heymann, B., and Gaub, H.E. (1997) *Science*, **275**, 1295.

1328 Fritz, J., Anselmetti, D., Jarchow, J., and Fernàndez-Busquets, X. (1997) *J. Struct. Biol.*, **119**, 165.

1329 Senden, T.J., de Meglio, J.M., and Auroy, P. (1998) *Eur. Phys. J. B*, **3**, 211.

1330 Shi, W., Zhang, Y., Liu, C., Wang, Z., Zhang, X., Zhang, Y., and Chen, Y. (2006) *Polymer*, **47**, 2499.

1331 Kratky, O. and Porod, G. (1949) *Recl. Trav. Chim. Pays-Bas*, **68**, 1106.

1332 Bustamante, C., Marko, J.F., Siggia, E.D., and Smith, S. (1994) *Science*, **265**, 1599.

1333 Marko, J.F. and Siggia, E.D. (1995) *Macromolecules*, **28**, 8759.

1334 Fleer, G.J., Cohen Stuart, M.A., Scheutjens, J.M.H.M., Cosgrove, T., and Vincent, B. (1993) *Polymers at Interfaces*, Chapman & Hall, London.

1335 Vologodskii, A. (1994) *Macromolecules*, **27**, 5623.

1336 Odijk, T. (1995) *Macromolecules*, **28**, 7016.

1337 Bouchiat, C., Wang, M.D., Allemand, J.F., Strick, T., Block, S.M., and Croquette, V. (1999) *Biophys. J.*, **76**, 409.

1338 Fischer, E.W. (1958) *Kolloid-Zeitschrift*, **160**, 120.

1339 Vrij, A. (1976) *Pure Appl. Chem.*, **48**, 471.

1340 Currie, E.P.K., Norde, W., and Cohen Stuart, M.A. (2003) *Adv. Colloid Interface Sci.*, **100–102**, 205.

1341 Netz, R.R. and Andelman, D. (2003) *Phys. Rep.*, **380**, 1.

1342 Kreer, T., Metzger, S., Müller, M., Binder, K., and Baschnagel, J. (2004) *J. Chem. Phys.*, **120**, 4012.

1343 Clayfield, E.J. and Lumb, E.C. (1966) *J. Colloid Interface Sci.*, **22**, 269.

1344 Meier, D.J. (1967) *J. Phys. Chem.*, **71**, 1861.

1345 Hesselink, F.T., Vrij, A., and Overbeek, J.T.G. (1971) *J. Phys. Chem.*, **75**, 2094.

1346 de Gennes, P.G. (1980) *Macromolecules*, **13**, 1069; (1987) *Adv. Colloid Interface Sci.*, **27**, 189.

1347 Milner, S.T., Witten, T.A., and Cates, M.E. (1988) *Macromolecules*, **21**, 2610.

1348 Karim, A., Satija, S.K., Douglas, J.F., Anker, J.F., and Fetters, L.J. (1994) *Phys. Rev. Lett.*, **73**, 3407.

1349 Habicht, J., Schmidt, M., Rühe, J., and Johannsmann, D. (1999) *Langmuir*, **15**, 2460.

1350 Biesalski, M. and Rühe, J. (2002) *Macromolecules*, **35**, 499.

1351 Dolan, A.K. and Edwards, S.F. (1974) *Proc. R. Soc. Lond. A*, **337**, 509.

1352 McLean, S.C., Lioe, H., Meagher, L., Craig, V.S.J., and Gee, M.L. (2005) *Langmuir*, **21**, 2199.

1353 Klein, J. and Luckham, P.F. (1984) *Macromolecules*, **17**, 1041.

1354 Israelachvili, J.N., Tirrel, M., Klein, J., and Almog, Y. (1984) *Macromolecules*, **17**, 204.

1355 Hadziioannou, G., Patel, S., Granick, S., and Tirrell, M. (1986) *J. Am. Chem. Soc.*, **108**, 2869.

1356 Marra, J. and Hair, M.L. (1988) *Macromolecules*, **21**, 2349.

1357 Marra, J. and Christenson, H.K. (1989) *J. Phys. Chem.*, **93**, 7180.

1358 Horn, R.G., Hirz, S.J., Hadziioannou, G., and Frank, C.W. (1989) *J. Chem. Phys.*, **90**, 6767.

1359 Taunton, H.J., Toprakcioglu, C., Fetters, L.J., and Klein, J. (1990) *Macromolecules*, **23**, 571.

1360 Raviv, U., Giasson, S., Kampf, N., Gohy, J.F., Jerome, R., and Klein, J. (2003) *Nature*, **425**, 163.

1361 O'Shea, S.J., Welland, M.E., and Rayment, T. (1993) *Langmuir*, **9**, 1826.

1362 Lea, A.S., Andrade, J.D., and Hlady, V. (1994) *Colloids Surf. A*, **93**, 349.

1363 Chatellier, X., Senden, T.J., and Joanny, J.F. di Meglio, J.M. (1998) *Europhys. Lett.*, **41**, 303.

1364 Butt, H.-J., Kappl, M., Müller, H., Raiteri, R., Meyer, W., and Rühe, J. (1999) *Langmuir*, **15**, 2559.

1365 Lyklema, H. and van Vliet, T. (1978) *Faraday Discuss. Chem. Soc.*, **65**, 25.

1366 Kenworthy, A.K., Hristova, K., Needham, D., and McIntosh, T.J. (1995) *Biophys. J.*, **68**, 1921.

1367 Witten, T.A. and Pincus, P.A. (1986) *Macromolecules*, **19**, 2509.

1368 Subramanian, G., Williams, D.R.M., and Pincus, P.A. (1996) *Macromolecules*, **29**, 4045.

1369 Murat, M. and Grest, G.S. (1996) *Macromolecules*, **29**, 8282.

1370 Milchev, A., Yamakov, V., and Binder, K. (1999) *Europhys. Lett.*, **47**, 675.

1371 Kelley, T.W., Schorr, P.A., Johnson, K.D., Tirrell, M., and Frisbie, C.D. (1998) *Macromolecules*, **31**, 4297.

1372 Block, S. and Helm, C.A. (2007) *Phys. Rev. E*, **76**, 030801.

1373 Pincus, P. (1991) *Macromolecules*, **24**, 2912.

1374 Balastre, M., Li, F., Schorr, P., Yang, J., Mays, J.W., and Tirrell, M.V. (2002) *Macromolecules*, **35**, 9480.

1375 Csajka, F.S., Netz, R.R., Seidel, C., and Joanny, J.F. (2001) *Eur. Phys. J. E*, **4**, 505.

1376 Forsman, J. (2006) *Curr. Opin. Colloid Interface Sci.*, **11**, 290.

1377 Hehmeyer, O.J. and Stevens, M.J. (2005) *J. Chem. Phys.*, **122**, 134909.

1378 Rühe, J. *et al.* (2004) *Adv. Polym. Sci.*, **165**, 79.

1379 Hayashi, S., Abe, T., Higashi, N., Niwa, M., and Kurihara, K. (2002) *Langmuir*, **18**, 3932.

1380 Block, S. and Helm, C.A. (2008) *Phys. Rev. E*, **112**, 9318.

1381 Kegler, K., Konieczny, M., Dominguez-Espinosa, G., Gutsche, C., Salomo, M., Kremer, F., and Likos, C.N. (2008) *Phys. Rev. Lett.*, **100**, 118302.

1382 Biggs, S. (1995) *Langmuir*, **11**, 156.

1383 Luckham, P.F. and Klein, J. (1984) *J. Chem. Soc., Faraday Trans., 1*, **80**, 865.

1384 Marra, J. and Hair, M.L. (1988) *J. Phys. Chem.*, **92**, 6044.

1385 Jusufi, A., Likos, C.N., and Ballauff, M. (2004) *Colloid Polym. Sci.*, **282**, 910.

1386 Klein, J. and Pincus, P. (1982) *Macromolecules*, **15**, 1129.

1387 Ingersent, K., Klein, J., and Pincus, P. (1986) *Macromolecules*, **19**, 1374.

1388 Bhattacharya, S., Milchev, A., Rostiashvili, V.G., Grosberg, A.Y., and Vilgis, T.A. (2008) *Phys. Rev. E*, **77**, 061603.

1389 Evers, O.A., Scheutjens, J.M.H.M., and Fleer, G.J. (1990) *Macromolecules*, **23**, 5221.

1390 Kumacheva, E., Klein, J., Pincus, P., and Fetters, L.J. (1993) *Macromolecules*, **26**, 6477.

1391 Israelachvili, J.N., Tandon, R.K., and White, L.R. (1980) *J. Colloid Interface Sci.*, **78**, 430.

1392 Pelssers, E.G.M., Cohen Stuart, M.A., and Fleer, G.J. (1989) *Colloids Surf.*, **38**, 15.

1393 Fan, A.X., Turro, N.J., and Somasundaran, P. (2000) *Colloids Surf. A*, **162**, 141.

1394 Luckham, P.F. and Klein, J. (1990) *J. Chem. Soc., Faraday Trans.*, **86**, 1363.

1395 Amiel, C., Sikka, M., Schneider, J.W., Tsao, Y.H., Tirrell, M., and Mays, J.W. (1995) *Macromolecules*, **28**, 3125.

1396 Ruths, M., Israelachvili, J.N., and Ploehn, H.J. (1997) *Macromolecules*, **30**, 3329.

1397 Klein, J. and Rossi, G. (1998) *Macromolecules*, **31**, 1979.

1398 Almog, Y. and Klein, J. (1985) *J. Colloid Interface Sci.*, **106**, 33.

1399 Dahlgren, M.A.G., Waltermo, A., Blomberg, E., Claesson, P.M., Sjostrom, L., Akesson, T., and Jonsson, B. (1993) *J. Phys. Chem.*, **97**, 11769.

1400 Poptoshev, E., Rutland, M.W., and Claesson, P.M. (1999) *Langmuir*, **15**, 7789.

1401 Ward, A.F.H. and Tordai, L. (1946) *J. Chem. Phys.*, **14**, 453.

1402 Langmuir, I. and Schaefer, V.J. (1937) *J. Am. Chem. Soc.*, **59**, 2400.

1403 Beer, M., Schmidt, M., and Muthukumar, M. (1997) *Macromolecules*, **30**, 8375.

1404 de Gennes, P.G. (1982) *Macromolecules*, **15**, 492.

1405 Scheutjens, J.M.H.M. and Fleer, G.J. (1985) *Macromolecules*, **18**, 1882.

1406 Ash, S.G. and Findenegg, G.H. (1971) *Trans. Faraday Soc.*, **67**, 2122.

1407 Black, A.P., Birkner, F.B., and Morgan, J.J. (1966) *J. Colloid Interface Sci.*, **21**, 626.

1408 Gaudreault, R., van de Ven, T.G.M., and Whitehead, M.A. (2005) *Colloids Surf. A*, **268**, 131.

1409 Klein, J. (1980) *Nature*, **288**, 248.

1410 Giesbers, M., Kleijn, J.M., Fleer, G.J., and Cohen Stuart, M.A. (1998) *Colloids Surf. A*, **142**, 343.

1411 Swenson, J., Smalley, M.V., and Hatharasinghe, H.L.M. (1998) *Phys. Rev. Lett.*, **81**, 5840.

1412 Granfeldt, M.K., Jönsson, B., and Woodward, C.E. (1991) *J. Phys. Chem.*, **95**, 4819.

1413 Huang, H. and Ruckenstein, E. (2004) *Adv. Colloid Interface Sci.*, **112**, 37.

1414 Chatellier, X., Senden, T.J., and Joanny, J.F. de Meglio, J.M. (1998) *Europhys. Lett.*, **41**, 303.

1415 Hugel, T., Grosholz, M., Clausen-Schaumann, H., Pfau, A., Gaub, H., and Seitz, M. (2001) *Macromolecules*, **34**, 1039.

1416 Cui, S.X., Liu, C.J., Wang, Z.Q., Zhang, X., Strandman, S., and Tenhu, H. (2004) *Macromolecules*, **37**, 946.

1417 Friedsam, C., Del Campo Bécares, A., Jonas, U., Seitz, M., and Gaub, H.E. (2004) *New J. Phys.*, **6**, 9.

1418 Long, J., Xu, Z., and Masliyah, J.H. (2006) *Langmuir*, **22**, 1652.

1419 Chatellier, X. and Joanny, J.F. (1998) *Phys. Rev. E*, **57**, 6923.

1420 Netz, R.R. and Joanny, J.F. (1999) *Macromolecules*, **32**, 9013.

1421 Hanke, F., Livadaru, L., and Kreuzer, H.J. (2005) *Europhys. Lett.*, **69**, 242.

1422 Waite, J.H. (2002) *Integr. Comp. Biol.*, **42**, 1172.

1423 Silverman, H.G. and Roberto, F.F. (2007) *Mar. Biotechnol.*, **9**, 661.

1424 Waite, J.H. and Tanzer, M.L. (1981) *Science*, **212**, 1038.

1425 Yu, M., Hwang, J., and Deming, T.J. (1999) *J. Am. Chem. Soc.*, **121**, 5825.

1426 Lee, H., Lee, B.P., and Messersmith, P.B. (2007) *Nature*, **448**, 338.

1427 Wang, J., Tahir, M.N., Kappl, M., Tremel, W., Metz, N., Barz, M., Theato, P., and Butt, H.-J. (2008) *Adv. Mater.*, **20**, 3872.

1428 Fantner, G.E., Hassenkam, T., Kindt, J.H., Weaver, J.C., Birkedal, H., Pechenik, L., Cutroni, J.A., Cidade, G.A.G., Stucky, G.D., Morse, D.E., and Hansma, P.K. (2005) *Nat. Mater.*, **4**, 612.

1429 Adams, J., Fantner, G.E., Fisher, L.W., and Hansma, P.K. (2008) *Nanotechnology*, **19**, 384008.

1430 Asakura, S. and Oosawa, F. (1954) *J. Chem. Phys.*, **22**, 1255; (1958) *J. Polym. Sci.*, **33**, 183.

1431 Rudhardt, D., Bechinger, C., and Leiderer, P. (1998) *Phys. Rev. Lett.*, **81**, 1330.

1432 Biben, T., Bladon, P., and Frenkel, D. (1996) *J. Phys.: Condens. Matter*, **8**, 10799.

1433 Roth, R., Evans, R., and Dietrich, S. (2000) *Phys. Rev. E*, **62**, 5360.

1434 Cowell, C., Li-In-On, R., and Vincent, B. (1978) *J. Chem. Soc., Faraday Trans., 1*, **74**, 337.

1435 de Hek, H. and Vrij, A. (1979) *J. Colloid Interface Sci.*, **70**, 592.

1436 Sperry, P.R., Hopfenberg, H.B., and Thomas, N.L. (1981) *J. Colloid Interface Sci.*, **82**, 62.

1437 Gast, A.P., Hall, C.K., and Russel, W.B. (1983) *J. Colloid Interface Sci.*, **96**, 251.

1438 Lekkerkerker, H.N.W., Poon, W.C.K., Pusey, P.N., Stroobants, A., and Warren, P.B. (1992) *Europhys. Lett.*, **20**, 559.

1439 Walz, J.Y. and Sharma, A. (1994) *J. Colloid Interface Sci.*, **168**, 485.

1440 Jönsson, B., Broukhno, A., Forsman, J., and Akesson, T. (2003) *Langmuir*, **19**, 9914.

1441 Feigin, R.I. and Napper, D.H. (1980) *J. Colloid Interface Sci.*, **75**, 525.

1442 Mao, Y., Cates, M.E., and Lekkerkerker, H.N.W. (1995) *Physica A*, **222**, 10.

1443 Richetti, P. and Kékicheff, P. (1992) *Phys. Rev. Lett.*, **68**, 1951.

1444 Milling, A.J. (1996) *J. Phys. Chem.*, **100**, 8986.

1445 Qu, D., Baigl, D., Williams, C.E., Möhwald, H., and Fery, A. (2003) *Macromolecules*, **36**, 6878.

1446 Sober, D.L. and Walz, J.Y. (1995) *Langmuir*, **11**, 2352.

1447 Crocker, J.C., Matteo, J.A., Dinsmore, A.D., and Yodh, A.G. (1999) *Phys. Rev. Lett.*, **82**, 4352.

1448 Klapp, S.H.L., Zeng, Y., Qu, D., and von Klitzing, R. (2008) *Phys. Rev. Lett.*, **100**, 118303.

1449 de Gennes, P.G. (1987) *C.R. Acad. Sci. Paris*, **305**, 1181.

1450 Ausserré, D. (1989) *J. Phys. (Paris)*, **50**, 3021.

1451 ten Brinke, G., Ausserre, D., and Hadziioannou, G. (1988) *J. Chem. Phys.*, **89**, 4374.

1452 Leermakers, F.A.M. and Butt, H.-J. (2005) *Phys. Rev. E*, **72**, 021807.

1453 van Alsten, J. and Granick, S. (1990) *Macromolecules*, **23**, 4856.

1454 Montfort, J.P. and Hadziioannou, G. (1988) *J. Chem. Phys.*, **88**, 7187.

1455 Hirz, S., Subbotin, A., Frank, C., and Hadziioannou, G. (1996) *Macromolecules*, **29**, 3970.

1456 Ruths, M. and Granick, S. (1999) *J. Phys. Chem. B*, **103**, 8711.

1457 Hu, H.W. and Granick, S. (1992) *Science*, **258**, 1339.

1458 Peanasky, J., Cai, L.L., Granick, S., and Kessel, C.R. (1994) *Langmuir*, **10**, 3874.

1459 Wang, J., Stark, R., Kappl, M., and Butt, H.-J. (2007) *Macromolecules*, **40**, 2520.

1460 Evmenenko, G., Mo, H., Kewalramani, S., and Dutta, P. (2006) *Polymer*, **47**, 878.

1461 Sun, G., Kappl, M., Pakula, T., Kremer, K., and Butt, H.-J. (2004) *Langmuir*, **20**, 8030.

Index

Surface and Interfacial Forces. Hans-Jürgen Butt and Michael Kappl
Copyright © 2010 Wiley-VCH Verlag GmbH & Co. KGaA
ISBN: 978-3-527-40849-8

:2

2,

7,

8,